Geophysical Monograph Series

Including
IUGG Volumes
Maurice Ewing Volumes
Mineral Physics Volumes

Geophysical Monograph 175

A Continental Plate Boundary:
Tectonics at South Island, New Zealand

David Okaya
Tim Stern
Fred Davey

Editors

American Geophysical Union
Washington, DC

Library of Congress Cataloging-in-Publication Data

A continental plate boundary : tectonics at South Island, New Zealand / David Okaya, Tim Stern, Fred Davey, editors.
    p. cm. -- (Geophysical monograph ; 175)
  ISBN 978-0-87590-440-5
1. Plate tectonics--New Zealand--South Island. 2. Continental margins--New Zealand--South Island. I. Okaya, David Akiharu. II.
Stern, Timothy A. III. Davey, Frederick J.

QE511.4C657 2007
551.1'3609937--dc22

                    2007045674

ISBN 978-0-87590-440-5
ISSN 0065-8448

Cover Photo:

The Huxley Valley leads towards Aoraki/Mt. Cook, New Zealand's highest peak in the Southern Alps (3754 m elevation). This glaciated valley follows structural trends associated with the transpressional Alpine fault, located only 30 km away, that separates the Pacific and Indo-Australian plates. Photograph courtesy of Simon Cox.

CONTENTS

SECTION III: Plate Boundary Dynamics

SECTION IV: Comparisons

PREFACE

Continental collision is a fundamental geologic process in plate tectonics that impacts on continental growth, continental deformation, the development of natural resources, and the occurrence of natural hazards. Convergent plate boundaries where continents collide are often broadly distributed and, where strain rates are high, result in major deformation of the continental lithosphere and the development of mountain ranges. Within mountain belts, compression, thrust faulting, and erosion can combine to generate uplift and overthickened crust, and exhume large sections of high-grade, once deeply buried rocks, associated faults, and sutures. Although the well-exposed surface geology of some mountain chains provides an important starting point for understanding the deformational processes acting at the plate boundary, well designed geophysical investigations are essential to understand the dynamics of a convergent continental plate boundary and constrain the deformational processes and their drivers within the deeper crust and lithospheric mantle.

The Pacific and Indo-Australian plates in South Island, New Zealand, are separated by the transpressional Alpine fault. Associated with this continental transform fault are the Southern Alps, a relatively simple and young orogen created by continental collision whose zone of deformation is laterally narrow and uncomplicated by subsequent tectonic overprinting. This mountain system offers the opportunity to understand how lithospheric rocks deform within a developing orogen and how this deformation may have changed with time. Because relative plate motions are oblique, the Alpine fault needs to accommodate both convergence and lateral slip. Topical questions regarding this plate boundary relate to whether strain is partitioned and whether it is localized or diffuse within the crust and/or lithospheric mantle. Furthermore, this transpressional plate boundary has often been compared to the Transverse Ranges of the San Andreas fault (California, USA) and contrasted to the Dead Sea transform (Israel-Jordan), the Northern Anatolian fault (Turkey), and the Denali fault (Alaska, USA). Continental collision at this plate boundary serves as a model to understand other

more complex active orogenic regions where deformation is broader or occurred over a much longer period, such as in the Cordillera of western North America where the width of deformation exceeds several hundreds of kilometers.

In the early 1990s an international collaboration formed between New Zealand and United States geoscientists in order to study continental collision at the Pacific/Indo-Australian plate boundary. The selection of this site in central South Island was attractive because of supposed similarities to the Transverse Ranges in southern California (of direct interest to the U.S. investigators); easy access well into the heart of the Southern Alps; a low population density which improved field experiment logistics, permitting, and signal-to-noise levels; a narrow island width, which allowed marine seismic methods to be applied onto both sides of the land-situated orogen; and an established foundation of scientific knowledge built by an excellent in-country scientific community. The initial collaboration, funded by the New Zealand Foundation for Research, Science, and Technology and the U.S.'s National Science Foundation (Continental Dynamics Program), was quickly joined by additional scientists primarily from New Zealand who performed their own relevant but independent studies. Thus a decade-long focus on the central South Island transpressional plate boundary ensued. More than fifty scientists and students participated in the overall set of studies, many of which involved substantial field observation experiments, and which has led to forty journal publications and a dozen graduate theses.

This volume represents a collection of papers which primarily summarizes the results that arose from our overall New Zealand-U.S. collaboration. The chapters cover a range of geological and geophysical investigations and provide further insight into the deformation and evolution of continental collision. The volume is divided into four sections. Preceding the first section, an inroductory chapter presents the scientific questions that motivated the collaborations and summarizes the key findings of the overall research. The first section presents the regional framework of the Pacific/Indo-Australian plate boundary within South Island; its six chapters summarize what is known of the geology and geophysics of what is essentially the far-field region relative to deformation at the plate boundary. The second section focuses on the plate boundary (Alpine) fault and the Southern Alps orogen; the two geological and two geophysical chapters characterize the near-field region of the plate boundary.

A Continental Plate Boundary: Tectonics at South Island, New Zealand
Geophysical Monograph Series 175
Copyright 2007 by the American Geophysical Union.
10.1029/175GM01

The third section contains three chapters which examine the dynamics of the plate boundary - how mechanically the plate boundary and its fault(s) accommodate both strike-slip and convergent motions. The final section presents three chapters that compare the Alpine fault-Southern Alps system to other transpressional and obliquely convergent plate boundaries. We have included a CD-ROM that contains both color versions of figures that are printed in black & white within the volume plus supplemental/oversize materials that provide added content to selected chapters.

As editors, we acknowledge and greatly appreciate the contributions provided by the authors and reviewers. This volume would not have been possible without the major efforts of the authors, who squeezed into their very full schedules the time to write summary and synthesis papers. Reviewers, as usual, have carried out an essential and often underappreciated task; we thank Duncan Agnew, Thora Arnadottir, Gary Axen, Geoff Batt, David Berryman, Tom Brocher, Tim Byrne, Ramon Carbonell, James Connolly, John Dewey, Donna Eberhart-Phillips, Susan Ellis, Andrew Gorman, David Gray, John Hole, Keith Howard, Leon Teng, Vadim Levin, Tim Little, Zhen Liu, John Louie, Peter Malin, Peter Molnar, Martin Reyners, David Rodgers, Martyn Unsworth, and five anonymous reviewers. We also thank Harm van Avendonk and Stuart Henrys for handling the editorial duties for two of the chapter papers.

We thank at AGU the book acquisition editors - initially Allan Graubard and subsequently Jeffrey Robbins. The patient support of the AGU books and special publications staff, particularly Dawn Seigler, administrative assistant, and Maxine Aldred, program coordinator, is greatly appreciated.

Funds to cover publication costs were provided by the U.S. National Science Foundation Continental Dynamics Program, the New Zealand Foundation for Research Science and Technology, GNS Science, and Victoria University of Wellington. We particularly thank Dr. Leonard Johnson, Program Director of NSF-CD, who also provided funds for a synthesis workshop held during June 2005, in Christchurch, New Zealand, that led directly to the organization and contents of this volume. We also direct special thanks to John McRaney, at the University of Southern California, who not only arranged for additional funds to defray publication costs but provided administrative and logistical support throughout our New Zealand-US projects that facilitated our large field experiments to take place.

Finally, we wish to dedicate this volume to three scientists who had profound influence on our interest in the South Island transpressional plate boundary and on our establishment of the New Zealand-U.S. collaboration: Prof. Richard I. Walcott (Victoria University of Wellington), Prof. Thomas L. Henyey (University of Southern California), and Prof. Thomas V. McEvilly (University of California, Berkeley). Dick Walcott was one of the first geophysicists in the 1970's to investigate the deformation of New Zealand using geodetic data; his studies of active deformation provided the foundations for many scientists' subsequent studies on the kinematics and dynamics of New Zealand's subduction and strike slip plate boundaries. Dick provided early guidance to our project scientific direction. His careful use of field observations to quantitatively constrain concepts was an approach that we applied throughout our South Island collaborations. Tom Henyey similarly valued field observations; his initial interest in New Zealand was in the collection of heat flow measurements in South Island lakes following a sabbatical in 1982 at the Department of Scientific and Industrial Research (DSIR) in Wellington. Tom's long-standing research in the tectonics of Southern California led to an early interest in the comparison between the transpressional San Andreas and Alpine faults. Discussions of this comparison with one of us (TS) beginning in 1988 led to concrete plans for a multinational geophysical examination of the Alpine fault plate boundary; Tom's leadership was instrumental in assembling our trans-Pacific collaboration and steering it through its original tenure. Tom McEvilly was the rare seismologist who knew how to design at scales ranging from the inner workings of seismometers all the way up to effective structures of national and international community organizations such as IRIS and FDSN. After providing sound advice at an early planning workshop in Wellington, New Zealand, Tom became an active project participant. He led our passive seismic experiments, in planning and with shovel in hand, and bird-dogged the seismic crew we contracted for seismic reflection profiling near Mount Cook. Tom, who passed away in 2002, had a wry sense of humor that could defuse the tension or enliven the dullness of any meeting. As a mentor he had a knack for correcting us in such a way so that we thought we solved mistakes on our own. We extend our gratitude to Professors Dick Walcott, Tom Henyey, and Tom McEvilly, who all led by their example how to accomplish high caliber research and gave us strong encouragement to carry out the collaborative research presented in this volume.

David Okaya
Tim Stern
Fred Davey

Continent-Continent Collision at the Pacific/Indo-Australian Plate Boundary: Background, Motivation, and Principal Results

David Okaya[1], Tim Stern[2], Fred Davey[3], Stuart Henrys[3], and Simon Cox[4]

BACKGROUND

Mountain belts are important and highly visible structural elements of the earth's continental crust. They impact societies by providing natural resources, hosting processes which produce natural hazards, and by their influence on weather and climate. One important mechanism in mountain formation is continent-continent collision at plate boundaries. Within mountain belts, compression, thrust faulting, and erosion can combine to generate uplift and over-thickened crust, and exhume large sections of high-grade, once deeply-buried rocks and associated faults, sutures, and folds. The well-exposed surface geology of mountain belts provides an important starting point or "ground truth" for geological and geophysical investigations of continental dynamics and the relevant processes in the deeper crust.

Continent-continent collision occurs at two major plate boundary settings. The first is at convergent plate boundaries where a continental lithospheric plate comes into contact with another such plate, often creating spectacular collisional orogens (e.g., the Himalayan system due to Indian-Eurasian plate collision and the Zagros belt due to Arabian-Eurasian plate interaction). The second major setting is at strike-slip plate boundaries, where a component of oblique convergence will often result in mountain building on either or both sides of the plate boundary. Notable cases are the Southern Alps in New Zealand and the Transverse

Ranges in southern California. The structure and kinematic regime of these transpressional orogens is thus, in all likelihood, more complex as a result of the need to accommodate large amounts of lateral slip, as well as convergence.

WHY STUDY NEW ZEALAND?

New Zealand is a largely submarine continent that lies across the boundary between the Pacific and Australian plates (Plate 1). Subduction of the Pacific plate beneath the Australian plate occurs at the eastern margin of North Island. Subduction of the Australian plate occurs along the Puysegur Trench and under the southwestern tip of South Island. Connecting these subduction zones of opposite polarity is a continental transform fault - the Alpine fault - which runs obliquely through South Island (Plate 2). Present day relative velocity between the two plates is about 38 mm/y. A component of compression has existed across the Alpine Fault for at least the past 5–10 million years [*Walcott*, 1998], which has resulted in the building of the Southern Alps with a maximum elevation of 3754 m.

Central South Island has long been considered a premier site to study oblique continent-continent convergence [*Molnar*, 1988]. Rates of erosion and concomitant vertical movement of crustal rock here are among the highest in the world (~10 mm/yr) [*Blythe*, 1998], and the Alpine fault has components of both strike-slip and dip-slip movement [*Norris et al.*, 1990]. The orogen is young, and the zone of deformation is relatively narrow (~80 km wide) and uncomplicated by subsequent tectonic overprinting (Plate 2). In many cases, such as in southeast Asia or the Cordillera of western North America, the width of the zone of deformation exceeds hundreds of kilometers and is often complicated by a lengthy history of tectonic episodes [*Dickinson*, 2003].

The rapid deformation of central South Island [*Walcott*, 1979; *Beavan et al.*, 1999] allows active processes to be studied thoroughly using Quaternary geology, geodesy, thermochronology, and seismology, building on established programs in these disciplines within New Zealand. The geology

[1]Department of Earth Sciences, University of Southern California, Los Angeles, California.

[2]School of Earth Sciences, Victoria University of Wellington, Wellington, New Zealand.

[3]GNS Science, Lower Hutt, New Zealand.

[4]GNS Science, Dunedin, New Zealand.

A Continental Plate Boundary: Tectonics at South Island, New Zealand
Geophysical Monograph Series 175
Copyright 2007 by the American Geophysical Union.
10.1029/175GM02

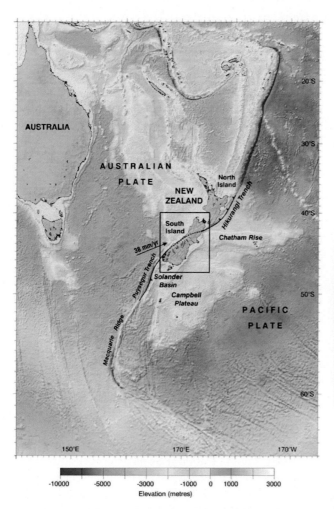

Plate 1. The plate tectonic setting of the New Zealand region. In South Island (box), the transpressional Alpine fault (dotted line) separates the Hikurangi and Puysegur trenches along the Pacific-Australian plate boundary. Morphology in color. [From *Davey et al.*, this volume].

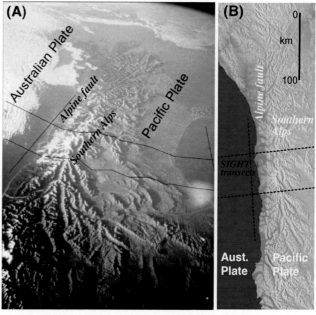

Plate 2. South Island, the Alpine fault and Southern Alps. (A) Oblique-view image of South Island with SIGHT lines superimposed. Visible is the Southern Alps orogen associated with the transpressional Alpine fault. Space Shuttle image STS59-229-017 courtesy of NASA. (B) Digital elevation image of South Island which reveals the linearity and sharpness of the Alpine fault. Elevation data collected by Shuttle Radar Topography Mission aboard NASA space shuttle. Image PIA06661 courtesy of NASA.

of the country (Plate 3A; see oversized version on the CD-ROM that accompanies this volume) permits the separation of older tectonic episodes from Cenozoic deformation, and makes quantifying the amounts of deformation by various mechanisms easier than in other areas of the world. This quantification is aided by the narrowness of the belt between two well-defined lithospheric plates. For example, the lower and middle crust has been turned up and exhumed along the Alpine fault (Figure 1) resulting in exposure of progressively deeper parts of the crust that can be sampled along a line only 20-30 km long [*Wellman*, 1979]. The combination of exceptionally rapid erosion [*Tippett and Kamp*, 1993; *Little et al.*, 2005] and youthful deformation allows the study of mountain building in its infancy through the simultaneous analysis of rocks deformed at the surface and those deeply exhumed in late Cenozoic time.

The relatively simple geology of New Zealand reveals deformation that undoubtedly has occurred elsewhere but simply cannot be seen through the complexity of the world's typical surface geology. The demonstration of 480 km of right-lateral slip on the Alpine fault [*Wellman*, 1955] preceded the recognition of roughly 300 km of slip on the San Andreas fault, in part because such displacements are far

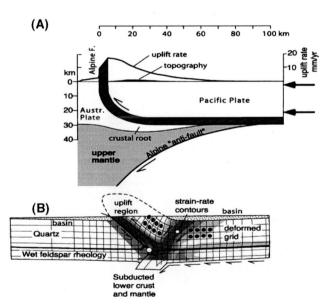

Figure 1. Early models of the Pacific-Australian plate boundary. (A) *Wellman*'s [1979] model for delamination at the Alpine fault. Crust and mantle separate. (B) Model for the geodynamic evolution of the Southern Alps [*Beaumont et al.*, 1994] made before the SIGHT experiments were conducted. Subduction of the lower crust and mantle occurs within this numerical model. Figure courtesy of *Little et al.* [2002].

more obvious from a geologic map of New Zealand than from one of California (Plate 3A). Moreover, the 480 km of slip on the Alpine fault is an underestimate for the relative displacement of the western North Island with respect to eastern South Island of New Zealand. Right-lateral shear parallel to the Alpine fault has been shown to be distributed over a relatively wide zone on South Island and perhaps over a wider zone on the North Island [e.g., *Norris*, 1979; *Sutherland*, 1999]. Reconstructions of plate motions suggest that as much as 800 km of right-lateral displacement has occurred [*Stock and Molnar*, 1982; *Sutherland*, 1999]. The difference between 450 km of slip on the Alpine fault and 800 km of total slip does not seem to have been absorbed by slip on secondary faults, but rather by distributed shear [*Molnar et al.*, 1999].

High quality geodetic measurements in New Zealand were central to the concepts of distributed deformation in continental regions [*Walcott*, 1984]. Differences between repeated geodetic surveys when integrated across South Island are consistent with both the rate and orientation of plate motion averaged over the last 2–5 Ma [*Walcott*, 1979, 1984, 1998; *Beavan et al.*, 1999]. Thus, there does not appear to be a need for postulating offshore slip or strain. Moreover, the geodetically measured deformation is not confined to faults but, like the permanent strain, is diffuse. This emphasizes the need to consider continuous deformation in kinematic and dynamic modeling [e.g., *Walcott*, 1979, 1984; *Moore et al.*, 2002].

Advances in subsurface geophysical methods and crustal-scale seismic exploration in late 1980s–early 1990s led to international NZ-US collaboration. Logistically, New Zealand offers some advantages not realizable in most orogenic belts. The Southern Alps is one of the few zones of continental convergence that can be studied on both sides from the sea [*Okaya et al.*, 2002]. The narrowness of South Island (Plate 3B; see oversized version on the CD-ROM that accompanies this volume) makes it possible to combine arrays of onshore and offshore seismic sources and receivers to greatly enrich the quality and quantity of data at modest cost. The compactness of the belt also minimizes the logistical difficulties for complementary geophysical experiments including gravity, magnetotellurics, and teleseismic seismology. With the exception of the highest part of the range, the Southern Alps are readily accessible via relatively straight rural farm roads and broad, deeply eroded, linear glacial valleys.

KEY QUESTIONS

By the early 1980's it was recognized that crust and mantle of the Pacific Plate beneath central South Island separated in some fashion and the crust was being obducted and rapidly

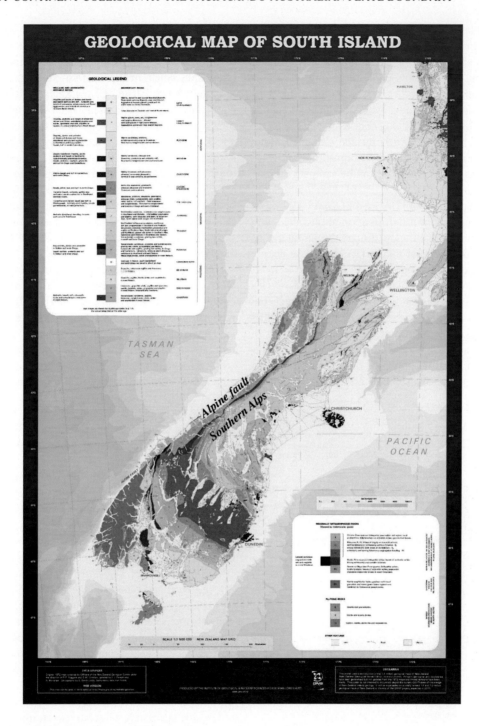

Plate 3. A. Geology map of South Island. Map produced from QMAP 1:250,000 geological database. This plate is a reduced version of the original oversized figure available on CD-ROM that accompanies this volume. Details of geological legend available on oversized figure. B. Location of South Island Geophysical Transects superimposed on geological map. (Top) Experimental observation locations shown for passive seismology (SAPSE), refraction and wide-angle reflection profiling (SIGHT), and petrophysical sample sites. Geological units identified on companion Plate 3A. (Bottom left) Land refraction (explosion source) profiles (SIGHT) superimposed on geology. (Bottom right) Seismic reflection profiles (CDP98) and magnetotelluric stations superimposed on digital topography. LP denotes Lake Pukaki, site of seismic reflection pilot study. This plate is a reduced version of the original oversized figure available on CD-ROM that accompanies this volume.

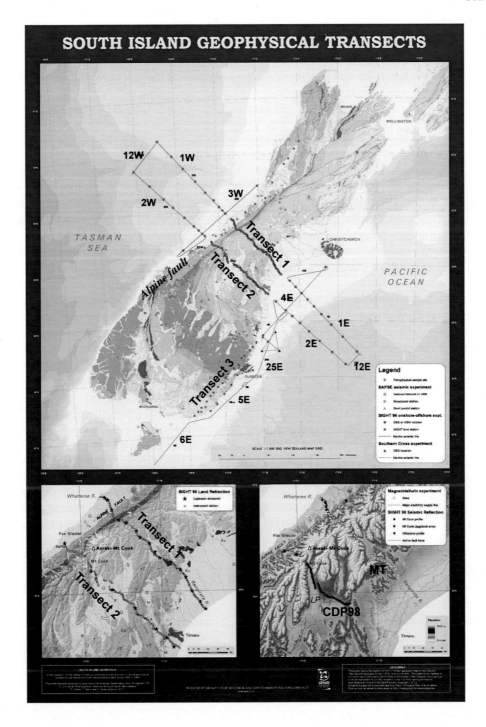

Plate 3. (Continued)

eroded (Figure 1a) [e.g., *Wellman*, 1979; *Adams*, 1980]. What was not clear at that time was the shape and position of the decollement between the crust and mantle. Moreover, it was not known how the thickened mantle lithosphere was being disposed of. Computer modeling in the early 1990's elaborated on Wellman's early conceptual models and defined their physical basis (Figure 1b). Shown in Figure 2 are conceptual models that were valid in 1993 to explain what may be happening beneath the Southern Alps. These are variations on the theme of decollement between crust and mantle, and/or lower crust-mantle; but most did not address the question of how the Alpine fault penetrates into the mantle.

The New Zealand-USA collaborative project SIGHT (South Island Geophysical Transect), formed in the early 1990's, set out to answer the following series of questions:

1. What is the expression of the plate boundary zone in both crust and mantle?

2. What are the processes (structures) in the lithosphere that accommodate the oblique convergence?

3. How are processes in the mantle coupled or related to processes in the crust?

4. How do the distributions of horizontal and vertical strains across the plate boundary differ?

5. Is strain generally partitioned along an oblique-convergent continent-continent plate boundary, or might New Zealand be different? How does the answer to this question relate to patterns of aseismic and seismic strain in New Zealand?

6. Do the reflection Moho and crustal reflectivity beneath South Island act as effective strain markers for Cenozoic deformation?

7. If so, does deformation appear to be manifested as ductile flow, brittle failure, or a broad elastic flexure?

8. What is the role of the Alpine fault in the convergent process? Is the apparent dip on the fault simply a surficial effect or does it persist well into the crust?

9. Has the 500 km or so of right lateral slip along the Alpine fault left an anisotropic seismic expression in the crust or mantle?

10. Can the mechanical processes of lateral slip along the Alpine fault and convergence across the southern Alps be considered independently in terms of structural expression or must they be treated as coupled?

11. How does the rapid erosion rate affect the mechanics of deformation?

12. When the seismic crustal image is combined with observed gravity data, how is the load of the Southern Alps supported? Is strength of the lithosphere important or is some dynamic support required?

These twelve key questions gave rise to a series of experimental targets as part of the science plan for the SIGHT project that could be directly tested by seismological and complementary geophysical work.

EXPERIMENTAL TARGETS

The above set of scientific questions coupled with the conceptual models (Figure 2) postulated various configurations of the crust-mantle system across the Southern Alps transpressional orogen. At the time of the project beginning, an integrated model of the crust and upper mantle beneath the Southern Alps including the Alpine fault remained to be proposed. The existing models (Figure 2) were each based on mainly one type of geological or geophysical approach; nevertheless they were illustrative of the range of hypotheses that could be directly tested by seismic reflection, refraction and ancillary geophysical work along the proposed transects.

In order to distinguish among these conceptual models, project field experiments set out to obtain the best possible information on the physical properties (e.g., rheology, composition) of the crust and lithospheric mantle and on the architecture of several key structural markers, including (1) the Moho, (2) lower-crustal reflectivity, (3) the Alpine fault and associated faults, and (4) other potential structural or compositional boundaries in the crust (e.g., top of a mafic lower crust). These structures served as important strain markers which, when combined with the results of geodetic, geobarometric/thermometric, and geological studies, produced new insights into the deformation that has accompanied transpressional orogeny in the Southern Alps. Imaging these targets provided constraints to address the above scientific questions; these constraints included:

- the rheology of the lithosphere, including the brittle/plastic transition zone,
- lithospheric composition, including continental vs. oceanic affinities, densities, and evidence for fluids,
- the dip and depth extent of the Alpine fault, especially its expression in the lower crust and mantle,
- the extent and configuration of antithetic thrusting in the eastern Alps,
- the pattern of uplift in the Alps,
- the configuration of the Moho and implications for isostatic compensation of the Alps,
- the velocity, reflectivity, and anisotropy of the upper mantle, including any evidence for shear strain accommodating plate motion,
- the underpinnings and configuration at depth of the Torlesse and Haast complexes, and

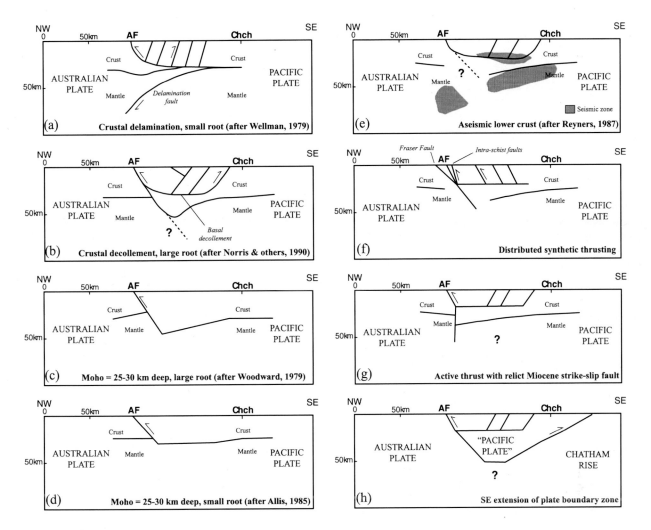

Figure 2. Schematic models of convergence at the transpressional plate boundary between the Pacific and Australian plates. From a historical standpoint, these models were valid in 1993 and represent testable hypotheses which formed the basis for the SIGHT research project. These models involve various combinations of Pacific crust exhumation, Pacific upper crust antithetic faulting, Pacific lower crust and/or mantle in subduction, and Australian plate as a backstop with a range of possible geometries. (a) Uplift of the Southern Alps using a process of delamination of the lithosphere into its crustal and mantle components; no Southern Alps root is necessary [*Wellman*, 1979; see Figure 1]. (b) Application of the critical wedge model to the Southern Alps suggesting an overthickened crust with frontal faults and the existence of a detachment in the ductile lower crust [*Koons*, 1989, 1990]. *Norris et al.* [1990] suggested that exposed fault mylonites from 20–25 km depth indicate the depth at which delamination, or detachment, is taking place. Large Moho root is present. (c) Moho topography with a large root under the Southern Alps [*Woodward*, 1979]. (d) Shallow Moho with no root under the Southern Alps [*Allis*, 1986]. (e) Intracrustal detachment with seismogenic zones [*Reyners*, 1987]. Moho which dips. (f) Distributed uplift along many synthetic faults as opposed to uplift along the primary Alpine fault (models #a, c, d) or in a two-sided critical wedge with antithetic faults (#b, e). (g) Is there evidence that the Alpine fault may be detached along a mid-crustal decollement as has been suggested for the San Andreas fault in both northern and southern California [e.g., *Hadley and Kanamori*, 1977; *Furlong et al.*, 1989; *Namson and Davis*, 1991], or perhaps stranded from its pre-Pliocene location in the lower crust and upper mantle as shown here? (h) Finally, what unforeseen Pacific/Indo-Australian convergence geometries might exist as suggested here or in #e or #g?

Table 1. Experimental targets and geophysical/geological methods which provide constraints. The experimental targets were used to discriminate among the project hypotheses and testable models (Figure 2). The methods provide observations which define how the experimental targets are applicable within South Island. In most cases an individual method can address several experimental targets.

	CDP/MCS Reflection	Onshore Refraction	Onshore/offshore Refract./reflect.	High-resolution CDP Reflection	Local seismicity	Teleseismic analysis
Rheology/Petrology of crust/mantle system	√	√	√		√	√
Configuration of Alpine fault	√	√	√	√	√	
Configuration of Moho	√	√	√			√
Pattern of seismicity in Southern Alps (S.A.)		√	√		√	
Relation of Alpine fault to thrusts in Westlands	√	√	√	√	√	
Extent & configuration of antithetic thrusts in eastern S.A.	√	√			√	
Pattern of uplift in S.A.	√	√	√	√	√	√
Isostatic compentation of S.A.	√	√	√			√
Sedimentation pattern around S.A.	√	√	√			
Underpinnings of Torlesse	√	√	√			
Velocity/reflectivity of lithospheric mantle	√	√	√		√	√
Extent/orientation of shear strain in mantle						√
Crustal anisotropy	√	√	√		√	√

• structural and compositional controls on seismicity along the Alpine fault.

Obtaining these images and physical property information required a multidisciplinary approach (Table 1). Individual targets needed information provided by different methods in order to produce well-constrained or well-imaged geometries and attributes. The core SIGHT project carried out experiments which included passive seismology, land refraction and double-sided seismic onshore-offshore profiling plus marine MCS profiling, high resolution seismic reflection profiling, magnetotellurics and petrophysical sampling. Additional experiments were performed including shallow seismics and gravity, regional geodetics, and numerous geological studies. Descriptions of these field efforts and their results are summarized in other chapters in this volume [e.g., *Davey et al.*, this volume; *Savage et al.*, this volume; *Stern et al.*, this volume] as well as in the numerous project publications.

THE SIGHT PROJECT AND GREATER SOUTH ISLAND OROGEN PROJECT

SIGHT, the South Island Geophysical Transect, refers to a range of field observation projects that were broadly focused on two transects across mid South Island in the period 1995-98 (Plate 3B). The main focus was active source seismic work but SIGHT also included magnetotellurics and petrophysical work. A passive seismic experiment - Southern Alps Passive Seismic Experiment (SAPSE) ran in parallel with SIGHT [*Leitner et al.*, 2001]. Data from both SAPSE and SIGHT were combined to build a tomographic image of the crust beneath the Southern Alps [*Eberhart-Phillips and Ban-*

Table 1. (Continued)

	Petrophysics	Magnetotellurics	Gravity	Flexural Modeling	Fission Track	Geological strip mapping
Rheology/Petrology of crust/mantle system	√	√	√	√		√
Configuration of Alpine fault		√	√	√	√	√
Configuration of Moho			√	√		
Pattern of seismicity in Southern Alps (S.A.)						
Relation of Alpine fault to thrusts in Westlands						
Extent & configuration of antithetic thrusts in eastern S.A.						
Pattern of uplift in S.A.			√	√	√	√
Isostatic compentation of S.A.			√	√		
Sedimentation pattern around S.A.			√			
Underpinnings of Torlesse	√		√			
Velocity/reflectivity of lithospheric mantle			√			
Extent/orientation of shear strain in mantle	√					
Crustal anisotropy	√					

nister, 2002]. Although these NSF-NZ Science Foundation-funded projects were mainly focused on geophysical data acquisition, additional studies in structural geology, geodesy and geodynamic modeling ran in parallel during the decade 1995–2005. Thus the term South Island Orogen Project is used to encompass all these projects. This book summarizes the results of the larger set of research under the South Island Orogen Project including the core SIGHT-SAPSE projects.

Primary participants on the South Island Orogen Project came from three New Zealand Institutions (Victoria University of Wellington, Otago University and GNS Science) and eight US institutions (U. Southern California, U. of California/Berkeley, MIT, San Diego State U., SUNY, U. of Utah, U. of Wisconsin, and U. of Wyoming). In addition there are some 25 outer-circle participants from other institutions. Fifteen students and postdocs completed studies based on just

the SIGHT/SAPSE projects. Funding came from one major NSF grant and several PGST grants from the New Zealand Science Foundation.

Previous Studies of Orogens

Prior to the period of the SIGHT experiment, seismic investigations of compressional orogens had been carried out in the Pyrenees [*Choukrone and team*, 1989], the European Alps [*Pfiffner et al.*, 1990], the Wind River Mountains [*Smithson et al.*, 1979] and in much lower and now eroded mountain chains like the Appalachians [*Cook et al.*, 1979] and the Grenville Front [*Green et al.*, 1988] and the Canadian Cordillera [*Cook and Varsek*, 1994]. These earlier studies focused on the geometries of thick and thin-skinned thrusting that occur at orogenic belts, the nature and thickness of the crust, and how regional isostatic compensation occurs. SIGHT focused on a relatively simple and youthful orogen with the intent of providing an important addition to these previous studies, which are in currently less active environments. At the time of SIGHT, knowledge of the geology of South Island was well known but that of the tectonics only moderately known with a distinct lack of knowledge of deeper structure and how deformation at the plate boundary was being accommodated.

WHAT WAS LEARNT?

Outcomes of the projects are many and varied and went well beyond that signaled in the original proposal. What follows are some of the outcomes that link to the key questions listed above. In-depth presentations of these outcomes are provided in the papers within this volume.

Field Methods

From an operational point of view one of the main successes of SIGHT was to demonstrate the power and efficiency of the "double-sided" onshore-offshore seismic exploration method to continental islands like South Island. Few global localities offer such a favorable setting for onshore-offshore exploration from both sides of a continental island. Specifically, the combination of a relatively unpopulated continental island, an island wide enough (160–200 km) to generate all crustal seismic phases from onshore shots, yet narrow enough that ship-generated air-gun waves could be detected with onshore seismic detectors from coast to coast (Plate 3B), and offshore waters where a seismic ship can operate without restriction. Thus one of the most distinctive outcomes of SIGHT was the acquisition of top quality crustal seismic data sets that, for example, gave rise to 600 km-long "supergathers" [*Okaya et al.*, 2002, 2003].

Crustal Structure

Crustal thickness. The most dominant structural feature discovered by SIGHT is the asymmetric crustal thickening of South Island that appears to be concentrated in the Pacific plate (eastern South Island) and is strongly asymmetric (Plate 4) [Kleffmann et al., 1998; Scherwath et al., 2003; van Avendonk et al., 2004]. Maximum crustal thickness is 44 ± 1.5 km beneath the highest portion of the Alps. This represents ~17 km of crustal thickening from both the east and west coast where well resolved refraction and reflection data put the Moho at 27 ± 1 km (Plate 4a and b).

On the third transect down eastern South Island (Plate 4c) a 600 km-long multichannel seismic and wide-angle seismic

reflection image was created [*Godfrey et al.*, 2001]. This profile highlighted the contrasting seismic fabric and thickness of the Eastern and Western Provinces [*Mortimer*, 2004]: the former being about 27 km thick and Permian-Cretaceous in age, and the latter 35 km thick and Paleozoic–Mesozoic in age (Plate 3B) [*Davey*, 2005]. The Eastern Province is the thin greywacke-schist crust that overlies a reflective lower oceanic crust and forms most of the crust of New Zealand. The Western Province is in effect a fragment of Gondwana continental crust that separated during the opening of the Tasman Sea [*Mortimer*, 2004].

On Transect 3 a serendipitous find was the deep crustal image of the mid-Miocene Dunedin volcano complex [*Godfrey et al.*, 2001]. Beneath the complex a domed area of underplating is interpreted near the Moho, and seismic velocities in both the crust and mantle are reduced by about 10–15%. Thus even though this volcano has been dormant for ~12 my there is still evidence of partial melts and/or volatiles in the upper mantle and lower crust.

Decollement surface beneath the Southern Alps. Central to many of the key questions listed above is the issue of how and where the mantle and crust separate beneath central South Island. A decollement at a depth between 35 and 15 km has been imaged and interpreted (Figure 3). A combination of wide-angle and migrated, high-resolution seismic reflection methods are used to show that the crust beneath the Southern Alps is split by this zone of broad and strong reflectivity (Figure 3). Below the decollement surface a crustal root has developed with rocks of seismic velocity 6.8 km/s, which are interpreted to be the oceanic crust on which the greywacke schists were deposited [*Scherwath et al.*, 2003]. Above the decollement the rocks are all of seismic velocity <6.3 km/s and are interpreted to be the greywacke–schist rocks (Plate 4). Interestingly, the lower crust delaminates and moves downward with the mantle lithosphere rather than attaching itself to the upper crust. It is unclear if this behavior is peculiar to just this orogen, or it is a more general phenomenon. If the latter, there are important implications for models of mantle instability, lithospheric detachments and continental evolution. Specifically, it is unclear of the role that the lower crust plays in mantle detachments: i.e., does it fully participate in a detachment [*Schott and Schmeling*, 1998] or is the buoyancy of even lower continental crust just too high to overcome [*Schott and Schmeling*, 1998; *Molnar and Houseman*, 2004]?

Flexural rigidity and strength estimates across the orogen. To the west of the Southern Alps the Australian plate is deformed by flexural bending under the load of the crust of the Pacific plate, which overthrusts the Australian plate at the

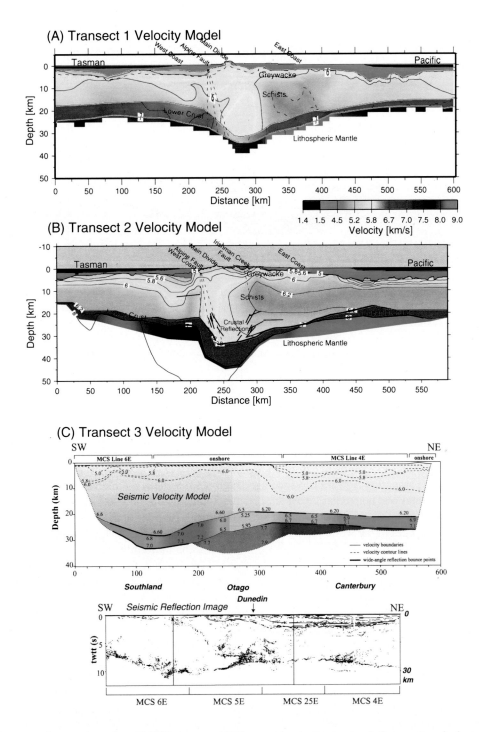

Plate 4. Crustal structure images along three SIGHT transects. (A) Transect 1 velocity structure [after *van Avendonk et al.,* 2004; see *Stern et al.,* this volume]. 5× vertical exaggeration. Low velocity zone in crust beneath Southern Alps in the hanging wall of the Alpine fault. (B) Transect 2 velocity structure [after *Scherwath et al.,* 2003; see *Stern et al.,* this volume]. 5x vertical exaggeration. Velocity structure similar to Transect 1, but with a deeper root of the Southern Alps. (C) Transect 3 velocity structure [from *Godfrey et al.,* 2001]. Underneath this seismic velocity structure is a line drawing of the marine seismic reflection profile along this transect.

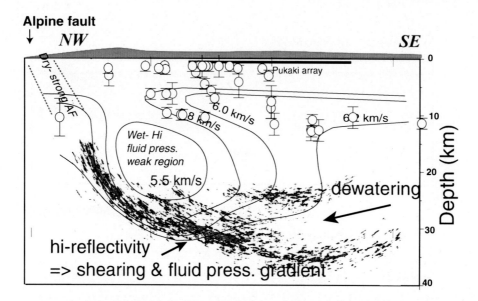

Figure 3. Migrated version of the western portion of the SIGHT seismic reflection profile. Superimposed on the migrated section are seismic velocity contours, earthquake locations. Interpreted regions of dewatering, shearing and high fluid pressure are shown [*Okaya et al.*, 2007].

Alpine fault [*Harrison*, 1999]. This bending can be tracked ~150 km west of the Alpine fault and implies an effective elastic thickness of 15 ± 5 km for the edge of the Australian plate. Beneath the Southern Alps, however, the flexural rigidity of the lithosphere is required to be reduced almost to zero. This latter result is based on an analysis of deformation of the Moho on Transect 1, assuming that prior to loading of the Southern Alps the Moho was horizontal [*Stern et al.*, 2002], and is in keeping with flexural rigidities estimated for other collisional mountain ranges of the world [*Stewart and Watts*, 1997].

The Alpine Fault zone. From the outset one of the key foci of the South Island Orogen project was the detailed structure of the Alpine fault zone, in particular, the search for geophysical properties from which we could learn about fault zone dynamics. Four different sub-disciplines have combined to contribute to this goal: structural geology, magnetotellurics, active source seismology and earthquake seismology. At depth the Alpine fault is most strongly manifested as a broad, southeast-dipping zone seen with electrical and seismic means (Plate 4 and Figure 3). It is characterized at lower crustal depths by seismic P-wave speeds that are 10% less than normal and electrical resistivities two orders of magnitude less than normal [see *Jiracek et al.*, this volume]. Combined, these observations suggest the presence of interconnected fluid and high fluid pressure [*Stern et al.*, 2001; *Wannamaker et al.*, 2002].

Mantle Structure

Thickened mantle lithosphere. Anomalous P-wave delays of up to 1 s were recorded for teleseismic ray-paths that have passed through the upper mantle beneath the Southern Alps [*Stern et al.*, 2000]. A high speed, and hence relative high-density, body is inferred directly beneath the root of the Southern Alps and with sides that are within 15 degrees of vertical. Because of the shape, dimensions (100 km deep, 80 km wide) and location of the anomalous body it is interpreted to be uniformly thickened mantle lithosphere [*Stern et al.*, 2000] rather than subducted lithosphere as originally proposed by, for example, *Wellman* [1979]. This high speed anomaly in the mantle is also evident in a regional tomography analysis [*Kohler and Eberhart-Phillips*, 2002].

Mantle deformation and the crustal root. A load in addition to topography is required to maintain the crustal root. The average topography in the central Alps is only 1600 m [*Koons et al.*, 1993] yet the observed root has an amplitude of about 17 km [*Scherwath et al.*, 2003]. Given a ~5:1 ratio of root to topography amplitude [*Fowler*, 1995], assuming local isostatic equilibrium, the Southern Alps root is nearly twice as large to that needed to just support topography. But in cross-sectional form the mass excess of the thickened mantle is similar in magnitude to that of surface topography. Together, these excess masses balance the negative buoyancy of the crustal root beneath the Southern Alps [*Stern et al.*,

2002]. Equal thickening of both lower crustal (7 km/s) and lower mid-crustal (6.1 km/s) rocks contributes to the root [*Scherwath et al.*, 2003].

Anisotropy and mantle deformation. Seismic anisotropy of crustal and mantle rocks is a major feature of the orogen (Figure 4). Strong shear-wave splitting in the mantle (SKS splitting) was identified at an early stage in the project [*Klosko et al.*, 1999] and have stimulated a variety of tectonic interpretations [*Molnar et al.*, 1999; *Little et al.*, 2002; *Moore et al.*, 2002; *Baldock and Stern*, 2005]. Because of

the excellent coverage of the onshore-offshore shooting upper mantle (*Pn*) anisotropy was also measured with active source methods [*Scherwath et al.*, 2002; *Baldock and Stern*, 2005]. The combination of SKS splitting and Pn anisotropy provides a powerful constraint on where the anisotropy and shearing in the mantle occurs.

All interpretations include an element of distributed deformation in the mantle lithosphere to account for the observed anisotropy. But arguably the most contentious issue faced by the project is in the second part of the first key question listed above: how is the plate boundary expressed in the mantle?

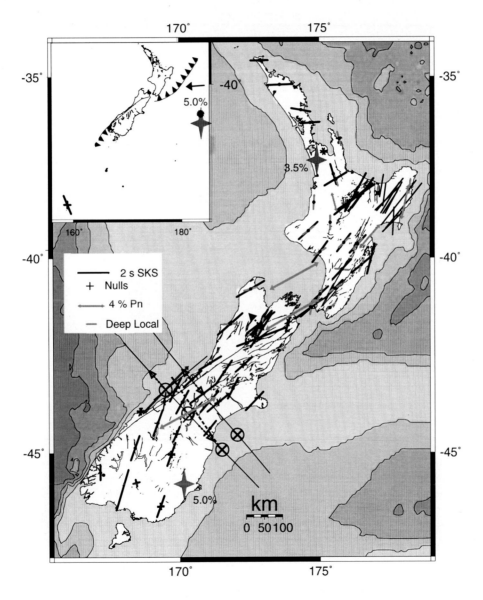

Figure 4. Map of shear wave splitting (anisotropy) measurements in New Zealand. Bars represent measurements of shear wave splitting. Direction of shear wave splitting is sub-parallel to the Pacific-Australian plate boundary. [From *Savage et al.*, this volume].

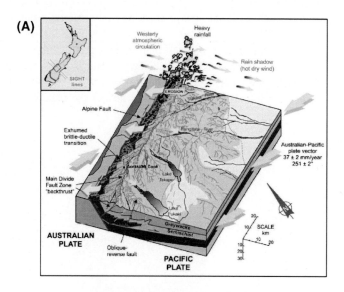

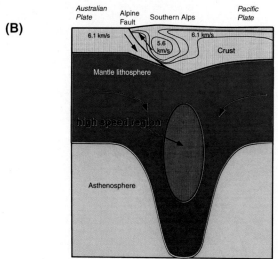

Plate 5. Integration of geological and geophysical results from SIGHT/SAPSE and the South Island Orogen Project. (a) block model of transpression along the Alpine fault [from *Cox and Sutherland*, this volume]. The Pacific plate upper and middle crust is exhumed along the Alpine fault zone. The lower crust accumulates in the root of the Southern Alps. Exhumation and climate-driven erosion laterally offsets the Southern Alps mountains from its root. (b) Crust-mantle structure inferred from SIGHT/SAPSE geophysical data [from *Stern et al.*, this volume]. A relatively cold, dense, and faster seismic wavespeed body is located under western South Island; the Southern Alps root is dynamically pulled downward by this mantle feature.

While evidence from seismic anisotropy and P-wave delay calls for distributed deformation in the mantle lithosphere rather than brittle faulting [*Molnar et al.*, 1999], the lateral extent of the mantle deformation is difficult to assess. Although measurements of anisotropy suggest that the deformation is at least as wide as South Island and even beyond [*Scherwath et al.*, 2002], this result is predicated on the assumption that the anisotropy is fresh and not inherited from a previous deformation episode. Modeling studies, on the other hand, are more comfortable with deformation in the mantle not extending beyond a width of 100 km [*Ellis et al.*, 2006]. This clearly is still an important issue as it bears on some fundamental issues of rock mechanics such as the temperature and pressure conditions under which dynamic recrystallization of olivine can occur [*Scherwath et al.*, 2002].

Seismic anisotropy in crustal rocks is not so well defined. Laboratory measurements on schist that form the bulk of the Pacific plate crustal rocks exhibit strong material-derived anisotropy [*Okaya, et al.*, 1995; *Christensen and Okaya*, this volume]. But the limited field measurements made orthogonally at a crustal scale indicate similar velocities and hence no seismically observed anisotropy signals [*Pulford et al.*, 2003].

Integrated View of the Plate Boundary

Plate 5 summarizes some plate boundary processes in central South Island. Relative plate motion is both dextral strike-slip movement along the Alpine fault (Australian plate to the north with respect to the Pacific plate), and convergent. The ratio of strike-slip to convergence is ~3.5:1. The Pacific upper and middle crust is exhumed along the Alpine fault or via backthrusts east of the fault; these mechanisms form the Southern Alps and its foothills. Lithospheric mantle appears to thicken uniformly beneath the Southern Alps, although it is not clear if it is Pacific or Australian plate mantle that is participating in the thickening. The thickened mantle is colder and hence denser than the surrounding asthenosphere and thus able to dynamically pull downward the overlying crust. This loading from the thickened mantle helps maintain a crustal root which is both laterally offset from the Southern Alps and excessively thick with respect to surface topography.

As the Pacific plate converges towards the plate boundary its lower crust accumulates into this drawn-down root. The overlying middle crust (semischists to schists) also deepens and undergoes metamorphic processes, which results in the release of fluids. These fluids aid in the backshear stair-step faulting which facilitates exhumation up the Alpine fault. These fluids that are transported along the fault produce the magnetotelluric anomaly and lower seismic velocities observed within the plate boundary fault zone. Exhumation rates are large but are balanced by climate-driven rapid erosion that keeps the height of the Southern Alps at modest levels.

SUMMARY

Benchmarks of success of large international projects come in many forms. Apart from the research that is summarized in this volume, in addition to papers published previously, there are two other measures of importance. First there is the manner in which a project involves the upcoming generation of earth scientists. Because all aspects of the South Island Orogen Project involved making new observations in the field; graduate students therefore formed the backbone of the field efforts. At least twenty graduate degrees have been granted, from both New Zealand and United States universities, linked to the South Island project with the first in 1995 and the most recent in 2007. The other mark of success is the degree to which this project has spawned new initiatives and research programs. In this regard the South Island Orogen Project has been highly successful. Since the results from the major data sets first started to appear at conferences, a plethora of geodynamical modeling studies that depend on these data have appeared [*Beaumont et al.*, 1996; *Batt and Braun*, 1999; *Gerbault et al.*, 2002; *Little et al.*, 2002; *Ellis et al.*, 2006; *Scherwath et al.*, 2006]. The focus of these studies is the kinematics and dynamics of continental tectonics. A new US-NZ GPS program to measure the vertical surface uplift rate of the Southern Alps began in 2000 [*Beavan et al.*, 2004]. Several new geophysical follow up projects are currently in the pipeline including an offshore study of SKS splitting using ocean-bottom broad-band seismographs, and a bore-hole seismograph array to search for microearthquakes in the Alpine fault zone.

New campaigns have been launched within the Alpine fault and its northward transition to subduction based on the success of SIGHT initial magnetotelluric work. Because of SIGHT, new projects for seismic profiling of other continental islands have been conducted and proposed. Seismic imaging studies in New Zealand's North Island recently were carried out [*Stratford and Stern*, 2006; *Henrys et al.*, 2006]. Double-sided onshore-offshore profiling patterned after SIGHT is scheduled to take place in Taiwan as part of a Taiwan-USA collaborative project. Within New Zealand students who gained field experience with SIGHT have moved into the local oil industry, where they have now applied onshore-offshore seismic methods to explore for difficult targets beneath or adjacent to shore lines.

Finally, our search to understand the dynamics of major continental transform faults has received a boost from the combined South Island Orogen Project. As this book goes to press a workshop will be held to examine the possibility of bringing the International Drilling Commission to New Zealand to drill the Alpine fault to a depth of ~5 km. Data from the South Island Orogen Project are central to formulate a science plan for the proposed drilling program. For example, because we can now show that the Alpine fault dips at a moderate angle into the crust, unlike at SAFOD in California [*Hickman et al.*, 2004] where costly directional drilling was required, deep drilling here will only have to deal with a vertical hole in order to intersect the Alpine plate boundary fault at depth.

Acknowledgements. Funding for the core SIGHT/SAPSE collaborations were provided by the New Zealand Foundation for Research Science and Technology, GNS Science, Victoria University of Wellington, and the U.S. National Science Foundation Continental Dynamics program (grants EAR-9219496 and EAR-9418530). Additional U.S. grants were provided by NSF Continental Dynamics (EAR-9418343 and EAR-9725883). Additional New Zealand support was provided by PGST and the Marsden Fund. We thank Belinda Smith Lyttle who used the N.Z. QMAP geological GIS database in order to make the maps of South Island (including the oversized versions).

REFERENCES

Allis, R. G. (1986), Mode of crustal shortening adjacent to the Alpine Fault, New Zealand, *Tectonics, 5*, 15–32.

Baldock, G., and T. Stern (2005), Width of mantle deformation across a continental transform: evidence from upper mantle (Pn) seismic anisotropy measurements, *Geology, 33*, 741v744.

Batt, G. E., and J. Braun (1999), The tectonic evolution of the Southern Alps from thermally coupled dynamic modelling, *Geophys. J. Int., 136*, 403–420.

Beaumont, C., P. J. J. Kamp, J. Hamilton, and P. Fullsack (1996), The continental collision zone, South Island, New Zealand: comparison of geodynamical models and observations, *J. Geophys. Res., 101*, 3333–3359.

Beavan, J., D. Matheson, P. Denys, M. Denham, T. Herring, B. Hager, and P. Molnar (2004), A vertical deformation profile across the Southern Alps, New Zealand, from 3.5 years of continuous GPS data, paper presented at The State of GPS Vertical Positioning Precision: Separation of Earth Processes by Space Geodesy, Cahiers de Centre Europeen de Geodynamique et seismologie.

Beavan, J., M. Moore, C. Pearson, M. Henderson, B. Parsons, S. Bourne, P. England, D. Walcott, G. Blick, D. Darby, and K. Hodgkinson (1999), Crustal deformation during 1994–1998 due to oblique continental collisionin the central Southern Alps, New Zealand, and implications for seismic potential of the Alpine Fault, *J. Geophys. Res., 104*, 25,233–25,255.

Blythe, A. E. (1998), Active tectonics and ultrahigh-pressure rocks, in *When Continents Collide: Geodynamics and Geochemistry of Ultrahigh-Pressure Rocks*, edited by B. R. Hacker and J. G. Liou, Kluwer Academic Publishers, Dordrecht, Netherlands, pp. 141–160.

Choukrone, P., and ECORS Team (1989), The ECORS Pyrenean deep seismic profile reflection data and overall structure of an orogenic belt, *Tectonics, 8*, 23–39.

Christensen, N. I., and D. A. Okaya (this volume), Compressional and shear wave velocities in South Island New Zealand rocks and their application to the interpretation of seismological models of the New Zealand crust.

Cook, F. A., D. Albaugh, L. Brown, S. Kaufman, J. Oliver, and R. Hatcher (1979), Thin skinned tectonics in the crystalline southern Appalachians: COCORP seismic reflection profiling of the Blue Ridge and Piedmont, *Geology, 7*, 563–567.

Cook, F. A., and J. L. Varsek (1994), Orogen-scale decollements, *Rev. Geophys., 32*, 37–60.

Cox, S. C., and R. Sutherland (this volume), Regional geological framework of South Island, New Zealand, and its significance for understanding the active plate boundary.

Davey, F. (2005), A Mesozoic crustal suture on the Gondwana margin in the New Zealand region, *Tectonics*, doi:10.1029/2004TC001719.

Davey, F., D. Eberhart-Phillips, M. Kohler, S. Bannister, G. Caldwell, S. Henrys, M. Scherwath, T. Stern, and H. van Avendonk (this volume), Geophysical structure of the Southern Alps Orogen, South Island, New Zealand.

Dickinson, W. R. (2003), The Basin and Range Province as a composite extensional domain, in *The George A. Thompson Volume: The Lithosphere of Western North America and Its Geophysical Characterization*, edited by S. L. Klemperer and W. G. Ernst, Geological Society of America, Boulder, Colorado, pp. 213–250.

Duclos, M., M. K. Savage, A. Tommasi, and K. R. Gledhill (2005), Mantle tectonics beneath New Zealand inferred from SKS splitting and petrophysics, *Geophys. J. Int., 163*, 760–774, doi:710.1111/j.1365-1246X.2005.02793.x.

Eberhart-Phillips, D., and S. C. Bannister (2002), Three-dimensional crustal structure in the Southern Alps region of New Zealand from inversion of local earthquake and active source data, *J. Geophys. Res.*, doi:10.1029/2001JB000567.

Ellis, S., J. Beavan, D. Eberhardt-Phillips, and B. Stockhert (2006), Simplified models of the Alpine fault seismic cycle: stress transfer in the mid-crust, *Geophys. J. Int., 166*, 386–402.

Furlong, K. P., W. D. Hugo, and G. Zandt (1989), Geometry and evolution of the San Andreas fault zone in northern California, *J. Geophys. Res., 94*, 3100–3110.

Gerbault, M., F. Davey, and S. Henrys (2002), Three-dimensional lateral crustal thickening in continental oblique collision: an example from the Southern Alps of New Zealand, *Geophys. J. Int., 150*, 770–779.

Godfrey, N. J., F. J. Davey, T. A. Stern, and D. Okaya (2001), Crustal structure and thermal anomalies of the Dunedin Region, South Island, New Zealand, *J. Geophys. Res., 106*, 30,835–30,848.

Green, A., B. Milkereit, A. Davidson, C. Spenser, D. R. Hutchinson, W. F. Cannon, M. W. Lee, W. F. Agena, J. C. Behrendt, and W. J. Hinze (1988), Crustal structure of the Grenville front and adjacent terranes, *Geology, 16*, 788–792.

Hadley, D. and H. Kanamori (1977), Seismic structure of the Transverse Ranges, California, *Geol. Soc. Am. Bull., 88*, 1469–1478.

Harrison, A. J. (1999), Multichannel seismic and flexural analysis of the Westland sedimentary basin, South Island, New Zealand, M.Sc. thesis, Victoria University of Wellington, Wellington, 177 pp.

Henrys, S., M. Reyners, I. Pecher, S. Bannister, Y. Nishimura, and G. Maslen (2006), Kinking of the subducted slab by escalator normal faulting beneath North Island of New Zealand, *Geology, 34*, 777–780.

Hickman, S., M. D. Zoback, and W. Ellsworth (2004), Introduction to special sections: preparing for the San Andreas Fault Observatory at Depth, *Geophys. Res. Lett., 31*, L12S01, doi:10.1029/2004GL020688.

Jiracek, G. J., V. M. Gonzalez, T. G. Caldwell, P. E. Wannamaker, and D. Kilb (this volume), Seismogenic, electrically conductive, and fluid zones at continental plate boundaries in New Zealand, Himalaya, and California-USA.

Klosko, E. R., F. T. Wu, H. J. Anderson, D. Eberhardt-Phillips, T. V. McEvilly, E. Audoine, and M. K. Savage (1999), Upper mantle anisotropy in the New Zealand region, *Geophys. Res. Lett., 26*, 1497–1500.

Kohler, M. D., and D. Eberhart-Phillips (2002), Three-dimensional lithospheric structure below the New Zealand Southern Alps, *J. Geophys. Res.*, 1074, 2225.

Koons, P. O. (1989), The topographic evolution of collisional mountain belts: a numerical look at the southern Alps, New Zealand, *Am. J. Sci., 289*, 1041–1069.

Koons, P. O. (1990), Two-sided orogen: collision and erosion from the Sandbox to the Southern Alps, New Zealand, *Geology, 18*, 679–682.

Leitner, B., D. Eberhart-Phillips, H. Anderson, and J. Nableck (2001), A focussed look at the Alpine fault, New Zealand: seismicity, focal mechanisms, and stress observations, *J. Geophys. Res., 106*, 2193–2220.

Little, T. A., S. Cox, J. K. Vry, and G. Batt (2005), Variations in exhumation level and uplift rate along the oblique-slip Alpine fault, central Southern Alps, *GSA Bull., 117*, 707–723.

Little, T. A., M. K. Savage, and B. Tikoff (2002), Relationship between crustal finite strain and seismic anisotropy in the mantle, Pacific-Australia plate boundary zone, South Island, New Zealand, *Geophys. J. Int., 151*, 106–116.

Molnar, P. (1988), Continental tectonics in the aftermath of plate tectonics, *Nature, 335*, 131–138.

Molnar, P., H. Anderson, E. Audoine, D. Eberhart-Philips, K. Gledhill, E. Klosko, T. McEvilly, D. Okaya, M. Savage, T. Stern, and F. Wu (1999), Continuous deformation versus faulting through the continental lithosphere of New Zealand, *Science, 286*, 516–619.

Molnar, P., and G. Houseman (2004), The effects of buoyant crust on the gravitational instability of thickened mantle lithosphere at zones of intracontinental convergence, *Geophys. J. Int., 158*, 1134–1150.

Moore, M., P. England, and B. Parsons (2002), Relation between surface velocity field and shear wave splitting in the South Island of New Zealand, *J. Geophys. Res., 107*, doi:10.1029/2000JB000093.

Mortimer, N. (2004), New Zealand's Geological Foundations, *Gondwana Res., 7*, 261–272.

Namson, J. S., and T. L. Davis (1991), Detection and seismic potential of blind thrusts in the Los Angeles, Ventura, and Santa Barbara areas and adjoining Transverse Ranges, Final Technical Report for USGS NEHRP Program, 31 pp.

Norris, R. J., P. O. Koons, and A. F. Cooper (1990), The obliquely convergent plate boundary in the South Island of New Zealand; implications for ancient collision zones, *J. Struct. Geol., 12*, 715–725.

Okaya, D., N. Christensen, D. Stanley, and T. Stern (1995), Crustal anisotropy in the vicinity of the Alpine Fault Zone, South Island New Zealand., *N. Z. J. Geol. Geophys., 38*, 579–584.

Okaya, D., S. Henrys, and T. Stern (2002), Double-sided onshore-offshore seismic imaging of a plate boundary: super-gathers across South Island of New Zealand, *Tectonophysics, 355*, 247–263.

Okaya, D., T. Stern, S. Holbrook, H. van Avendonk, F. Davey, and S. Henrys (2003), Imaging a plate boundary using double-sided onshore-offshore seismic profiling, *Leading Edge, 22*, 256–260.

Okaya, D., T. A. Stern, and S. H. Henrys (2007), Seismic image of continental collision at a transform plate boundary, New Zealand, *Science* (submitted).

Pfiffner, O. A., and others (1990), Crustal shortening in the Alpine orogen: results from deep seismic reflection profiling in the eastern Swiss Alps, line NFP 20-East, *Tectonics, 9*, 1327–1355.

Pulford, A., M. K. Savage, and T. A. Stern (2003), Absent anisotropy: the paradox of the Southern Alps orogen, *Geophys. Res. Lett.*, doi:10.1029/2003GL017758.

Reyners, M. (1987), Subcrustal earthquakes in the central South Island, New Zealand, and the root of the Southern Alps, *Geology, 15*, 1168–1171.

Savage, M., M. Duclos, and K. Marson-Pidgeon (this volume), Seismic anisotropy in South Island, New Zealand.

Scherwath, M., T. Stern, F. Davey, and R. Davies (2006), Three-dimensional lithospheric deformation under oblique continental collision from gravity analysis in South Island, New Zealand, *Geophys. J. Int., 167*, 906–916.

Scherwath, M., T. Stern, A. Melhusih, and P. Molnar (2002), Pn anisotropy and distributed upper mantle deformation associated with a continental transform fault, *Geophys. Res. Lett.*, doi:10.1029/2001GL014179.

Scherwath, M., T. A. Stern, F. J. Davey, D. Okaya, W. S. Holbrook, R. Davies, and S. Kleffmann (2003), Lithospheric structure across oblique continental collision in New Zealand from wide-angle P-wave modeling, *J. Geophys. Res.*, doi:10.1029/2002JB002286.

Schott, B., and H. Schmeling (1998), Delamination and detachment of a lithospheric root, *Tectonophysics, 296*, 225–247.

Smithson, S. B., J. A. Brewer, S. Kaufman, J. E. Oliver, and C. Hurich (1979), Structure of the Larimide uplift from COCORP deep reflection data and gravity data., *J. Geophys. Res., 84*, 5955–5972.

Stern, T., S. Kleffmann, D. Okaya, M. Scherwath, and S. Bannister (2001), Low seismic wave-speeds and enhanced fluid pressure beneath the Southern Alps, New Zealand, *Geology, 29*, 679–682.

Stern, T. A., P. Molnar, D. Okaya, and D. Eberhart-Phillips (2000), Teleseismic P-wave delays and modes of shortening the mantle beneath the South Island, New Zealand, *J. Geophys. Res., 105,* 21,615–21,631.

Stern T., D. Okaya, S. Kleffmann, M. Scherwath, S. Henrys, and F. Davey (this volume), Geophysical exploration and dynamics of the Alpine fault zone.

Stern, T., D. Okaya, and M. Scherwath (2002), Structure, deformation and strength of a continental transform from seismic observations, *Earth Planets Space, 54,* 1011–1021.

Stewart, J., and A. B. Watts (1997), Gravity anomalies and spatial variations of flexural rigidity at mountain ranges, *J. Geophys. Res., 102,* 5327–5352.

Stratford, W. R., and T. A. Stern (2006), Crustal and upper mantle structure of a continental back-arc: Central North Island, New Zealand, *Geophys. J. Int., 166,* 469–484.

Sutherland, R. (1999), Basement geology and the tectonic development of the greater New Zealand region: an interpretation from regional magnetic data, *Tectonophysics, 308,* 341–362.

Tippett, J. M., and P. J. J. Kamp (1993), Fission track analysis of the Late Cenozoic vertical kinematics of continental Pacific crust, South Island, New Zealand, *J. Geophys. Res., 98,* 16,119–16,148.

van Avendonk, H., W. S. Holbrook, D. Okaya, J. Austin, F. Davey, and T. Stern (2004), Continental crust under compression: a seismic reflection study of South Island Geophysical Transect 1, South Island, New Zealand, *J. Geophys. Res., 109,* doi:10.1029/2003JB002790.

Walcott, R. I. (1979), Plate motion and shear strain rates in the vicinity of the Southern Alps, in *The Origin of the Southern Alps,* R. Soc. N. Z. Bull., vol. 18, edited by R. I. Walcott and M. M. Cresswell, Royal Society of New Zealand, Wellington, pp. 5–12.

Walcott, R. I. (1984), The kinematics of the plate boundary zone through New Zealand: a comparison ofshort and long term deformation, *Geophys. J. R. Astron. Soc., 79,* 613–633.

Walcott, R. I. (1998), Modes of oblique compression: late Cenozoic tectonics of the South Island of New Zealand, *Reviews of Geophysics, 36,* 1–26.

Wannamaker, P. E., G. R. Jiracek, J. A. Stodt, T. G. Caldwell, V. M. Gonzalez, J. D. McKnight, and A. D. Porter (2002), Fluid generation and pathways beneath an active compressional orogen, the New Zealand Southern Alps, inferred from magnetotelluric data, *J. Geophys. Res., 107,* 2117, doi:10.1029/2001JB000186.

Wellman, H. W. (1955), New Zealand quaternary tectonics, *Geol. Runds., 43,* 248–257.

Wellman, H. W. (1979), An uplift map for the South Island of New Zealand, *Bull. R. Soc. N. Z., 18,* 13–20.

Woodward, D. J. (1979), The crustal structure of the Southern Alps, New Zealand, as determined by gravity, *in The Origin of the Southern Alps, R. Soc. N. Z. Bull.,* vol. 18, R. I. Walcott and M. M. Cresswell, Royal Society of New Zealand, Wellington, pp. 95–98.

S. Cox, GNS Science, Dunedin, New Zealand.

F. Davey and S. Henrys, GNS Science, Lower Hutt, New Zealand.

D. Okaya, Department of Earth Sciences, University of Southern California, Los Angeles, CA, USA.

T. Stern, School of Earth Sciences, Victoria University of Wellington, Wellington, New Zealand.

Regional Geological Framework of South Island, New Zealand, and its Significance for Understanding the Active Plate Boundary

Simon C. Cox

GNS Science, Dunedin, New Zealand

Rupert Sutherland

GNS Science, Avalon, Lower Hutt, New Zealand

New Zealand basement is composed of distinct volcano-sedimentary terranes, intruded by batholiths and overprinted by metamorphism, that were accreted to the Pacific margin of Gondwana during the Paleozoic and Mesozoic. Extension and sea floor spreading during the Cretaceous and Paleogene thinned and isolated a large fragment of Gondwana, most of which still remains submerged. A sedimentary section comprising rifted (oldest), passive and convergent (youngest) margin episodes was deposited unconformably on basement, changing locally in character as the Australian-Pacific plate boundary developed during the Neogene. Terranes were offset by up to 470 km on the Alpine Fault, bent dextrally by distributed deformation, and now constrain the maximum plate motion to be around 850 km. Displacement of the basal unconformity (Waipounamu erosion surface) and laterally offset Tertiary sediments constrain long-term plate boundary deformation. New Zealand's landscape evolved during the Pleistocene, and has been profoundly affected by asymmetric rainfall and erosion. This has affected geomorphology, glaciation and glacial cycles, uplift, exhumation and rock distribution across the Southern Alps, controlled the first order shape of the orogen, and probably the distribution of deformation. Pleistocene glacial cycles shaped the landscape and left a fragmentary record of moraines, outwash surfaces, alluvial terraces and fans. Fluvio-glacial landforms have been offset by active faults and constrain the rate and location of much late Quaternary deformation, but evidence of active deformation is absent from the most mountainous region, and the geological record Australian-Pacific plate motion during the late Quaternary is incomplete.

INTRODUCTION

A Continental Plate Boundary: Tectonics at South Island, New Zealand
Geophysical Monograph Series 175
Copyright 2007 by the American Geophysical Union.
10.1029/175GM03

New Zealand straddles the Australian-Pacific plate boundary (Figure 1) and has been widely referred to as a "natural laboratory" for the study of earth deformation and plate boundary processes. Formation of its main crustal components occurred by accretion onto the Gondwana plate margin

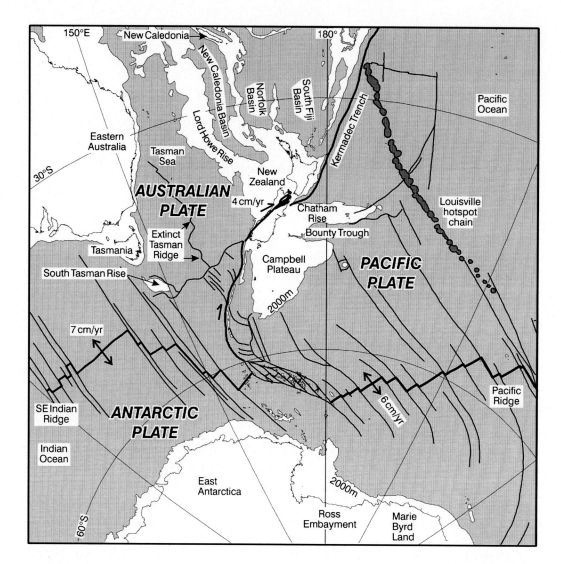

Figure 1. Map of the New Zealand region showing the coastline and bathymetry <2000 m (white) and >2000 m (grey) approximating areas of continental crust and oceanic crust, respectively. The relative motions of the Pacific, Australian and Antarctic plates, spreading ridges and fracture zones are also shown [*DeMets et al.* 1994].

during the Paleozoic and Mesozoic. Later rifting and subsidence resulted in deposition of widespread cover sequences during the Cretaceous and Cenozoic. The Australian-Pacific plate boundary developed during the Paleogene and Neogene and involved marked changes in basin sedimentation. Rates of plate motion are now sufficiently high that the effects of deformation can be directly observed. Transpression has exposed the South Island landmass, built mountain ranges of sufficient elevation to host glaciers, developed a strongly orographic climate with rapid erosion rates, and exposed a variety of basement rocks and cover sequences with their constituent histories. Perhaps not surprisingly, New Zealand hosts many geological features from which key observations have changed the world's understanding of tectonics.

Following a brief description of New Zealand's plate motion history, this paper describes the basement, sedimentary cover, and surficial geology in "geological" order. It contains no new data and is largely a descriptive summary which aims to provide a brief overview of plate boundary processes recorded by these rocks and deposits, and highlight some of the key geological features that have been used to constrain the dynamics of the Australian-Pacific plate boundary.

PLATE MOTION HISTORY OF NEW ZEALAND

New Zealand's tectonic history is most simply divided into three phases: Paleozoic to Early Cretaceous Gondwana margin events (Rangitata orogeny) (Plate 1A); Late

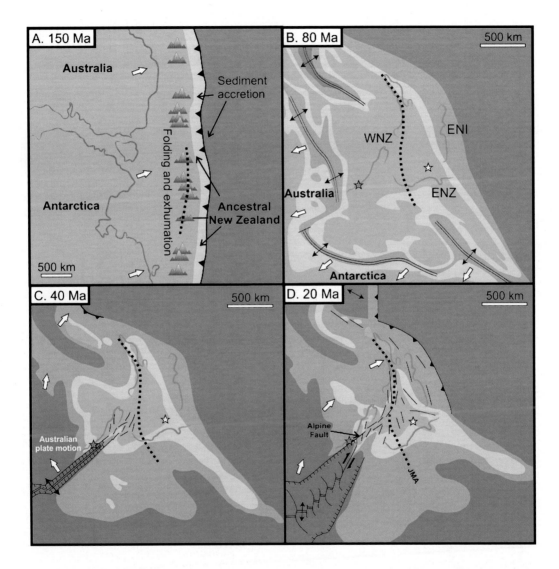

Plate 1. Cartoon reconstructions of the plate motion and tectonic setting of New Zealand at 150, 80, 40 and 20 Ma. Stars show the locations of points that eventually became juxtaposed either side of the Alpine Fault at Franz Josef village (see also Figure 3). The Junction Magnetic Anomaly is represented schematically with a dashed line. WNZ = Western Province, ENZ = Eastern Province, ENI = eastern North Island. A more-detailed model from 40 Ma to present is provided in *King* [2000].

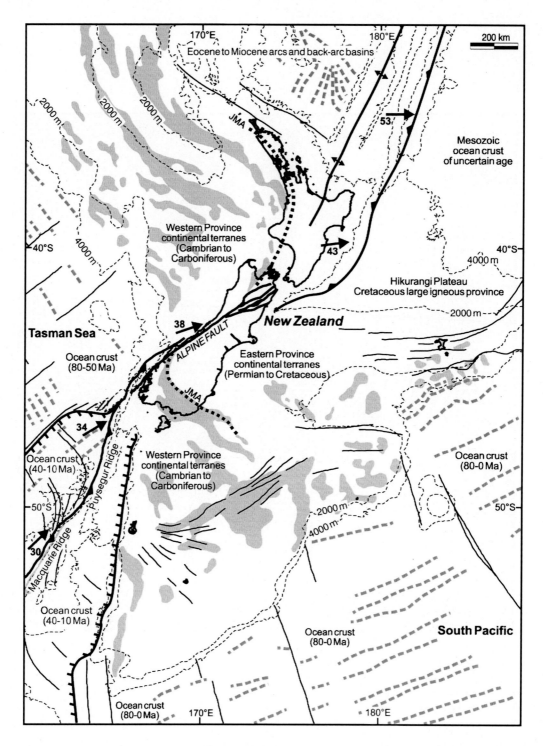

Figure 2. Summary of the current plate tectonic setting of New Zealand showing: the age and nature of crust in the region; oceanic magnetic lineations (dashed grey lines) [*Cande et al.*, 1989; *Sutherland*, 1999]; the Junction Magnetic Anomaly (JMA - dashed black line) [*Hunt,* 1978]; and large positive magnetic anomalies in continental crust (grey shaded areas) [*Sutherland,* 1999].

Cretaceous rifting and Gondwana break-up (Plate 1B); and Eocene to Quaternary evolution of the plate boundary that is still active (Kaikoura orogeny) (Plate 1C, 1D; Figure 2). The magnetic timescale of *Cande and Kent* [1995] is used in the following discussion.

It is difficult to quantify Paleozoic to Early Cretaceous plate tectonic events because this time predates most of the ocean crust that remains on Earth, and the New Zealand evolution took place along a convergent margin (Plate 1A). The remains of the convergent margin are preserved along the northern margin of the Chatham Rise, where the Hikurangi Plateau, a large igneous province that was probably contemporaneous (122 Ma) with the Ontong Java and Manihiki plateau [*Mortimer and Parkinson*, 1996; *Wood and Davy*, 1994], is underthrust. Paleomagnetism and deep-sea drilling results indicate it is likely that Early Cretaceous dextral strike-slip motion accompanied the final phase of convergence [*Sutherland and Hollis*, 2001]. Details of the Paleozoic to Early

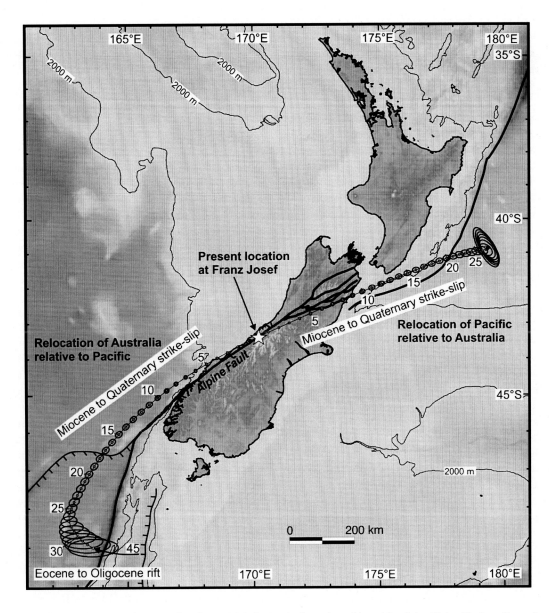

Figure 3. Reconstruction of the plate motion histories at 1 Ma intervals for points either side of the Alpine Fault at Franz Josef during the past 45 Ma, using plate rotations determined by *Cande and Stock* [2004] and the Eocene reconstruction of *Sutherland* [1995]. The trajectories highlight differences in tectonic setting of two points that are now juxtaposed by the Alpine Fault. Ellipses show interpolated 95% confidence regions. (See also colour version on the CDROM which accompanies this volume).

Cretaceous history of New Zealand are primarily inferred from geological observations (discussed below).

The Gondwana break-up phase was associated with widespread rifting, basin formation, and eventual formation of ocean crust at the Pacific and Tasman ridges (Plate 1B). The oldest ocean crust along the southern continental margin of New Zealand (Figure 2) is spatially correlated with negative magnetic anomalies [*Sutherland*, 1999a] that are continuous southward with a clearly recognisable alternating sequence of Cretaceous seafloor-spreading anomalies [*Gaina et al.*, 1998; *Molnar et al.*, 1975]; hence, widespread seafloor spreading is inferred to have started during chron 33r (79–83 Ma) and dates the time that New Zealand separated from Australia and Antarctica (Plate 1B). Seafloor spreading ceased in the Tasman Sea at about chron 24 (52 Ma), but continues today at the Pacific ridge [*Cande et al.*, 1995; *Gaina et al.*, 1998].

Substantial changes in global plate motions and plate boundary geometries occurred in the western Pacific during the Early and Middle Eocene (55–37 Ma) [*Steinberger et al.*, 2004]. New Zealand was split and passive margins formed as the Eocene plate boundary developed through New Zealand (Plate 1C, Figure 2): seafloor spreading was occurring south of New Zealand by chron 18 (38–40 Ma) [*Wood et al.*, 1996]. However, nappes were emplaced in New Caledonia as Late Eocene convergence started [*Aitchison et al.*, 1995] (Figure 1; Plate 1C), and Eocene-Oligocene deformation through most of New Zealand was very minor and restricted to low rates of rifting in the far south. Plate motions through New Zealand can be precisely determined by closure of the Australia-Antarctica-Pacific plate-motion circuit [*Cande and Stock*, 2004] and provide an explanation for the apparent enigma that a through-going plate boundary formed ca. 20 Myr later in New Zealand than elsewhere in the western Pacific. The Eocene and Oligocene Australia-Pacific stage rotations had poles that were very close to New Zealand: the entire Australia-Pacific plate boundary was pivoting about New Zealand, and the rate of motion within New Zealand remained very low (Figure 3; also on the CDROM which accompanies this volume).

During Late Oligocene time (29–24 Ma), the pole of Australia-Pacific relative rotation migrated southeast: rates of motion increased within New Zealand and the direction of motion changed to make the plate boundary predominantly strike-slip in character. The relative motion is illustrated in Figure 3 by relocating a point on the Alpine Fault (at Franz Josef). Both Australia and Pacific frames of reference are considered, and the difference is significant. A difference arises because points on either side of the Alpine Fault were previously separated by several hundred km; the total plate motion is ca. 800 km. A point on the west side of the fault experienced significant rifting and passive margin formation, followed by a period of transtension, and then transpression (Figure 3). In contrast, a point on the eastern side of the Alpine Fault experienced almost no relative plate motion until about 26 Ma, and then the motion has been at a slightly increasing rate and a relatively constant direction since then (Figure 3).

The interplay between Cenozoic plate boundary geometry and relative plate motion has produced the current architecture of South Island. Plate motions were such that the Eocene rifted margin of the Australian plate formed parallel to what would become the Miocene-Quaternary plate motion direction. The contrast in lithospheric buoyancy on either side of the Eocene passive margin, and its geometry, have been a primary control on the evolution of the South Island plate boundary and geometry of the Eocene rift margin adequately explains how and why the Alpine Fault formed [Sutherland et al., 2000].

BASEMENT

Terranes

The basement geology of New Zealand is composed of a series of tectonostratigraphic units (terranes) and igneous suites that are a rifted fragment of Gondwana [*Mortimer*, 2004] (Plate 2). Unlike most of the major constituent blocks of Gondwana, New Zealand contains no exposed Precambrian strata or cratonic rocks. Instead, the New Zealand fragment formed during Paleozoic and Mesozoic subduction-accretion processes along the Gondwana Pacific margin, that subsequently separated into a distinct sub-continent of relatively thinned continental crust by Cretaceous to Paleogene rifting and extension. The constituent basement terranes have been grouped into a Western and an Eastern Province due to their distinctive geological histories and separation by a long-lived, ca. 375–110 Ma, composite, subduction related batholith (The Median Batholith) [*Mortimer et al.*, 1999a, 1999b]. Terranes and igneous rocks are now arranged in relatively elongate belts marking the local orientation of the ancient Gondwana margin, which has been truncated at a high angle by the current Australian-Pacific plate boundary.

Western Province basement is dominated by Late Cambrian to Late Ordovician quartzose sediments of the Buller Terrane that have been regionally metamorphosed [*Nathan*, 1976]. The Takaka Terrane is volumetrically less significant, containing a diverse assemblage of deformed and metamorphosed Paleozoic sedimentary and volcanic rocks [*Cooper*, 1989]. The two terranes were amalgamated, possibly in the earliest Devonian, and subsequently intruded by plutons of the Karamea, Paparoa and Hohonu Batholiths during the Devonian and Carboniferous [*Cooper and Tulloch*, 1992].

Eastern Province terranes are dominated by lithic and feldspathic metagreywackes, but include volcanic, intrusive and ophiolitic assemblages that were accreted onto

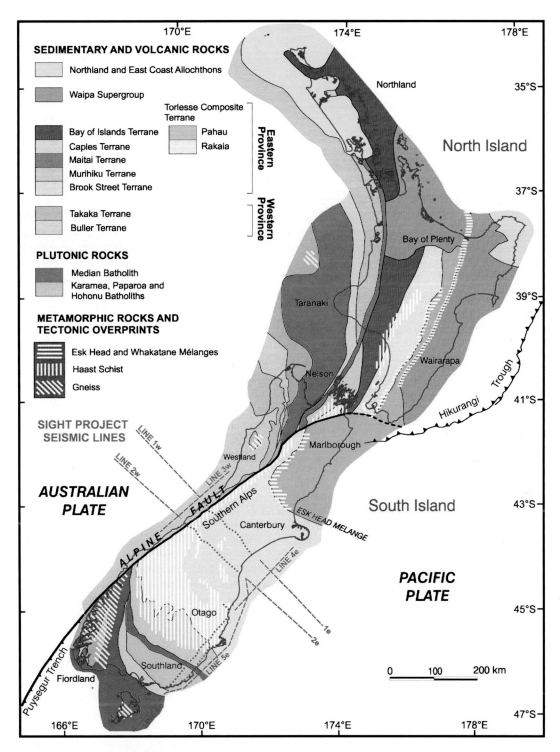

Plate 2. Simplified basement geology of onland and near-shore New Zealand. Post-110 Ma sedimentary rocks are not shown. The map illustrates the main terranes that were accreted against the margin of Gondwana during the Paleozoic and Mesozoic. These were intruded by batholiths, overprinted by metamorphism and deformation, then bent and offset during the Neogene by the Australian-Pacific plate boundary. The Northland and East Coast Allocthons, and Waipa Supergroup, have been overthrust and deposited on Eastern Province terranes, respectively, to form basement in north and east of North Island (adapted from *Mortimer*, 2004). The location of the SIGHT project geophysical transects are shown for reference [*Davey et al.*, this volume].

the margin of Gondwana [*Bishop et al.*, 1985; *Coombs et al.*, 1976; *Gray and Foster*, 2004; *MacKinnon*, 1983; *Mortimer*, 1995]. From inboard to outboard, now approximately southwest to northeast, they are the Brook Street, Murihiku, Matai, Caples, Bay of Islands (part of the former Waipapa), Rakaia (older Torlesse) and Pahau (younger Torlesse) Terranes [*Mortimer*, 2004]. Their depositional ages mostly range from Permian to Jurassic, but the Pahau Terrane contains Late Jurassic to Early Cretaceous sedimentary rocks that are partly recycled from the Rakaia Terrane [*MacKinnon*, 1983; *Wandres et al.*, 2004, 2005].

In northern and eastern North Island basement rocks comprise allochthonous Early Cretaceous-Paleogene ophiolitic, volcanic and sedimentary rocks that were thrust over North Island at the end of the Oligocene (Northland and East Coast Allochthons). A blanket of Median Batholith-derived volcaniclastic sedimentary rocks (Waipa Supergroup) [*Kear and Mortimer*, 2003] was deposited across the Eastern Province terranes during the Late Jurassic to Early Cretaceous and now forms basement in central and eastern North Island (Plate 2).

Gondwana Convergent Margin - Metamorphism and Deformation

A well-established convergent margin involved accretion and continental growth of Gondwana throughout much of the Mesozoic, continuing until the late Early Cretaceous (Plate 1A). There was probably considerable complexity during this extended period of active tectonics, but the spatial and temporal history is not well-recorded by the rocks and is only beginning to be resolved [e.g., *Gray and Foster*, 2004; *Mortimer*, 2004]. A commonly adopted tectonic model for New Zealand involves convergence of the Permian–Jurassic Rakaia and Pahau Terranes with arc–forearc basin sequences (Brook Street, Murihiku and Caples Terranes) caused by subduction of Permian oceanic crust beneath the Gondwana margin. Oblique-dextral transpression is suggested by paleomagnetism and some sedimentary-source reconstructions [*Adams and Kelley*, 1998; *Sutherland and Hollis*, 2001; Mortimer, 2004]. Regional metamorphism is attributed to accretion-related structural thickening during the Jurassic – Early Cretaceous, while subduction-related gabbro-noritic to granitic plutonic rocks were emplaced in the Median Batholith [*Mortimer et al.*, 1999a, 1999b]. Adakitic magmatism in the western margin of the batholith during the late Early Cretaceous (125–105 Ma) was possibly associated with crustal thickening.

The basement terranes have been weakly metamorphosed to at least zeolite or prehnite-pumpellyite facies, but most lack penetrative foliations. Gneissic fabrics are developed in some Western Province rocks (Plate 2) as a result of amphibolite-granulite facies recrystallisation during either

the Devonian or Cretaceous [*Mortimer*, 2004]. A significant regional metamorphic overprint was developed across parts of the Rakaia (older Torlesse quartzofeldspathic greywacke) and Caples (volcaniclastic greywacke) Terranes during the Jurassic to Early Cretaceous, resulting in the Haast Schist (Plate 2). In Otago the schist forms a 150 km-wide northwest-trending structural anticlinorium, with subhorizontally foliated greenschist facies (garnet-biotite-albite zone) rocks exposed in the axial region, and lower-grade semi-schist (prehnite-pumpellyite facies) on its moderate- to steeply-dipping flanks [*Mortimer*, 2003]. Recumbent folds, nappes, shear zones and prominent stretching lineations were developed within the schist [*Mortimer*, 1993]. Schist in Marlborough also forms a two-sided belt, containing correlative structures that were originally contiguous with those in Otago [*Little et al.*, 1999; *Little and Mortimer*, 2001; *Mortimer and Johnston*, 1993]. Alpine schist exposed in a narrow belt through the Southern Alps (Plate 2) linking schists in Otago and Marlborough, has a distinctly different history, being the product of Late Cretaceous amphibolite-facies peak metamorphism with a Neogene ductile-brittle overprinting deformation during their exhumation in the hanging wall of the Alpine Fault [see below; *Little et al.*, 2005; *Little et al.*, this volume; *Mortimer*, 2000; *Vry et al.*, 2004]. The names Otago, Alpine and Marlborough schist are used widely in New Zealand literature, but boundaries between them are undefined, and geologists have generally avoided formal subdivision of metamorphic overprints and Haast Schist.

The Pahau Terrane (younger Torlesse quartzofeldspathic greywacke) was deposited during the Late Jurassic to Early Cretaceous, while deformation and metamorphism that formed Haast Schist were occurring further inboard in the Gondwana accretionary prism [*Gray and Foster*, 2004; *Little et al.*, 1999]. The Pahau Terrane contains recycled Rakaia sediments [*MacKinnon*, 1983; *Wandres et al.*, 2004, 2005]. It is now separated from its older equivalent by the strongly sheared Esk Head Mélange (Plate 2) [*Bradshaw*, 1973; *Silberling et al.*, 1988], which is a belt of rocks interpreted to reflect shallow-level accretion-related deformation. The Esk Head Mélange extends across Canterbury-Marlborough, where it is rotated and offset by the Marlborough faults and truncated by the Alpine Fault. Zones of mélange and broken formation exposed in North Island are thought to be correlatives of the Esk Head Mélange [*Begg and Johnston*, 2000], separated laterally across the plate boundary (discussed further below).

Terrane Offset and Oroclinal Bending

The Maitai Terrane contains the distinctive Dun Mountain ophiolite assemblage, which has been important to the

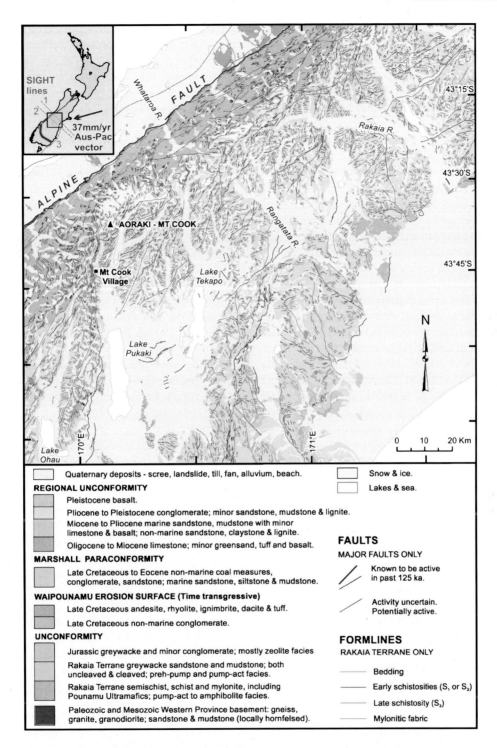

Plate 3. Orientation of bedding in Rakaia Terrane greywacke, or schistosity in semischistose and schistose rocks, across central South Island (Figure generated from QMAP 1:250,000 Geological Map of New Zealand dataset - *Cox and Barrell*, 2007). Formlines mark the trace of these structures on topography and are a good approximation of strike because dips are mostly moderate or steep. Bedding has a predominantly NW-SE strike in the east, but changes to predominantly NE-SW in the west, and there are numerous smaller (1–4 km scale) variations about steeply plunging folds. All known faults in the region are shown, with those known to have been active in the past 125 ka shown in red.

worldwide understanding of plate boundary deformation. Harold Wellman discovered the Alpine Fault [*Wellman and Willett*, 1942] and suggested in 1949 that it offset equivalent rock-types in Nelson and South Westland by about 300 miles [*Benson*, 1952] (Plate 2). The observation substantially contributed to the notion that large amounts of strike-slip motion may occur on transcurrent faults, which has since gained worldwide acceptance [*Nathan*, 2005]. The Maitai Terrane rocks are spatially associated with magnetic anomalies that include the laterally-continuous Junction Magnetic Anomaly (Figure 2). The Matai Terrane and Junction Magnetic Anomaly have been traced across the New Zealand subcontinent and provide an important marker for plate tectonic reconstruction and calculation of strain [e.g., *King*, 2000; *Little et al.*, 2002; *Sutherland*, 1999a].

Basement terranes and structures within the terranes are distinctly curved close to the Alpine Fault (Plate 2). In southeast South Island, >60 km from the Alpine Fault, they strike NW but closer to the fault their strike swings to N or NNE. Terranes in southern South Island have been truncated and displaced in a dextral sense to northern South Island, where they strike NNE and curve through N to NE along the western side of North Island. The recurved Z-shaped arc is consistent with distributed dextral bending, but exactly when this "oroclinal bending" occurred and what proportion is related to the current plate boundary, has been one of the most widely debated topics in New Zealand geology [e.g., *Kamp*, 1987; *Kingma*, 1959; *Suggate*, 1963; *Little and Mortimer*, 2001; *Sutherland*, 1999; and references therein]. Some studies attribute the curvature to Mesozoic deformation. In contrast, review of seafloor-based plate reconstruction, rotated post-Oligocene structures, and paleomagnetic data support late Cenozoic bending. The debate has important implications for calculated rates of displacement on the Alpine Fault, its age of inception, and the extent of distributed "off-fault" plate deformation.

The orientation of bedding in Rakaia Terrane greywacke is highly variable (Plate 3) [*Cox and Barrell*, 2007]. In general, bedding dips are steep (60-90°), and rocks contain numerous changes in strike at 1 to 4 km intervals about steeply plunging folds and across faults. Gently dipping beds (<30°), locally overturned and isoclinally folded, occur only in semischists along the Main Divide and the Two Thumb Range of Canterbury. There is a broad regional swing in bedding strike across South Island, from predominantly NW in the southeast, through N to NNE in the northwest (Plate 3). The swing matches the oroclinal bend of terranes elsewhere in South Island (Plate 2), caused by distributed dextral bending and fault offsets near the Australian-Pacific plate boundary.

The Maitai, Murihiku and Caples Terranes are separated by 440–470 km across the Alpine Fault, depending on which geological marker is selected. This separation is the maximum possible strike-slip displacement on the fault. The true strike-slip component is probably less than this separation, now exaggerated by shortening across the plate boundary with a component of apparent offset caused by overthrusting and erosion [*Kingma*, 1959]. The Esk Head Mélange in northeastern South Island (Plate 2), for example, is less than 200 km separated from its North Island equivalent [*Mazengarb*, 1994]. In addition to the 440-470 km separation on the Alpine Fault, the Maitai Terrane shows dextral bending of 120 km across South Island, and potentially 240 km in North Island. Plate reconstructions constrained by magnetic anomalies and fracture zones through plate-circuit closure are most plausibly reconciled with New Zealand's surface geology if the Maitai and other terranes had a smooth and continuous geometry with little bending or offset prior to the Eocene [*Sutherland*, 1999b]. If the 440-470 km separation of terranes across the Alpine Fault is late Eocene-Recent offset, then it accounts for ca. 55% of the plate displacement, with ca. 45% accommodated by distributed deformation [*Sutherland*, 1999b]. Bending probably involves transpressional rather than simple shear deformation [*Little et al.*, 2002c]. A degree of coupling between the upper crust and the lithospheric mantle is suggested by coincidence of calculated maximum finite strain azimuths with fast polarisation azimuths of seismic waves [*Little et al.*, 2002c].

Uplift and Exhumation of Basement

In the late Early Cretaceous the convergent tectonic regime was replaced by one of extension and crustal rifting that led to the separation of the New Zealand sub-continent from Gondwana [*Bradshaw*, 1989]. Metamorphic core complexes developed gneissic fabrics in the Western Province [*Gibson et al.*, 1988; *Tulloch and Kimbrough*, 1989], whereas reactivated shear zones, faulting, and regional uplift occurred in the Eastern Province [*Deckert et al.*, 2002; *Gray and Foster*, 2004]. Coarse, locally derived sediments initially filled fault-controlled depocentres on basement [*Laird*, 1993]. Erosion and marine transgression then began to cut a time-transgressive unconformity across basement rocks, the Waipounamu erosion surface [*LeMasurier and Landis*, 1996], which forms a conspicuous onshore geomorphic feature in southeast South Island (Figure 4).

Neogene transpression across the Australian-Pacific plate boundary (Figure 3; also on the CDROM which accompanies this volume) has further uplifted basement rocks, pushing the Southern Alps upwards into the path of a westerly-dominated air circulation pattern and setting up a strongly orographic climatic regime. Asymmetric rainfall and erosion are reflected in the first-order pattern of uplift, exhumation and rocks exposed southeast of the Alpine Fault (Plates 4, 5; see also below). Uplift has been less pronounced in the east,

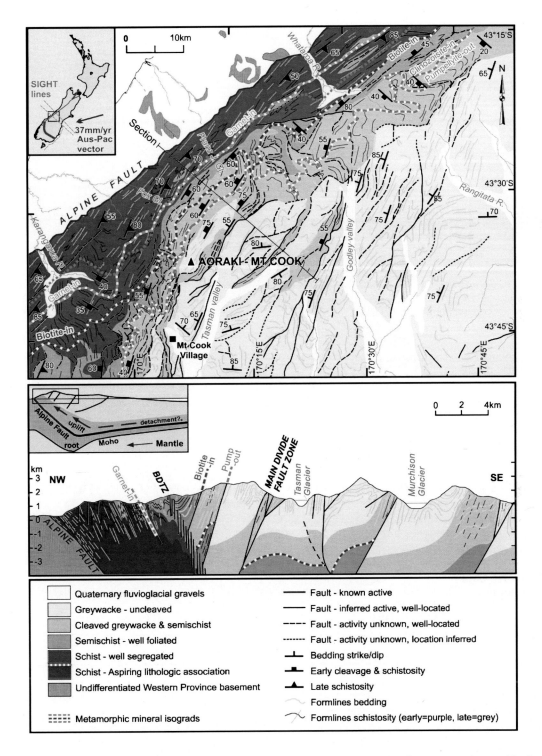

Plate 4. Geological map of the central Southern Alps illustrating the northwestward transition from greywacke to schist in the Rakaia Terrane, together with generalised orientations of bedding and schistosity (both dip/strike symbols and formlines are used). Schistosity is mostly oblique to the Alpine Fault, and metamorphic mineral isograds are not parallel to textural changes. The geological cross-section (note expanded scale) shows greater detail than the map, highlighting the relationship between northwest-dipping Main Divide Fault Zone and other oblique-reverse backthrusts to the Alpine Fault. (BDTZ = brittle ductile transition zone).

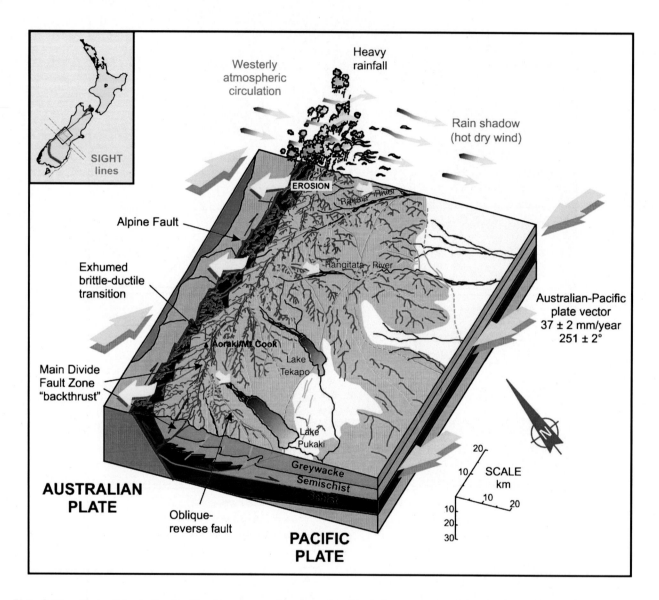

Plate 5. The effects of Australian-Pacific plate transpression shown in a block diagram summarising dominant climatic pattern, erosion, geology, structure and tectonics of the Southern Alps and central South Island.

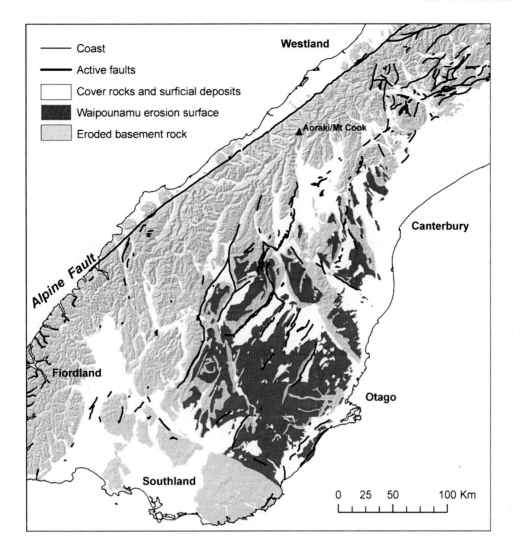

Figure 4. Map of southern South Island showing the distribution of the Waipounamu Erosion surface relative to known active faults. The erosion surface is locally buried beneath areas of Late Cretaceous-Cenozoic sedimentary rocks and surficial deposits (white), and elsewhere has been eroded (grey). It has variable relief of up to 2000 m and continues beneath the Canterbury plains and offshore where it has been mapped by seismic studies. (Figure generated from the QMAP 1:250,000 Geological Map of New Zealand dataset, GNS Science).

where the Waipounamu erosion surface serves as a marker of Neogene convergent deformation [e.g., *Litchfield*, 2001; *Stirling*, 1990, 1991; *Yeats*, 1987]. In the Otago region, the erosion surface has been offset into a series of parallel flat-topped mountain ridges (15 to 20 km-wavelength) with little or no sediment cover. The surface is locally modified and dissected by streams on the ridges, but the unconformity is well-preserved beneath Cenozoic cover in elongate valleys beneath the ridges (Figure 4). Plate convergence has resulted in up to 2000 m relief, currently expressed by episodic fault activity and cumulative slip rates of ca. 2–3 mm/yr [*Beanland and Barrow-Hurlbert*, 1988; *Beanland and Berryman*, 1989;

Norris and Nicolls, 2004; *Norris et al.*, 2005]. Processes in fault-fold growth and interaction, including the notion that separate fault-fold segments have coalesced into single continuous ridges during Quaternary time, have been inferred from drainage patterns, displacement of the erosion surface, and exposure-age dating [*Jackson et al.*, 1996; *Markley and Norris*, 1999; *Norris and Nicolls*, 2004]. A similar exhumed erosion surface situated on the other side of the Australian-Pacific plate boundary in the northwest Nelson region, has also been used as a reference to constrain Neogene deformation [*Rattenbury et al.*, 1998; *Suggate*, 1978 p.347; *Wellman*, 1940].

A near-continuous mid-upper crustal section is exposed across the Southern Alps in the uplifted hanging wall southeast of the Alpine Fault [*Grapes*, 1995, *Grapes and Watanabe*, 1992; *Little et al.*, 2005] (Plates 4, 5). Mid-crustal mylonites and amphibolite facies Alpine schist adjacent to the fault contain evidence of a Neogene ductile deformation overprint that constructively reinforced and reoriented the pre-existing Mesozoic metamorphic fabrics [*Norris and Cooper*, 2003, this volume; *Little et al.*, 2002a, this volume]. An exhumed, fossil, brittle-ductile transition zone (Plate 4 – BDTZ) [*Little et al.*, 2002b; *Wightman and Little*, this volume] separates Alpine schist in the central Southern Alps from relatively undeformed eastern greywacke (Rakaia Terrane – older Torlesse) and semischist that were metamorphosed during the Mesozoic, but suffered only brittle effects during the Neogene Australian-Pacific plate transpression [*Cox and Findlay*, 1985; *Cox et al.*, 1997].

The belt of Alpine schist (Plate 4) is not constant in width beside the Alpine Fault, but varies from 8 to 26 km [*Cox and Barrell*, 2007; *Little et al.* 2005]. It contains multiple generations of metamorphic fabrics, folds, and syn- to post-metamorphic quartz veins. Details of Alpine schist structure are provided elsewhere in this volume [see *Little et al.*, this volume; *Wightman and Little*, this volume]. The main fabrics are early schistosity (or schistosities) and slaty cleavage subparallel to remnant bedding and metamorphic isograds, and a later, near-vertical, schistosity or crenulation cleavage parallel to fold axial planes. Metamorphic zone boundaries (isograds) are slightly oblique to boundaries in the textural development of schistosity and cleavage (isotects) (Plate 4) [*Little et al.*, 2005]. Alpine schist is seismically anisotropic along its schistosity [*Godfrey et al.*, 2002], but this fabric is not parallel to the Alpine Fault. In the central Southern Alps where the schist belt is narrow (8 km), tight to isoclinal folding has aligned the early schistosity with 035–045° strike weakly oblique to the 055° Alpine Fault. Elsewhere, the folding is open, early schistosity strikes nearly perpendicular to the Alpine Fault, and the schist belt is wider (up to 26 km). Beneath the Southern Alps, the prominent early schistosity is probably subhorizontal, similar to the orientation of fabrics in Otago schist. Original sedimentary features are almost entirely obscured in Alpine schist, but the westernmost part beside the Alpine Fault contains a relatively high proportion of pelitic schist and mafic metavolcanic greenschist, with only minor psammitic schist. The name Aspiring lithologic association is used to describe this assemblage of rock derived from distal oceanic sediments, which was first recognised in Otago [*Craw*, 1984; *Mortimer*, 1993] but is now extended to northern South Island [*Cox and Barrell*, 2007; *Nathan et al.* 2002]. Discontinuous lenses of serpentinite, gabbro and metabasite (Pounamu Ultra-

mafics) were tectonically incorporated within the Aspiring lithologic association prior to the schist-forming deformation and regional metamorphism [*Cooper and Reay*, 1983]. Pounamu Ultramafics are interpreted to be imbricated slices of ocean crust on which the Rakaia terrane was deposited, and may correlate with mid-crustal seismic reflections imaged east of the Southern Alps beneath the Canterbury region [*Kleffman et al.* 1998].

Amounts and rates of basement exhumation have been modelled by thermochronological and petrological studies [e.g., *Adams*, 1979, 1981; *Batt and Braun*, 1999; *Batt et al.*, 2000; *Grapes and Watanabe*, 1992, 1994; *Little et al.*, 2005; *Tippett and Kamp*, 1993a, 1993b]. Thermochronological models are simplified by the lack of constraint on both past and present geothermal gradients, and the use of one dimensional uplift trajectories [see *Walcott*, 1998], but they clearly highlight differences between tectonic (total rock) uplift, exhumation/denudation, and surface uplift [*Kamp and Tippett*, 1993]. Vertical exhumation rates are broadly consistent with tectonic uplift rate estimates [e.g., *Adams*, 1980a, 1980b; *Norris and Cooper*, 2001; *Simpson et al.*, 1994; *Wellman*, 1979] varying by an order of magnitude across the Southern Alps from ca. 1 mm/yr in the middle of South Island, to around 10 mm/yr and 20–30 km of total exhumation where the rainfall and erosion rates are highest immediately adjacent to the Alpine Fault. Thermal annealing, partial annealing, and gas retention zones can be mapped with metamorphic mineral zones and isograds, and help differentiate otherwise monotonous greywacke and schist sequences across the Alpine Fault hanging wall [*Batt et al.*, 2000; *Kamp*, 1997, 2001; *Tippett and Kamp*, 1993a, 1993b]. A major backthrust near the drainage divide crest of the Southern Alps, the Main Divide Fault Zone, juxtaposes greywackes with different fission track ages that indicate significant Neogene (<5 Ma) vertical throw on the structure [*Cox and Findlay*, 1995]. Elsewhere, evidence is lacking in the Southern Alps for differential fault throws that have significance greater than the uncertainty (around ± 2 km) of fission track methods [*Tippett and Kamp*, 1993a].

Thermochronology along the plate boundary suggests there is only a short 20 km segment of the ~700 km long Southern Alps with potential to have reached exhumational steady state [*Little et al.*, 2005]. Amphibolite-facies schists have been completely exhumed from >500°C in the Neogene, in a 20 km long central Southern Alps region, but elsewhere were partially uplifted and cooled during the Paleogene [*Batt et al.*, 2000]. Along-strike progressive exhumation has also been demonstrated by low temperature thermochronological data across Fiordland, reflecting northward migration of the leading edge of the Australian plate subducted beneath the Pacific plate [*House et al.*, 2002; *House et al.*, 2005]. Detailed models of valley evolution in the Southern Alps are currently being constrained and tested using low temperature thermochronology on basement rocks [e.g., *Batt et al.*, 2000; *Herman et al.*, 2007].

SEDIMENTARY COVER SEQUENCES

General

The Cretaceous-Cenozoic sedimentary succession is composed of major transgressive-regressive megasequences, punctuated by a number of regional unconformities, now preserved in and between a series of basinal depocentres on the New Zealand sub-continent [*Carter*, 1988; *King et al.*, 1999 and references therein]. Sequences typically begin with Cretaceous non-marine and marginal marine rocks, changing up section through marine clastics to carbonates by the end of the Paleogene. The Neogene sequence typically comprises marine clastics passing up into non-marine coarse lithic sediments. Basin evolution can generally be classified into syn-rift (oldest), passive margin, and active margin (youngest) episodes. These coincide closely with the predicted plate-tectonic setting and evolution during Gondwana breakup, sea floor spreading between New Zealand and Gondwana, initiation of the proto Australian-Pacific plate boundary, then development of the modern plate boundary.

The stratigraphic architecture of New Zealand's sedimentary basins reflects the interplay of tectonic, eustatic, and oceanographic processes, the collective effects of which controlled accommodation space and cyclicity of deposition through time. In detail, basins exhibit multiple phases of structural evolution and depositional fill [*Carter*, 1988; *King et al.*, 1999]. Marked differences occur in basins originating on the eastern and western sides of New Zealand [e.g., *Field and Browne* et al., 1989; *King et al.*, 1999; *Nathan et al.*, 1986], in part reflecting differences in their location relative to the Australia-Pacific pole of rotation (e.g., Figure 3; also on the CDROM which accompanies this volume). Facies variations, depositional cycles and unconformities appear to be dominated by local tectonic activity, but influenced by regional tectonics, eustatic sea level changes and ocean currents [e.g., *King et al.*, 1999; *Norris and Carter*, 1982; *McMillan and Wilson*, 1997]. The evolution of the basins, deduced from their geology, is commonly used to develop a local structural history then related to plate boundary models and regional dynamics [e.g., *Norris et al.*, 1978; *Norris and Carter*, 1982; *Sircombe and Kamp*, 1998]. It requires systematic overviews, however, to fully differentiate between the effects of local, regional and global processes in the sedimentary record [e.g., *Kamp*, 1986; *King et al.*, 1999].

Basin Development

Break-up of Gondwana (Late Cretaceous) and formation of the Tasman Sea and South Pacific Ocean was initially associated with widespread continental extension and then isolation of the New Zealand sub-continent by mid-ocean ridges (Plate 1B). In the east, this period involved deposition of coarse graben-fill deposits, formation of the Waipounamu erosion surface, localised volcanism and passive margin transgression over the erosion surface. Sedimentary facies were similar in the west, although rift basins subsided more rapidly, possibly linked to the Tasman Ridge in a transform zone [*Carter and Norris*, 1976; *Kamp*, 1986; *Laird*, 1981, 1994]. Seafloor spreading in the Tasman Sea ceased during the Early Eocene. Changes in structural and sedimentation style occurred in southern and western basins during the Middle Eocene-Oligocene, with rapid subsidence associated with oblique sea floor spreading south of New Zealand [*Norris et al.*, 1978] (Plate 1C).

Carbonate deposition was widespread during Oligocene time, reflecting a reduction in clastic detritus and a period when most of the New Zealand sub-continent was submerged. A number of unconformities, disconformities and depositional hiatuses are recorded within the carbonate sequence [*McMillan and Wilson*, 1997], the most significant of which is the Marshall Paraconformity [*Carter*, 1985; *Carter and Landis*, 1972; *Fulthorpe et al.*, 1996; *Nelson et al.*, 2004]. This surface developed across the eastern side of the New Zealand subcontinent as a depositional hiatus near the Oligocene-Miocene boundary, and represents widespread submarine erosion, winnowing and non-deposition as the Antarctic circumpolar current strengthened at a time of glacioeustatic sea level fall [*Loutit et al.*, 1988; *Carter*, 1985]. The Marshall Paraconformity is regionally significant, has been identified and mapped offshore, and can be correlated between basins. It has potential for constraining subsequent Neogene deformation, albeit complicated (c.f. Waipounamu erosion surface) by effects of differential sedimentary compaction and paleo water depth.

Plate Boundary Initiation

Major changes in sedimentation, volcanism, and tectonism occurred at the beginning of the Miocene throughout most of South Island (Plate 1D). Principally, there was a change from epicontinental transgressive sedimentary patterns to active fault-localised sedimentation and regression, which coincided with increasing sedimentation rates. The increased tempo of tectonics probably reflects strike-slip motion as the Australian-Pacific plate boundary developed through the region (Plate 1C, D). The Alpine Fault developed as a strike-slip link between west-dipping subduction beneath the North Island and spreading and oblique extension southwest of South Island [*Lebrun et al.*, 2003] (Figures 1, 2), but there is no direct evidence for the exact age of the fault. Latest Oligocene [*Carter and Norris*, 1976] or earliest Miocene [*Kamp*, 1986] inception is inferred on the basis of sedimentation changes and paleogeographic reconstruction of basins, possibly coinciding with a swarm of latest Oligocene to Early Miocene lamprophyre dykes in west Otago [*Cooper et al.*, 1987]. Uplift of ranges occurred in both North and South Islands, and some basins were everted exposing the sedimentary sequence. By the Late Miocene, sedimentation almost everywhere in South Island was dominated by clastic material [*King et al.*, 1999], much of it sourced from basement rocks. Terrigenous sediments were eroded from the rising ranges and thick sequences of clastic (locally volcaniclastic) sediments were deposited in subsiding basins and across stable platform areas, resulting in progradation of sand wedges across the continental shelf. Back-arc extension and volcanism also occurred in the North Island during the Miocene and Pliocene, whereas intraplate volcanics became locally prominent in South Island, including large centres at Dunedin and Banks Peninsula.

In Pliocene and Pleistocene time the plate boundary developed as a continental collision zone. Detritus shed from rising mountain systems built extensive piedmont gravel fans and plains, while deposition of marine mudstone continued offshore [*Field and Browne et al.*, 1989; *Nathan et al.*, 1986]. A change in the direction of plate motion and increased convergence is predicted to have occurred between about 12 and 6 Ma [*De Mets et al.*, 1994; *Cande and Stock*, 2004; *Sutherland*, 1995; *Walcott*, 1978, 1998]. The Southern Alps were rapidly uplifted, a crustal root developed, and enormous volumes of schist and greywacke detritus spread across South Island. Concomitantly, there was increase in the offshore sedimentation rate adjacent to the Southern Alps, and an influx of lithic-dominated material over previously quartz-dominated sedimentation [*Field and Browne et al.*, 1989; *Sircombe and Kamp*, 1986].

Cover Sequence Constraints on Plate Motion

The development of sedimentary sources and depositional centres is largely governed by vertical motion of land relative to sea level. Fault block interaction within a transform system such as the Australian-Pacific plate boundary inevitably results in vertical motion, although the total amount of horizontal motion may be considerably greater [*Norris and Carter*, 1982]. The vertical throw on the Alpine Fault, for example, has had profound physiographic and sedimentary effects, but is considerably less than the 440–470 km horizontal separation. Where source areas are distinctive, the detritus composition and distribution can be used to provide information on the relative vertical motion of the source areas. Material in fault-controlled basins of southwestern South Island, for example, records vertical motion and shedding of detritus from compositionally distinct basement terranes at various times during the Eocene-Miocene [*Manville*, 1996; *Norris and Carter*, 1982]. The present elevation of exposed marine sedimentary rocks, relative to inferred paleowater depths of formation, have provided some crude rates of uplift in South Westland since the Miocene [*Sutherland* et al., 1995]. Clasts of oligoclase zone Alpine schist in late Pliocene conglomerates indicates the time by which schist rocks of this grade must have been exhumed in the hanging wall of the Alpine Fault [*Sutherland*, 1996; *Mortimer et al.*, 2001].

Lateral fault displacements of sediments away from distinctive source areas are less commonly recognised, but have potential to place important constraints on displacement rate averaged since the depositional age of the sediment. Younger sediments (e.g., Pleistocene-Quaternary) need sufficient displacement for deposits to have been isolated from their source areas and catchment drainage, and rate calculations are commonly hampered by a lack of precise age determination. Studies of older sediments (e.g., Miocene-Pliocene) lack knowledge of paleogeography so fault displacements can be difficult to conclusively distinguish from sedimentary transport, but their age has allowed time for greater offset which improves precision of displacement rate calculations. Several studies have utilised the distinctiveness of basement terranes forming the New Zealand sub-continent and their high-angle to the direction of cross-cutting Australian-Pacific plate boundary displacement. Neogene clastic sedimentary deposits are preserved in several basins on Western Province basement, containing significant quantities of non-local Eastern Province detritus that have been transported across and offset by strike-slip movement on the Alpine Fault and other faults [*Cutten*, 1979; *Rose*, 1996; *Smale*, 1991; *Sutherland*, 1996]. Relocation of a Fiordland-derived Pliocene conglomerate that is currently exposed >100 km from its source area, for example, indicates the average displacement rate on the Alpine fault has been >27 ± 4 mm/yr since the Pliocene [*Sutherland*, 1994].

SURFICIAL DEPOSITS AND LANDSCAPE EVOLUTION

Landscape Evolution

New Zealand's landscape is young and vigorously evolving, and has been strongly influenced by effects of global climatic fluctuations, local weather patterns, and tectonic activity. Convergence across the Australian-Pacific plate boundary in South Island, and consequences of Pacific plate subduction in North Island, have pushed otherwise thinned and submerged crust upward into the path of a westerly atmospheric circulation pattern (Plate 5) [e.g., *Koons*, 1990; *Shulmeister et al.*, 2004]. A dramatic orographic climate regime results from the disturbance of a westerly atmospheric circulation by the Southern Alps. Heavy rainfall (>10 m/yr) occurs on the western slopes, whereas in the east precipitation is much lighter (≤1 m/yr) and a rainshadow is developed [*Griffiths and McSaveney*, 1983; *Henderson and Thompson*, 1999]. In New Zealand's mid-latitude setting, the mountains receive a significant proportion of their precipitation as snow and glaciers have formed.

Episodes of glaciation probably began in the Late Pliocene [*Suggate*, 1990]. As mountains continued to develop during the Pleistocene, glacial processes significantly modified the landscape in central and southern South Island, leaving an extensive and complex record of erosion and deposition. Large outwash surfaces and alluvial fans were produced down-valley of the ice limits during glacial episodes. In contrast, interglacial periods were characterised by downcutting of rivers and lake formation behind moraine loops. Periglacial conditions prevailed during the Pleistocene in eastern and northern South Island, and central and southern North Island, where loess accumulation, glacial-climate river aggradation, and interglacial degradation were widespread. Glacio-eustatic sea level fluctuation combined with tectonic uplift to leave a flight of marine terraces around the coast of New Zealand that have been used to constrain uplift rates [*Kim and Sutherland*, 2004; *Suggate*, 1965; *Pillans*, 1990]. By the Late Pleistocene, the topographic setting and landscape of South Island were broadly similar to that of the present day.

Glacial Cycles and Glacial Landforms

New Zealand's glaciers are affected by local as well as global climatic fluctuations [*Oerlemans*, 1997; *Shulmeister et al.*, 2004]. The resultant landforms and glacial deposits provide a record that is fragmentary in time and space, yet fundamental for constraining tectonic deformation across South Island. 5% of South Island is currently covered with ice; there are over 3000 glaciers ranging from tiny (10^4 m^2) snowfields to the largest (10^8 m^2) Tasman Glacier [*Chinn*, 2001]. Glacier morphology varies across the Southern Alps, with steep, highly active glaciers that have rapid (5–8 years) response to climatic events in the west, and gentler, larger valley glaciers with slower (up to 100 year) response times in the east [*Chinn*, 2001; *Gellatly et al.*, 1988]. The re-

sponse of different catchments to climatic variation is, and has been, dependent principally on the amount of catchment area above or below the varying snow accumulation/ablation equilibrium line [*Porter*, 1975]. Net gains/losses of ice are currently influenced by the strength of westerly atmospheric circulation [e.g., *Chinn*, 1995; *Fitzharris et al.*, 1992; *Porter*, 1975; *Shulmeister et al.*, 2004].

New Zealand's record of glaciation stretches back to the Late Pliocene with two separate, poorly preserved sequences of deposits interbedded with outwash gravels on the western side of the Southern Alps [*Gage*, 1945; *Suggate*, 1990]. Onshore evidence of glaciation is also patchy for most of the early to mid-Pleistocene, with increasingly better preservation of glacial deposits and landforms from the mid-late Pleistocene.

Late Pleistocene glaciers were much more extensive than the present glaciers in the Southern Alps [*Barrell et al.*, 2005; *Suggate*, 1990; *Suggate and Almond*, 2005]. They coalesced in the main mountain valleys to form piedmont lobes in the west and extended through the foothills to outwash plains in the east (Figure 5). Correlation of glacial sequences from one catchment to another is complicated and is still being refined, particularly as more numerical dates are obtained via radiocarbon, luminescence and surface-exposure cosmogenic methods. A series of relatively well-defined moraines mark the extent of ice during the Last Glacial Maximum (LGM) at about 26 ka (cal. yr. BP). There was then a retreat, formation of a series of moraines at or inside the full LGM extent from 23.5–19 ka, followed by rapid retreat and major mass wasting of the glaciers. Lakes in the centre of South Island now occupy glacial troughs that were formed by this glacial retreat (Plates 3, 5). Well-defined moraine loops and till ridges nearer the headwaters of many catchments mark a "late-glacial" ice advance in the period 11.4–14 ka that coincides approximately with the Younger Dryas and Antarctic Cold Reversal events [*Barrell et al.* 2005, *Denton and Hendy*, 1994; *Denton et al.*, 2005; *McGlone*, 1995; *Suggate*, 1990].

At least six minor glacial advances occurred during the Holocene [*Gellatly et al.*, 1988]. However, the last 100 years since the end of the "Little Ice Age" have been characterised by significant ice-loss (ca. 25% decrease by area) and the disappearance of many snowfields [*Bishop and Forsyth*, 1988; *Chinn*, 1996; *Sara*, 1970], despite periodic local increases in mass-balance due to annual changes in atmospheric flow patterns [*Chinn*, 1995; *Fitzharris et al.*, 1997].

Geomorphology, Erosion and Tectonics

The crest of the Southern Alps mountain chain lies 15–30 km southeast of the Alpine Fault, reaching a maximum elevation

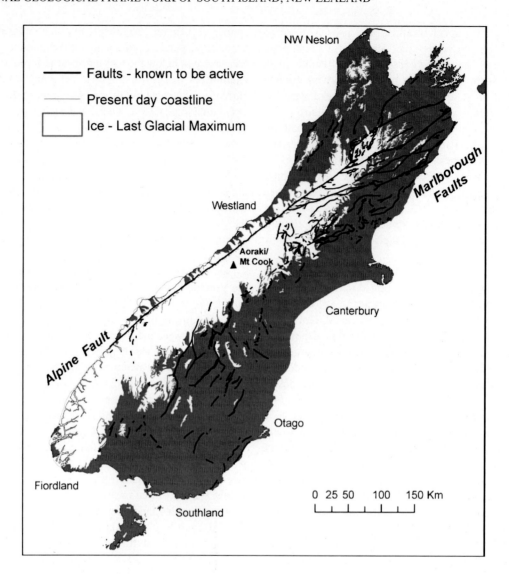

Figure 5. Late Quaternary fault displacements across the South Island plate boundary zone are insufficient to account for plate tectonic calculations of Australian-Pacific motion. This map shows all faults known to be active (displacement in last 125 ka) compared with the extent of ice during the Last Glacial Maximum c. 26 ka [after *Barrell et al.*, 2005]. Although there are numerous faults beneath the area recently covered in ice (see Plates 3, 4), there is a conspicuous absence of evidence for Quaternary displacement on all but the Alpine and Marlborough faults. Deformation probably occurs within the Southern Alps but is difficult to observe due to the youthful landscape, high rates of erosion, and a lack of offset features that have been preserved.

of 3754 m at Aoraki-Mount Cook, and is nearly everywhere coincident with the drainage divide (Main Divide). Geomorphology changes eastward across the Southern Alps from intensely dissected in the west, to strongly glaciated along the Main Divide, to more subdued eastern landscapes. There is an eastward increase in the age of the landforms that reflects variation in exhumation rates, rock lithology and relief.

Erosion appears to correlate strongly with precipitation, and is concentrated on the "windward" western side of the range [*Adams*, 1980a; *Adams*, 1980b; *Hicks et al.*, 1996]. A number of studies have utilised suspended sediment yields in rivers to estimate erosion [e.g., *Griffiths*, 1979; *Milliman and Meade*, 1983; *Soons*, 1986]. Yields vary across the Southern Alps from ca. 15,000 t/km/yr for the western side of the range, ca. 1600 t/km/yr for large basins draining the east side of the divide, ca. 600 t/km/yr for basins in the eastern foothills, to ca. 100 t/km/yr for dry intermontane basins in the eastern rainshadow. In terms of mean ground lowering, values of the around 4–7 mm/yr are obtained for major valleys draining the western side of the Southern Alps which, despite well-developed forest cover, exceed those determined for most other parts of the world by an order of magnitude [Soons, 1986]. Smaller catchments in the west have around 1–4 mm/yr denudation, falling at the upper end of worldwide values. Despite considerable uncertainties, the highest erosion rate estimates are broadly equivalent to rock uplift rates and Alpine Fault dip-slip rates [Adams, 1980a; Norris and Cooper, 2001], and order-of-magnitude variations in erosion rate across the Southern Alps corroborate expected variations based on geology and geomorphology [Whitehouse, 1986].

Fluvial and mass movement processes dominate erosion throughout most of South Island. Active floodplains, alluvial fans and debris cones are prominent landscape features. Rainfall appears to be the most important variable controlling sediment yield [Whitehouse, 1986], with short-term fluctuations from events such as earthquakes and floods. Material is carried predominantly in rivers as suspended load, with dissolved load and bed load both <5% [Adams, 1980a]. Present yields from glaciated and non-glaciated basins do not appear to differ significantly [Griffiths, 1979; Griffiths, 1981], leading to a hypothesis that erosion rates may be equal or higher during deglaciation than when valleys are protected by a cover of ice and receive a greater proportion of their precipitation as snow [Whitehouse, 1986]. Gravitational mass wasting appears to be the primary mode of erosion within the Southern Alps. Rock avalanches, rockfalls and landslides are widespread and frequent in the Southern Alps, particularly in areas of fault-zone weakened bedrock [Cox and Findlay, 1995; Korup, 2004; McSaveney, 2002; Whitehouse and Griffiths, 1983]. Landslides west of the Main Divide account for an average 9 ± 4 mm/yr downwasting [Hovius et al., 1997], involve as much as 1000 years worth of contemporary sediment yield, and play an important role in the erosion process as temporary storers of sediment [Korup, 2005].

The high erosion rates across the Australian-Pacific plate boundary are thought to have a profound effect on tectonic processes, and to have shaped the Southern Alps into an asymmetric orogen [Koons, 1990; Koons 1994; Plate 5]. A coupled system is proposed whereby removal of Pacific plate material by erosion enhances uplift, and uplift of rocks into the westerly airstream concentrates erosion [e.g., McSaveney, 1979; Koons, 1990]. Modelling of tectonics and topography suggests the Southern Alps may have been dynamically maintained at near steady-state elevation for the past million years [Koons, 1989; Koons 1995], and it

has been proposed that the width and shape of the orogen, and potentially the distribution of deformation and nature of fault behaviour, may be influenced by changes in climate-related erosion [Koons, 1994]. In detail, the geomorphology of valleys and ridges appears unlikely to reach steady-state, due to the influence of cycles in fluvial erosion and glaciation [Herman and Braun, 2006].

Offshore Sedimentation and Volcanic Deposition

Compared with the onshore glacial deposits, much more continuous records of Quaternary glaciation and paleo-environmental conditions appear to be preserved in marine sediments. Large volumes of sediment have been carried offshore [Adams, 1980a; Adams, 1980b] and deposited in offshore fan systems [e.g., Carter et al., 2004]. Fluvial sediment supply into these fan systems is estimated at ca.109 Mt/yr [Hicks and Shankar, 2003] corresponding to a very significant ~0.5% of sediment input into the world ocean for such a small landmass. Times of maximum supply are attributed to lowstands (glaciations), whereas supply is thought to be reduced during highstands [see Carter et al., 2004] due to increased sediment storage on land (e.g., glacial lakes) and shelf sedimentation.

DSDP 594 drillhole, situated on the margins of the Bounty fan 300 km offshore of South Island east coast, contains alternating units of pelagic ooze and hemipelagic ooze containing terrigenous material. These are interpreted to represent respective local warm period deposition interlayered with glacially-derived material from the Southern Alps, with evidence of at least 12 major periods of Southern Alps glaciation during the late Pleistocene [Nelson et al., 1985]. Comparison of the offshore and onshore geology implies that much onshore evidence of glaciation is missing, probably due to uplift/erosion or reworking by subsequent glaciations [see Suggate, 1990]. Marine core MD97 2121, recovered from a slope basin in the forearc east of North Island, contains an important high-resolution paleo-environmental record of the past ca.30,000 years [Carter et al., 2002]. Elsewhere, the Wanganui Basin evolved in a back-arc setting in western North Island with one of the most complete Plio-Pleistocene marine stratigraphic records in the world [Carter and Naish, 1998a].

While glaciation dominated evolution of South Island landscape during the Pleistocene, subsidence and volcanic activity prevailed in the north. The Taupo Volcanic Zone evolved as one of the most violent and productive silicic volcanic regions in the world, and numerous other volcanic centres developed from Taranaki to Northland. Erupted magma volume from the Taupo Volcanic Zone since 1.6 Ma is estimated at a massive 0.28 m^3/s [Wilson, 1996]. Tephra sourced from North Island were widely spread, as both airborne and waterborne deposits, and have been preserved throughout New Zealand's Quaternary geological record [e.g., *Carter et al.*, 1995]. They provide distinctive and useful chronostratigraphic marker horizons for subdivision and correlation of Quaternary deposits [e.g., *Almond*, 1996; *Barrell et al.*, 2005; *Carter and Naish*, 1998b].

Quaternary Deformation

The historical effects of earthquakes in New Zealand have been documented since the Marlborough earthquake of 1848, with the 1855 M8+ Wairarapa earthquake being the most severe earthquake felt in New Zealand since European settlement [*Downes*, 1995]. One of the earliest records of fault displacement related to earthquakes was made by Alexander McKay, who documented dextral offsets on fence lines crossing the Hope Fault as a result of the 1888 M7–7.3 North Canterbury earthquake [*Cowan*, 1991; *McKay*, 1890]. There are now well over one hundred studies providing assessment and refinements of Quaternary fault displacements and calculated displacement rates in New Zealand. Examples are too plentiful to list individually, and many will be cited in papers of this volume [see *Berryman et al.*, 1992; *Norris and Cooper*, 2001; *Pettinga et al.*, 2001; *Sutherland et al.*, 2005 for some summaries]. Horizontal displacement rates are generally harder to observe than vertical, because they require the identification and dating of a linear feature [*Norris and Cooper*, 2001]. These may be geomorphological, such as moraine ridges, terrace risers, stream or river channels, or geological deposits with a distinctive source, such as moraines or fluvial sediments. Vertical displacement rates involve recognition of an offset surface, such as marine or alluvial terraces. Vertical uplift/subsidence rates may be determined from terraces or lake shore-lines, or sedimentary deposits where the original position is known, raised above a reference surface at the time of formation such as sea level [*Pillans*, 1990; *Wellman*, 1979]. Dip-slip rates can be obtained where fault dip is known and slip-rates calculated if the actual fault slip-vector is known (e.g., from striations). In practice, determining displacement rates and slip-rates is not straight forward. Offset features need to be preserved in an inherently active and unstable tectonic environment. Uncertainties are introduced in both measuring the true fault-related offset and assigning an age to the feature [*Norris and Cooper*, 2001]. Large gaps remain in our knowledge of where Quaternary Australian-Pacific plate motion has been accommodated, and the scale of single-event displacements and hazard posed by various faults [e.g., see *Sutherland et al.*, this volume]. Whether displacements are the result of seismic, slow-rupture earthquakes, or aseismic creep, and encompassing any post-seismic after slip, is virtually unknown but has been the subject of much controversy [see *Walcott*, 1998; *Sutherland et al.*, this volume].

Direct evidence of late Quaternary deformation is poorly preserved in the rapidly evolving landscape of the Southern Alps, but late Quaternary deposits and landforms that have been offset by active faults or buckled by active folds are preserved in more stable lowland areas [e.g., *Beanland and Berryman*, 1989; *Jackson et al.* 1996; *Norris and Nicolls*, 2004]. By far the greatest rate of late Quaternary slip has occurred on the Alpine Fault. It has a relatively constant strike-slip displacement rate that is well constrained by offset late Quaternary deposits to be 23 ± 2 mm/yr in the southwestern onshore section [*Sutherland et al.*, 2006], and less well-constrained at about 27 ± 5 mm/yr through central South Island [*Berryman et al.*, 1992; *Norris and Cooper*, 2001, this volume]. Dip-slip rates are variable along the fault, being close to zero in the southwest, and rising to a maximum of >10 mm/yr locally in central South Island [*Norris and Cooper*, 2001, this volume]. The Pacific plate motion relative to the Australian plate, calculated from sea floor spreading data, is 37 ± 2 mm/year on a bearing of 251 ± 2° in central South Island [*DeMets et al.* 1994; see also *Cande and Stock*, 2004]. Strike-slip displacement on the Alpine Fault represents about 50-80% of fault-parallel plate motion, which is consistent with dislocation models of geodetic strain [*Beavan et al.*, 1999, this volume; *Wallace et al.*, 2007].

A large number of other faults have been mapped in South Island, but within the area overlain by ice during the LGM there is little evidence that can be used to determine late Quaternary displacement rates (Figure 5). The Alpine and Marlborough faults are the only major faults within this area confirmed to be "active" with displacement in the last 125 ka. Beyond the LGM ice extent, structures require displacement rates >0.3 mm/yr to have produced offset features that are preserved or are easily identifiable in the landscape. Although displacement rates are not precisely known, the total deformation distributed on these structures is insufficient to account for the 20-50% of plate motion that is not accommodated on the Alpine Fault. Up to about 10 mm/yr strike-slip motion appears to be "missing" [*Sutherland et al.*, 2006]. Global Positioning System (GPS) surveying since the early 1990's indicates plate motion is being converted into elastic strain (<1 mm per kilometre each year) which is concentrated in the Southern Alps [*Beavan et al.*, 1999, this volume; *Wallace et al.*, 2007], but how, where and when this energy might be released is not known. The presence of elevated topography and a concentration of geodetic deformation and earthquakes [Leitner et al., 2001] provide indirect, yet compelling, evidence that the "missing" component of plate boundary displacement probably occurs in the Southern Alps.

Whereas glaciers may have buried or swept the onshore alpine region clear of evidence of fault displacements, an extensive post-glacial surface was developed offshore and provides a marker of deformation. Recent studies using high-resolution seismic reflection methods and multi-beam swath mapping across the continental shelf are a reminder that deformation does occur offshore [e.g., *Barnes et al.*, 2005; *Collet et al.*, 1995; *Lamarche and Barnes*, 2005, *Nodder et al.*, 2005]. The presence of active folding and distributed 'off-fault' deformation is also becoming increasingly recognised onshore [e.g., *Barrell et al.*, 1996; *Jackson et al.* 1996; *Nicol and van Dissen*, 2002]. The extent to which "missing deformation" may occur (i) within the Southern Alps, (ii) locally along known active faults as distributed deformation (e.g., as fault-folds), (iii) offshore, or (iv) partly related to uncertainties in plate tectonic models, remains an important unresolved issue for New Zealand seismic hazard assessment.

SUMMARY

South Island of New Zealand is a fragment of Gondwana that was formed by Paleozoic and Mesozoic subduction-accretion processes at the paleo-Pacific margin; it split from Gondwana during the Late Cretaceous; and was rent by dextral strike-slip displacement on the Alpine Fault and distributed deformation during the Cenozoic. The kinematics of each part of each tectonic phase (including the present) was partially controlled by inherited faults and fabrics, but also involved the formation of new faults and fabrics, and the deposition of a sedimentary record [*Carter*, 1988; *King et al.*, 1999].

Pre-Late Cretaceous basement rocks are a series of volcano-sedimentary accreted terranes, batholiths that intrude these terranes, and schist and gneiss metamorphic overprints on the sedimentary and plutonic rocks (Plate 2) [*Mortimer*, 2004]. The paleo-Pacific plate boundary left pervasive fault, fold and metamorphic fabrics that are commonly sub-parallel to terrane boundaries. Widespread extension started at 110–100 Ma and New Zealand finally separated from Gondwana at 83–79 Ma when seafloor spreading initiated in the Tasman Sea and Southern Ocean (Plate 1). Cretaceous rifting caused reactivation of the strong Gondwana-margin fabric and new normal faults formed at moderate or high angles to terrane boundaries, to accommodate the imposed extensional plate motion; the 80–70 Ma extension direction is evident from orientations of fracture zones of marginal Late Cretaceous ocean crust (Figure 2).

The Cenozoic Australia-Pacific plate boundary initiated during Eocene-Oligocene time [*Carter and Norris*, 1976; *Kamp*, 1986], but very little plate motion occurred during this initial phase, because South Island was close to the pole of relative plate rotation. Minor rift reactivation occurred and new passive margins formed adjacent to a spreading ridge farther south (Plate 1). Although little intra-continental deformation occurred at this time, the western passive margin that was formed would eventually have a profound impact on the continental kinematics: the passive margin was progressively displaced into South Island during Miocene-Quaternary time and caused the Alpine Fault to form (Plate 1, Figure 2) [*Sutherland et al.*, 2000]. Basement terranes represent elongate markers that are now deformed by movement at the Australia-Pacific plate boundary, and are distinctly curved adjacent to the Alpine Fault with geometry sympathetic with distributed dextral shear across the plate boundary. The broad curvature of basement terranes, combined with the apparent Alpine Fault offset of ca. 470 km, is consistent with the inference based upon marine magnetic anomalies that around 800 to 850 km of Australia-Pacific plate motion occurred during the Cenozoic (Figures 2, 3) [*King*, 2000; *Sutherland*, 1999b].

An influx of lithic-dominated sediment into basins marks the onset of strike-slip and local convergence across the plate boundary, and the first-order coarsening-upward sequence records progressive growth of the Southern Alps during Neogene time [*Carter* 1988; *Carter and Norris*, 1976; *King et al.*, 1999; *Youngson et al.*, 1998]. Regional-scale unconformities at the base of the sedimentary rock sequence [*LeMasurier and Landis*, 1996], and within it [*Carter*, 1985], have been folded and offset, with throws of several kilometres preserved outside the zone of high exhumation near the Alpine Fault (Plate 4). The high angle between the plate boundary and the strike of distinctive basement terranes enables displaced Miocene and Pliocene sediments in a few instances to be linked back to their sedimentary source, and they are shown to be consistent with displacement rates of 20-30 mm/yr on the Alpine Fault [e.g., *Sutherland*, 1994].

The Southern Alps create an orographic climate with asymmetric rainfall and erosion that affects the pattern of geomorphology, glaciation, uplift, erosion, exhumation, and rock-types exposed across South Island (Plate 5). Pleistocene glacial cycles carved and shaped the land [*Suggate*, 1990], and there has been extensive modification of the landscape in the central Southern Alps during the Holocene. Records of early Quaternary tectonic events were largely wiped, but numerous active faults have been recognised beyond the extent of glaciers, with displacement rates constrained by offset late Quaternary features. Offset moraines and hillslopes in southwestern South Island show that the Alpine Fault has pure strike-slip displacement at a rate of 23 ± 2 mm/yr, and in central South Island, where local uplift rates may exceed 10 mm/yr, the dextral-oblique slip rate is constrained to be 27 ± 5 mm/yr [*Norris and Cooper*, 2001; *Sutherland et al.*, 2006]. Folding and distributed deformation southeast of the Alpine Fault is becoming more widely recognised but a deficit of around 10 mm/yr remains when the sum of geologically-determined fault-slip rates are compared with the plate-motion rate of 38 ± 2 mm/yr [*DeMets et al.*, 1994].

Acknowledgements. This review incorporated ideas and logic from a huge variety of sources that are well beyond the comparative specialist expertise of the authors. Hopefully the reference list provides due acknowledgement, as the sources are too numerous to be named individually. Numerous discussions with David Barrell, and consultations with Tim Little, Vernon Manville, Nick Mortimer, Richard Norris and Andy Tulloch, helped clarify the relative importance and details of various geological features in our geological history and the evolution of the Australian-Pacific plate boundary. Material provided by Nick Mortimer, David Barrell, and the QMAP 1:250,000 GIS database have been used to generate the diagrams. The review was prepared under GNS Science's 'Impacts of Global Plate Tectonics in and around New Zealand Programme' (PGST Contract C05X0203). Jane Forsyth, David Barrell and two anonymous referees are thanked for providing critical and careful reviews.

REFERENCES

Adams, C. J. D. (1979), Age and origin of the Southern Alps, in *The Origin of the Southern Alps*, edited by R. I. Walcott and M. M. Cresswell, pp. 73–78, Royal Society of New Zealand Bulletin 18.

Adams, C. J. D. (1981), Uplift rates and thermal structure in the Alpine Fault Zone and Alpine Schists, Southern Alps, New Zealand, in *Thrust and nappe tectonics,* edited by K. R McClay, and N. J. Price, pp. 211–222. Oxford: Blackwell. Special Publication Geological Society of London 9.

Adams, C. J. and S. Kelley (1998), Provenance of Permian-Triassic and Ordovician metagraywacke terranes in New Zealand: Evidence from Ar/Ar dating of detrital micas, *Geol. Soc. Am. Bull.*, 110, 422–432.

Adams, J. (1980a) Contemporary uplift and erosion of the Southern Alps, New Zealand: Summary, *Geol. Soc. Am. Bull.*, 91, *Part I*, 2–4.

Adams, J. (1980b) Contemporary uplift and erosion of the Southern Alps, New Zealand: Part II, *Geol. Soc. Am. Bull.*, 91, *Part II*, 1–114.

Aitchison, J., G. L. Clarke, S. Meffre, and D. Cluzel (1995), Eocene arc-continent collision in New Caledonia and implications for regional Southwest Pacific tectonic evolution, *Geology*, 23(2), 161–164.

Almond, P. C. (1996), Loess, soil stratigraphy and Aokautere Ash on Late Pleistocene surfaces in South Westland: interpretation and correlation with the glacial stratigraphy, *Quat. Int.*, 34–36, 163–176.

Barrell, D. J. A., B. V. Alloway, J. Shulmeister, and R. M. Newnham (editors) (2005), Towards a climate event stratigraphy for New Zealand over the past 30,000 years. Institute of Geological and Nuclear Sciences Limited, Lower Hutt, *Institute of Geological and Nuclear Sciences Science Report 2005/07*. 12 p. + 1 poster.

Barrell, D. J. A., P. J. Forsyth, and M. J. McSaveney (1996), Quaternary geology of the Rangitata Fan, Canterbury Plains, New Zealand, *Institute of Geological and Nuclear Sciences Science Report 96/23*.

Barnes, P. M., R. Sutherland, and J. Delteil (2005), Strike-slip structure and sedimentary basins of the southern Alpine Fault, Fiordland, New Zealand, *Geol. Soc. Am. Bull.*, 117(3/4), 411–435.

Batt, G. E., J. Braun, B. P. Kohn, and I. McDougall (2000), Thermochronological analysis of the dynamics of the Southern Alps, New Zealand, *Geol. Soc. Am. Bull.*, 112, 250–266.

Batt, G. E., and J. Braun (1999), The tectonic evolution of the Southern Alps, New Zealand: insights from thermally coupled dynamical modelling, *Geophys. J. Int.*, 136, 403–420.

Beanland, S., and S. A. Barrow-Hurlbert (1988), The Nevis-Cardrona Fault System, Central Otago, New Zealand, Late Quaternary tectonics and structural development, *N. Z. J. Geol. Geophys.*, 31, 337–352.

Beanland, S., and K. R. Berryman (1989), Style and episodicity of late Quaternary activity on the Pisa-Grandview Fault Zone, Central Otago, New Zealand, *N. Z. J. Geol. Geophys.*, 32(4), 451–461.

Beavan, J., M. Moore, C. Pearson, M. Henderson, B. Parsons, S. Bourne, P. England, R.I. Walcott, G. Blick, D. Darby, and K. Hodgkinson, (1999), Crustal deformation during 1994–98 due to oblique continental collision in the central Southern Alps, New Zealand, and implications for seismic potential of the Alpine Fault, *J. Geophys. Res.*, 104 (B11), 25,233–25,255.

Beavan, J., S. Ellis, L. Wallace, P. Denys (this volume) Kinematic Constraints from GPS on Oblique Convergence of the Pacific and Australian Plates, Central South Island, New Zealand.

Begg, J. G., and M. R. Johnstone (2000), Geology of the Wellington area, Institute of Geological and Nuclear Sciences 1:250,000 geological map 10. 1 sheet +64p.

Benson, W. N. (1952), Meeting of the Geological Division of the Pacific Science Congress in New Zealand, February 1949, Interim Proceedings of the Geological Society of America Feb 1950, 1, 11–13.

Berryman, K. R., S. Beanland, A. F. Cooper, H. N. Cutten, R. J. Norris, and P. R. Wood (1992), The Alpine Fault, New Zealand: variation in Quaternary structural style and geomorphic expression, *Ann. Tecton.*, *Spec. Issue—Suppl.*, 6, 126–163.

Bishop, D. G., J. D. Bradshaw, and C. A. Landis (1985), Provisional terrane map of the South Island, New Zealand, in *Tectonostratigraphic terranes of the circum-Pacific region*, edited by D. G. Howell, A.A.P.G. Circum-Pacific Council for Energy and Mineral Resources Earth Science Series 1, 515–521.

Bishop, D. G., and P. J. Forsyth (1988), Vanishing Ice – An introduction to glaciers based on a study of the Dart Glacier, 56 pp. Dunedin, New Zealand: John McIndoe and New Zealand Geological Survey.

Bradshaw, J. D. (1973), Allochthonous Mesozoic fossil localities in melange within the Torlesse rocks of north Canterbury, *J. R. Soc. N. Z.*, *32*, 169–181.

Bradshaw, J. D. (1989), Cretaceous geotectonic patterns in the New Zealand region, *Tectonics*, *8*, 803–820.

Cande, S. C., J. L. LaBrecque, R. L. Larson, W. C. Pitman III, X. Golovchenko, W. F. Haxby (1989), Magnetic lineations of the worlds ocean basins (map), American Association of Petroleum Geologists, Tulsa, OK, USA.

Cande, S. C., and J. M. Stock (2004), Pacific-Antarctic-Australia motion and the formation of the Macquarie Plate, *Geophys. J. Int.*, *157*, 399–414.

Cande, S. C., and D. V. Kent (1995), Revised calibration of the geomagnetic polarity timescale for the Late Cretaceous and Cenozoic, *J. Geophys. Res.*, *100*(4), 6093–6095.

Cande, S. C., C. A. Raymond, J. M. Stock, and W. F. Haxby (1995), Geophysics of the Pitman Fracture Zone, *Science*, *270*, 947–953.

Carter, L., B. Manighetti, M. Elliot, N. A. Trustrum, and B. Gomez (2002), Source, sea level and circulation effects on the sediment flux to the deep ocean over the past 15 ka off eastern New Zealand, *Global Planet. Change*, *33*(3/4), 339–355.

Carter, L., C. S. Nelson, H. L. Neil and P. C. Froggatt (1995), Correlation, dispersal, and preservation of the Kawakawa Tephra and other late Quaternary tephra layers in the Southwest Pacific Ocean, *N. Z. J. Geol. Geophys.*, *38*(1), 29–46.

Carter, L., R. M. Carter, and I. N. McCave, (2004), Evolution of the sedimentary system beneath the deep Pacific inflow off eastern New Zealand, *Mar. Geol.*, *205*(14): 9–27.

Carter, R. M., and C. A. Landis (1972), Correlative Oligocene unconformities in southern Australasia, *Nat. Phys. Sci.*, *237*, 12–13.

Carter, R. M., and R. J. Norris (1976), Cainozoic history of southern New Zealand: an accord between geological observations and plate tectonic predictions, *Earth Planet. Sci. Lett.*, *31*(1), 85–94.

Carter, R. M. (1985), The mid-Oligocene Marshall paraconformity, New Zealand: coincidence with global eustaic sea-level fall or rise?, *J. Geol.*, *93*, 359–371.

Carter, R. M. (1988), Post-breakup stratigraphy (Kaikoura Synthem: Cretaceous-Cenozoic) of the continental margin off southeastern New Zealand, *N. Z. J. Geol. Geophys.*, *31*(4), 405–429.

Carter, R. M., and T. R. Naish (1998a), A review of Wanganui Basin, New Zealand: global reference section for shallow marine, Plio-Pleistocene (2.5–0 Ma) cyclostratigraphy, *Sediment. Geol.*, *122*(1/4), 37–52.

Carter, R. M., and T. R. Naish (1998b), Have local stages outlived their usefulness for the New Zealand Pliocene-Pleistocene?, *N. Z. J. Geol. Geophys.*, *41*(3), 271–279.

Chinn, T. J. (1995), Glacier fluctuations in the Southern Alps of New Zealand determined from snowline elevations, *Arctic Alpine Res.*, *27*(2), 187–198.

Chinn, T. J. (1996), New Zealand glacier responses to climate change of the past century, *N. Z. J. Geol. Geophys.*, *39*(3), 415–428.

Chinn, T. J. (2001), Distribution of the glacial water resources of New Zealand, *J. Hydrol. N. Z. 40* (2), 139–187.

Collot, J.-Y., J. Delteil, R. H. Herzer, R. Wood, K. B. Lewis, J.-C. Audru, B. Mercier de Lepinay, M. Popoff, E. Ruellan, M. Sosson, P. Barnes, G. Lamarche, J.-F. Lebrun, B. Pontoise, B., Toussaint, S. Calmant, B. Pelletier, F. Chanier, J. Ferriere, E. Chaumillon, M. Coffin, B. W. Davy, C. I. Uruski, D. Christoffel, S. Lallemand, A. Mauffret, A. Orpin, and R. Sutherland (1995), Sonic imaging reveals new plate boundary structures offshore New Zealand, *EOS*, *76*(1):1, 4–5.

Coombs, D. S., C. A. Landis, R. J. Norris, J. M. Sinton, D. J. Borns, and D. Craw (1976), The Dun Mountain Ophiolite Belt, New Zealand, its tectonic setting, constitution and origin, with special reference to the southern portion, *Am. J. Sci.*, *276*, 561–603.

Cooper, A. F., B. A. Barriero, D. L. Kimbrough, and J. M. Mattinson (1987), Lamprophyre dike intrusion and the age of the Alpine Fault, New Zealand, *Geology*, *15*, 941–944.

Cooper, A. F., and A. Reay (1983), Lithology, field relationships, and structure of the Pounamu Ultramafics from the Whitcombe and Hokitika Rivers, Westland, New Zealand, *N. Z. J. Geol. Geophys.*, *26*(4), 359–379.

Cooper, R. A. (1989), Early Paleozoic terranes of New Zealand, *J. R. Soc. N. Z.*, *19*, 73–112.

Cooper, R. A., and A. J. Tulloch (1992), Early Palaeozoic terranes in New Zealand and their relationship to the Lachlan Fold Belt, *Tectonophysics*, *214*, 129–144.

Cowan, H. A. (1991), The North Canterbury earthquake of September 1, 1888, *J. R. Soc. N. Z.*, *21*(1), 13–24.

Cox, S. C., and D. J. A. Barrell, Geology of the Aoraki area. Institute of Geological and Nuclear Sciences 1:250,000 geological map 15, 1 sheet + 71 pages. Lower Hutt, New Zealand: GNS Sciences.

Cox, S. C., and R. H. Findlay (1995), The Main Divide fault zone and its role in formation of the Southern Alps, New Zealand, *N. Z. J. Geol. Geophys.*, *38*, 489–499.

Cox, S. C., D. Craw, and C. P. Chamberlain, (1997), Structure and fluid migration in a late Cenozoic duplex system forming the Main Divide in the central Southern Alps, New Zealand, *N. Z. J. Geol. Geophys.*, *4*, 359–374.

Craw, D. (1984), Lithologic variations in Otago Schist, Mt Aspiring area, northwest Otago, New Zealand, *N. Z. J. Geol. Geophys.*, *27*(2), 151–166.

Cutten, H. N. C. (1979), Rappahannock Group: Late Cenozoic sedimentation and tectonics contemporaneous with Alpine Fault movement, *N. Z. J. Geol. Geophys.*, *22*(5), 535–553.

Davey, F. J., D. Eberhart-Phillips, M. Kohler, S. Bannister, G. Caldwell, S. Henrys, M. Scherwath, T. Stern, and H. van Avedonk (this volume), 3D Structure of the Southern Alps Orogen, South Island, New Zealand.

Deckert, H., U. Ring, and N. Mortimer (2002), Tectonic significance of Cretaceous bivergent extensional shear zones in the Torlesse accretionary wedge, central Otago Schist, New Zealand, *N. Z. J. Geol. Geophys.*, *45*(4): 537–547.

DeMets, C., R. G. Gordon, D. F. Argus, and S. Stein (1994), Effect of recent revisions to the geomagnetic time scale on estimates of current plate motions, *Geophys. Res. Lett.*, *21*, 2191–2194.

Denton, G. H., and C. H. Hendy (1994), Younger Dryas age advance of Franz Josef Glacier in the Southern Alps of New Zealand, *Science*, *264*(5164), 1434–1437.

Denton, G. H., R. B. Alley, G. C. Comer, and W. Broeker (2005), The role of seasonality in abrupt climate change, *Quat. Sci. Rev.*, *24*, 1159–1182.

Downes, G. L. (1995), Atlas of isoseismal maps of New Zealand earthquakes. Institute of Geological and Nuclear Sciences monograph 11, 304 p., Lower Hutt, New Zealand: Institute of Geological and Nuclear Sciences.

Field, B. D., and G. H. Browne (1986), Lithostratigraphy of Cretaceous and Tertiary rocks, southern Canterbury, New Zealand. Wellington, New Zealand Geological Survey Record 14.

Field, B. D., and G. H. Browne, et al. (1989), Cretaceous and Cenozoic sedimentary basins and geological evolution of the Canterbury Region, South Island, New Zealand, New Zealand Geological Survey Basins Studies 2.

Fitzharris, B. B., J. E. Hay, and P. D. Jones (1992), Behaviour of New Zealand glaciers and atmospheric circulation changes over the past 130 years, *Holocene*, *2*(2), 97–106.

Gaina, C., D. R. Mueller, J.-Y. Royer, J. M. Stock, J. L. Hardebeck, and P. Symonds (1998), The tectonic history of the Tasman Sea; a puzzle with 13 pieces, *J. Geophys. Res.*, *103*(6), 12,413–12,433.

Gage, M. (1945), The Tertiary and Quaternary geology of Ross, Westland, *Trans. R. Soc. N. Z.*, *75*, 138–159.

Gellatly, A. F., T. J. H. Chinn, and F. Rothlisberger (1988), Holocene glacier variations in New Zealand: A review, *Quat. Sci. Rev.*, *7*, 227–242.

Gibson, G. M., I. McDougall, and T. R. Ireland (1988), Age constraints on metamorphism and the development of a metamorphic core complex in Fiordland, southern New Zealand, *Geology*, *16*(5): 405–408.

Godfrey, N. J., N. I. Christensen, and D. A. Okaya (2002), The effect of crustal anisotropy on reflector depth and velocity determination from wide-angle seismic data: a synthetic example based on South Island, New Zealand, *Tectonophysics*, *355*, 145161.

Grapes, R. H. (1995), Uplift and exhumation of Alpine Schist in the Franz Josef - Fox Glacier area of the Southern Alps, New Zealand, *N. Z. J. Geol. Geophys. 38*, 525–533.

Grapes, R. H., and T. Wantanabe (1992), Metamorphism and uplift of Alpine Schist in the Franz Josef-Fox Glacier area of the Southern Alps, New Zealand, *J. Metamorph. Geol.*, *10*, 171–180.

Grapes, R. H., and T. Wantanabe (1994), Mineral composition variation in Alpine Schist, Southern Alps, New Zealand: Implications for recrystallisation and exhumation, *Island Arc*, *3*, 163–181.

Gray, D. R., and D. A. Foster (2004), [40]Ar/[39]Ar thermochronologic constraints on deformation, metamorphism and cooling/exhumation of a Mesozoic accretionary wedge, Otago Schist, New Zealand, *Tectonophysics*, *385*, pp. 181–210.

Griffiths, G. A. (1979), High sediment yields from major rivers of the western Southern Alps, New Zealand, *Nature*, *282*, 61–63.

Griffiths, G. A. (1981), Some suspended sediment yields from South Island catchments, New Zealand, *Water Resour. Bull.*, *17*(4), 662–671.

Griffiths, G. A., and M. J. McSaveney (1983), Distribution of mean annual precipitation across some steepland regions of New Zealand. *N. Z. J. Sci.*, *26*, 197–209.

Henderson, R.D., and S.M. Thompson (1999), Extreme rainfalls in the Southern Alps of New Zealand, *J. Hydrol. N. Z.*, *38*(2), 309–330.

Herman, F., and J. Braun (2006), Fluvial response to horizontal shortening and glaciations: A study in the Southern Alps of New Zealand, *J. Geophys. Res.*, *111*, F01008, doi:10.1029/2004JF000248.

Herman F., J. Braun, and W. J. Dunlap (2007), Tectonomorphic scenarios in the Southern Alps of New Zealand, *J. Geophys. Res.*, *112*, B04201, doi:10.1029/2004JB003472.

Hicks, D. D., J. Hill, and U. Shankar (1996), Variation of suspended sediment yields around New Zealand: the relative importance of rainfall and geology, *IHAS Publ.*, *236*:149–156.

Hicks, D. D., and U. Shankar (2003), Sediment from New Zealand Rivers, *NIWA Chart, Miscellaneous Series*, *79*.

Hovius, N., C. P. Stark, and P. A. Allen (1997), Sediment flux from a mountain belt derived by landslide mapping, *Geology*, *25*, 231–234.

House, M. A., M. Gurnis, P. J .J. Kamp, and R. Sutherland (2002), Uplift in the Fiordland region, New Zealand: implications for incipient subduction, *Science*, *297*(5589), 2038–2041.

House, M. A., M. Gurnis, R. Sutherland, and P. J. J. Kamp (2005), Patterns of Late Cenozoic exhumation deduced from apatite and zircon U-He ages from Fiordland, NZ, *Geochem. Geophys. Geosyst.*, *6*(9), Q09013, doi:10.1029/2005GC000968.

Hunt, T., 1978. Stokes Magnetic Anomaly System, *N. Z. J. Geol. Geophys.*, *21*, 595–606.

Jackson, J., R. Norris, and J. Youngson (1996), The structural evolution of active fault and fold systems in central Otago, New Zealand: evidence revealed by drainage patterns, *J. Struct. Geol.*, *18*, 217–234.

Kamp, P. J. J. (1986), The mid-Cenozoic Challenger Rift system of western New Zealand and its implications for the age of Alpine Fault inception, *Geol. Soc. Am. Bull.*, *97*(3): 255–281.

Kamp, P. J. J. (1987), Age and origin of the New Zealand orocline in relation to Alpine Fault movement, *J. Geol. Soc. (Lond.)*, *144*, 641–652.

Kamp, P. J. J. (1997), Paleogeothermal gradient and deformation style, Pacific front of the Southern Alps Orogen: constraints from fission track thermochronology, *Tectonophysics*, *271*, 37–58.

Kamp, P. J. J., and J. M. Tippett (1993), Dynamics of Pacific plate crust in the South Island (New Zealand) zone of oblique continent-continent convergence, *J. Geophys. Res.*, *98*(B9) 16, 105–16,118.

Kamp, P. J. J. (2001), Possible Jurassic age for part of Rakaia Terrane: implications for tectonic development of the Torlesse accretionary prism, *N. Z. J. Geol. Geophys.*, *44*(2), 185–203.

Kear, D., and Mortimer, N. (2003), Waipa Supergroup, New Zealand: a proposal, *J. R. Soc. N. Z.*, *33*(1): 149–163

Kim, K. J. and Sutherland, R. (2004), Uplift rate and landscape development in southwest Fiordland, New Zealand, determined using [10]Be and [26]Al exposure dating of marine terraces, *Geochim. Cosmochim. Acta*, *68*(10): 2313–2319.

King, P. R. (2000), Tectonic reconstructions of New Zealand 40 Ma to the present, *N. Z. J. Geol. Geophys. 43*(4), 611–638.

King, P. R., T. R. Naish, G. H. Browne, B. D. Field, and S. W. Ed-brooke (1999), Cretaceous to Recent sedimentary patterns in New Zealand, Lower Hutt: Institute of Geological and Nuclear Sciences, Institute of Geological and Nuclear Sciences folio series 1.

Kingma, J. T. (1959), The tectonic history of New Zealand, *N. Z. J. Geol. Geophys.*, *2*, 1–55.

Kleffman, S., F. Davey, A. Melhuish, D. Okaya, and T. Stern (1998), Crustal structure in the central South Island from the Lake Pukaki seismic experiment, *N. Z. J. Geol. Geophys.*, *41*, 39–49.

Koons, P. O. (1989), The topographic evolution of collisional mountain belts: a numerical look at the Southern Alps, New Zealand, *Am. J. Sci.*, *289*, 1041–1069.

Koons, P. O. (1990), Two-sided orogen: collision and erosion from the sandbox to Southern Alps, New Zealand, *Geology*, *18*, 679–682.

Koons, P. O. (1994), Three-dimensional critical wedges: tectonics and topography in oblique collisional orogens, *J. Geophys. Res.*, *99*, 12,301–12,315.

Koons, P. O. (1995), Modelling the topographic evolution of collisional belts, *Annu. Rev. Earth Planet Sci.*, *23*, 375–408.

Korup, O. (2004), Geomorphic implications of fault zone weakening: slope instability along the Alpine Fault, South Westland to Fiordland, *N. Z. J. Geol. Geophys.*, *47*(2), 257–267.

Korup, O. (2005), Distribution of landslides in southwest New Zealand, *Landslides*, *1*, 43–51, doi:10.1007/s10346-004-0042-0.

Laird, M. G. (1981), The late Mesozoic fragmentation of New Zealand segment of Gondwana, in *Gondwana Five*, edited by M. M. Cresswell, and P. Vella, Proceedings of the fifth international Gondwana Symposium Wellington, February 1980, 311–318. Balkema, Rotterdam.

Laird, M. G. (1993), Cretaceous Continental Rifts: New Zealand Region, in *South Pacific Sedimentary Basins. Sedimentary Basins of the World 2*, edited by P. F. Ballance, Elsevier, Amsterdam.

Laird, M. G. (1994), Geological aspects of the opening of the Tasman sea, in *Evolution of the Tasman Sea Basin*, edited by G. J. van der Lingen, K. M. Swanson, and R. J. Muir, A.A. Balkema, Rotterdam.

Lamarche, G., and P .M. Barnes (2005), Fault characterisation and earthquake source identification in the offshore Bay of Plenty, in *Programme and abstracts, Geological Society of New Zealand 50th Annual conference, Kaikoura, New Zealand,* edited by J. R. Pettinga, and A. M. Wandres, *Geol. Soc. N. Z. Misc. Publ.*, *119A*, p. 42.

LeBrun, J. F., G. Lamarche, and J. Y. Collot (2003), Subduction initiation at a strike-slip plate boundary: The Cenozoic Pacific-Australian plate boundary, south of New Zealand, *J. Geophys. Res.*, *108*(B9), 2453, doi:10.1029/2002JB002041.

LeMasurier, W. E., and C. A. Landis (1996), Mantle-plume activity recorded by low-relief erosion surfaces in West Antarctica and New Zealand, *Geol. Soc. Am. Bull.*, *108*, 1450–1466.

Litchfield, N. J. (2001), The Titri Fault System: Quaternary-active faults near the leading edge of the Otago reverse fault province, *N. Z. J. Geol. Geophys.*, *44*(4), 517–534.

Little, T. A., N. Mortimer, and M. McWilliams (1999), An episodic Cretaceous cooling model for the Otago-Marlborough Schist, New Zealand, based on [40]Ar/[39]Ar white mica ages, *N. Z. J. Geol. Geophys.*, *42*, pp. 305–343.

Little, T. A., and N. Mortimer (2001), Rotation of ductile fabrics across the Alpine Fault and Cenozoic bending of the New Zealand orocline, *J. Geol. Soc. Lond.*, *15*, 745–756.

Little, T. A., R. J. Holcombe, and B. R. Ilg (2002a), Ductile fabrics in the zone of active oblique convergence near the Alpine Fault, New Zealand: identifying the neotectonic overprint, *J. Struct. Geol.*, *24*, 193–217.

Little, T. A., R. J. Holcombe, and B. R. Ilg (2002b), Kinematics of oblique continental collision and ramping inferred from micro-stuctures and strain in middle crustal rocks, central Southern Alps, New Zealand, *J. Struct. Geol.*, *24*, 219–239.

Little, T. A., M. K. Savage, and B. Tikoff (2002c), relationship between crustal finite strain and seismic anisotropy in the mantle, Pacific-Australia plate boundary zone, South Island, New Zealand, *Geophys. J. Int. 151*, 106–116.

Little, T. A., S. C. Cox, J. K. Vry, and G. Batt (2005), Variations in exhumation level and uplift-rate along the oblique-slip Alpine Fault, central Southern Alps, New Zealand, *Geol. Soc. Am. Bull.*, *117*(5), 707–723.

Little, T A, R. Wightman, R. J. Holcombe, and M. Hill (this volume), Transpression models and ductile deformation of the lower crust of the Pacific Plate in the central Southern Alps, a perspective from structural geology.

Long, D. T., S. C. Cox, S. Bannister, M. C. Gerstenberger, and D. Okaya (2001), Upper crustal structure beneath the eastern Southern Alps and the MacKenzie Basin, New Zealand, derived from seismic reflection data, *N. Z. J. Geol. Geophys.*, *46*(1), 21–39.

McGlone, M.S. (1995), Late glacial landscape and vegetation change and the younger dryas climatic oscillation in New Zealand, *Quat. Sci. Rev.*, *14*(9):867–881.

McKay, A. (1890), On the Earthquakes of September 1888, in the Amuri and Marlborough districts of the South Island, *N. Z. Geol. Surv. Rep. Geol. Explor.*, *20*,1–16.

McMillan, S. G. C., and G. J. Wilson (1997), Allostratigraphy of coastal south and east Otago: a stratigraphic framework for interpretation of the Great South Basin, New Zealand, *N. Z. J. Geol. Geophys.*, *40*, 91–107.

McSaveney, M. J. (1979), Is the Kaikoura Orogeny a climatic event: An influence of weather on a plate boundary. Ministry of Works and Development, Christchurch, New Zealand, Report No. WS73.

McSaveney, M. J. (2002), Recent rockfalls and rock avalanches in Mount Cook National Park, New Zealand, *Geol. Soc. Am. Rev. Eng. Geol.*, *v. XV*, 35–70.

MacKinnon, T. C. (1983), Origin of the Torlesse terrane and coeval rocks, South Island, *New Zealand. Geol. Soc. Am. Bull.*, *94*, 967–985.

Manville, V. (1996), Sedimentology and stratigraphy of Prospect Formation, Te Anau Basin, western Southland, New Zealand, *N. Z. J. Geol. Geophys.*, *39*(3), 429–444.

Markley M., and R. J. Norris (1999), Structure and neotectonics of the Blackstone Hill Antiform, Central Otago, New Zealand, *N. Z. J. Geol. Geophys.*, *42*, 205–218.

Mazengarb, C. (1994), Estimating actual displacement along the Alpine Fault: time for a test, *Geol. Soc. N. Z. Newslett.*, *105*(58).

Mazengarb, C., and D. H. M. Harris (1994), Cretaceous stratigraphic and structural relations of Raukumara Peninsula, New

Zealand: stratigraphic patterns associated with the migration of a thrust system, *Ann. Tecton., 8*(2): 100-108.

Milliman, J. D., and J.D. Meade (1983), World-wide delivery of river sediment to the oceans, *J. Geol., 91,* 1–22.

Molnar, P., T. Atwater, J. Mammerickx, and S. M. Smith (1975), Magnetic anomalies, bathymetry and the tectonic evolution of the south Pacific since the late Cretaceous, *Geophys. J. R. Astron. Soc., 40,* 383–420.

Mortimer, N. (1993), Jurassic tectonic history of Otago Schist, New Zealand, *Tectonics, 12,* 237–224.

Mortimer, N. (1995), Origin of the Torlesse Terrane and coeval rocks, North Island, New Zealand, *Int. Geol. Rev., 3,* 891–910.

Mortimer, N., R. Sutherland, and S Nathan (2001), Torlesse greywacke and Haast Schist source for Pliocene conglomerates near Reefton, New Zealand, *N. Z. J. Geol. Geophys., 44*(1), 105–111.

Mortimer, N., and M. R. Johnstone (1990), Discovery of a new Rangitata structure offset by the Alpine Fault: enigmatic 350km-long synform within the Caples-Pelorous terrane, *Geol. Soc. N. Z. Misc. Publ., 50A*(99).

Mortimer, N. (2000), Metamorphic discontinuities in orogenic belts: example of the garnet-biotite-albite zone in the Otago schist, New Zealand, *Int. J. Earth Sci., 89*(2), 295–306.

Mortimer, N. (2003), A provisional structural thickness map of the Otago Schist, New Zealand, *Am. J. Sci., 303,* pp. 603–621.

Mortimer, N. (2004), New Zealand's geological foundations, *Gondwana Res., 7*(1), 262–272.

Mortimer, N., and D. Parkinson (1996), Hikurangi Plateau; a Cretaceous large igneous province in the Southwest Pacific Ocean, *J. Geophys. Res., 101*(B1), 687–696, doi:10.1029/95JB03037.

Mortimer, N., A. J. Tulloch, R. N. Spark, N. W. Walker, E. Ladley, A. Allibone, and D. L. Kimbrough (1999a), Overview of the Median Batholith, New Zealand: a new interpretation of the geology of the Median Tectonic Zone and adjacent rocks, *J. Afr. Earth Sci., 29,* 257–268.

Mortimer, N., P. Gans, A. Calvert, and N. Walker (1999b), Geology and thermochronology of the east edge of the Median Batholith (Median Tectonic Zone): a new perspective on Permian to Cretaceous crustal growth in New Zealand, *Island Arc, 8,* 404–425.

Nathan, S., H. J. Anderson, R. A. Cook, R. H. Herzer, R. H. Hoskins, J. I. Raine, D. Smale (1986), Cretaceous and Cenozoic sedimentary basins of the West Coast Region, South Island, New Zealand, *N. Z. Geol. Surv. Basins Stud., 1.*

Nathan, S.; Rattenbury, M.S.; Suggate, R.P. (2002), Geology of the Greymouth area. Institute of Geological and Nuclear Sciences 1:250,000 geological map 12: Lower Hutt, Institute of Geological and Nuclear Sciences Limited.

Nathan, S. (1976), Geochemistry of the Greenland group (Early Ordovician), New Zealand, *N. Z. J. Geol. Geophys., 19,* 683–700.

Nathan, S. (2005), Harold Wellman - A man who moved mountains. Victoria University Press, Wellington, New Zealand. ISBN 0-86473-506-5, 272 pages.

Nicol, A., and R. J. Van Dissen (2002), Up-dip partitioning of displacement components on the oblique-slip Clarence Fault, New Zealand, *J. Struct. Geol., 24,* 1521–1535.

Nelson, C. S., C. H. Hendy, G. R. Jarrett, and A. M. Cuthbertson (1985), Near-synchroneity of New Zealand alpine glaciations and Northern Hemisphere continental glaciations during the past 750 k yr, *Nature, 318,* 361–363.

Nelson, C. S., D. Lee, P. Maxwell, R. Maas, P. J. J. Kamp, S. Cooke (2004), Strontium isotope dating of the New Zealand Oligocene, *N. Z. J. Geol. Geophys., 47*(4), 719–730.

Nodder, S. D., G. Lamarche, J. N. Proust (2005), Active faulting and paleoseismicity of the offshore Kapiti-Manawatu fault system, southern North Island, New Zealand, in *Programme and abstracts, Geological Society of New Zealand 50th Annual Conference, Kaikoura, New Zealand,* edited by J. R. Pettinga and A. M. Wandres, p. 61, Geological Society of New Zealand Miscellaneous Publication 119A.

Norris, R. J. (2004), Strain localisation within ductile shear zones beneath active faults: The Alpine Fault contrasted with the adjacent Otago fault system, New Zealand, *Earth Planets Space, 56,* 1095–1101.

Norris, R. J., R. M. Carter, and I. M. Turnbull (1978), Cainozoic sedimentation in basins adjacent to a major continental transform boundary in southern New Zealand, *J. Geol. Soc. (Lond.), 135,* 191–205.

Norris, R. J., and R. M. Carter (1982), Fault-bounded blocks and their role in localising sedimentation and deformation adjacent to the Alpine Fault, southern New Zealand, *Tectonophysics, 87*(1–4), 11–23.

Norris, R. J., and A. F. Cooper (2001), Late Quaternary slip rates and slip partitioning on the Alpine Fault, New Zealand, *J. Struct. Geol., 23,* 507–520.

Norris, R. J, and A. F. Cooper (2003), Very high strains recorded in mylonites along the Alpine Fault, New Zealand: Implications for deep structure of plate boundary faults, *J. Struct. Geol., 25,* 2141–2257.

Norris, R. J, and A. F. Cooper (this volume), The Alpine Fault, New Zealand: Surface Geology and Field Relationships.

Norris, R. J., and R. Nicolls (2004), Strain accumulation and episodicity of fault movements in Otago, *EQC Research Report 01/445,* p. 145.

Norris, R. J., E. Bennett, J. Youngson, J. Jackson, G. Raisbeck, and F. Yiou (2005), The growth of anticlinal ranges in an active fold-thrust belt, Central Otago, New Zealand, in *Programme and abstracts, Geological Society of New Zealand 50th Annual Conference, Kaikoura, New Zealand,* edited by J. R Pettinga. and A. M. Wandres, p. 62, Geological Society of New Zealand Miscellaneous Publication 119A.

Pettinga, J.R., M. D. Yetton, R. J. Van Dissen, and G. L. Downes (2001), Earthquake source identification and characterisation for the Canterbury region, South Island, New Zealand, *Bull. N. Z. Soc. Earthq. Eng., 34*(4), 282–317.

Pillans, B. (1990), Pleistocene marine terraces in New Zealand: a review, *N. Z. J. Geol. Geophys., 33,* 219–231.

Porter, S. C. (1975), Equilibrium-line altitudes of Late Quaternary glaciers in the Southern Alps, New Zealand, *Quat. Res., 5,* 27–47.

Rattenbury M. S., R. A. Cooper, and Johnston M. R. (compilers) (1998), Geology of the Nelson area. Institute of geological and Nuclear Sciences 1:250,000 geological map 9. 1 sheet + 67 pages. Lower Hutt, New Zealand: Institute of Geological and Nuclear Sciences Limited.

Rose, R.V. (1996), Summary of Miocene to Recent conglomerate provenance and gold content and plate boundary tectonics of the West Coast region, p.82–109, in *The Changing Face of West Coast Mining*, The Australasian Institute of Mining and Metallurgy, New Zealand Branch, 29th annual conference 1996, Wellington, New Zealand Branch.

Sara, W.A. (1970), Glaciers of Westland National Park. Wellington: DSIR, *Department of Scientific and Industrial Research Information Series, 75.* 47 p.

Shulmeister, J. (2005), Warming up the Last Glaciation, *Geol. Soc. N. Z. Newslett., 137,* 40.

Shulmeister, J.; I. Goodwin, J. Renwick, K. Harle, L. Armand, M. S. McGlone, E. Cook, J. Dodson, P. P. Hesse, P. Mayewski, and M. Curran (2004), The Southern Hemisphere westerlies in the Australasian sector over the last glacial cycle: a synthesis, *Quat. Int., 118/119,* 23–53.

Silberling, N. J., K. M. Nichols, J. D. Bradshaw, and C. D. Blome (1988), Limestone and chert in tectonic blocks from Esk Head subterrane, South Island, New Zealand, *Bull. Geol. Soc. Am., 100,* 1213–1223.

Simpson, G. D. H., A. F. Cooper, R. J. Norris (1994), Late Quaternary evolution of the Alpine Fault zone at Paringa, South Westland, New Zealand, *N. Z. J. Geol. Geophys., 37,* 49–58.

Sircombe, K. N., and P. J. J. Kamp (1998), The South Westland Basin: seismic stratigraphy, basin geometry and evolution of a foreland basin within the Southern Alps collision zone, New Zealand, *Tectonophysics, 300*(1/4), 359–387.

Smale, D. (1991), Provenance changes and movement on the Alpine Fault indicated by heavy minerals from Cretaceous-Cenozoic sediments in south Westland, *J. R. Soc. N. Z., 21*(2), 151–160.

Soons, J. M. (1986), Erosion rates in a superhumid environment, *Int. Geomorphol., Part 1,* 885–896.

Steinberger, B., R. Sutherland, and R. J. O'Connell (2004), Prediction of Emperor-Hawaii seamount locations from a revised model of global plate motion and mantle flow, *Nature, 430,* 167–173.

Stirling, M.W. (1990), The Old Man Range and Garvie Mountains: Tectonic geomorphology of the Central Otago peneplain, New Zealand, *N. Z. J. Geol. Geophys., 33*(2), 233–243.

Stirling, M.W. (1991), Peneplain modification in an alpine environment of Central Otago, New Zealand, *N. Z. J. Geol. Geophys., 34*(2), 195–201.

Suggate, R. P. (1963), The Alpine Fault, *Trans. R. Soc. N. Z., 2,* 105–109.

Suggate, R. P. (1965), Late Pleistocene geology of the northern part of the South Island, New Zealand, *N. Z. Geol. Surv. Bull., 77.*

Suggate, R. P. (1978), The late mobile phase: Cretaceous. Pp346–351 In: R. P. Suggate, G. R. Stevens, and M. T. TePunga (Editors), *The Geology of New Zealand.* Government Printer, Wellington. 2 volumes, 820 p.

Suggate, R. P. (1990), Late Pliocene and Quaternary glaciations of New Zealand, *Quat. Sci. Rev., 9,* 175–197.

Suggate, R. P., and P. C. Almond (2005), The Last Glacial Maximum (LGM) in western South Island, New Zealand: implications for the global LGM and MIS 2, *Quat. Sci. Rev., 24,* 1923–1940.

Sutherland, R. (1994), Displacement since the Pliocene along the southern section of the Alpine Fault, New Zealand, *Geology, 22,* 327–330.

Sutherland, R. (1995), The Australia-Pacific boundary and Cenozoic plate motions in the SW Pacific: some constraints from Geosat data, *Tectonics, 14*(4): 819–831.

Sutherland, R. (1996), Transpressional development of the Australia-Pacific boundary through southern South Island, New Zealand: constraints from Miocene-Pliocene sediments, Waiho-1 borehole, South Westland, *N. Z. J. Geol. Geophys., 39*(2), 251–264.

Sutherland, R. (1999a), Basement geology and tectonic development of the greater New Zealand region: an interpretation from regional magnetic data, *Tectonophysics, 308*(3), 341–362.

Sutherland, R. (1999b), Cenozoic bending of New Zealand basement terranes and Alpine Fault displacement: a brief review, *N. Z. J. Geol. Geophys., 42*(2), 295–301.

Sutherland, R., and C. Hollis (2001), Cretaceous demise of the Moa plate and strike-slip motion at the Gondwana margin, *Geology, 29,* 279–282.

Sutherland, R., S. Nathan, I. M. Turnbull, and A. G. Beu (1995), Pliocene-Quaternary sedimentation and Alpine Fault related tectonics in the lower Cascade valley, South Westland, New Zealand, *N. Z. J. Geol. Geophys., 38*(4), 431–450.

Sutherland, R., F. Davey, and J. Beavan (2000), Kinematics of plate boundary deformation in South Island, New Zealand, is related to inherited lithospheric structure, *Earth Planet. Sci. Lett., 177,* 141–151.

Sutherland, R., K. R. Berryman, and R. J. Norris (2006), Quaternary slip rate and geomorphology of the Alpine Fault: implications for kinematics and seismic hazard in southwest New Zealand, *Geol. Soc. Am. Bull., 118,* 464–474.

Sutherland, R., D. Eberhart-Phillips, R. A. Harris, T. Stern, J. Beavan, S. Ellis, S. Henrys, S. Cox, R. J. Norris, K. R. Berryman, J. Townend, S. Bannister, J. Pettinga, B. Leitner, L. Wallace, T. A. Little, A. F. Cooper, M. Yetton, and M. Stirling (this volume), Do great earthquakes occur on the Alpine Fault in central South Island, New Zealand?

Tippett, J. M., and P. J. J. Kamp (1993a), The role of faulting in rock uplift in the Southern Alps, *N. Z. J. Geol. Geophys., 36,* 497–504.

Tippett, J. M., and P. J. J. Kamp (1993b), Fission track analysis of the late Cenozoic vertical kinematics of continental Pacific crust, South Island, New Zealand, *J. Geophys. Res., 98,* 16,119–16,148.

Tulloch, A. J., and D. L. Kimbrough (1989), The Paparoa Metamorphic Core Complex, Westland-Nelson, New Zealand: Cretaceous extension associated with fragmentation of the Pacific margin of Gondwana, *Tectonics, 8*(6), 1217–1234.

Vry J. K., J. Baker, R. Maas, T. A. Little, R. Grapes, and M. Dixon (2004), Zoned (Cretaceous and Cenozoic) garnet and the timing of high grade metamorphism, Southern Alps, New Zealand, *J. Metamorph. Geol., 22,* 137–157.

Walcott, R. I. (1978), Present tectonics and Late Cenozoic evolution of New Zealand, *Geophys. J. R. Astron. Soc., 52*(1), 137–164.

Walcott, R. I. (1998), Modes of oblique compression: late Cenozoic tectonics of the South Island, New Zealand, *Rev. Geophys., 36,* 1–26.

Wallace, L. M., R. J. Beavan, R. McCaffrey, K. R. Berryman, and P. Denys (2007), Balancing the plate motion budget in the South Island, New Zealand using GPS, geological and seismological data, *Geophys. J. Int.*, *168*(1), 332–352, doi:10.1111/j.1365-246X.2006.03183.x

Wandres, A. M., J. D Bradshaw, and T. Ireland (2005), The Paleozoic–Mesozoic recycling of the Rakaia Terrane, South Island, New Zealand: sandstone clast and sandstone petrology, geochemistry, and geochronology, *N. Z. J. Geol. Geophys.*, *4*, 229–245.

Wandres, A. M., J. D.Bradshaw, S. D. Weaver, R. Maas, T. R. Ireland, and G. N. Eiby (2004), Provenance analysis using conglomerate clast lithologies: a case study from the Pahau Terrane of New Zealand, *Sediment. Geol.*, *167*, 57–89.

Wellman, H. W. (1940), The Otataran peneplain of northwest Nelson. Unpublished MSc thesis, University of New Zealand (Copy held in library, Victoria University of Wellington).

Wellman, H. W. (1979), An uplift map for the South Island of New Zealand and a model for uplift of the Southern Alps, in *The Origin of the Southern Alps*, edited by R. I. Walcott, and M. M. Cresswell, *R. Soc. N. Z. Bull.*, *1*, 13–20.

Wellman, H.W. and R.W. Willett (1942), The geology of the west coast from Abut Head to Milford Sound, part I, *Trans. R. Soc. N. Z.*, *71*, 282.

Whitehouse, I. E. (1986), Geomorphology of a compressional plate boundary, Southern Alps, New Zealand, *Int. Geomorph.*, Part 1, 897–924.

Whitehouse, I. E., and G. A. Griffiths (1983), Frequency and hazard of large rock avalanches in the central Southern Alps, New Zealand, *Geology*, *11*, 331–334.

Wightman, R., and T. A. Little (this volume), Deformation of the Pacific Plate above the Alpine Fault ramp and its relationship to expulsion of metamorphic fluids: An array of backshears.

Wilson, C. J. N. (1996), Taupo's atypical arc, *Nature*, *379*, 27–28.

Wood, R., and B. Davy (1994), The Hikurangi Plateau, *Mar. Geol.*, *118*, 153–173.

Wood, R. A., G. Lamarche, R. H. Herzer, J. Delteil, and B. Davy (1996), Paleogene seafloor spreading in the southeast Tasman Sea, *Tectonics*, *15*, 966–975.

Yeats, R. S. (1987), Tectonic map of Central Otago based on Landsat imagery, *N. Z. J. Geol. Geophys.*, *30*, 261–271.

Youngson, J. H., D. Craw, C. A. Landis, and K. R. Scmitt (1998), Redefinition and interpretation of late Miocene-Pleistocene terrestrial stratigraphy, Central Otago, New Zealand, *N. Z. J. Geol. Geophys.*, *41*(1), 5–68.

Geophysical Structure of the Southern Alps Orogen, South Island, New Zealand

F. J. Davey[1], D. Eberhart-Phillips[2], M. D. Kohler[3], S. Bannister[1], G. Caldwell[1], S. Henrys[1], M. Scherwath[4], T. Stern[5], and H. van Avendonk[6]

The central part of South Island of New Zealand is a product of the transpressive continental collision of the Pacific and Australian plates during the past 5 million years, prior to which the plate boundary was largely transcurrent for over 10 My. Subduction occurs at the north (west dipping) and south (east dipping) of South Island. The deformation is largely accommodated by the ramping up of the Pacific plate over the Australian plate and near-symmetric mantle shortening. The initial asymmetric crustal deformation may be the result of an initial difference in lithospheric strength or an inherited suture resulting from earlier plate motions. Delamination of the Pacific plate occurs resulting in the uplift and exposure of mid-crustal rocks at the plate boundary fault (Alpine fault) to form a foreland mountain chain. In addition, an asymmetric crustal root (additional 8 – 17 km) is formed, with an underlying mantle downwarp. The crustal root, which thickens southwards, comprises the delaminated lower crust and a thickened overlying middle crust. Lower crust is variable in thickness along the orogen, which may arise from convergence in, and lower lithosphere extrusion along, the orogen. Low velocity zones in the crust occur adjacent to the plate boundary (Alpine fault) in the Australian and Pacific plates, where they are attributed to fracturing of the upper crust as a result of flexural bending for the Australian plate and to high pressure fluids in the crust derived from prograde metamorphism of the crustal rocks for the Pacific plate.

INTRODUCTION

The Pacific-Australian plate boundary crosses the length of South Island, New Zealand (Plate 1), where it is dominantly a transpressive boundary linking the east-dipping subduction zone southwest of South Island from west-dipping subduction east of North Island. South Island's existence is largely the result of the plate tectonic deformation of the region along the Pacific-Australian plate boundary since its inception during the mid-Cenozoic. Prior to this time, the land mass of South Island was extremely low lying, with only a small proportion of the present landmass above sea level [*Suggate et al.*, 1978; *Balance*, 1993]. This was due to extension and thinning of the New Zealand micro-continent lithosphere during late Cretaceous extension, and erosion and subsidence (cooling) associated with the break-up of

[1]GNS Science, Gracefield, Lower Hutt, New Zealand.

[2]GNS Science, Dunedin, New Zealand.

[3]Center for Embedded Networked Sensing, University of California, Los Angeles, California.

[4]Leibniz-Institute of Marine Sciences, IFM-GEOMAR, Kiel, Germany.

[5]School of Earth Sciences, Victoria University of Wellington, Wellington, New Zealand.

[6]Institute of Geophysics, University of Texas, Austin, Texas.

A Continental Plate Boundary: Tectonics at South Island, New Zealand
Geophysical Monograph Series 175
Copyright 2007 by the American Geophysical Union.
10.1029/175GM04

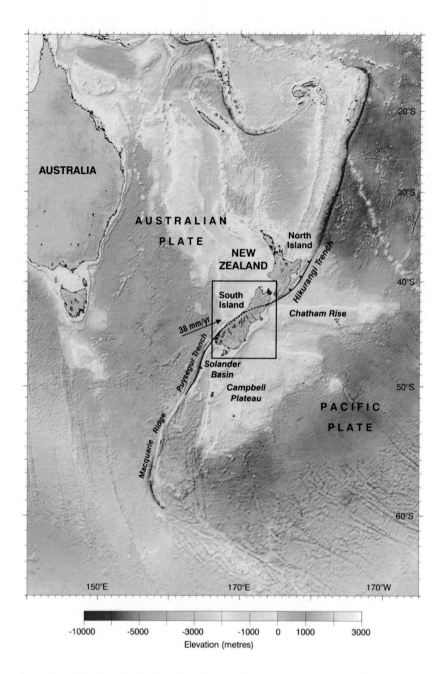

Plate 1. The plate tectonic setting of the New Zealand region, showing the morphology (color scale) and plate boundary through the region (solid line). Subduction boundaries (Hikurangi and Puysegur) are marked by solid line and triangles, transcurrent boundary (Alpine fault) by a dotted line, and convergence direction and rate by the annotated arrow. The location of Figures 1, 10, 13 and 14 are shown by the box.

Gondwana and onset of seafloor spreading between New Zealand, Australia and Antarctica. About 45 Ma ago the Pacific-Australian plate boundary developed, extending from the rifted margin off southwest Campbell Plateau, through the axis of the present South Island and along the convergent margin off east and north North Island to the old Cretaceous convergent margin off east Australia [*Sutherland*, 1999]. Initially this plate margin was dominantly transtensional, but it developed mainly as a transcurrent boundary until, at about 5 Ma, a significant component of convergence commenced leading to the thickening of the lithosphere and the emergence of South Island and the Southern Alps in particular. The structure and development of this orogen and the adjacent region during this transpressive phase is the focus of this chapter.

We first review the range of geophysical data available for South Island. These data are then used to define the main features of the three-dimensional structure of the orogen and provide the context for the detailed structure and development of the plate boundary zone as defined by the Alpine fault. The chapter closely complements the previous chapter on the geological data base for South Island, and uses those and other data to constrain its inferences. The conclusions of the analysis give models that provide a baseline of the general features for continental transpressive plate boundaries as discussed in the latter part of this volume [Cox and Sutherland, this volume].

GEOPHYSICAL CHARACTERISTICS OF THE SOUTH ISLAND COLLISIONAL OROGEN

i)Topography - The morphology of South Island is well known with several digital elevation models available for the region (e.g., Etopo2, Land Information New Zealand). These models typically have resolutions of about 25 m. Figure 1 shows the morphology of South Island based on a grid of 1000 m (sub-sampled from a 25 m grid), adequate to bring out the dominant features of the morphology. In general terms, the morphology of South Island consists of an axial mountain range, the Southern Alps, that is narrow and high in the central part but wider and lower to the north and south, and is flanked by coastal plains. The high central part of the Southern Alps has five main zones from west to east: i) a narrow coastal plain in the west, delineated at its eastern margin by a linear range front (Alpine fault), ii) a steep, west to northwest dipping mountain front that rises to the main divide at over 3000 m elevation in its central part (see below), iii) an intermontane plateau at about 1000 m, iv) a steep slope on the eastern side of the mountains, leading into v) a broad, gently east-dipping coastal plain to the east coast (Figure 2b).

This simple two-dimensional morphology is modified in the north and south of South Island [see also *Cox and*

Sutherland, this volume]. In the north, where splay faulting from the Alpine fault to the east starts, the mountains become lower but wider, and extend to the east coast, thus terminating the northward extent of the eastern coastal plain (Figure 2a). Their maximum height is about 2500 m. The major splay faults coincide with distinct linear morphological features (Figure 1). Elevated morphology occurs west of the Alpine fault trace in the northern region. In the south, the mountains are also lower, reaching maximum elevations of about 2500 m, and broader, covering most of southern South Island (Figure 2c). The morphology of the region is less two-dimensional and reflects the grain of the geology that swings around from a NNE to a ESE trend. The Alpine fault is a more subdued morphological feature, but it is still remarkably linear until it runs offshore at Milford Sound.

The region of maximum elevation of the Southern Alps coincides with the narrowest part of the mountain belt where the highest uplift rates of up to 17 mm/y occur [*Adams*, 1979; *Wellman*, 1979]. Erosion however, is taking place at about the same rate [*Adams*, 1980] so that the mountain ranges are in approximate dynamic equilibrium.

ii) Seismicity distribution - The major features of seismicity distribution in New Zealand have been known for some time. *Hatherton* [1980] reported on shallow seismicity for the period 1956-75. Subsequently *Reyners* [1989] reported on shallow and deep seismicity for the period 1964-87. The upgrade of the New Zealand National Seismograph Network (NZNSN) from analogue to digital instruments that took place between 1986 and the mid 1990s allowed a more comprehensive analysis to be made [*Anderson and Webb*, 1994]. In Figure 3 we plot the distribution of seismicity up to 2002. The distribution of seismicity falls into three distinct zones that correspond in general to the plate tectonic context of South Island. In the north, the influence of the southern end of the Hikurangi subduction system is dominant. Deep seismicity in this region traces the subducting Pacific lithosphere along the eastern margin of northern South Island to depths of about 200 km (Figure 3c,d). The southwest termination of this deep seismicity is sharp and coincides closely with the region where the Wairau and Awatere faults splay off eastwards from the Alpine fault. Shallow seismicity (Figure 3b) is extensive and intense in this same region, and delineates some of the major faults including the eastern splay faults. Shallow seismicity is also intense west of the Alpine fault in northwest South Island, but no unequivocal causative faults have been identified. In the south of South Island, the influence of the Puysegur-Fiordland subduction system dominates seismicity (Figure 3c,d). The intermediate-depth seismicity is restricted in extent defining a narrow, near-vertical zone up to 140 km deep under central-northern

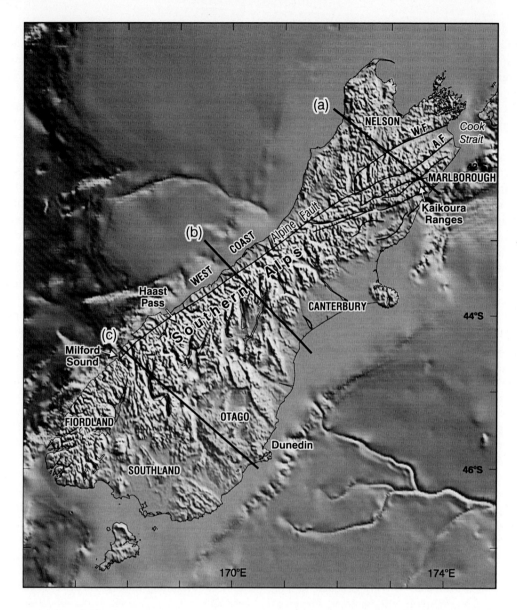

Figure 1. Locations and topography of the South Island region (grey scale illuminate from the northeast). The Alpine fault and Marlborough faults marked by the thin lines. The topographic profiles (a, b and c) in Figure 2 are located by the thick lines. AF - Awatere fault, WF - Wairau fault.

Fiordland that traces the subduction of the Australian plate north-eastwards from Puysegur Trench under southwest South Island. The detailed geometry of the subducted plate is controversial, as the part of the plate from the Puysegur trench to well under southern South Island is not well constrained by seismicity. This may partially arise from limited station coverage. The shallow seismicity in the region is concentrated above the deep seismicity, and presumably relates to the subduction process.

The region between the two subduction zones is nearly devoid of earthquakes below 25-km depth (Figure 3c,d), but

detailed studies of parts of the region [e.g., *Reyners*, 1987, 1988] show abundant shallow seismicity and a low level of deeper seismicity. A major improvement in the knowledge of seismicity in South Island, particularly under the Southern Alps, has resulted from the Southern Alps Passive Seismic Experiment [SAPSE, *Anderson et al.*, 1997], where an array of 26 broadband and 14 short-period recording seismographs was deployed over South Island, concentrated on the central Southern Alps for up to 12 months. The array recorded 5491 earthquakes and most of the shots of the active source seismic shooting of SIGHT [*Davey et al.*, 1998]. These data

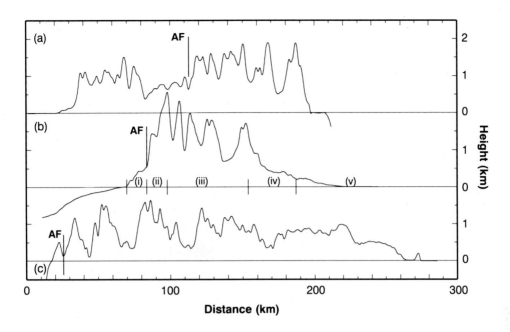

Figure 2. Topographic profiles across South Island: a) northern, b) central Southern Alps, c) southern. AF marks the location of the Alpine fault. Regions i–v in (b) are discussed in the text. Note the broadening of the high topography in the northern and southern regions of South Island, plus the northern region's steep eastern margin.

have allowed much more accurate seismicity locations (Figure 4) with depth errors <3 km, as well as the production of a 3D tomographic model of P and S-wave velocity for South Island (see next section and Plate 2) [*Eberhart-Phillips and Bannister*, 2002].

A major feature of the crustal seismicity distribution is its restriction to depths of less than 12 ± 2 km indicating the depth of brittle crust under the Southern Alps [*Leitner et al.*, 2001]. The maximum depth decreases to about 8 km under the region of maximum uplift in the central Southern Alps indicating localised elevated temperatures east of the Alpine fault. The rate of seismicity is subdued in a similar region about 10–20 km southeast of the Alpine fault. Distributed seismicity is observed throughout a roughly 80-km wide region southeast of the Alpine fault, with the greatest moment release in the vicinity of the Alpine fault. The Southern Alps seismicity does not cluster along specific faults, indicating that a wide range of structures are producing slip. This was borne out in the Mw 6.7 1994 Arthur's Pass earthquake sequence (Figure 4) which activated a broad zone of seismicity with varied focal mechanisms, following the primary reverse faulting event [*Abercrombie et al.*, 2000]. Several tens of intermediate-depth earthquakes (30–97 km) have been recorded in central South Island [*Reyners*, 1987; *Kohler and Eberhart-Phillips*, 2003] some of which align on a zone dipping at about 18° generally west to northwest, down to about

60 km depth [*Reyners*, 1987, 2005]. The earthquakes are not associated with crustal subduction but all lie within or on the margins of thickened crust or uppermost mantle seismic high-velocity anomalies [*Kohler and Eberhart-Phillips*, 2003]. The earthquakes are generally small ($M_L \leq 4.0$) but *Reyners* [2005] infers a subcrustal depth (50 km) for the 1943 Lake Hawea M_W 5.9 earthquake.

iii) 3D tomography, earthquake travel-time and receiver function studies - Recordings of earthquakes and shots from the active source seismic experiment (SIGHT) were used for the derivation of 3D models of P-wave velocity and Vp/Vs ratios, constrained for the deeper part of the models by the gravity data coverage of the region [*Eberhart-Phillips and Bannister*, 2002]. The models (Plate 2), accurate to 0.2 km/s in P-wave velocity, were derived using nodes with 8–40 km spacing horizontally and 3–10 km spacing vertically, with velocities allowed to vary smoothly between nodes and hypocentral parameters included in the inversion. The models show a broad downwarp of the 7.5 km/s velocity contour (indicative of Moho) to depths of about 35 km under the central Southern Alps that shallows towards the coast. The deepest part trends to the east away from the highest part the Southern Alps following the axis of the gravity low (see below). A low-velocity layer occurs in the middle crust under the south-eastern part of the Southern Alps (Plate 2b,

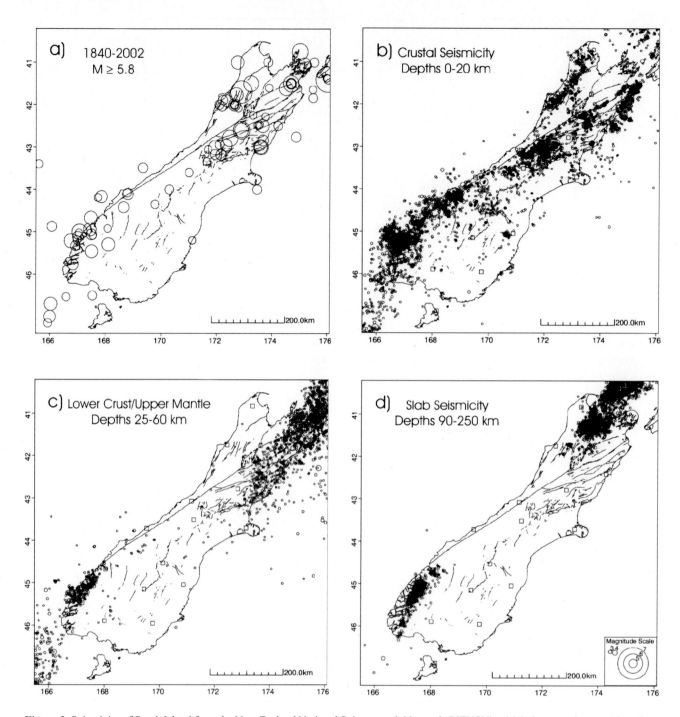

Figure 3. Seismicity of South Island from the New Zealand National Seismograph Network (NZNSN). a) Moderate-to-large earthquakes, $M_L \geq 5.8$, 1840–2002. $M_L \geq 3.0$, 1990–2002: b) crustal seismicity, 0–20 km depth; c) lower crust/upper mantle, 25–60 km depth; d) slab, 90–250 km depth. Squares = NZNSN stations.

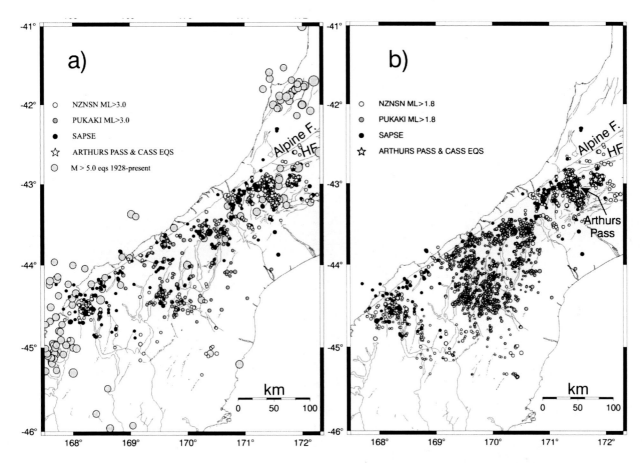

Figure 4. Shallow central South Island seismicity [*Leitner et al.*, 2001]. Maps showing quality locations of SAPSE (solid circles), Pukaki (shaded circles) and NZNSN (open circles) earthquakes. (a) $M_L > 3$. Large shaded circles are $M_L > 5$ earthquakes since 1920. Note that larger earthquakes follow the seismicity patterns for the smaller earthquakes. (b) $M_L > 1.8$. HF marks the Hope fault.

see also *Reyners and Cowan*, 1993; *Kleffmann*, 1999]. Low-velocity zones are imaged along the Alpine fault, and are southeast-dipping in the central oblique-slip region and vertical in the southern strike-slip region (Plate 2). Given the sparse data and coarse grid, it is not possible to fully define the subsurface fault zone. There are clear differences in the low-velocity zone between the central and southern sections, and the character at 14-km depth compared to 6-km depth (Plate 2a-b). At 14-km depth, the low-velocity zone appears as a distinct feature trending to the southeast away from the Alpine fault, such that it does not join up with the southern Alpine fault low-velocity zone Plate 2b). The variation in the thickness of the incoming Pacific plate is shown in the 30-km depth velocity slice (Plate 2c). The crust is thinner in the Canterbury (east-central South Island) region, and thickens by 5–10 km towards the southeast. The shape of the crustal root is shown by the large low-velocity region (Plate 2c). This extends well south of the oblique-slip fault section, with the most pronounced low velocity occurring 80

km south of Mt. Cook. The crustal root is also seen in the cross-sections (Plate 2d–g), where both vertical and dipping lines are shown for the Alpine fault, since the true variation of fault dip with depth is not known.

The tomographic model to the north [*Eberhart-Phillips and Reyners*, 1997] shows a relatively narrow velocity low throughout the crust associated with the Awatere fault, a major splay fault of eastern South Island (northern white line, Plate 2a–b). In cross-section, the subducted slab is shown in both seismicity and velocity, with relatively low velocities between 40 and 100 km depth, reflecting the continental nature of the subducted crust in this region (Figure 5). Strong P-wave anisotropy of up to 12% is found in the brittle crust in the region of the Awatere and Wairau faults [*Eberhart-Phillips and Henderson*, 2004] and is aligned with faults (Figure 5). Different anisotropic azimuths are found in the more ductile region below 20-km depth in a zone of progressive coupling and partial subduction. They are more aligned with the relative plate motion and associated with crustal

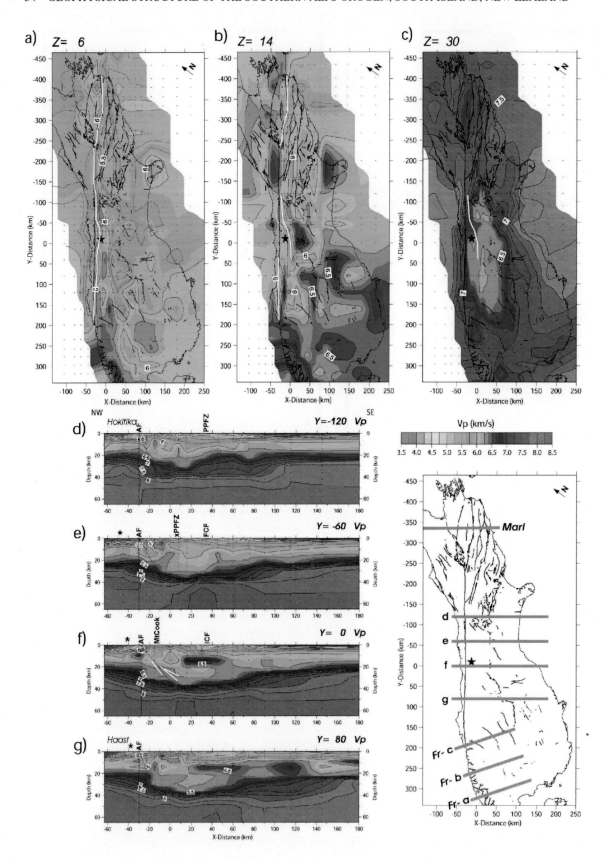

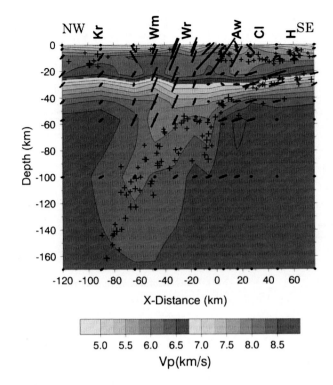

Figure 5. Cross-section of Marlborough P-velocity from inversion including azimuthal (hexagonal) anisotropy, with shear-wave splitting initial model [*Eberhart-Phillips and Henderson*, 2004], location indicated in Plate 2. Isotropic component of Vp is shown by grey-scale and contoured at 0.25 km/s. Bars at each node show the magnitude and fast azimuth of azimuthal anisotropy. These are 2-D azimuths even on cross-sections, i.e., vertical bar = north azimuth and horizontal bar = east azimuth. The maximum magnitude in each plot is specified. Hypocenters are shown by pluses. Abbreviations are as follows: Aw, Awatere fault; Cl, Clarence fault; H, Hope fault; Kr, Karamea fault; Wm, Waimea fault; and Wr, Wairau fault. Note no significant crustal thinning across the Wairau Fault.

thickening (Plate 2c). A detailed receiver function study along a profile across the Wairau fault to the Clarence fault [*Wilson et al.*, 2004] inferred a southeast (along profile) dipping Moho (24 km deep north of the Wairau fault to 34 km just south of the fault) under the region with per-

vasive anisotropy of unspecified orientation in the middle crust (15–20 km) of the region. In the Fiordland subduction region in the south, tomographic velocity models [*Eberhart-Phillips and Reyners*, 2001] show high-velocity Pacific mantle below 80-km depth, which impinges on the subducted Australian slab and causes it to bend to near vertical (Figure 6).

Delays in teleseismic P-wave arrivals from distant earthquakes, detected by the detailed receiver arrays (about 80 seismographs each) along the two main transects of the SIGHT project, compared to a standard earth model have been used to define a high-velocity anomaly in the upper mantle under the crustal root [*Stern et al.*, 2000]. The delays across the array show a major advance (reduction in delay) across South Island [see also *Stern et al.*, this volume]. The relative delays were modeled by correcting for the crustal velocity model derived from the active source experiment and assigning the residual delays to a broad vertical positive velocity anomaly in the upper mantle constrained by ray tracing. The preferred model for the velocity anomaly (maximum 7% wrt 8.1 km/s) is centred below the crustal root and is about 80 km wide, extending from 60 to 170 km in depth (dashed contours in Figure 7). The high density of seismographs along each transect and detailed ray tracing allowed *Stern et al.* [2000] to closely constrain the geometry and magnitude limits of the anomalous body in the upper mantle. They show that at least its western margin is within 15° of vertical. This model gives a local residual delay that *Stern et al.* [2000] assign to a low-velocity crust within the hanging wall of the Alpine fault.

A three-dimensional uppermost mantle seismic velocity model below South Island has also been derived by teleseismic P-wave travel-time inversion using waveform data from the SAPSE experiment and the New Zealand National Seismograph Network [*Kohler and Eberhart-Phillips*, 2002]. The velocity images show a near-vertical, northwest dipping, high-velocity (2–4% relative to regionally average uppermost mantle velocities - 8.1 km/s) structure in the uppermost mantle (gray scale contours in Figure 7) that also directly underlies thickened crust along the NNE-SSW axis of the Southern Alps (east of the Alpine fault) (Figure 8).

Plate 2. (Opposite) South Island 3-D P-velocity model incorporating Southern Alps [Eberhart-Phillips and Bannister, 2002], Marlborough [Eberhart-Phillips and Reyners, 1997] and Fiordland [Eberhart-Phillips and Reyners, 2001]. Map views at (a) 6-km depth, (b) 14-km depth, and (c) 30-km depth. White lines show low-velocity zone, M = high-velocity underlying Mackenzie basin (Fig 5b). Orientation is parallel to Alpine fault, with coastline and active faults shown, star = Mt Cook. Location map marks location of cross sections. Line marked "Marlborough" locates profile in Figure 5 across Marlborough. Lines marked "Fr-a" through "-c" locate profiles shown in Figure 6. WF - Wairau fault, AF - Awatere fault. (d)–(g) Y cross-sections. Dip of Alpine fault is not known. For reference both vertical and dipping lines (generally following a crustal low velocity zone) are shown. Locations are indicated for the Alpine fault (AF), Porters Pass fault zone (PPFZ), Forest Creek fault (FCF), Mt. Cook, Irishman's Creek fault (ICF). White lines in (f) = reflectors from SIGHT98 CDP line [see Stern at al., this volume].

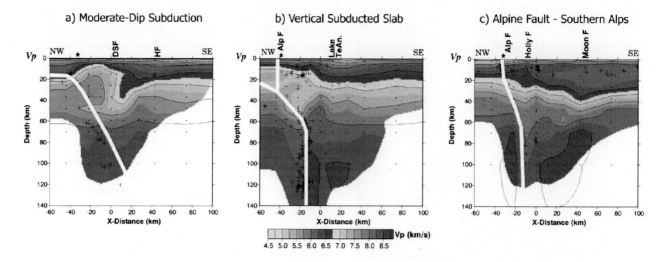

Figure 6. Tectonic interpretation of 3-D velocity model [*Eberhart-Phillips and Reyners*, 2001], cross-sections as indicated in Plate 2. a) Moderate-dip subduction occurs in the southern part of Fiordland. b) North of Doubtful Sound, the subducted slab is bent to vertical by a high-velocity body in the Pacific mantle and partitioned slip occurs on the shallow plate interface and the Alpine fault. c) North of Milford Sound, the Alpine fault becomes the lithospheric boundary to the Australian plate, and oblique convergence forms the Southern Alps. Locations are indicated for: the Alpine fault, Hollyford fault, Moonlight fault, Dusky fault (DSF), Hauroko fault (HF), and Lake Te Anau. * - coastline.

It extends from about 40 km to 200 km in depth from the northern to the southern end of South Island. The seismic anomaly varies from 60 to 100 km in width. The narrowest portion corresponds to the narrowest region of crustal deformation in central and northern South Island, and the widest portion corresponds to the wider plate boundary deformation zone in Otago (Figure 8). The velocity anomaly does not appear to increase its dip with depth like a subducted oceanic lithospheric slab and its geometry can be visualized as a long sheet descending into the mantle. The anomaly is broadly similar to that derived by forward modeling of teleseismic P-wave delays noted earlier (Figure 8) but is less well constrained in dip and magnitude due to data distribution and smoothing involved in the tomographic inversion. Other tests indicated a partial (40%) recovery of anomaly amplitude that is best at shallower depths, and a northwest dip bias of about 10° for a vertical body.

The high upper mantle velocities obtained by *Kohler and Eberhart-Phillips* [2002] extend from the 100-km thick sheet in northwest Nelson to southern South Island. In the north, the high-velocity anomaly (3–5%) corresponds to the descending slab of the Hikurangi subduction zone, with a depth extent corresponding to the deepest seismicity. The anomaly indicates a gradual change from a slab environment associated with deep seismicity in the oceanic lithosphere to continental lithospheric collision conditions which are characterized by no deep seismicity. The images suggest that continental mantle lithospheric deformation takes over

where subduction stops, but they do not provide evidence for a sharp oceanic lithospheric slab edge where the deep seismicity ends. In the south, the high-velocity structure at 100 km depth under eastern Fiordland is continuous with the Southern Alps high-velocity anomaly but extends deeper, to at least 220 km. Neither the teleseismic nor local earthquake studies of Fiordland find high velocities associated with the subducted Australian plate. It is relatively young (20–40 ma) and may be thinner, warmer, and less dense than older slabs [*Kohler and Eberhart-Phillips*, 2002].

Regional earthquake data have been used for Common Mid-Point (CMP) stacking of arrivals from local earthquakes recorded by the SIGHT transect recorders. They give an estimate of crustal thickness for South Island, indicating that the Southern Alps crustal root thins and narrows north of the SIGHT transects [*Wilson and Eberhart-Phillips*, 1998]. Aftershocks of Fiordland earthquakes observed on a profile along the Southern Alps indicate high Pn velocities (8.6 ± 0.1 km/s) and a crustal thickness increasing southwards to about 49 ± 6 km under southern South Island [*Bourguignon et al.*, this volume].

Teleseismic data have also been used for the calculation of shear-wave splitting using SKS phases. The results demonstrate strong anisotropy of the upper mantle under South Island where the fast component is NNE oriented over most of South Island with a possible change to a more north orientation over the southern half of the island [*Klosko et al.*, 1999; see also *Savage et al.*, this volume].

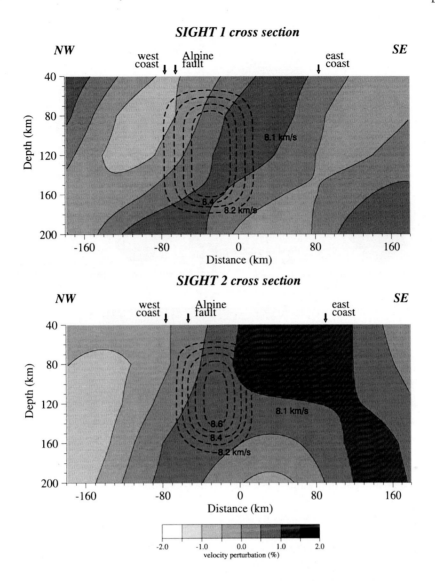

Figure 7. Upper mantle high-velocity anomaly under South Island. Vertical section of the mantle velocity anomaly along SIGHT transects 1 and 2. The grey shaded contours are of % velocity perturbation from a regionally averaged upper mantle velocity, after *Kohler and Eberhart-Phillips* [2002], and are overlain by dashed contours of the mantle velocity model (0.1 km/s intervals with a base level of 8.1 km/s) after *Stern et al.* [2000]. The Kohler and Eberhart-Phillips "slab" dips steeply northwest through the centre of the *Stern et al.* "blob".

iv) Wide angle active source/velocity models - Early active source seismic refraction and wide-angle measurements in South Island have provided preliminary velocity models across central South Island [*Smith et al.*, 1995], for the Lake Tekapo area [*Kleffmann et al.*, 1998] and for the Fiordland area [*Davey and Broadbent*, 1980]. The latter showed lower crustal velocities of about 7.3 km/s within 3 km of the surface associated with the large positive gravity anomaly in Fiordland. The results of the SIGHT project have superseded those of the former two projects (Figure 9). The SIGHT

project recorded active source seismic measurements along two major transects (Transect 1 - T1 and Transect 2 - T2) across South Island, and two main tie lines (Transect 3 - T3, and 3W) parallel to the coast [Figure 10, *Stern at al.*, 1997; *Davey et al.*, 1998]. Measurements were made along the two main transects using explosive charges on land recorded by an array of 400 recorders placed on each transect across South Island. These were supplemented by recording of offshore shots (airgun at 50 m spacing) by arrays of about 200 recorders on each line across the island and 17 ocean bottom

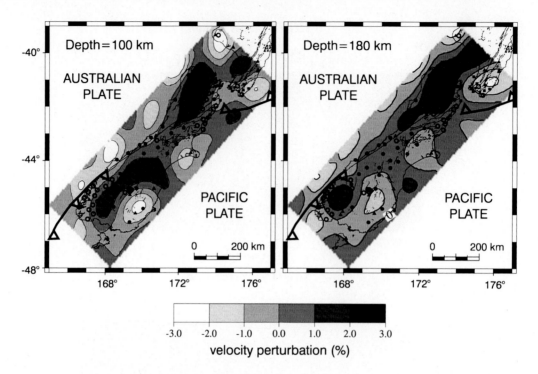

Figure 8. Depth slices at 100 km and 180 km depth of velocity anomaly, after *Kohler and Eberhart-Phillips* [2002], showing the continuity of the high-velocity slab from southern North Island to off the central south coast of South Island and its position relative to the surface position of the Alpine fault.

seismographs offshore along each transect. Offshore vertical incidence reflection data constrained the sedimentary section along the offshore part of the transects. Seismic velocities along the northern transect, T1, were modelled (Figure 10a) by a tomographic inversion code [*van Avendonk et al.*, 2004] using dense parameterization that is adjusted to give the smoothest model with minimum residuals. T2 [*Scherwath et al.*, 2003] and tie lines T3 off the east coast [*Godfrey et al.*, 2001] and Line 3w of the west coast [*Melhuish et al.*, 2004] have been modelled using the forward modelling and inversion code of *Zelt and Smith* [1992]. This method uses sparse parameterization and minimum perturbation from the starting model to minimise residuals. The different modelling techniques give rise to differences in the derived models (such as the broader crustal root in T1). The velocity models are shown in Figures 11b–d respectively. Ray coverage was good for most of T1 and T2 over the plate boundary. The deeper part of the crustal root is not well constrained due to poorer ray coverage.

The two main transects that cross the plate boundary have a number of features in common: similar crustal velocity structures under the coastal regions and further offshore, a low-velocity region around the Alpine fault (plate boundary), and a thickened crustal section under the Southern Alps

(although T2 was about 5 km thicker than T1, indicating changes along strike). T2 defined a low-velocity upper-mid crustal zone west of the west coast that has been attributed to flexural stress in the region [*Scherwath et al.*, 2003]. A low-velocity lower-mid crust in the Pacific Plate was detected on T2, similar to that detected on the passive seismic tomography models noted above. Upper mantle velocities on both T1 and T2 are generally above 8 km/s, however, a zone of reduced (7.6–7.7 km/s) uppermost mantle velocities exist around the plate boundary, extending to about 100 km offshore in the west. Its eastern limit is poorly resolved due to reduced data coverage below the crustal root.

The tie lines 3w and T3 (Figures 10 and 11) were designed to image the velocity structure of the South Island crust away from the region of present deformation and thus provide a crustal velocity model of the lithosphere that probably entered the deformation zone. Line 3w resulted in a fairly simple velocity model of thin crust with a well-defined reflective lower crust. The uppermost mantle velocities are high (8.5 km/s) at the southern end. Tie line T3 along the east coast of South Island has a fairly simple crustal model where it crosses the main transects T1 and T2, showing a thick upper crust and very thin lower crust. The uppermost mantle velocities for the intersecting east coast transects are

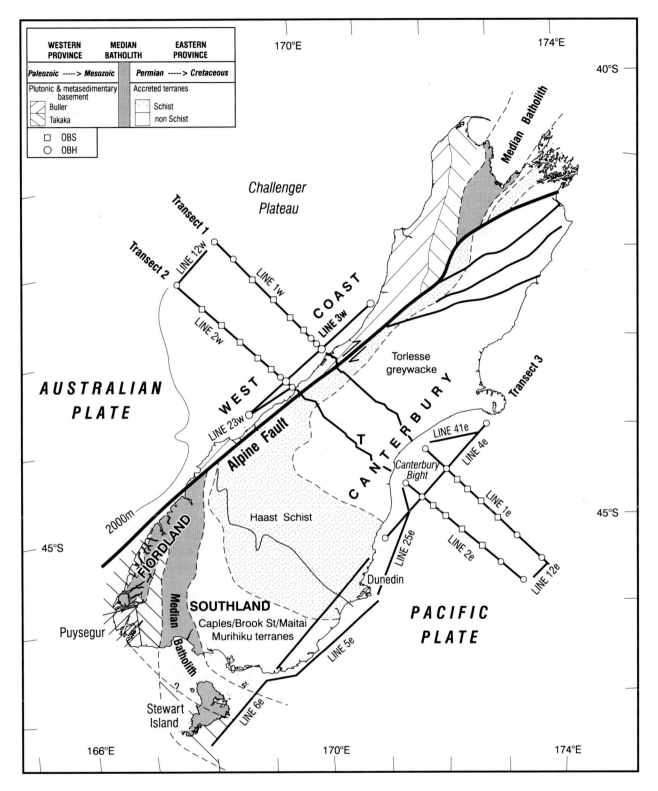

Figure 9. The location of profiles of the SIGHT project overlain on a generalised geology of South Island. The two main transect of SIGHT each comprise a land transect and their extension offshore to east and west. Tie lines are formed by Transect 3 (line 4e, a land component and line 6e) and line 3w. The Alpine fault and Marlborough splay faults to the north are also shown.

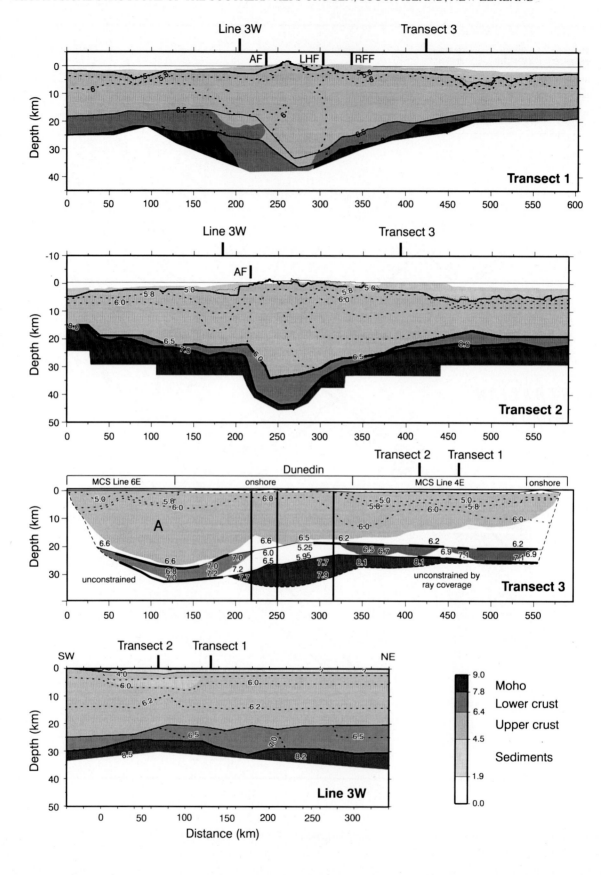

all similar (8.1–8.3 km/s). The velocity model for T3 becomes more complex to the south probably resulting from the past deformation of multiple accreted terranes [*Mortimer et al.*, 2002]. There are significant changes in crustal thickness and a distinct localised low-velocity lower crust and uppermost mantle near Dunedin that was interpreted as remnant hot rock related to the Miocene Dunedin volcano. The crustal sections for all four transects agree to within 0.1–0.2 km/s and 1 km in depth where they cross.

To investigate crustal anisotropy in the vicinity of the Alpine fault, *Pulford et al.* [2003] modelled SmS phases from transect T1 data for shear-wave splitting. A maximum time difference of 0.08 s was measured indicating little or no anisotropy above 15 km for the direction of ray paths. However, significant upper mantle anisotropy occurs. The uppermost mantle P-wave velocities on line 3w are high (8.2–8.5 km/s) in contrast to velocities of about 7.6–7.8 km/s on the cross transects T1 and T2, implying significant uppermost mantle anisotropy (11 ± 5% for T1-3w) in this region [*Scherwath et al.*, 2002]. The uppermost mantle velocities for the intersecting east coast transects are all similar, with velocities of 8.1–8.3 km/s for the intersections of T3 with T1 and T2 [*Godfrey et al.*, 2001; *Scherwath et al.*, 2003; *van Avendonk et al.*, 2004] implying little or no upper mantle anisotropy off the east coast.

v) MCS/CDP images - Crustal near-vertical incidence seismic reflection data (CDP) are limited to the coastal region of South Island and along the main transects of the SIGHT project. In 1996, marine MCS reflection data were recorded to 16 s twt in association with the shots for the off-onshore experiments [*Godfrey et al.*, 2001; *Mortimer et al.*, 2002; *Greenroyd et al.*, 2003; *Melhuish et al.*, 2004; *Davey*, 2005]. In 1998 CDP data were recorded overland along the central part of T2 and the west coast part of T1 [*Henrys et al.*, 1998]. Extensive industry data exist for the east coast of South Island [*Field et al.*, 1989] and to a lesser extent for the West Coast [*Nathan et al.*, 1986]. They are useful for linking the SIGHT images to geological control at petroleum wells, but are limited in depth mainly to basement.

Interpreted Moho is well-recorded on all SIGHT96 reflection data, apart from the data corresponding to the far offshore (>100 km from the coast) of the two trans-South Island transects (T1 and T2). Off the West Coast, the seismic images typically show a distinct base to the lower crustal reflectivity at about 9 s twt that is interpreted to be the Moho and corresponds very closely to refraction Moho. Above this is a highly reflective lower crust that is about 2 s twt thick and corresponds closely to a seismic refraction lower crust with velocities of 6.5–7.0 km/s, a relatively acoustically transparent upper crust with a consistent velocity of about 5.9 to 6.1 km/s, and a sedimentary section of variable thickness and character. Uppermost mantle dipping reflector packages have been imaged on the MCS data and are inferred to show old lithospheric convergent structures [*Melhuish et al.*, 2004]. Dipping mantle reflectors have been recorded on the wide angle data at apparent depths (ranges) of up to 100 km, but no precise locations have been derived.

The images off the east coast of South Island are more variable. Moho varies in character from a single reflector to the base of a highly reflective zone [*Mortimer et al.*, 2002] and its depth varies strikingly from about 7.5 s twt to 11 s twt off southeast South Island (Figure 11). Off central South Island (Canterbury) the reflective lower crust is either thin or non-existent, corresponding in general to the wide-angle refraction results. It is overlain by an upper crust showing sparse, subdued and gently dipping reflectors that conform with the metamorphic grade of the greywackes and schists forming the crust in the south. The sedimentary section in this region is thick. Further south, the sediments thin and the crustal section becomes more complex [*Mortimer et al.*, 2002] with a conspicuous, highly reflective lower crust of restricted lateral extent near Dunedin that is coincident with low lower crustal and upper mantle seismic velocities [*Godfrey et al.*, 2001].

Off the southern part of South Island, the seismic reflection images are complex. Moho is generally well-imaged at the base of a reflective lower crust with variable thickness. Dipping uppermost mantle events cross and displace the Moho, and are inferred to be major faults and paleosutures [*Sutherland and Melhuish*, 2000; *Davey*, 2005]. The crustal sections consist of a reflective lower crust with a generally transparent upper crust overlain by a sedimentary section of variable thickness [*Melhuish et al.*, 1999; *Sutherland and Melhuish*, 2000; *Davey*, 2005]. In the southwest, off Fiordland, seismic data show only the sedimentary section, and crustal reflectivity is poor [*Wood et al.*, 2000]. The seismic data image accretionary sediment packages along the highly obliquely convergent Fiordland margin [*Barnes et al.*, 2005]. A continental-oceanic plate boundary is inferred under the deep water of the upper continental rise [*Wood et al.*, 2000].

The coverage of seismic reflection data on land in South Island is limited. Surveys for petroleum or groundwater

---

Figure 10. (Opposite)   Velocity models along a) Transect 1 [*van Avendonk et al.*, 2004], b) Transect 2 [*Scherwath et al.*, 2003], c) Transect 3 [*Godfrey et al.*, 2001], d) Line 3w [*Melhuish et al.*, 2005]. Velocity contours are in km/s and also shown by a broad grey scale. Note the distinct asymmetry of the orogen. AF - Alpine fault, LHF - Lake Heron fault, RFF - Range Front fault.

exploration have been undertaken for the main sedimentary basins of West Coast, Canterbury, and Southland and are restricted to basement depths or shallower. A small (27 km) crustal survey was carried out in central South Island region (Tekapo) and used to image the base of the crust and possibly the Alpine fault [*Kleffmann et al.*, 1998]. A more extensive survey, SIGHT98, comprised a 75-km long profile across the central part of the South Island orogen and a 35-km profile across the Alpine fault and the west coastal plain. The shallow part of the data records a complex structure of faults and dipping strata [*Long et al.*, 2003] in the schists of the central mountains. Although complex, the image helps to understand the character of the inferred antithetic faulting of the convergent orogen. The deeper data [*Henrys et al.*, 1998; also in *Stern et al.*, this volume] were used to image two gently west-dipping reflectors within the mid-lower crust under the eastern part of the mountains. The lower one was inferred to mark the detachment zone between the upper crust that is being exposed at and to the east of the Alpine fault, and the lower crust that is being "subducted" into a crustal root under the Southern Alps. The data were also used to image three steeply-dipping reflectors under the western part of the Southern Alps, corresponding closely to the inferred position of the Alpine fault [*Stern et al.*, 2001].

vi) potential field and other data - Good regional gravity coverage exists for South Island with stations at an average of 5 km spacing [*Reilly and Whiteford*, 1979]. Free air, Bouguer and isostatic anomaly compilations are available. Magnetic data coverage is poor, with about 60% of South Island, primarily the northern and southern thirds of the island, covered by high altitude (3.3 km) aeromagnetic surveys [*Woodward and Hatherton*, 1975; *Hunt*, 1978]. Gravity and magnetic measurements have been made over the coastal offshore region by ships and satellite [e.g., *Sutherland*, 1996; *Sandwell and Smith*, 1997].

Gravity data have been used extensively for interpreting crustal and upper mantle structure of the Southern Alps and Alpine fault [e.g., *Reilly*, 1962; *Woodward*, 1979; *Allis*, 1986; *Stern*, 1995] and also for constraining seismic models [*Eberhart-Phillips and Bannister*, 2002]. The Bouguer gravity field of South Island has a number of major features (Figure 12). The two dominant features are the broad negative anomaly associated with the Southern Alps, and the large local gravity high (200 mgal) associated with the Fiordland margin. The anomaly associated with the Southern Alps reaches a minimum Bouguer anomaly of −100 mgal with the −80 mgal contour extending over a region of 200 km by 50 km. It is broader in the south and terminates east of Fiordland, although a narrow low continues to the south coast. Although the gravity anomaly appears to correspond closely with the topography, the mountains are not in isostatic equilibrium [*Reilly and Whiteford*, 1979]. The lowest negative isostatic anomaly of −30 mgal is offset about 10 km to the

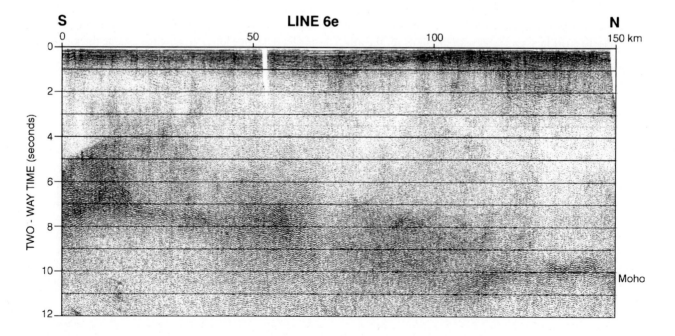

Figure 11. Seismic reflection profile SIGHT line 6e off southeast South Island, showing the highly variable thickness of the reflective lower crust, the sharp Moho and the large changes in its depth.

southeast of the maximum topography [*Woodward*, 1979], implying that the corresponding crustal root is offset to the southeast and/or some other forces are operating. In addition, the axis of the gravity anomaly trends off to the southeast from the plate boundary trend [e.g., *Gerbault et al.*, 2002]. These factors have been explained in several ways including offset of the Southern Alps root to the southeast [e.g., *Allis*, 1986; *Stern*, 1995], the presence of a cold, high-density mantle blob under the Southern Alps pulling down the lithosphere [*Stern et al.*, 2000; *Scherwath et al.*, 2006], with subducting lithospheric roll back [*Waschbusch et al.*, 1998], and lower crustal ductile flow towards the south [*Gerbault et al.*, 2002].

There are no magnetic data over the central part of the Southern Alps orogen. The coverage does, however, define the trace of the Stokes magnetic anomaly/Dun Mountain ophiolite [*Woodward and Hatherton*, 1975; *Hunt*, 1978].

This anomaly corresponds to an obducted Permian oceanic crustal element in the Maitai terrane. It forms a major marker for the horizontal deformation of the New Zealand micro-continent since the late Mesozoic that defines the transcurrent offset on the Alpine Fault.

Heat flow data in the region are few (Figure 13). In general, heat flow is about 45–60 mW/m^2 but high heat flow values of about 80–90 mW/m^2 have been measured along the Southern Alps and in the Dunedin region [*Funnell and Allis*, 1996; *Shi et al.*, 1996; *Cook et al.*, 1999]. Along the Southern Alps a very high heat flow value of 190 mW/m^2 has been obtained in the region of maximum uplift near Franz Josef by *Shi et al.* [1996], who use this to constrain a model of the uplift and thermal state of the Southern Alps. In the Dunedin region, a high heat flow anomaly of about 90 mW/m^2 coincides with a mantle helium anomaly of up to 88% [*Hoke et al.*, 2000].

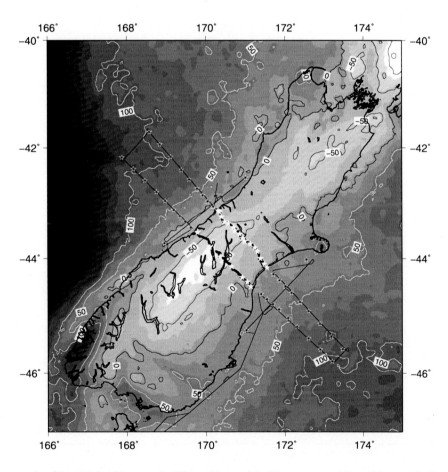

Figure 12. Bouguer anomaly of South Island (grey scale (25 mgal intervals) with contours at 50 mgal intervals) after *Reilly and Whiteford* [1979], showing the large negative Bouguer anomaly associated with the Southern Alps, that trends away from the Alpine fault, and the large positive anomaly associated with the Fiordland region (southwest South Island). The two main SIGHT transects (white stars - shot locations) and the location of the offshore profiles (black stars - OBS locations) are marked.

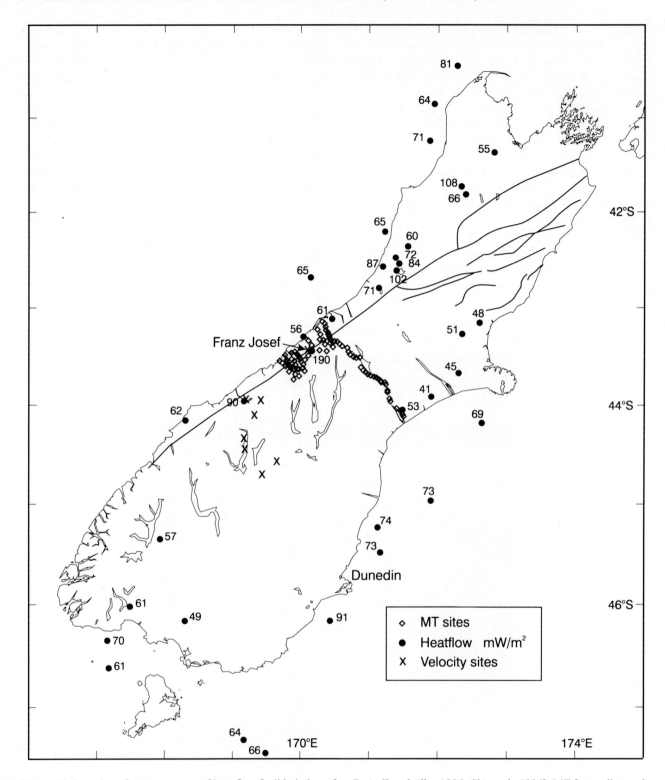

Figure 13. Location of measurements of heat flow [solid circles, after *Funnell and Allis*, 1996; *Shi et al.*, 1996], MT [open diamonds, after *Wannamaker et al.*, 2002; *Caldwell*, personal communication] and detailed petrophysics data [crosses after *Okaya et al.*, 1995] in South Island. Heat flow values are annotated. Note high heat flow associated with the Alpine fault culminating with a maximum value of 190 mW/m^2 in the central region.

Crustal electrical conductivity measurements have been made at several sites in the Southern Alps including a profile along transect T1 (Figure 13) using the magnetotelluric technique [*Ingham and Brown*, 1998; *Caldwell et al.*, 1999; *Wannamaker et al.*, 2002; see also *Jiracek et al.*, this volume]. 2D inversion of the MT data along the transect produced models showing a concave-upwards, middle to lower crustal conductivity body under the Southern Alps extending from near the Alpine fault to about 50 km to the southeast [*Wannamaker et al.*, 2002]. In the northwest, the conductive body rises towards the Alpine fault and becomes near-vertical about 5 km east of the fault. In the southeast the body rises to the surface coincident with a major back thrust of the orogen. Limited data to the northeast and southwest indicate that the conductive body extends along the axis of the Southern Alps.

Physical properties of rocks are available for a wide range of samples across the South Island [PETLAB data base, *Mortimer*, 2005; see also *Christensen and Okaya*, this volume]. These are largely density and chemical properties plus acoustic ultrasonic velocity measurements. Detailed velocity measurements on a set of samples from the Southern Alps (Figure 13) showed that higher-grade schistose rocks of the Southern Alps have significant velocity anisotropy with values of up to 17% for P waves and up to 24% for S waves [*Okaya et al.*, 1995; *Godfrey et al.*, 2000].

STRUCTURE OF THE OROGEN

The development of the orogen is the result of a combination of convergent lithospheric tectonics and climate-related erosion [e.g., *Adams*, 1980; *Koons*, 1989]. Lithospheric structure provides clues about the dynamic relationships between mountain uplift, lithospheric deformation, and associated uppermost mantle flow, and hence our understanding of tectonic processes. Central South Island provides an important example of the main structures and processes forming an orogen at a continental collisional plate boundary. It is the result of a relatively simple transpressive continental plate boundary that has been in existence for only the last 5–6 Ma.

i) Broad structure (crustal, upper mantle) - In central South Island there is a simple transpressive continental plate boundary. To the north there is the influence of the Hikurangi subduction system and the associated splay faulting through northeastern South Island. To the south there is the influence of the Fiordland subducting margin, the highly uplifted Fiordland lithospheric block, and the earlier oroclinal bending of the South Island terranes. The structures, processes and influence of these regions will be discussed later. The prime data sets are the two lithospheric transects across central South Island (T1 and T2) that formed the main focus of the SIGHT experiment [*Scherwath et al.*, 2003; *van Avendonk et al.*, 2004]. These two transects show a number of common features (Figures 11 and 15). A prime observation is that the orogen is distinctly asymmetric at all lithospheric levels, with deformation being focussed under the eastern Pacific plate. In detail, deformation tends to be asymmetric in the crust, but symmetric within the mantle lithosphere in terms of anisotropy with respect to the plate boundary and, assuming the *Stern et al.* [2000] model, accommodation of mantle shortening. However, there is also evidence here and in other transpressional plate boundary regions that the descending mantle lithosphere may be undergoing asymmetric deformation [see *Fuis et al.*, this volume].

a) surface morphology. In the west, there is a narrow continental shelf, and a narrow coastal plain before the surface expression of the plate boundary - the Alpine fault or the plate boundary thrust fault of crustal extent - is reached. This marks the base of a steep mountain front that rises to the main divide at an altitude of over 3000 m, a high intermontane region at a height of about 1500 m, largely between the main plate boundary fault and the antithetic back thrusts of the orogen [e.g., Lake Heron fault, *Cox*, 1995]. A major back thrust [Canterbury Range Front, *Cox*, 1995] marks the eastern limit of the uplifted orogen and is followed to the east by a broad outwash coastal plain and a broad continental shelf before deep water is again reached (Figures 2 and 11).

b) basement morphology-sedimentary cover. Thick (4 km) sediments underlie the western coastal strip and narrow continental shelf. They are the result of thrust loading of the Australian plate by the Pacific plate ramping up at the Alpine fault, and deposition from high erosion of the western flanks of the Southern Alps due to high precipitation associated with the dominant westerly winds. The sedimentary stratigraphy is that of a classic foreland thrust belt [*Sircombe and Kamp*, 1998]. To the east of the Southern Alps, thick (up to 4.5 km), broad, outwash fan deposits originating in the high alps underlie the broad coastal plain and continental shelf [*Field et al.*, 1989].

c) crust - The crust is about 20 km thick at the extremities of the two major transects where it was thinned by rifting during the development of the continental-oceanic boundary. At the coast it thickens to about 25–30 km and then increases in thickness again to form an asymmetric crustal root, about 200 km wide and 40–50 km thick under the Southern Alps. Upper-middle crustal rocks of the Australian plate typically have seismic velocities of about 5.6 km/s at the surface ranging up to 6.1 km/s at depth. Lower crust is thin—about 1–4 km thick—with a seismic velocity of about 6.9 km/s. In contrast the crust of the Pacific plate has a seismic velocity of about 6.0 km/s, rising to a mid crust high-velocity layer of about 6.3 km/s at a depth of 15 km. Its lower crust is variable in thickness, thinning from about 5 km at the outboard end of the transects to almost zero at the coast, then thickening to about 10 km to form the lower part of the crustal root. Seismic velocities for the lower crust are about 6.9–7.0 km/s. The crustal root is formed almost equally by thickened middle crust and thickened lower crust [*Scherwath et al.*, 2003]. Immediately east of the plate boundary, the western margin of the Pacific crust is a broad (40 km wide) low-velocity zone stretching from near the surface to the base of crust [*Stern et al.*, 2001; *Scherwath et al.*, 2003; *van Avendonk et al.*, 2004]. The low-velocity zone is located in the region where the crust is being upturned along the Alpine fault and where the highly anisotropic schists are inferred to form most of the upper-mid crust. This suggests that the low-velocity crust may be the result of anisotropy of the schists. However, teleseismic residual modelling [*Stern et al.*, 2000] also indicates low-velocity rocks in this region. The orientation of the ray paths for the teleseismic events and for the active source survey are perpendicular, indicating isotropic rocks. *Stern et al.* [2001] propose over-pressured fluids derived from prograde metamorphism as the cause of the low velocities. MT data [*Wannamaker et al.*, 2002] indicate that the low-velocity zone and mid-crust decollement are of high conductivity, and also suggest fluids derived from prograde metamorphism as the cause. Limited MT data along the strike of the Southern Alps suggest the along-axis continuation of the high-conductivity zone is a feature of the convergent plate boundary. The velocity structure in the crust is remarkably consistent among all four transects, suggesting little regional crustal anisotropy.

Transects T1 and T2 also differ in important details (see Figure 10). Major differences can be seen in the depth of the root (deeper in T2), the dip on the western side of the root (steeper in T2), and the thickness of the lower crust in the root zone. These differences may partially be due to the different analysis techniques used; for example the dip of the Moho on the western side of the root is much less in T1 but wide-angle CDP stacked images along both profiles suggest that the dip on T1 is steeper than as modelled on the wide-angle velocity models [*Henrys et al.*, 2004]. The thickness of lower crust is important for defining how convergence is partitioned between upper and lower crust. The increased thickness of lower crust in T2 can be modelled by the obduction and erosion of upper-middle crust at the Alpine fault and by the accumulation of lower crust and lower-middle crust into the crustal root. Such a mechanism does not appear possible for the velocity structure of T1, as lower crustal material is of near-constant thickness across the crustal root region. This change can be explained by lower crustal flow towards the south [*Gerbault et al.*, 2002]. The overall increase in the depth of the base of the crustal root to the south is consistent with the gravity data.

Differences are apparent between the active source 2D models and the 3D first-arrival tomography, particularly in the detailed structure (Plate 2, Profiles d and e, and Figure 10, Transects 1 and 2). The 2-D profiles have layered structure determined by forward modelling of arrivals from surface sources, and have additional features derived from secondary arrivals. This technique produces a detailed Moho and smooth crustal velocities within layers. The 3-D gridded inversion includes seismic sources throughout the volume and thus has more information on heterogeneity within the crust, although the Moho will be a smooth gradient over 3–10 km. The 3-D inversion uses a wide range of raypath azimuths and thus the results will show the average velocity of the volume surrounding the node, rather than an anisotropic velocity for a specific azimuth. The results of the two methods are similar, within the constraints of the techniques, which show the reliability of the results. For example, the 7.5 km/s contour would give an approximate 3-D model Moho of 40-42 km depth on the transects, similar to the forward models. The high-velocity body within the Pacific plate crust shown by the 3-D inversion is similar to that seen in the 2-D model, but it is a smoother feature with the higher velocity smeared throughout the crust. The 3-D model can thus be interpreted together with the 2-D models to examine how the Southern Alps structure varies north and south of the transects.

d) upper mantle - Uppermost mantle velocities are about 8.0 km/s under the Australian plate at the western end of the transects and about 8.1–8.3 km/s under the Pacific plate at the eastern end of the transects (Figure 10). For about 100 km on either side of the plate boundary, the uppermost mantle velocities are low - 7.8 km/s - along transects T1 and T2 corresponding to strong mantle rock anisotropy of up to about 11% [*Scherwath et al.*, 2002]. Under T3 off the east coast, little or no upper mantle anisotropy exists, suggesting that the eastern edge of anisotropic upper mantle is under eastern South Island in this region. The degree and direction of anisotropy for the west coast are consistent with that obtained from shear-wave splitting (SKS) studies [*Klosko et al.*, 1999]. The anisotropy arises from the large degree of horizontal deformation [850 km over a 400 km wide zone, *Sutherland*, 1999] that occurred in this part of New Zealand during the Cenozoic associated with the development of the Alpine Fault.

Teleseismic data recorded along Transect 1 have been used to model a high-velocity, vertically-oriented mantle body centered about 120 km under the crustal root with its western margin immediately below the Alpine Fault [*Molnar et al.*, 1999; *Stern et al.*, 2000]. This high-velocity region (maximum velocity anomaly of +7%) has a width of about 80 km and a vertical extent of about 100 km (Figure 7), and is suggested to result from uniform thickening of mantle lithosphere. Gravity modelling is consistent with the mantle body being of higher-density rock than the surrounding lithospheric mantle, symmetrically disposed with respect to the crustal root [*Stern et al.*, 2000], and extending well to the north and south of the transects [*Scherwath et al.*, 2006]. Uniform thickening of mantle lithosphere directly below a crustal root is consistent with growth of a Rayleigh-Taylor type of instability beneath the Southern Alps [*Houseman and Molner*, 1997]. In contrast, the centre of the high-velocity upper mantle body of *Kohler and Eberhart-Phillips* [2002] lies about 40 km east of the crustal root and dips steeply to the northwest which may indicate origin from the Pacific continental mantle (Figure 7). It extends from the Hikurangi subducted slab in the north to the southern margin of South Island (Figure 8), more extensive than the major Southern Alps gravity anomaly, suggesting it may be caused, in part, by a different mechanism for accommodating mantle lithospheric shortening to that inferred by *Stern et al.* [2000]. On the basis of limited sub-crustal seismicity in the central Southern Alps, *Reyners* [2005] proposes that the uppermost mantle beneath the crustal root of the Southern Alps is rigid enough to store sufficient shear strain to produce a $M_W 5.9$ earthquake and that the pattern of seismicity is consistent with the subduction of rigid mantle. Modeling studies by *Gerbault et al.* [2003] also require a strong (~200 MPa) mantle lithosphere with an intermediate rate of strain softening. The deep (long range) reflectors recorded by the wide angle experiment off the West Coast and noted earlier, may be reflections from this upper mantle structure.

A single model for the deformation of the lithospheric mantle which can integrate the interpretations of *Kohler and Eberhart-Phillips* [2002], *Reyners* [2005], and *Stern et al.* [2000] is likely a combination of both brittle and viscous deformation – plate-like behaviour in the upper part of the lithospheric mantle with distributed thickening and viscous instability in the lower part, as suggested by the modeling studies of *Pysklywec et al.* [2002].

ii) Constraints from other data - The depth of the brittle crust can be constrained by heat flow modelling and by seismicity. Based on limited data, modelling of the crustal thermal state under the Alps and the associated geotherms indicates a depth to the base of the crustal brittle layer (300° geotherm) of about 15 km under the east coast, thickening to about 25 km under the Southern Alps and rising to about 10 km adjacent to the Alpine Fault [*Shi et al.*, 1996]. However, a better constraint on the depth of the base of the brittle layer is defined by the maximum seismicity depths of 12 km on the flanks and 8 km under the central Southern Alps [*Leitner et al.*, 2001]. Thermal modelling also suggests the occurrence of brittle mantle seismicity to depths of about 60 km under the central Southern Alps [*Shi et al.*, 1996], consistent with the seismicity results of *Reyners* [1987, 2005] and *Kohler and Eberhart-Phillips* [2003].

The axis of the major central South Island gravity anomaly lies at an angle of about –15° to the plate boundary (Alpine fault), similar to the eastwards displacement of the crustal root in the south derived from seismic tomography [*Eberhart-Phillips and Bannister*, 2002]. The data are interpreted to show a thickening and broadening of the crustal root to the south [*Scherwath et al.*, 2006], consistent with the results of *Bourguignon et al.* [this volume] and an increase in the gravity anomaly associated with the dense upper mantle body [*Scherwath*, 2002]. This is compatible with a model of extrusion of highly ductile lower crustal material to the south [*Gerbault et al.*, 2002]. The model of south-directed extrusion is also consistent with plate tectonic models [*Walcott*, 1998; *Cande and Stock*, 2004] that propose greater shortening across the plate boundary in the north of the region than in the south whereas the crustal root thins to the north [*Scherwath*, 2002]. Thickening of the lower lithosphere towards the south occurs largely in the lower crust and upper mantle and can be attributed to lower lithospheric flow arising from the transcurrent movement at the plate boundary [*Gerbault et al.*, 2002; *Scherwath et al.*, 2006].

iii) Regional variation (north and south) - The simple 2D structure observed in central South Island does not continue to the north and south. This is because the simple transpressive continental tectonics are modified by other processes and inherited structures as the plate boundary changes from a transform plate boundary to west-dipping subduction in the north and east-dipping subduction in the south. In the north, the major change occurs near Arthur's Pass where the first of the northeast-trending Marlborough faults splays off to the east from the Alpine fault and the "big bend" in the Alpine fault trace commences. In the south, the morphology of the Southern Alps undergoes a major change in the Haast Pass region where they broaden and have reduced elevation. The plate boundary, as delineated by the Alpine fault, becomes a subdued morphological feature. At times it shows minor uplift to the west, eventually running offshore at Milford Sound at the northern end of the Fiordland block where plate-boundary deformation comprises highly oblique, oceanic-continental subduction. These changes are reflected in the geology and gravity which show a swing in structure from northeast-southwest - parallel to the Alpine fault and the present plate boundary - to west-northwest-east-south-east and the termination of the major gravity anomaly. In northern Fiordland, the Alpine fault becomes aligned with the subducted slab, about 80 km northwest of the crustal root (Figure 6c). The continuation to the south of the high-velocity, near-vertical body in the upper mantle [*Kohler and Eberhart-Phillips*, 2002] lies east of the subducted plate under Fiordland, and is aligned with the Solander Basin and Macquarie Ridge to the south. Whether it continues along the Macquarie Ridge strike-slip plate boundary or terminates is unknown, but it appears to be unrelated to the Fiordland subducting plate.

a) Marlborough region. The northeastern end of the plate boundary through South Island is formed by the Marlborough Fault System (Plate 1), a system of faults that splay off to the east from the Alpine fault and along which transcurrent plate motion is transferred across Cook Strait [*Lewis*

et al., 1994] and into the back-arc region of the North Island (Hikurangi) subduction system. The region coincides with the southern end of the zone of intermediate to deep seismicity [*Anderson and Webb*, 1994] associated with the Hikurangi subduction zone. *Reyners and Robertson* [2004] propose that the subducting Pacific plate does not stop abruptly here but instead becomes aseismic, possibly due to a change in dehydration conditions in the slab. The region also coincides with the position where the subducting slab changes from oceanic to continental as the Chatham Rise is

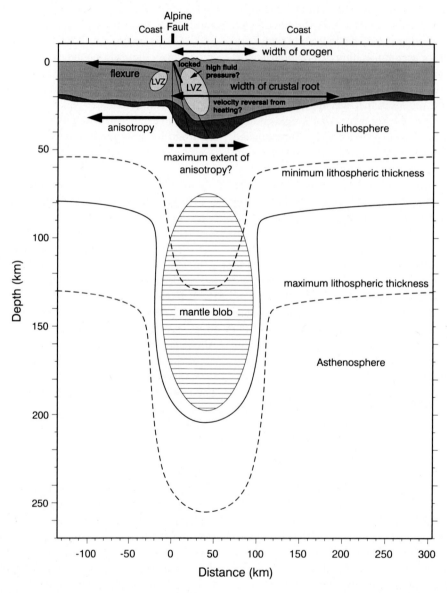

Figure 14. Lithospheric cross section and upper mantle velocity anomaly across the Southern Alps orogen [after *Scherwath*, 2002, Figure 6.7], showing the major structural/tectonic features of the orogen.

brought into the subducting region. *Eberhart-Phillips and Reyners* [1997] present 3D velocity images for the region derived from the tomographic inversion of earthquake data. In the uppermost mantle they image a low-velocity zone associated with seismicity. They infer this to be related to the continental nature of the subducted plate with the increase in the amplitude of the velocity anomaly to the southwest associated with the increasing thickness of the continental crust being subducted. They suggest that subduction is relatively minor and the plate interface may be locked, leading to the intense deformation at the coastal region. As a result of this attempt to subduct buoyant continental crust, the overriding plate has been compressed leading to the development of the Kaikoura Ranges (Inland and Seaward) with their very sharp onset at the coast (Figure 2). This overlying plate in a region that extends as far to the northwest as the Awatere fault, is also associated with a low seismic velocity and high seismicity. It is inferred to be weak, probably as a result of elevated pore pressures caused by fluids derived from the dehydration of the subducted continental crust. Deformation in the region is accommodated by the east-northeast-striking dextral strike-slip Marlborough Fault System, with early (Miocene and later) movement largely concentrated on the Wairau fault in the west, where *Wilson et al.* [2004] infer significant crustal thinning, although this is not seen on the tomographic images of *Eberhart-Phillips and Reyners* [1997] perhaps because of lack of resolution in the latter. Late Quaternary slip rates on the faults increase from west to east [*van Dissen and Yeats*, 1991] with most of the present relative plate motion taken up on a belt of distributed shear closer to the Hikurangi Trough [*Bibby*, 1981]. The fault azimuths change from about 70° in the south to 50° in the north as a result of increasing thrust faulting in the northeast and rapid uplift of the coastal ranges.

b) Fiordland region. The southern end of the plate boundary system runs offshore at the northern end of the Fiordland region and continues along the coast to the southern end of South Island into Puysegur Trench. Oceanic crust of the Australian plate lies to the west of the plate boundary and highly oblique subduction occurs along it. The juxtaposition of oceanic crust with very high grade metamorphic rocks onshore, along a sharp lithospheric boundary with a component of convergence, has given rise to a strong positive gravity anomaly onshore and a complementary negative anomaly offshore. The onshore part of this region also coincides with a restricted zone of intermediate depth seismicity where lower crustal seismic velocities occur very near the surface [*Woodward*, 1972; *Davey and Broadbent*, 1980; *Davey and Smith*, 1983; *Eberhart-Phillips and Reyners*, 2001] (Figures 5b and 7a).

Recent data from a 24-station temporary array provide accurate hypocentres, focal mechanisms and 3-D velocity structure for the region. They document the change towards the southwest from lithospheric thickening under the Southern Alps to near-vertical subduction of oceanic Australian plate under Fiordland, and the transition from this near-vetical subduction to inferred lower-angle subduction of the Australian plate at the Puysegur trench south of South Island [*Eberhart-Phillips and Reyners*, 2001; *Reyners et al.*, 2002]. The data were used to image a high-velocity body in the Pacific mantle which they infer has caused the subducting Australian oceanic slab to bend progressively steeper to the northeast (Figure 6b). This interpretation follows the ploughshare or twisted subducted plate model, originally proposed by *Christoffel and Van der Linden* [1972], invoking a break in the subducted plate to achieve the steep dip and tight "bending" also suggested by *Davey and Smith* [1983]. The bending is consistent with subducted plates losing most of their strength inboard of the trench [*Billen and Gurnis*, 2005]. *Malservisi et al.* [2003] propose a tear along the Fiordland margin to achieve a similar effect. The present tectonics of the region are characterized by the unusual subduction environment along the Fiordland coast and the broad extensional environment to the east where subdued basin and range deformation of the pre-Cenozoic basement occurs.

CONCLUSIONS

The Pacific-Australian plate boundary through South Island is a major transcurrent fault joining west-dipping oceanic subduction under the North Island in the north, and east-dipping, highly oblique subduction off Fiordland in the southwest. This transcurrent deformation has experienced an increasingly large amount of convergence during the past 5 Ma or so. The central part of South Island has been taken to represent a simple, 2D convergent continental plate boundary, but more detailed investigation demonstrates that the deformation is 3D in nature, resulting from the transpressive nature of the relative plate motion. Research over the central, quasi 2D sector of the plate boundary identifies some of the major features of transpressive continental plate boundaries. The SIGHT transects illustrate these features well (Figure 14). The deformation is accommodated by the ramping up of the upper part of the Pacific plate over the Australian plate. This initially asymmetric deformation may be defined by an initial difference in lithospheric strength [Australian plate stronger than the Pacific plate, e.g., *Koons*, 1990; or v.v *Scherwath*, 2002], or an inherited suture resulting from earlier plate motions [*Sutherland et al.*, 2000]. Evidence supports both explanations and the real cause may be some combination of both [e.g., *Gerbault et al.*, 2003]. The ramping up of the Pacific crust is accompanied by the delamination of the Pacific plate with the result of the uplift and exposure of mid-crustal rocks at the plate boundary fault

(Alpine fault) forming a foreland mountain chain that has developed a particular climatic response: increased precipitation and the development of increased uplift resulting from the positive feedback loop of erosion and isostatic unloading of the underlying rocks. This is accompanied by the formation of a thick crustal root (additional 8–17 km) formed by the delaminated lower crust and a thickening of the overlying middle crust. Lower crust is variable in thickness both along and across the orogen. The variation may arise from pre-transpression extension away from the focus of the orogen, and convergence in and along the orogen with lower lithosphere extrusion being proposed for the latter. Low-velocity zones in the crust occur adjacent to the plate boundary (Alpine fault) in both the Australian and Pacific plates. Fracturing of the upper crust as a result of flexural bending is proposed as the cause for the low-velocity crust in the Australian plate. The low-velocity zone in the Pacific plate at the plate boundary is suggested to be caused by high-pressure fluids in the crust derived from prograde metamorphism of the crustal rocks as they are being exhumed. This explanation is supported by high conductivity measurements in the same region from MT observations. The ramp fault and the delaminating decollement are well-imaged on seismic reflection data. First-order calculations indicate a conservation of mass within the system [*van Avendonk et al.*, 2004; *Henrys et al.*, 2004]. The accommodation of lithospheric mantle shortening is probably a combination of both brittle and viscous deformation. The structure of the orogen forming the Southern Alps conforms to theoretical and data-based models showing an uplifted region constrained by the major ramp fault and antithetic faulting.

Acknowledgements. We gratefully acknowledge the tremendous support of the people involved in the field work that provided the basis for this paper, in particularly in the SIGHT and SAPSE experiments. We also acknowledge funding support from NSF (grants EAR-9805224 (MDK)) and the NZ Foundation for Research Science and Technology. Reviews by Martin Reyners and Andrew Gorman were greatly appreciated.

REFERENCES

Abercrombie, R. E., T. H. Webb, R. Robinson, P. J. McGinty, P. J. Mori, and J. R. Beavan (2000), The engima of the Arthur's Pass, New Zealand, earthquake 1: reconciling a variety of data for an unusual earthquake sequence, *J. Geophys. Res.*, *105*, 16,119–16,137.

Adams, J. (1979), Vertical drag on the Alpine Fault, New Zealand, in *The Origin of the Southern Alps, Bull. R. Soc. N. Z.*, vol. 8, edited by R. I. Walcott and M. M. Cresswell, Royal Society of New Zealand, Wellington, New Zealand, pp. 47–54.

Adams, J. (1980), Contemporary uplift and erosion of the Southern Alps, New Zealand, *Geol. Soc. Am. Bull.*, *91*(1), Pt1, 2–4; Pt2, 1–114.

Allis, R. G. (1986), Mode of crustal shortening adjacent to the Alpine fault, New Zealand, *Tectonics*, *5*, 15–32.

Anderson, H. and T. Webb (1994), New Zealand seismicity: patterns revealed by the upgraded National Seismograph Network, *N. Z. J. Geol. Geophys.*, *37*, 477–493.

Anderson, H., D. Eberhart-Phillips, T. McEvilly, F. Wu, and R. Uhrhammer (1997), Southern Alps passive seismic experiment. *Institute of Geological and Nuclear Sciences Ltd Science Report 97/21*, Institute of Geological and Nuclear Sciences Ltd., Lower Hutt.

Balance, P. F. (1993), The Paleo-Pacific, post-subduction, passive margin thermal relaxation sequence (Late Cretaceous-Paleogene) of the drifting New Zealand continent, in *South Pacific Sedimentary Basins, Sedimentary Basins of the World*, *2*, edited by P. F. Balance, Elsevier, Amsterdam, pp. 93–110.

Barnes, P. M., R. Sutherland, and J. Delteil (2005), Strike-slip structure and sedimentary basins of the southern Alpine Fault, Fiordland, New Zealand, *Geol. Soc. Am. Bull.*, *117*(3/4), 411–435.

Bibby, H. (1981), Geodetically determined strain across the southern end of the Tonga-Kermadec-Hikurangi subduction zone, *Geophys. J. R. Astron. Soc.*, *66*(3), 513–533.

Billen M. I., and M. Gurnis (2005), Constraints on subducting plate strength within the Kermadec trench, *J. Geophys. Res.*, *110*, B05407, doi:10.1029/2004JB003308.

Bourguignon, S., M. K. Savage, and T. Stern (this volume), Crustal thickness and Pn anisotropy beneath the Southern Alps oblique collision, New Zealand.

Cande, S. C. and J. M. Stock (2004), Pacific-Antarctic-Australia motion and the deformation of the Macquarie Plate, *Geophys. J. Int.*, *167*, 399–414.

Caldwell G. T. (1999), 1998 South Island magnetotelluric transect, *Newslett. N. Z. Geophys. Soc.*, *49*, 32–33.

Christensen and Okaya (this volume).

Christoffel, D. A., and W. J. M. van der Linden (1972), Macquarie Ridge, New Zealand Alpine Fault transition, II: the Australian-New Zealand sector, in *Antarctic Oceanology, Antarctic Res. Ser.*, vol. 19, edited by D. E. Hayes, American Geophysical Union, Washington, DC, pp. 235–242.

Cook, R. A., R. Sutherland, and H. Zhu (1999), Cretaceous-Cenozoic geology and petroleum systems of the Great South Basin, New Zealand, *Institute of Geological and Nuclear Sciences Monograph*, 188 pp.

Cox, S. C. (1995), Geological transect across the Southern Alps of New Zealand, *Institute of Geological and Nuclear Sciences science report 95/30*, Institute of Geological and Nuclear Sciences Ltd, Lower Hutt, New Zealand, 32 pp.

Cox, S. C., and R. Sutherland (this volume), Regional geological framework of South Island, New Zealand, and its significance for understanding the active plate boundary.

Davey F J., and E. G. C. Smith (1983), The tectonic setting of the Fiordland region, *Geophys. J. R. Astron. Soc.*, *72*, 23–28.

Davey, F. J., and M. Broadbent (1980), Seismic refraction measurements in Fiordland, southwest New Zealand, *N. Z. J. Geol. Geophys.*, *23*, 395–406.

Davey, F. J., T. Henyey, W. S. Holbrook, D. Okaya, T. A. Stern A. Melhuish, S. Henrys, D. Eberhart-Phillips, T. McEvilly, R. Urhammer, H. Anderson, F. Wu, G. Jiracek, P. Wannermaker,

G. Caldwell, and N. Christensen (1998), Preliminary results from a geophysical study across a modern continent-continent collisional plate boundary—the Southern Alps, New Zealand, *Tectonophysics*, *288*, 221–235.

Davey, F. J. (2005), A Mesozoic crustal suture on the Gondwana margin in the New Zealand region, *Tectonics*, *24*(4), TC4006, doi:1029/2004TC001719.

Eberhart-Phillips, D., and C. M. Henderson (2004), Including anisotropy in 3-D velocity inversion and application to Marlborough, New Zealand, *Geophys. J. Int.*, *156*(2), 237–254.

Eberhart-Phillips, D.; and M. E. Reyners (1997), Continental subduction and three-dimensional crustal structure: the northern South Island, New Zealand, *J. Geophys. Res.*, *102*(B6), 11,843–11,861.

Eberhart-Phillips, D. and M. E. Reyners (2001), A complex, young subduction zone imaged by three-dimensional seismic velocity, Fiordland, New Zealand, *Geophys. J. Int.*, *146*(3), 731–746.

Eberhart-Phillips, D. and S. C. Bannister (2002), Three-dimensional crustal structure in the Southern Alps region of New Zealand from inversion of local earthquake and active source data, *J. Geophys. Res., Solid Earth*, *107*(B10), doi:10.1029/2001JB000567.

Field, B. D., G. H. Browne, B. W. Davy, R. H. Herzer, R. H. Hoskins, J. I. Raine, G. J. Wilson, R. J. Sewell, D. Smale, and W. A. Watters (1989), Cretaceous and Cenozoic sedimentary basins and geological evolution of the Canterbury region, South Island, New Zealand. Lower Hutt: New Zealand Geological Survey, *N. Z. Geol. Surv. Basin Stud.*, *2*, 94 pp.

Fuis, G. S., M. D. Kohler, M. Scherwath, U. ten Brink, H. J. A. van Avendonk, and J. M. Murphy (this volume), A comparison between the transpressional plate boundaries of South Island, New Zealand, and Southern California, USA: the alpine and San Andreas fault systems.

Funnell, R. H., and R. G. Allis (1996), Hydrocarbon maturation potential of offshore Canterbury and Great South Basins, in *1996 New Zealand Petroleum Conference Proceedings*, vol. 1, Ministry of Commerce, Wellington, New Zealand, pp. 22–30.

Gerbault, M., F. J. Davey, and S. A. Henrys (2002), Three-dimensional lateral crustal thickening in continental oblique collision; an example from the Southern Alps, New Zealand, *Geophys. J. Int.*, *150*(3), 770–779.

Gerbault, M., S. A. Henrys, and F. J. Davey (2003), Numerical models of lithospheric deformation forming the Southern Alps of New Zealand, *J. Geophys. Res.*, *108*(B7), 2341, doi:10.1029/2002JB001716.

Godfrey, N. J., N. I. Christensen, and D. A. Okaya (2000), Anisotropy of schists: contributions of crustal anisotropy to active-source seismic experiments and shear-wave splitting observations, *J. Geophys. Res.*, *105*, 27, 991–28, 007.

Godfrey, N. J., F. Davey, T. A. Stern, and D. A. Okaya (2001), Crustal structure and thermal anomalies of the Dunedin region, South Island, New Zealand, *J. Geophys. Res., B, Solid Earth Planets*, *106*(12), 30,835–30,848.

Greenroyd, C., J. Yu, A. Melhuish, J. M. Ravens, F. Davey, G. Maslen, and SIGHT WORKING GROUP (2003), New Zealand South Island GeopHysical Transect (SIGHT): marine active-source seismic component - a processing summary. *Institute of Geological and Nuclear Sciences science report 2003/04*.

Hatherton, T. (1980), Shallow seismicity in New Zealand 1956-75, *Journal of Royal Society of New Zealand*, *10*, 19–25.

Henrys, S. A., D. Okaya, A. Melhuish, T. A. Stern, and S. Holbrook (1998), Near-vertical seismic images of a continental transpressional plate boundary: Southern Alps, New Zealand, *Eos*, *79*(45, Suppl.), F901.

Henrys, S. A., D. J. Woodward, D. Okaya, and J. Yu (2004), Mapping the Moho beneath the Southern Alps continent-continent collision, New Zealand, using wide-angle reflections, *Geophy. Res. Lett.*, *31*(17), L17602, doi:10.1029/2004GL020561.

Hoke, L., R. Poreda, A. Reay, S. D. Weaver (2000), The subcontinental mantle beneath southern New Zealand characterised by helium isotopes in intraplate basalts and gas-rich springs, *Geochim. Cosmochim. Acta*, *64*, 2489–2507.

Houseman, G A., and P. Molnar (1997), Gravitational (Rayleigh-Taylor) instability of a layer with non-linear viscosity and convective thinning of continental lithosphere, *Geophys. J. Int.*, *128*, 125–150.

Hunt, T. (1978), Stokes magnetic anomaly system, *N. Z. J. Geol. Geophys.*, *21*(5), 595–606.

Ingham, M., and C. Brown (1998), A magnetotelluric study of the Alpine Fault, New Zealand, *Geophys. J. Int.*, *135*(2), 542–552.

Jiracek, G. R., V. M. Gonzalez, T. G. Caldwell, P. E. Wannamaker, and D. Kilb (this volume), Seismogenic, electrically conductive, and fluid zones at continental plate boundaries in New Zealand, Himalaya, and California-USA.

Kleffmann, S. (1999), Crustal structural studies of a transpressional plate boundary: the central South Island of New Zealand, PhD thesis, Victoria University of Wellington.

Kleffmann, S., F. Davey, A. Melhuish, D. Okaya, and T. Stern (1998), Crustal structure in central South Island from the Lake Pukaki seismic experiment, *N. Z. J. Geol. Geophys.*, *41*, 39–49.

Klosko, E. R., F. T. Wu, H. J. Anderson, D. Eberhart-Phillips, T. V. McEvilly, E. Audoine, M. K. Savage, and K. R. Gledhill (1999), Upper mantle anisotropy in the New Zealand region, *Geophy. Res. Lett.*, *26*(10), 1497–1500.

Kohler, M. D. and D. Eberhart-Phillips (2002), Three-dimensional lithospheric structure below the New Zealand Southern Alps, *J. Geophys. Res.*, *107*(B10), 2225, doi:10.1029/2001JB000182.

Kohler, M. D. and D. Eberhart-Phillips (2003), Intermediate-depth earthquakes in a region of continental convergence: South Island, New Zealand, *Bull. Seis. Soc. Am.*, *93*, 85–93.

Koons, P. O. (1989), The topographic evolution of collisional mountain belts: a numerical look at the Southern Alps, New Zealand, *Am. J. Sci.*, *289*, 1041–1069.

Koons, P. O. (1990), Two-sided orogen: collision and erosion from the sandbox to the Southern Alps, New Zealand, *Geology*, *18*, 679–682.

Leitner, B., D. Eberhart-Phillips, H. Anderson, and J. L. Nabelek (2001), A focused look at the Alpine Fault, New Zealand: seismicity, focal mechanisms, and stress observations, *J. Geophys. Res.*, *106*, 2193–2220.

Lewis, K. B., L. Carter, and F. J. Davey (1994), The opening of Cook Strait: interglacial tidal scour and aligning basins at a subduction to transform plate edge. *Mar. Geol.*, *116*(3/4), 293–312.

Long, D. T., S. C. Cox, S. C. Bannister, M. C., Gerstenberger, and D. Okaya (2003), Upper crustal structure beneath the eastern

Southern Alps and the Mackenzie Basin, New Zealand, derived from seismic reflection data. *N. Z. J. Geol. Geophys.*, *46*(1), 21–39.

Malservisi, R., K. P. Furlong, and H. Anderson (2003), Dynamic uplift in a transpressional regime: numerical model of the subducted area of Fiordland, New Zealand, *Earth Planet. Lett.*, *206*, 349–364.

Melhuish A., W. S. Holbrook, F. Davey, D. A. Okaya, and T. Stern (2004), Crustal and upper mantle seismic structure of the Australian Plate, South Island, New Zealand, *Tectonophysics*, *395*, 113–135, doi:10.1016/j.tecto2004.09.005.

Melhuish, A., R. Sutherland, F. J. Davey, and G. Lamarche (1999), Crustal structure and neotectonics of the Puysegur oblique subduction zone, New Zealand, *Tectonophysics*, *313*(4), 335–362.

Molnar, P., H. J. Anderson, E. Audoine, D. Eberhart-Phillips, K. R. Gledhill, E. R. Klosko, T. V. McEvilly, D. Okaya, M. K. Savage, T. Stern, and F. T. Wu (1999), Continuous deformation versus faulting through the continental lithosphere of New Zealand, *Science*, *286*, 516–519.

Mortimer, N. (2005), PETLAB: New Zealand's rock and geoanalytical database, *Newslett. Geol. Soc. N. Z.*, *136*, 27–31.

Mortimer, N., F. J. Davey, A. Melhuish, J. Yu, and N. J. Godfrey (2002), Geological interpretation of a deep seismic reflection profile across the Eastern Province and Median Batholith, New Zealand: crustal architecture of an extended Phanerozoic orogen, *N. Z. J. Geol. Geophys.*, *45*, 349–363.

Nathan, S., H. J. Anderson, R. J. Cook, R. H. Herzer, R. H. Hoskins, J. I. Raine, and D. Smale (1986)., Cretaceous and Cenozoic sedimentary basins of the West Coast region, South Island, New Zealand, *N. Z. Geol. Surv. Basin Stud.*, *1*, New Zealand Geological Survey, Lower Hutt.

Okaya, D., N. Christensen, D. Stanley, and T. Stern (1995), Crustal anisotropy in the vicinity of the Alpine Fault Zone, South Island, New Zealand, *N. Z. J. Geol. Geophys.*, *38*(4), 579–583.

Pulford, A., M. Savage, and T. Stern (2003), Absent anisotropy: the paradox of the Southern Alps orogen, *Geophy. Res. Lett.*, *30*(20), 2051, doi:10.1029/2003GL017758.

Pysklywec, R. N., C. Beaumont, and P. Fullsack, Lithospheric deformation during the early stages of continental collision: numerical experiments and comparison with South Island, New Zealand, *J Geophys. Res.*, *107*(B7), doi:10.1029/2001JB000252.

Reyners, M., (1987), Subcrustal earthquakes in the central South Island, New Zealand, and the root of the Southern Alps, *Geology*, *15*, 1168–1171.

Reyners, M. (1988), Reservoir-induced seismicity at Lake Pukaki, New Zealand, *Geophys. J. R. Astron. Soc.*, *93*(1), 127–135.

Reyners, M. (1989), New Zealand seismicity 1964–87: an interpretation, *N. Z. J. Geol. Geophys.*, *32*, 307–315.

Reyners, M. (2005), The 1943 Lake Hawea earthquake—a large subcrustal event beneath the Southern Alps of New Zealand, *N. Z. J. Geol. Geophys.*, *48*(1), 147–152.

Reyners, M. and H. Cowan (1993), The transition from subduction to continental collision: crustal structure in the North Canterbury region, New Zealand, *Geophys. J. Int. 115*(3), 1124–1136.

Reyners, M. E., R. Robinson, A. Pancha, and P. J. McGinty (2002), Stresses and strains in a twisted subduction zone: Fiordland, New Zealand, *Geophys. J. Int.*, *148*(3), 637–648.

Reyners, M., and E. Robertson (2004), Intermediate depth earthquakes beneath Nelson, New Zealand, and the southwest termination of the subducted Pacific plate, *Geophy. Res. Lett.*, *31*, L04607, doi:10.1029/2003GL019201.

Reilly, W. I. (1962), Gravity and crustal thickness in New Zealand, *N. Z. J. Geol. Geophys.*, *5*(2), 228–233.

Reilly, W. I., and C. M. Whiteford (1979), South Island: Bouguer anomalies, *Gravity Map of New Zealand 1:1,000,000*, DSIR, Wellington, New Zealand, 1 map.

Sandwell, D. T. and W. H. F. Smith (1997), Marine gravity anomaly from Geosat and ERS 1 satellite altimetry, *J. Geophys. Res., B, Solid Earth Planets*, *102*(5), 10,039–10,054.

Savage, M. K. (1999), Seismic anisotropy and mantle deformation: what have we learned from shear wave splitting?, *Rev. Geophys.*, *37*, 65–106.

Savage et al. (this volume).

Scherwath, M. (2002), Lithospheric structure & deformation in an oblique continental collision zone, South Island, New Zealand, PhD thesis, Victoria University of Wellington, Wellington, New Zealand.

Scherwath, M., T. Stern, A. Melhuish, and P. Molnar (2002), Pn anisotropy and distributed upper mantle deformation associated with a continental transform fault, *Geophys. Res. Lett.*, *29*(8), doi:10.1029/2001GLO141–79.

Scherwath, M., T. Stern, F. Davey, D. Okaya, W. S. Holbrook, R.. Davies, and S. Kleffmann (2003), Lithospheric structure across oblique continental collision in New Zealand from wide-angle P wave modelling, *J. Geophys. Res., B, Solid Earth Planets*, *108*(B12), doi:10.1029/2002JB002286.

Scherwath, M., T. Stern, F. Davey, and R. Davies (2006), Three-dimensional lithospheric deformation and gravity anomalies associated with oblique continental collision in South Island, New Zealand, *Geophys. J. Int.*, *167*, 906–916, doi:10.1111/j.1365-246X.2006.03085.

Shi, Y., R. Allis, and F. Davey (1996), Thermal modeling of the Southern Alps, New Zealand, *Pure Appl. Geophys.*, *146*(3/4), 469–501.

Sircombe, K. N., and P. J. J. Kamp (1998), The South Westland Basin: seismic stratigraphy, basin geometry and evolution of a foreland basin within the Southern Alps collision zone, New Zealand, *Tectonophysics*, *300*(1/4), 359–387.

Smith, E. G., C. Smith, T. Stern, and B. O'Brien (1995), A seismic velocity profile across the central South Island, New Zealand, from explosion data, *N. Z. J. Geol. Geophys.*, *38*, 565–570.

Stern, T. A. (1995), Gravity anomalies and crustal loading at and adjacent to the Alpine Fault, New Zealand, in *Origin of the Southern Alps II, Origin South. Alps Symp.*, vol. 38(4), 593–600.

Stern, T. A., P. E. Wannamaker, D. Eberhart-Phillips, D. Okaya, F. J. Davey, and S. I. P. W. Group (1997), Mountain building and active deformation studied in New Zealand, *Eos Trans. AGU*, *78*, 329, 335–336.

Stern, T., P. Molnar, D. Okaya, and D. Eberhart-Phillips (2000), Teleseismic P wave delays and modes of shortening the mantle lithosphere beneath South Island, New Zealand, *J. Geophys. Res.*, *105*, 21,615–21,631.

Stern, T., D. Okaya, S. Kleffmann, M. Scherwath, S. Henrys, and F. Davey (this volume), Geophysical exploration and dynamics of the Alpine fault zone.

Stern, T., S. Kleffmann, D. Okaya, M. Scherwath, and S. Bannister (2001), Low seismic-wave speeds and enhanced fluid pressure beneath the Southern Alps of New Zealand, *Geology*, *29*(8), 679–682.

Suggate, P. R., G. R. Stevens, and M. T. Te Punga (eds.) (1978), *The Geology of New Zealand*, New Zealand Government Printer, Wellington, New Zealand, 2 vols.

Sutherland, R. (1996), Gravity anomalies and magnetic lineations in the South Pacific, 1:15,000,000, *Geophysical Map/Institute of Geological and Nuclear Sciences*, *10*, Institute of Geological and Nuclear Sciences, Lower Hutt.

Sutherland, R., (1999), Cenozoic bending of New Zealand basement terranes and Alpine Fault displacement: a brief review, *N. Z. J. Geol. Geophys.*, *42*, 295–301.

Sutherland, R., F. Davey, and J. Beavan (2000), Plate boundary deformation in South Island, New Zealand, is related to inherited lithospheric structure, *Earth Planet. Sci. Lett.*, *177*, 141–151.

Sutherland, R. and A. Melhuish (2000), Formation and evolution of the Solander Basin, southwestern South Island, controlled by a major fault in continental crust and upper mantle, *Tectonics*, *19*, 44–61.

Townend, J. (1997), Estimates of conductive heat flow through bottom-simulating reflectors on the Hikurangi margin and southwest Fiordland, New Zealand, *Mar. Geol.*, *141*, 209–220.

van Avendonk, H. J. A., W. S. Holbrook, D. Okaya, J. K. Austin, F. Davey, and T. Stern (2004), Continental crust under compression: a seismic refraction study of SIGHT Transect I, South Island, New Zealand, *J. Geophys. Res.*, *B, Solid Earth Planets*, *109*(B6), doi:10.1029/2003JB002790.

Van Dissen, R., and R. S. Yeates (1991), Hope fault, Jordan thrust, and uplift of the Seaward Kaikoura Range, New Zealand, *Geology*, *19*, 393–396.

Walcott, R. I. (1998), Modes of oblique compression; late Cenozoic tectonics of the South Island of New Zealand, *Rev. Geophys.*, *36*(1), 1–26.

Wannamaker, P. E., G. R. Jiracek, J. A. Stodt, T. G. Caldwell, V. M. Gonzalez, J. D. McKnight, and A. D. Porter (2002), Fluid generation and pathways beneath an active compressional orogen, the New Zealand Southern Alps, inferred from magnetotelluric data, *J. Geophys. Res.*, *B, Solid Earth Planets*, *107*(6), doi:10.1029/2001JB000186.

Waschbusch, P., G. Batt, and C. Beaumont (1998), Subduction zone retreat and recent tectonics of the South Island of New Zealand, *Tectonics*, *17*(2), 267–284.

Wellman, H. W. (1979), An uplift map for the South Island of New Zealand, and a model for uplift of the Southern Alps, in *The Origin of the Southern Alps, Bull. R. Soc. N. Z.*, vol. 18, edited by R. I. Walcott and M. M. Cresswell, Royal Society of New Zealand, Wellington, New Zealand, pp. 13–20.

Wilson, C. K., C. H. Jones, P. Molnar, A. F. Sheehan., and O. S. Boyd (2004), Distributed deformation in the lower crust and upper mantle beneath a continental strike-slip fault zone: Marlborough fault system, South Island, New Zealand, *Geology*, *32*(10), 837–840, doi:10.1130/G20657.

Wilson, D., and D. Eberhart-Phillips (1998), Estimating crustal thickness in the central South Island, *Institute of Geological and Nuclear Sciences science report 98/27*, Institute of Geological and Nuclear Sciences, Lower Hutt, 61 pp.

Wood, R. A., R. H. Herzer, R. Sutherland, and A. Melhuish (2000), Cretaceous-Tertiary tectonic history of the Fiordland margin, New Zealand, *N. Z. J. Geol. Geophys.*, *43*(2), 289–302.

Woodward, D. J. (1972), Gravity anomalies in Fiordland, southwest New Zealand, *N. Z. J. Geol. Geophys.*, *15*(1), 22–32.

Woodward, D. J. (1979), The crustal structure of the Southern Alps, New Zealand, as determined by gravity, *The Origin of the Southern Alps, Bull. R. Soc. N. Z.*, vol. 18, edited by R. I. Walcott and M. M. Cresswell (eds.), Royal Society of New Zealand, Wellington, pp. 95–98 .

Woodward, D. J., and T. Hatherton (1975), Magnetic anomalies over southern New Zealand, *N. Z. J. Geol. Geophys.*, *18*(1), 65–82.

Zelt, C. A. and R. B. Smith (1992), Seismic traveltime inversion for 2-D crustal velocity structure, *Geophys. J. Int.*, *108*, 16–34.

S. Bannister, G. Caldwell, F. J. Davey, S. Henrys, GNS Science, Gracefield, Lower Hutt, New Zealand. (f.davey@gns.cri.nz)

D. Eberhart-Phillips, GNS Science, Dunedin, New Zealand.

M. D. Kohler, Center for Embedded Networked Sensing, University of California, Los Angeles, California, USA.

M. Scherwath, Leibniz-Institute of Marine Sciences, IFM-GEOMAR, Kiel, Germany.

T. Stern, School of Earth Sciences, Victoria University of Wellington, Wellington, New Zealand.

H. van Avendonk, Institute of Geophysics, University of Texas, Austin, Texas, USA.

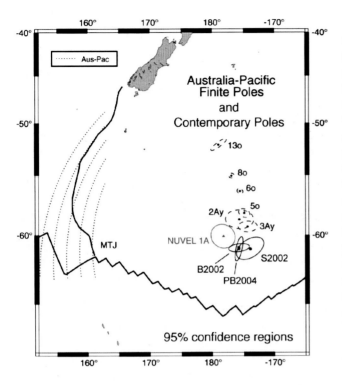

Figure 2. Australia-Pacific finite pole locations (dashed uncertainty ellipses) from *Cande and Stock* [2004], NUVEL-1A (grey), and contemporary GPS (solid black). B2002: *Beavan et al.* [2002]; S2002: *Sella et al.* [2002]; PB2004: *Prawirodirdjo and Bock* [2004]. The GPS poles differ significantly from Cande and Stock's 2Ay and 3Ay poles if the formal 95% confidence regions are used in the significance test. The GPS poles also differ from NUVEL-1A, but not as strongly.

et al. [2002] Euler vector has been confirmed in later work (with additional data) by *Prawirodirdjo and Bock* [2004]. The GPS pole of *Sella et al.* [2002] is even further southeast, but the fact that they find internal deformation in the Australian plate that is not found in the other studies makes their result questionable. Taken together, the GPS-derived Euler vectors are significantly different from the NUVEL-1A vector at the 95% confidence level. The GPS pole is, however, very close to an estimate of *Spitzak and DeMets* [1996] using fracture zones observed by satellite altimetry, but the Spitzak and DeMets rotation rate is some 3% faster [*Beavan et al.*, 2002]. More recently, an Euler vector of *Cande and Stock* [2004] using both magnetic anomalies and fracture zones, and including the motion since ~6 Ma of the previously-unrecognized Macquarie plate, give a Pacific-Australia pole since chron 2Ay (2.58 My) that is some 250 km northeast of NUVEL-1A and a similar distance north of the GPS poles, but again with a very similar rotation rate (Figure 2).

The implications of these different Euler vectors for dextral motion and shortening integrated across South Island are significant. Evaluated at Fox Glacier (43.5°S, 170°E) on the central section of the Alpine Fault (Table 1) the Cande and Stock shortening component since 2.58 Ma is 6.5±1.8 mm/yr, compared to 9.1±1.5 mm/yr from contemporary GPS, where the uncertainties are quoted at 95% confidence. These rates are barely different at 95% confidence, but there are potential problems with the lower rate when compared to geological evidence. First, the geologically determined dip-slip rate on the central Alpine Fault is as much as 8-12 mm/yr [*Norris and Cooper*, 2001], though these maximum dip-slip rates appear to be confined to a ca. 50-km long zone west of the highest part of the Southern Alps; rates decrease to ca. 6 mm/yr to the north and taper to zero to the south [*Norris and Cooper*, 2001]. *Little et al.* [2005] discuss the higher uplift rates, exhumation, and narrower width of the central Southern Alps compared to surrounding regions, and suggest that a restraining bend along the Alpine Fault ramp may enhance local convergence and uplift rates there. However, even if the lower dip-slip rate (north of the central region) of 6 mm/yr is taken as representative, neotectonic studies in the eastern foothills of the southern Alps indicate additional shortening there of >1 mm/yr [e.g., *Van Dissen et al.*, 1994; *Pettinga et al.*, 2001; *Upton et al.*, 2004]. Second, *Cande and Stock*'s [2004] stage poles imply that only about 40±15 (95%) km of shortening is available to build the central Southern Alps over the past ~6 My, substantially smaller than the ~70 km inferred at the same latitude by *Walcott* [1998] (though in both cases there is some shortening at a slower rate prior to 6 Ma). Most modeling studies of long-term development of the Southern Alps have used shortening rates of ~10 mm/yr and total convergence of ~100 km to obtain consistency with various observed phenomena. However, it is likely that by adopting relatively modest changes to the thermal, rheological, and erosion-rate assumptions, such models could also fit the observed phenomena with slower rates and smaller total shortening. Such models are therefore unlikely to provide a way to differentiate between shortening rates in the 4–10 mm/yr range, nor total convergence in the 40–100 km range.

The higher dextral rate of the GPS estimate means that ~3 mm/yr more dextral motion must be accommodated off the Alpine Fault. The kinematic inversion of the GPS velocities by *Wallace et al.* [2006] show that it is possible to accommodate this motion without violating any geological constraints. If the *Wallace et al.* [2006] inversion is repeated using the Cande and Stock 2Ay Pacific-Australia Euler vector, and omitting GPS data from the rigid interior of the Pacific plate (to avoid introducing present-day data in conflict with the Cande and Stock plate motion boundary condition), the results for Alpine Fault slip rate and locking distribution are almost unchanged from those in *Wallace et al.* [2006]. However, the slip rate is significantly reduced on the faults along the eastern foothills of the Alps. Though this result is in better accord with geological estimates for slip on these faults, the overall model has significantly worse agreement with the GPS velocity data in South Island, with the goodness-of-fit statistic, χ_n^2, averaging 50% higher in eastern South Island.

We conclude that either the Australia-Pacific plate motion rate has accelerated over the past 2.6 My, or that there are significant short-term fluctuations in plate-motion rates (so that present-day GPS between plate interiors does not measure the plate motion averaged over tens or hundreds of thousands of years), or that there is a systematic error in GPS velocity estimation, or that there is a systematic error somewhere in the series of assumptions and inferences through which Cande and Stock's Euler vectors are derived. Of these possibilities, it seems unlikely that a systematic error of the size necessary could be present in the GPS data, as the site velocities at all sites and on both plates are generated in a consistent way, and the distribution of sites across the plates is reasonably uniform.

3. DEFORMATION IN PLATE BOUNDARY ZONES

The measurement of deformation in plate boundary zones has been revolutionized since the 1980s by the application of space geodetic methods, principally GPS because of its relatively low cost and ease of use. Surveys repeated at intervals ranging from months to years and, more recently, continuously-operating GPS stations, enable the velocities of survey marks to be measured with mm/yr precision in the horizontal, and about a factor of three worse in the vertical. Plate tectonic velocities, which range up to tens of mm/yr or higher, are thus easily resolved. More importantly, their spatial distribution and temporal variability can be measured with sufficient accuracy to potentially provide constraints on competing models of plate boundary deformation processes.

Continuous velocity and strain-rate fields can be generated from GPS site velocities by a number of methods. The method most commonly used for New Zealand data is derived from the work of *Haines* [1982] and *Haines and Holt* [1993], which treats the continental crust as if it deforms like a thin viscous plate. The method was adapted for GPS data by *Beavan and Haines* [2001] who derived deformation-rate maps based on nationwide data collected through 1998.

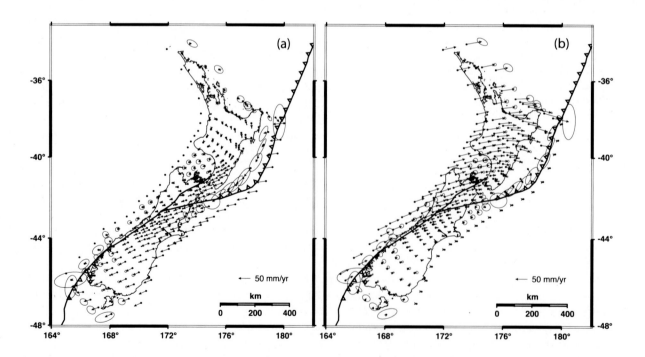

Figure 3. Contemporary horizontal velocity field relative to (a) Australian Plate, (b) Pacific Plate, obtained by modeling repeated GPS observations between 1991 and 2005 using the method of *Beavan and Haines* [2001]; c.f. Figure 5 of that paper. The velocities at the corners of the model grid cells are plotted. 95% confidence ellipses are shown, based on a white noise model scaled so that the χ_n^2 of the model fit to the input velocity data is about 3 (this is a crude way of allowing for colored noise in the GPS data).

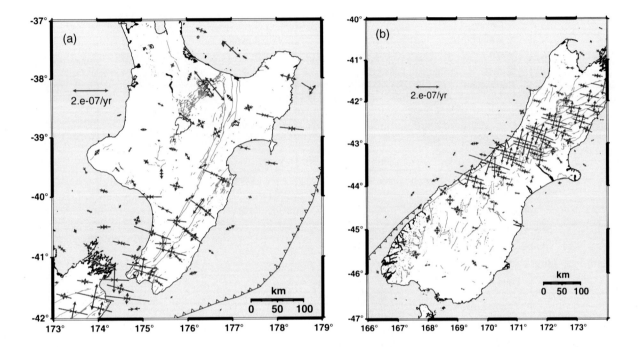

Plate 1. Principal strain rates from the model of Figure 3, but without relative plate velocity constraints on the boundaries of the model. The strain rates are therefore due entirely to the GPS velocity observations and the smoothing imposed by the condition that average strain rate be minimized within each grid cell of the model. The balance between fitting input velocities and minimizing average strain rates is chosen so that the χ_n^2 of the model fit to the data is about 3. It is chosen to be >1 to allow for the fact that the uncertainties on the input velocity data are calculated from a white noise model. The strain rates are averages over each grid cell and are plotted at the center of each cell.

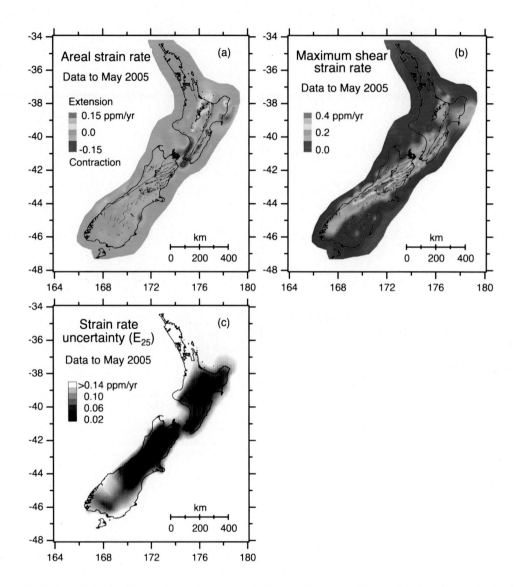

Plate 2. Images of (a) areal strain rate, (b) maximum (engineering) shear strain rate, and (c) strain-rate uncertainties based on a 25 km length scale. These correspond to the principal shear strain rates of Plate 1, and are updated from Plate 3 of *Beavan and Haines* [2001].

Since then, substantially more data have been collected, and updated versions of the velocity and strain-rate maps are given in Figures 3 and Plates 1 and 2. Figure 3 [cf. Figure 5 of *Beavan and Haines*, 2001] shows the velocity field with plate motion constraints from *Beavan et al.* [2002] applied on the boundaries of the model. Plates 1 and 2 [cf. Figure 6 and Plate 3 of *Beavan and Haines*, 2001] show the strain rates with free boundaries, so these strains result entirely from the velocity data and smoothness constraints. If plate motion boundary conditions are applied, there is little difference to the strain rates within the onshore GPS network, but the model requires additional straining offshore. This is an indication of present-day deformation offshore of New Zealand, which is also found from other methods of interpreting the GPS data. Plates 1 and 2 show the total strain rate at the present time. During interseismic intervals this is likely to include a substantial contribution from elastic strain due to locking or partial coupling between the surface and some depth on major faults. (We use the terms locking and coupling as a purely kinematic description of the present-day interseismic slip rate, or lack of slip, on fault surfaces; see *Wang and Dixon*, 2004, for further discussion.) There may also be strain transmitted through the elastic crust from viscously-deforming material below. Some workers [e.g., *Walcott*, 1979, 1998] have suggested that the upper crust can also deform steadily and aseismically in the long term, and such strains, if present, would also be recorded in the present-day strain-rate field.

The strain-rate maps show more detail and have lower uncertainties than those in *Beavan and Haines* [2001] because of the near doubling of the time period of GPS observations and the substantial improvement in spatial density and distribution of site velocities (more than 800 sites compared to fewer than 400). Many features between the new and old strain maps are similar, but appear with improved spatial resolution. As well as the clear WNW-ESE extension in the Taupo Volcanic Zone (Figure 1b), there is a zone of dilatation continuing SSW in the direction of Wanganui. There is a clear zone of WNW-ESE contraction running up the eastern North Island as far as Gisborne, and a small region of extension further north on the Raukumara Peninsula. The pronounced contraction in the southernmost North Island and across Cook Strait remains, and the regions of approximately east-west contraction in northwest Nelson and along much of the South Island east coastal region south of Christchurch are accentuated in the new maps. Deformation associated with the Alpine Fault is highlighted by a <50 km wide zone of high shear strain (>0.3 ppm/yr). South of Haast, the intensity of shearing diminishes; this is in part because the straining becomes less localized in this region and in part because of a lack of GPS data close to the southern Alpine Fault.

Many studies of coseismic and interseismic displacement fields have successfully explained the observed deformation under the assumption that the Earth's crust behaves as an elastic medium. The results of these studies in New Zealand are reviewed later. One model that has proven particularly useful in recent years, in large part because it interprets the velocity field over a 2-D surface rather than along 1-D profiles, is the rotating elastic block model. Numerous geological studies indicate that the majority of permanent deformation at the ground's surface in plate boundary zones can be accounted for by slip on faults. These faults bound what are often approximated as elastic crustal blocks that remain coherent over long time periods (i.e., many earthquake cycles). This "elastic block" assumption has been successfully implemented in the interpretation of GPS velocity fields from plate boundary zones around the world [e.g., *McClusky et al.*, 2001; *McCaffrey*, 2002, 2005; *McCaffrey et al.*, 2000; *Meade and Hager*, 2005; *Wallace et al.*, 2004, 2005]. The long-term rotation of these blocks is detected by GPS methods, but this motion is modulated by the effects of elastic strain fields due to variable coupling on the block boundaries (faults). If the block-bounding faults creep steadily at the long-term fault slip rate, as is observed at a very limited number of sites around the world, there will be no elastic strain affecting the GPS velocities. More usually, the faults are locked or partially coupled for some "interseismic" time period, then slip suddenly in large earthquakes. During the interseismic interval, elastic strain accumulates in the region surrounding the fault, leading to reduced (relative to the long-term) motion between the two sides of the fault. The coupling on a fault surface is well described by the coupling ratio, ϕ, which takes the value 0 where the fault is currently slipping at its long-term rate, and 1 where the fault is currently not slipping at all. ϕ may vary both along strike and down dip; for most faults it is 1 at the surface and decreases to zero at some depth (though values outside this range are possible if a fault is temporarily slipping faster than, or in the opposite direction to, its long-term rate).

In such a model, the long-term motions and elastic strains may be separated by their different behaviors: the long-term motion is described everywhere on the block by an angular velocity vector defining the rigid-body rotation of the block, while the velocity differences due to the elastic strain field reduce with distance from the fault. The GPS velocity field can thus be mathematically inverted to estimate simultaneously the distribution of locking on the fault surfaces and the angular velocities describing the relative rotation of tectonic blocks [e.g., *McCaffrey*, 2002].

The calculated rotation axis may be within or outside the block. When the axis is within or close to the block, the rotation of that block relative to nearby blocks is generally obvious in the velocity field. When the axes of two adjacent blocks are each far from the blocks, the relative motion between the two blocks will appear as a translation. For blocks with a horizontal dimension comparable to the fault locking depth, there are significant correlations between the fault slip rate and the block rotation rate, so it is not possible to uniquely separate the two signals without additional information.

During earthquakes, the strain accumulation in the crust surrounding the fault is, suddenly, more or less reversed. Following large earthquakes, postseismic deformation is often observed and is variously attributed to afterslip on parts of the fault plane (within or adjacent to the original rupture area) that did not rupture in the main earthquake, viscous flow in mid-lower crustal or deeper rocks, or poro-elastic effects due to fluid flow (usually of water) in the changed stress field. These signals are all observable at the surface by GPS methods, and it is a significant challenge to use the observed data to identify the causative mechanism.

In recent years "silent earthquakes" have also been identified in plate boundary zones. In these, fault slip occurs at rates higher than normal plate motion but much slower than normal earthquake slip rates so that fault motion is accomplished without destructive ground shaking. To date, the vast majority of these events have been observed in subduction zone settings, so they may not be an important feature of central South Island.

How the deformation observed at the surface relates to the deformation occurring at depth is controversial [e.g., *Savage*, 2000]. For example, in the context of the elastic block model, do the crustal block boundaries penetrate into the mantle as discrete shear zones, are they converted into wide deformation zones at depth, or are they decoupled from motion in the mantle below? And what forces drive the motion of the crustal blocks? Is it a general rule that the driving forces are edge forces due to far-field plate motion? Or that the forces are due to viscous coupling from below? Or is there a range of different behavior, with each tectonic situation needing to be analyzed on a case-by-case basis? And what is the role of gravitational forces due to density gradients and topography?

4. FAULT LOCKING AND SLIP RATES FROM HORIZONTAL DEFORMATION

We review the results of geodetic studies in the central Southern Alps since the early 1990s. We do not review studies in the Marlborough region [e.g., *Bourne et al.*, 1998] as these are in a complex transition region north of the area studied in the SIGHT project, which is a primary focus of this volume.

Pearson et al. [1995] compared shear strain rates derived from a 1992 GPS survey and a 1978 terrestrial survey along a ~40 km wide transect (Arthur's Pass; Figure 1b) projected onto a profile just south of the Hope Fault-Alpine Fault junction. Forward modeling with the Alpine Fault dipping 50° SE and a locking depth of 12 km provided a fair fit to the shear strain rates associated with strike-slip motion and a poor fit to those associated with dip-slip motion. The spatial resolution and uncertainty of the shear strain rate estimates were such that a wide range of model locking depths and slip rates would have provided equally good fits. The closeness of the survey to the Marlborough faults, even though these were crudely allowed for in the analysis, also makes the interpretation of a linear profile problematic in this region. An important result from this study was that only about 2/3 of the shear strain rate expected from NUVEL-1A relative plate motion is accommodated in the vicinity of the Alpine Fault, with substantial shear strain being accumulated further to the east, most probably in the Porter's Pass-Amberley fault zone (PPAFZ, Figure 1b).

Beavan et al. [1999] interpreted the velocity field from a number of GPS surveys between 1994 and 1998 in a ~100-km wide swath across South Island and central Southern Alps between Jacobs River and Waitaha River (Central South Island network; Figure 1b), a region where the characteristics of the Alpine Fault were believed to be fairly uniform. The velocity field was found to be fairly uniform along strike of the Alpine Fault, so was projected onto a single strike-normal profile before interpretation in order to increase spatial resolution in this direction. The use of GPS-GPS rather than GPS-terrestrial surveying provided a large increase in precision, and allowed inversion of the surface velocities for fault slip rates and locking depths, which was considered superior to forward modeling. Interpretation of the data as uniform slip below the locking depth on a single fault plane was not favored because this gave significantly larger misfits between model and data than other models. It was clear, as in the Arthur's Pass case [*Pearson et al.*, 1995], that significant strain is accumulating well east of the Alpine Fault. *Beavan et al.* [1999] modeled this using a two-fault model, one being the Alpine Fault and the other dipping in the opposite direction with a deeper locking depth and reminiscent of *Wellman*'s [1979] conceptual idea, later investigated in models of Southern Alps deformation by *Beaumont et al.* [1996] and *Gerbault et al.* [2003]. Beavan et al. acknowledged that this second fault was only one way of matching the broader surface deformation, and other approaches were investigated by *Moore* [1999] and *Moore et al.* [2002]. The results for the Alpine Fault were strike-slip rates rather lower than those inferred geologically, dip-slip rates rather higher, and a very shallow locking depth of 5–8 km. This locking depth was considered plausible because small and moderate earthquakes in the region are generally located shallower than 12 km [*Leitner et al.*, 2001; see Plate 4] and the high heat flow in the immediate vicinity of the Alpine Fault [*Allis and Shi*, 1995] could make the locking depth on the fault even shallower than this.

Pearson et al. [2000] modeled surface velocities from two GPS surveys in 1995 and 1998 that cross the Southern Alps in a ~20-km wide transect (Hawea-Haast; Figure 1b) just north of where the fault is believed to become vertical and the dip-slip motion ceases. The survey network trends significantly southward to the southeast of the Alps, so there is some question about the along-strike uniformity of the observations when they are projected onto a strike-normal profile for interpretation. The way of defining the reference frame, by holding a single station in Dunedin fixed between surveys, meant that a component of reference frame rotation was included in the modeled velocity field; the effect of this, however, is likely to be small. The modeling was done by grid search on slip rate and locking depth with other parameters held fixed, first for a one-fault model and then for a two-fault model similar to the central Southern Alps case. The fault slip results were similar to those in the central Southern Alps [*Beavan et al.*, 1999] though less well constrained because of fewer data, larger uncertainties, and poorer network geometry. A one-fault model implied a locking depth of ~20 km, greater than the 12-km maximum depth of small and moderate earthquakes, so a two-fault model similar to the Central South Island case was preferred. In this model, the Alpine Fault was best fit with a 10±2 km locking depth, strike-slip rate slightly slower than geological estimates and dip-slip rate rather faster.

The previous three studies all suffer from interpreting the velocity or strain-rate field projected onto a profile. It is particularly unlikely that the assumption of a uniform along-strike velocity field is satisfied in the Arthur's Pass or Hawea-Haast cases, as these are close to significant along-strike changes in the Alpine Fault and Southern Alps. However, it is also possible that there is along-strike variation in the velocity field in the Central South Island network. With the collection of substantial additional GPS data since the pioneering studies, *Wallace et al.* [2006] interpreted the 2-D velocity field of South Island in a single 3-D model that solves for long-term rotations of crustal blocks, coupling distribution along the faults forming the block boundaries, and optionally uniform strain rates within the blocks [*McCaffrey*, 2002, 2005] As well as the GPS data themselves, geological fault slip rate and azimuth data and earthquake slip vectors were included in the inversion. Some of the geological slip rates were applied as hard constraints; this was particularly necessary in regions like Marlborough,

where the presence of several adjacent, sub-parallel faults makes the non-uniqueness of the inversion problem more severe than usual. Both along-strike and down-dip variations in coupling were solved for, with the only constraints being that the coupling ratio decreases monotonically with depth and becomes zero at the bottom of the model fault. Importantly, the analysis was done in an Australia-fixed reference frame with the Pacific relative plate motion [from *Beavan et al.*, 2002] applied as a boundary condition on the eastern side of the model, so that the model is fully consistent with present-day Pacific-Australia relative plate motion.

For the ~200 km of the Alpine Fault between the Haast and Kokatahi rivers (Figure 1b), *Wallace et al.*'s [2006] model uses a 45°–55° eastward dipping Alpine Fault as one block boundary, and a distributed network of faults following the eastern foothills of the Alps as another. Unlike *Beavan et al.* [1999] and *Pearson et al.* [2000], they do not explicitly use a dislocation on a deep northwest dipping fault to model the long wavelength deformation. An assumption of the block model is that some sort of steady deformation at depth takes up the relative motion between the blocks, and that this does not cause elastic deformation at the surface. In their best fitting model (which we will refer to as W1, and has $\chi_n^2 = 1.16$), *Wallace et al.* [2006] infer a coupling-ratio distribution for the Alpine Fault that averages 70–85% between the surface and 18 km depth, with the smaller coupling values occurring in the central 80 km of the fault (Plate 3a). Strike-slip rates are at the high end of geological estimates and dip-slip rates are similar to geological estimates. Even if it is assumed (not entirely accurately) that an average 70% locking to 18 km depth gives surface deformation equivalent to 100% locking to 12 km depth, this locking depth is substantially greater than the 5–8 km inferred by *Beavan et al.* [1999], though it is similar to the maximum depth of small and moderate earthquakes [*Leitner et al.*, 2001]. *Wallace et al.* [2006] also tested for distributed strain within the Southern Alps, to determine whether some slip could be accommodated on structures similar to the Main Divide fault zone (MDFZ, Figure 1b) [*Cox and Findlay*, 1995; the MDFZ itself is now thought to be inactive, S. Cox, personal communication, 2005] or other faults within the Southern Alps [e.g., *Little*, 2004], or indeed by some sort of aseismic deformation within that region. They found that up to 5 mm/yr of dextral motion could be accommodated in this way. Furthermore, this had the result of markedly reducing the coupling ratio on the Alpine Fault, particularly in the central 80 km between Jacob's River and Harihari (Figure 1b), and of bringing the inferred slip rate closer to the mid point of geological estimates (Plate 3b). This model, which we refer to as W2, is more consistent with the shallow locking depth of *Beavan et al.* [1999] but is a poorer fit ($\chi_n^2 = 1.63$) to the GPS data than model

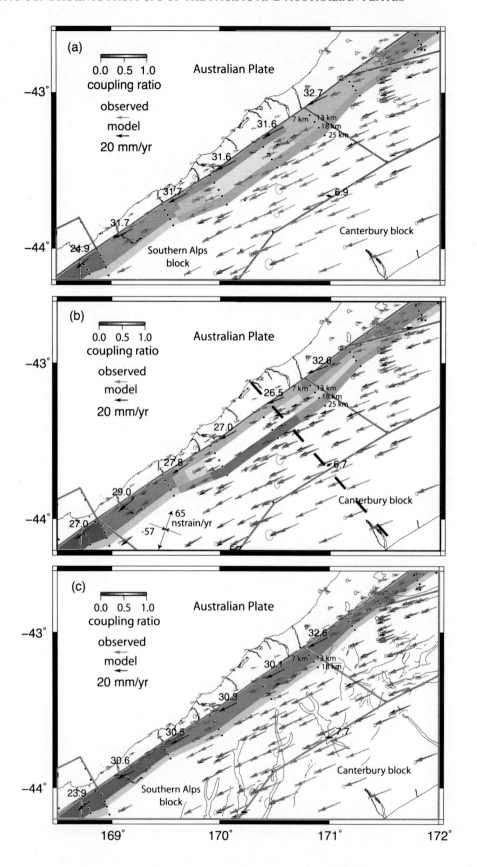

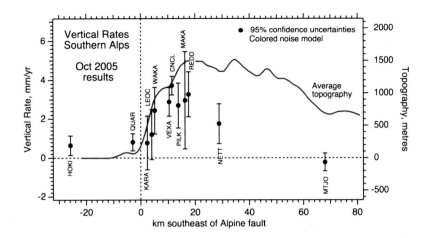

Figure 4. Vertical rates measured by continuous and semi-continuous GPS in a profile across the Southern Alps near Karangarua River (Figure 1b). The 95%-confidence error bars are based on a power-law + white noise model [*Williams et al.*, 2004; *Beavan*, 2005]. Only relative velocities are measured accurately by GPS so the vertical rates can all be adjusted higher or lower by a constant value. The zero on this plot was chosen by setting the average vertical velocity of two east coast sites to zero (these sites are in Christchurch and Dunedin and are off-scale to the right on this figure).

W1 – though the poor χ_n^2 is largely due to a few sites at the northern end of the straining Southern Alps block, perhaps indicating a model deficiency in that region.

The *Wallace et al.* [2006] study provides the most credible results to date for the distribution of coupling ratio and slip rate on the Alpine Fault for three reasons: (1) the superior data set available for analysis; (2) the along-strike variations in the surface velocity field being taken into account; and (3) the simultaneous estimation of coupling distribution on faults and long-term block motions (and therefore long-term fault slip rates). By allowing for variable coupling with depth and a small amount of distributed strain within the Southern Alps, they obtain Alpine Fault slip rates comparable to the geological estimates and coupling-ratio distributions comparable to the depth of small and moderate seismicity, without the explicit assumption of a deeper (>25 km) northwest-dipping fault beneath the Southern Alps. The *Wallace et al.* [2006] results are strong evidence in favor of long-term strain accumulation on the Alpine Fault that is

released by occasional large earthquakes. The nature of the few mm/yr distributed strain within the Southern Alps is unclear, although we believe it is most likely to be due to strain accumulation on a number of small faults [perhaps similar to the MDFZ or to the shear zone of *Little*, 2004] that will eventually fail in moderate earthquakes. But there is no way at present to rule out a few mm/yr of aseismic deformation at the surface in this region.

5. VERTICAL DEFORMATION

Vertical coordinates from GPS are subject to a variety of systematic errors, including satellite geometry, atmospheric and ionospheric effects, and electromagnetic environment around the antenna. In addition, vertical rates due to tectonic deformation are generally much less than 10 mm/yr, while horizontal rates are often an order of magnitude faster. To reduce systematic errors in campaign GPS vertical velocities, it is necessary to keep the antenna environment as

Plate 3. (Opposite)   Results from rotating, elastic block model of central South Island for three cases: (a) where Southern Alps block undergoes no long-term permanent strain (model W1); (b) where permanent distributed straining of the Southern Alps block is solved for as shown near bottom left of map (model W2); (c) similar to W1, but forcing the coupling ratio to trend to zero between 13 and 18 km depth, rather than between 18 and 25 km (model B3, which gives results virtually identical to model B1). Black arrows with numbers alongside them show the inferred long-term slip rate (mm/yr) on block-bounding faults in the central South Island. The red-white-blue image shows the distribution of coupling ratio (φ value, from 0 to 1) on the Alpine Fault. The slip rate deficit is φ multiplied by the relative block motion. Node locations in the model are shown as black dots. The fit of the GPS velocity data (red arrows with uncertainty ellipses) to the model (black arrows, often hidden beneath the red ones) is also shown. The dashed line in (b) shows the approximate location of the MT profile of Plate 4. Models W1 and W2 are from Figures 5 and 10 of Wallace et al. [2006].

constant as possible between surveys. This includes using the same antenna (or at least the same antenna type), setting the antenna the same height above the ground during each survey, and also analyzing the data using the same software with the same elevation cut-offs, mapping functions, and tropospheric models. For various historical reasons some of these precautions have not been taken with GPS campaigns in New Zealand, except for a repeat survey across Otago recently initiated by Otago University. Therefore, New Zealand GPS campaigns have not so far provided useful interseismic vertical rates (though useful coseismic vertical displacement data have been obtained for the 1994 Arthur's Pass and 2003 Fiordland earthquakes; see Section 8).

Continuous GPS avoids the worst of these pitfalls because the antenna environment remains much more constant. Apparent offsets at the times of equipment (e.g., antenna) changes can also be estimated and corrected from the daily coordinate time series available with continuous GPS. A profile of continuous GPS stations has operated across the central Southern Alps in the vicinity of Karangarua River (Figure 1b) since early 2000, and has been supplemented by "semi-continuous" GPS where 3 receivers are shared between 6 stations and data are collected for at least 3 months before the receiver is moved to another station. In this mode, the same antenna is always used at the same station, insofar as this is possible. Preliminary results from this network are given by *Beavan et al.* [2004], who found contemporary vertical rates up to 5-6 mm/yr relative to the east coast with uncertainties somewhat over 1 mm/yr at 95% confidence, using a power-law plus white noise model. A more recent analysis (Figure 4) using 5.5 years of data (rather than 3.5 years) gives slower vertical rates, the maximum being about 4 mm/yr relative to the east coast with uncertainties below 1 mm/yr at 95% confidence. The highest rates are several km west of the main divide of the Southern Alps, and the pattern of variation across the Alps is consistent with observed horizontal rates if a simple locked Alpine Fault model is adopted. The rates are also reasonably consistent with results from numerical models of Southern Alps evolution [*Gerbault et al.*, 2003; *Ellis et al.*, 2006a]. As more data are collected from this network, we expect reduced uncertainties in vertical rates and better ability to discriminate between competing models of tectonic deformation.

Vertical rates from absolute gravity would supplement those from GPS, and would provide additional information on the dynamics of the mountain building process. For example, are the mountains in isostatic equilibrium? Is the crustal root beneath the Alps still growing? A first-epoch absolute-gravity survey was made in 2000 at several of the CGPS sites, but has not been repeated at this writing.

6. UPPER MANTLE/LOWER CRUST BEHAVIOR

There is considerable debate as to whether distributed or localized deformation occurs in the lower crust and mantle lithosphere beneath central South Island [e.g., *Molnar et al.*, 1999 vs. *Gerbault et al.*, 2003; *Batt and Braun*, 1999; *Beavan et al.*, 1999]. Surface velocities from GPS cannot by themselves be inverted to provide a unique constraint on the behavior of the lower lithosphere [e.g., *Beavan et al.*, 1999; *Moore et al.*, 2002]. However, they can be used to place limits on length scales over which the lithosphere is presently deforming, and when combined with other geological and geophysical evidence they may help discriminate between competing models.

6.1 Style of Deformation in the Mantle Lithosphere

Both the normal and strike-slip components of surface velocity change significantly over a width of ca. 100 km on the eastern side of the Alpine Fault in central South Island [*Beavan et al.*, 1999; *Wallace et al.*, 2006]. High strain rates are focused within ~30 km of the central Alpine Fault [*Beavan and Haines*, 2001; Plate 2], but the zone of highest strain rates broadens and moves further east of the fault as the junction with the Marlborough faults is approached. *Ellis et al.* [2006a] determined under what conditions the various proposed styles of lithosphere behavior correctly predicted the surface velocity distribution normal to the central Alpine Fault. Although surface velocities could not be used to discriminate between deformation along a narrow shear zone ("mantle subduction" type models) and diffuse deformation within the mantle, they found that there is an upper limit of ca. 100 km for the width over which significant present-day mantle lithosphere deformation can be taking place. If mantle lithosphere were deforming over a significantly greater width than this, the decrease in surface velocities across the plate boundary would occur over a broader region than is currently observed.

A similar conclusion was reached on different grounds by *Wallace et al.* [2006], who used a block rotation approach to invert a combination of GPS, seismological and geological data. They observe anticlockwise vertical axis rotation of tectonic blocks in eastern South Island, opposite to the clockwise rotation that would be predicted if the crust were responding to significant present-day dextral mantle deformation reaching hundreds of kilometers east of the Alpine Fault. These conclusions are valid unless either (1) crust is almost completely detached from mantle, or (2) mantle lithosphere is significantly weaker than the crust [e.g., *Jackson*, 2002]. The <100 km width for deforming mantle predicted from GPS is compatible with the dimensions of a high-speed zone imaged directly beneath the Southern Alps from teleseismic P-wave delays [*Stern et al.*, 2000; c.f. Plate 4; *Stern et al.*, this volume]. In contrast, studies of shear-wave and Pn-anisotropy in the mantle (e.g., *Klosko et al.*, 1999; *Savage et al.*, 2004; *Baldock and Stern*, 2005; *Bourguignon et al.*, this volume) show anisotropy far to the east of the Alpine Fault and have been interpreted in terms of mantle deformation distributed over a wide region, >200 km and even as much as 400 km. The discrepancy may result from considering two entirely different timescales: the seismic anisotropy is measuring the total strain over the 20+ My history of deformation of the Alpine Fault in central South Island; the GPS is measuring the present-day strain that is accumulating. To fully understand what the anisotropy is telling us, it must be interpreted in terms of the thermo-mechanical evolution of the orogen [e.g., *Savage et al.*, this volume]. Anisotropy becomes saturated ("frozen in") with increasing strain, and this can lead to the interpretation that strain is distributed over a wide zone when in fact the bulk of the present-day strain is occurring in a narrower region. It is also possible that strain has become more localized in the lower lithosphere in the past few My [e.g., *Little et al.*, 2002], and that the anisotropic fabric is preferentially "sampling" the deep, weaker part of the mantle where deformation may indeed be more diffuse [*Pysklywec et al.*, 2002].

6.2 Style of Deformation in the Lower Crust

Similar analyses of GPS data to those described above have been used to demonstrate that the lower crust cannot be deforming over a diffuse shear zone more than 30 km wide [*Moore et al.*, 2002]. Other evidence constrains this even further. Seismic velocity anomalies imaged by SIGHT [*Stern et al.*, 2001] and magnetotelluric phase anomalies interpreted as regions of interconnected fluid [*Wannamaker et al.*, 2002] extend down-dip from the Alpine Fault into the mid-lower crust. Rocks exhumed from depths >20 km up the Alpine Fault show a localized mylonite zone and associated hanging-wall structures that appear to have absorbed most of the plate boundary strain over a width of <4 km (and possibly <1 km) [*Norris and Cooper*, 2003; *Little et al.*, 2005]. It is therefore likely that a large proportion of plate boundary strain in the lower crust is taken up in a narrow shear zone representing a down-dip continuation of the Alpine Fault. The observable strain and topography on the eastern side of South Island, however, does indicate a small amount of diffuse deformation, which may be transmitted via a sub-horizontal region of shear in the lower crust [e.g., *Norris et al.*, 1990].

7. COMBINING LONG-TERM EVOLUTION OF PLATE BOUNDARY COLLISION WITH THE EARTHQUAKE CYCLE

Repeated GPS observations measure the interseismic strain currently accumulating across central South Island. This must be combined with paleoseismological data on fault studies, and integrated over several seismic cycles, in order to understand how it leads to accumulation of finite strain and development of topography and crustal thickening. Many geodynamic models of the Southern Alps exist that do not include elastic processes [e.g., *Beaumont et al.*, 1996; *Batt and Braun*, 1999], and these cannot be compared directly to interseismic GPS observations. Geodynamic models that include both elastic and inelastic properties of the crust as well as embedded faults can be used to integrate deformation through one or many seismic cycles. There are two different approaches: (1) A long-term model is constructed and run for millions of years; fault motion is approximated as steady frictional creep, and in some cases a short-term seismic cycle may be imposed on the longer-term model (where the fault is locked) in order to investigate interseismic deformation to compare with GPS measurements [e.g., *Gerbault et al.*, 2003]; (2) The current crustal structure and fault configuration of the orogen can be used as a starting condition on which the seismic cycle is imposed [e.g., *Chéry et al.*, 2001 (San Andreas Fault); *Vergne et al.*, 2001 (Himalayas); *Ellis et al.*, 2006b (Southern Alps)].

Gerbault et al. [2003] modeled the 2D evolution of the central Southern Alps over 7 My of collision in order to investigate the effects of material behavior, boundary conditions, and surface processes. The Alpine Fault was modeled as a frictionally weak zone (rather than a discrete surface) so could not be "locked" to mimic an interseismic interval. Instead, the average surface velocities over one timestep (30 years) at the end of the 7 My interval were compared to the GPS velocity profiles of *Beavan and Haines* [2001]. The qualitative trend in the GPS data showing a drop-off in convergent velocity over ca. 100 km to the east of Alpine Fault was reproduced, although the total convergence rate was different.

Ellis et al. [2006b] started from the present-day configuration of the central Southern Alps and modeled both normal and strike-slip components of motion. The models were explicitly designed to investigate each part of the seismic cycle, with the Alpine Fault represented by a dipping contact surface with a prescribed frictional strength. Seismic cycles were simulated by locking the fault for 500 years and then releasing it for a short interval to simulate the static effects of an earthquake. The model fault experienced stress drops of several MPa along its upper (brittle) surface during a model earthquake, while at the same time stresses increased in the mid-crust, loading the mid- and lower-crust. The effect was to enhance ductile strain localization beneath the brittle fault, extending downwards towards the base of the crust. Predicted slip magnitudes along the fault during the seismic interval were compatible with paleoseismological estimates, and predicted interseismic velocities were in good agreement with those measured by GPS. The utility of the model was that it provided insight into the relative contributions of elastic vs. inelastic deformation during the seismic cycle. Despite many simplifying assumptions (particularly regarding thermal structure, which is still poorly constrained), this approach holds considerable promise for integrating GPS results with other data at a variety of time scales.

8. COSEISMIC DEFORMATION

Two earthquakes with significant coseismic signals have been observed in central South Island in the GPS era, as well as one in Fiordland [*Reyners et al.*, 2003]. The 1994 M_w 6.7 Arthur's Pass and 1995 M_w 6.2 Cass earthquakes occurred in the region where the PPAFZ trends towards the Alpine Fault (Figure 1b), and they perhaps represent part of the process by which the PPAFZ is becoming the next fault in the Marlborough fault system.

There was only one pre-earthquake GPS survey prior to the 1994 event, 1.5 years earlier in December 1992. Immediate post-seismic observations were limited to part of the network because of midwinter conditions. This meant that corrections for interseismic strain accumulation depended on assumptions rather than observation, and assessment of early postseismic deformation was limited. However, these effects are probably small compared to the maximum observed surface displacement of 500 mm. *Arnadottir et al.* [1995] modeled the horizontal and vertical displacement by slip on a NNW-trending left-lateral "cross fault" between the PPAFZ and the Alpine Fault. The majority of slip was on the northern part of the fault plane near the epicentral area. Later inversions of seismic data found convincing evidence that at least the early part of the rupture was on a NE-trending thrust. It has been possible to partially reconcile the GPS and seismic inversions by a model consisting of slip on a NE-trending thrust followed within seconds by slip on an adjacent NNW-trending strike-slip zone, but the relative proportion of slip on the two fault planes remains a difficulty [*Abercrombie et al.*, 2000]. Aftershock locations by *Robinson and McGinty* [2000], now refined using double-differencing techniques by *Bannister et al.* [2006], accentuate a series of short, ENE-trending, en-echelon fault segments rather than a continuous NNW-trending strike slip zone. It

is therefore possible that the apparent left-lateral fault slip was in fact due to a series of short, right-lateral en-echelon faults. Whichever fault plane model is chosen, however, the horizontal projection of the P-axis is consistent at 115°–120° on all fault segments, suggesting that the earthquake is a response to a consistent regional stress field. A poor quality (because of topography and vegetation among other issues) interferogram was obtained from JERS radar images before and after the earthquake [*Pearson et al.*, 1999] but this was unable to provide a useful constraint on the main shock. It did however, indicate that the largest (M_L 6.1) aftershock was due to right-lateral slip on the ENE-trending Harper Fault near the southern end of the aftershock zone.

Pre-earthquake GPS data existed in only one quadrant of the 1995 Cass earthquake, and the coseismic displacement observations were therefore unable to provide positive constraints on the faulting parameters. However, the GPS displacements were consistent with the fault parameters determined from seismic data [*Gledhill et al.*, 2000]: a thrust on a north-trending, west-dipping plane. The P-axis was consistent with that of the Arthur's Pass earthquake.

9. DISCUSSION

None of the GPS observations or results described herein were part of the NZ/US SIGHT project, but there has been a close interest from all investigators in the results from the various ongoing geological and geophysical experiments in South Island. In this section we attempt to bring together results where there are strong synergies between the GPS and other disciplines, and where the GPS may provide important constraints on future work.

Plate 4 summarizes some of the magnetotelluric (MT) and seismic results from SIGHT, overlaid with the faults used in the GPS block models and some geological information. The high conductivity regions inferred from the MT data have previously been interpreted as due to fluids arising from prograde metamorphism in a thickening crust, with the conductivity enhanced by connected pore space due to shear deformation and, nearer the surface, by fracturing [*Wannamaker et al.*, 2002]. The high conductivity regions correlate well with the block boundaries used in the GPS-based model, and these boundaries were chosen without reference to the MT results. We discuss this further below.

9.1 Alpine Fault

Wallace et al.'s [2006] best-fitting model (W1; Plate 3a) finds a long-term slip rate on the central section of the Alpine Fault of about 32 mm/yr, with a reduction in coupling ratio on the fault from 100% at the surface to about 70% at

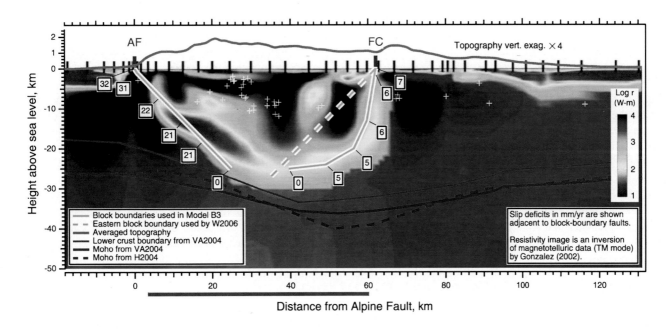

Plate 4. Summary of the GPS block model fault coupling results overlaid on seismicity (white crosses) and some of the magnetotelluric and active seismology results from the SIGHT project. The MT image [*Gonzalez*, 2002] assumes conductivity is isotropic beneath the Southern Alps. The red lines are the estimated Moho from *van Avendonk et al.* [2004] (VA2004; solid line) and *Henrys et al.* [2004] (H2004; dashed line), showing the root of the Southern Alps displaced well to the east of the highest topography. The brown line is the top of the lower crust from VA2004. AF shows the surface outcrop of the Alpine Fault, and FC the surface outcrop of the Forest Creek Fault. The boundaries of the Southern Alps block used by *Wallace et al.* [2006] (W2006) to interpret the GPS velocity data are shown as green lines on white. On the east side of the block, W2006 used a constant-dip boundary (dashed line). We have remodeled their best-fitting model (W1) by the same method, but using a boundary (solid line; model B3) that follows the high conductivity region; this fits the data just as well as the constant-dip boundary. The numbers beside these lines show the slip deficit rates in mm/yr that are causing elastic strain to build up in the surrounding crust; these are the total slip deficit rates, so can be viewed as the vector sum of the down-dip and along-strike components. The values shown for the Alpine Fault assume that the Southern Alps block undergoes little or no long-term permanent strain. If uniform permanent strain is allowed (model W2), the sub-surface values change from 31,22,21,21 mm/yr to 27,14,14,14 mm/yr. The choice of eastern boundary does not change the inferred distribution of coupling ratio or slip rate on the Alpine Fault. The earthquakes shown are from *Leitner et al.* [2001], and are those within 30 km of the MT profile. The purple line beneath the horizontal axis shows the approximate location of the high velocity body in the mantle ($v_p > 8.4$ km/s; ca. 80-180 km depth) inferred by *Stern et al.* [2000].

7 km depth. Between 7 and 18 km depth, adjacent to the highest conductivity region, the coupling ratio remains at about 70%. Below this the coupling ratio falls to zero, though this is constrained by the assumption of zero coupling (i.e., the fault or shear zone is slipping steadily at its long term rate) at 25 km depth. If the Southern Alps block is allowed to permanently strain (due to aseismic deformation or to earthquakes on known and unknown faults within the block) the long-term slip rate reduces to 27 mm/yr and the coupling-ratio values reduce from 70% to 50% (model W2; Plate 3b).

The *Wallace et al.* [2006] conclusions are somewhat different from those of *Beavan et al.* [1999] who modeled the GPS velocity data as a 2-D profile and assumed the Alpine Fault to be coupled from the surface to a certain locking depth, then slipping freely below that depth; Beavan et al. also allowed the along-strike and down-dip components of the coupling ratio to be different from each other. *Beavan et al.* [1999] determined the locking depth to be between 5 and 8 km. In both studies, there is a significant reduction in coupling below about 5–8 km depth, which implies that the top 5–8 km of the central Alpine Fault is likely to slip in the major earthquakes that recur every few hundred years with 4–8 m offsets (see review of previous work in *Sutherland et al.*, this volume).

The models of *Wallace et al.* [2006] imply that considerable additional elastic strain energy (50–70% of the long-term motion) is stored on the Alpine Fault down to 18 km depth. If the deeper part of the fault were to slip coseismically in Alpine Fault earthquakes, the earthquake magnitude would be substantially larger than in models where only the top 5-10 km of the fault is assumed to slip (see *Sutherland et al.*, this volume). However, because of the presence of high temperatures and high fluid pressures (see *Sutherland et al.*, this volume) we think it is likely that a substantial part of the strain stored on the deeper part of the fault would be released as afterslip following a major earthquake, rather than coseismically.

We investigate how well the *Wallace et al.* [2006] GPS data can be fit if the locking depth is forced to be shallower than in their best-fitting model, W1. In the W1 model, the coupling ratio is forced to trend linearly to zero between 18 and 25 km depth. We repeat the Wallace et al. analysis, forcing the coupling to trend to zero between (1) 13 and 18 km (Model B1), and (2) 7 and 13 km (Model B2). For model B1 the fit is not markedly worse, with the slip rate on the Alpine Fault decreasing slightly, that on the eastern boundary increasing slightly, the overall χ_n^2 increasing by 10%, and the χ_n^2 of the Eastern South Island block increasing by 30%. Model B2 violates geological constraints by requiring 8–9 mm/yr slip rate on the eastern boundary; also, the overall χ_n^2 increases by 30% and the χ_n^2 of the Eastern South Island block increases by 50%. For model B1 the coupling ratio is

0.85 between 1 and 13 km depth, slightly higher than in the W1 model. These results show that model B1, with fairly complete coupling to 13 km depth and a taper to zero coupling at 18 km depth, provides an acceptable fit to the GPS data – nearly as good a fit as model W1. Model B1 is also more consistent than model W1 with the observed distribution of moderate seismicity (Plate 4). This shows that with the presently available GPS data, we can achieve reasonable consistency with more-or-less full coupling to about 12 km depth on the Alpine Fault by at least two methods: (1) forcing the coupling to be shallower, as in our model B1; (2) allowing strain within the Southern Alps Block, as in model W2 from *Wallace et al.* [2006].

Other interpretations are possible for the deep locking depths obtained in the model W1 inversion. For a dislocation model that uses an elastic layer over a viscoelastic half-space (which may be a better model of the real Earth than an elastic half-space), the surface deformation pattern tends to spread out with time during the interseismic period. When interpreting a late-stage deformation pattern using an elastic half-space model, the true locking depth can be overestimated, perhaps by a factor of two or more for a vertical strike-slip fault [*Savage and Lisowski*, 1998]. However, *Ellis et al.*'s [2006b] seismic cycle models find this effect to be small for the Alpine Fault.

Finally, we note that the constant Alpine Fault dip assumed in the model is not well constrained by the GPS data. It is based on geological observations at the surface [e.g., *Berryman et al.*, 1992] and reflections from an inferred shear zone at about 25 km depth [*Kleffmann et al.*, 1998]. Modest variations in the shape of the fault as a function of depth provide equally good fits to the GPS data, and the conclusions on long-term slip rate and locking distribution do not change significantly.

9.2 Eastern Foothills

On the east side of the Southern Alps block *Wallace et al.* [2006] used a constant-dip boundary (shown dashed in Plate 4) that coincides at the surface with a network of discontinuous, low slip-rate faults that follow the eastern foothills of the Southern Alps. These are the Forest Creek Fault in the region of the MT transect, the Lake Heron Fault to the north, the Irishman's Creek Fault to the south, and a number of others (Figure 1b). (The Irishman's Creek Fault is known to dip SE, but could be a splay off a deeper NW-dipping fault; *Long et al.*, 2003]. We use a single fault surface to represent these known discontinuous faults because the spatial density of GPS velocities is insufficient to discriminate multiple nearby faults. In fact, we expect that there are additional faults within this fault zone that are yet to be discovered. The known faults are well expressed geologically with ca. 1 mm/yr shortening, but possible strike slip motion on the faults is harder to detect. Wallace et al.'s best-fitting model (W1) gives the requisite shortening but also implies ca. 5 mm/yr of right-lateral strike slip within the fault zone. Since the individual faults in the fault zone tend to be oriented counter-clockwise of the zone itself (towards the north-south direction), some of the inferred strike slip could be expressed as thrusting on these faults; nevertheless, the rates are still significantly higher than is presently recognized geologically.

The MT results show a distinct sub-vertical region of high conductivity beneath the fault zone. One interpretation of this is the presence of a fracture network that allows easy passage of fluids from depth [*Wannamaker et al.*, 2002]. Whether the fractures were caused by hydrofracturing due to the rise of fluids, or by fracturing due to earthquakes on upper-crustal faults, the two processes working together are likely to produce a zone of weakness where both faulting and fluid flow are concentrated. A sub-vertical high-conductivity region also occurs ca 5–10 km south-east of the Alpine Fault (Plate 4). This is interpreted by *Wannamaker et al.* [2002] as escape of fluids at lithostatic pressure along a fault-fracture mesh at the locus of the brittle-ductile transition. We do not place our western block boundary through this feature, as our boundary is constrained to the surface expression of the Alpine Fault. However, such a fault-fracture mesh could contribute to the internal strain in the Southern Alps block as discussed in Section 9.1.

Using the same approach as *Wallace et al.* [2006], we have remodeled the GPS data using an eastern boundary of the Southern Alps block that follows the high conductivity zone at depth (Plate 4; Model B3). As expected (due to the relatively low slip rate on this boundary and the relative sparseness of GPS sites in its vicinity) the data can be fit just as well with this boundary as with the planar boundary, and the inferred coupling distribution and slip rate on the Alpine Fault are not changed.

9.3 Crustal Deformation and Geodynamic Models

We are therefore finding a relatively consistent story that is also in line with many geodynamic models of the Southern Alps that predict a doubly-vergent zone of crustal deformation [e.g., *Norris et al.*, 1990; *Beaumont et al.*, 1996; *Gerbault et al.*, 2003]. Incoming Pacific upper-crust (and mid-crust?) is mainly transported up the ramp of the Alpine Fault, but there is also a westward-dipping distributed zone of bending and faulting near the eastern foothills of the mountains. As crustal material passes into this zone new faults are formed and the earlier ones pass through and reduce in activity (or, deformation may continue on these faults if they accommodate distributed deformation within the Southern Alps

themselves). Because of their limited lifetime the eastern foothills faults are therefore unlikely to develop into major through-going features like the Alpine Fault. Most geodynamic models have considered only the shortening component of deformation. The much larger along-strike motion that is present in central South Island takes advantage of the existing zones of weakness at the Alpine Fault and in the eastern foothills to accommodate this motion.

9.4 Lower Crust and Mantle

The nature of deformation in the lower crust and mantle is far less accessible to the surface deformation observations made by GPS. However, numerical models of the earthquake cycle superimposed on long-term deformation can only be made consistent with GPS surface velocity observations if the width of deformation zones in the mantle and lower crust satisfy certain constraints. The width of present-day shearing must be less than ~100 km in the upper mantle [*Ellis et al.*, 2006a], and less than ~30 km in the lower crust [*Moore et al.*, 2002].

10. CONCLUSIONS

Present-day AUS-PAC plate motion from GPS differs from "geological" estimates by a few mm/yr, and this difference is significant at the 95% confidence level. This discrepancy needs to be investigated further.

The GPS and other geophysical survey results (especially MT) across the central Alpine Fault and Southern Alps, together with geological observations, provide a fairly consistent story on the origin of the present-day surface deformation field. Furthermore, this is consistent with numerical models of the long-term evolution of the Southern Alps continental collision zone.

The present-day deformation is also consistent with the geological history of major Alpine Fault earthquakes every few hundred years. Coseismic slip on the central Alpine Fault during a major earthquake could well go significantly deeper than the 5–10 km assumed as a maximum in some recent work (*Sutherland et al.*, this volume). It is also likely that significant afterslip will occur to depths of 18 km or more in order to relieve the accumulated strains due to the partial coupling at these depths.

GPS is best suited to providing information on upper crust deformation, but some constraints on the width of contemporary deformation in the lower crust and upper mantle are possible.

Acknowledgments. The preparation of this paper was funded by the New Zealand Foundation for Research, Science and Technology. The GPS data since 1999 have been collected principally by

GNS Science and Otago University. We acknowledge our colleagues at Lamont-Doherty Earth Observatory, Oxford University, the Department of Survey and Land Information (now Land Information New Zealand) and Victoria University of Wellington, who were important participants in South Island GPS projects from 1992–1998. Thank you to Peter Molnar, Brad Hager and Tom Herring for allowing publication of Figure 4 using data from the SAGENZ US-NZ joint project, and to Donna Eberhart-Phillips for providing the seismicity data in Plate 4. Also to Grant Caldwell and Rupert Sutherland for discussions, and Grant Caldwell and Donna Eberhart-Phillips for their reviews of an earlier draft of the manuscript. We also thank Duncan Agnew and Thora Arnadottir for reviews that enabled us to significantly improve the paper.

REFERENCES

Abercrombie, R. E., T. H. Webb, R. Robinson, P. J. McGinty, J. J. Mori, and R. J. Beavan (2000), The enigma of the Arthur's Pass, New Zealand, earthquake 1. Reconciling a variety of data for an unusual earthquake, *J. Geophys. Res.*, *105*(B7), 16,119–16,137.

Allis, R. G., and Y. Shi (1995), New insights to temperature and pressure beneath the central Southern Alps, New Zealand, *N. Z. J. Geol. Geophys.*, *38*, 585–592.

Arnadottir, T., J. Beavan, and C. Pearson (1995), Deformation associated with the 18 June 1994 Arthur's Pass earthquake, New Zealand, *N. Z. J. Geol. Geophys.*, *38*, 553–558.

Baldock, G., and T. Stern (2005), Width of mantle deformation across a continental transform: Evidence from upper mantle (Pn) seismic anisotropy measurements, *Geology*, *33*(9), 741–744, doi: 10.1130/G21605.1.

Bannister, S., C. Thurber, and J. Louie (2006), Detailed fault structure highlighted by finely relocated aftershocks, Arthur's Pass, New Zealand, *Geophys. Res. Lett.*, *33*, L18315, doi:10.1029/2006GL027462.

Batt, G., and J. Braun (1999), The tectonic evolution of the Southern Alps, New Zealand: Insights from fully thermally coupled dynamical modeling, *Geophys. J. Int.*, *136*, 403–420.

Beaumont, C., P. J. Kamp, J. Hamilton, and P. Fullsack (1996), The continental collision zone, South Island, New Zealand: Comparison of geodynamical models and observations, *J. Geophys. Res.*, *101*(B2), 3333–3359.

Beavan, J. (2005), Noise Properties of Continuous GPS Data From Concrete-Pillar Geodetic Monuments in New Zealand, and Comparison With Data From U.S. Deep Drilled Braced Monuments, *J. Geophys. Res.*, *110*(B8), B08410, doi:10.1029/2005JB003642.

Beavan, J., and J. Haines (2001), Contemporary horizontal velocity and strain-rate fields of the Pacific-Australian plate boundary zone through New Zealand, *J. Geophys. Res.*, *106*(B1), 741–770.

Beavan, J., M. Moore, C. Pearson, M. Henderson, B. Parsons, S. Bourne, P. England, D. Walcott, G. Blick, D. Darby, and K. Hodgkinson (1999), Crustal deformation during 1994–1998 due to oblique continental collision in the central Southern Alps, New Zealand, and implications for seismic potential of the Alpine fault, *J. Geophys. Res.*, *104*(B11), 25,233–25,255.

Beavan, J., P. Tregoning, M. Bevis, T. Kato, and C. Meertens (2002), The motion and rigidity of the Pacific Plate and implications for plate boundary deformation, *J. Geophys. Res.*, *107*(B10), 2261, doi:10.1029/2001JB000282.

Beavan, J., D. Matheson, P. Denys, M. Denham, T. Herring, B. Hager, and P. Molnar (2004), A vertical deformation profile across the Southern Alps, New Zealand, from 3.5 years of continuous GPS data, in *Proceeding of the Workshop, The State of GPS Vertical Positioning Precision: Separation of Earth Processes by Geodesy, Cahiers du Centre Européen de Géodynamique et de Séismologie*, vol. 23, edited by T. van Dam and O. Francis, pp. 111–123, Centre Européen de Géodynamique et de Séismologie, Luxembourg.

Berryman, K. R., S. Beanland, A. F. Cooper, H. N. Cutten, R. J. Norris, and P. R. Wood (1992), *Ann. Tecton. Spec. Issue (Suppl.)*, *VI*, 126–163.

Bourguignon S., M. K. Savage, and T. Stern, Crustal thickness and Pn anisotropy beneath the Southern Alps oblique collision, New Zealand, *this volume*.

Bourne, S. J., T. Arnadottir, J. Beavan, D. Darby, P. C. England, B. Parsons, R. I. Walcott, and P. R. Wood (1998), Crustal deformation of the Marlborough fault zone in the South Island of New Zealand: Geodetic constraints over the interval 1982–1994, *J. Geophys. Res.*, *103*(B12), 30,147–30,165.

Cande, S., and J. Stock (2004), Pacific-Antarctic-Australia motion and the formation of the Macquarie Plate, *Geophys. J. Int.*, *157*, 399–414.

Chéry, J., M. D. Zoback, and R. Hassani (2001), An integrated mechanical model of the San Andreas fault in central and northern California, *J. Geophys. Res.*, *106*, 22,051–22,066.

Cox, S., and R. H. Findlay (1995), The Main Divide Fault Zone and its role in formation of the Southern Alps, New Zealand, *N. Z. J. Geol. Geophys.*, *38*, 489–499.

DeMets, C., R. G. Gordon, D. F. Argus, and S. Stein (1994), Effect of recent revisions to the geomagnetic reversal time scale on estimates of current plate motions, *Geophys. Res. Lett.*, *21*, 2191–2194.

Ellis, S., J. Beavan, J., and D. Eberhart-Phillips (2006a), Bounds on the width of mantle lithosphere flow derived from surface geodetic measurements: Application to the central Southern Alps, New Zealand, *Geophys. J. Int.*, doi:10.1111/j.1365-246X.2006.02918.x.

Ellis, S., J. Beavan, D. Eberhart-Phillips, and B. Stöckhert (2006b), Simplified models of the Alpine Fault seismic cycle: Stress transfer in the mid-crust. *Geophys. J. Int.*, doi:10.1111/j.1365-246X.2006.02917.x.

Gerbault, M., S. A. Henrys, and F. J. Davey (2003), Numerical models of lithospheric deformation forming the Southern Alps of New Zealand, *J. Geophys. Res.*, *108*(B7), 2341, doi:10.1029/2001JB001716.

Gledhill, K., R. Robinson, T. Webb, R. Abercrombie, J. Beavan, J. Cousins, and D. Eberhart-Phillips (2006), The M_w 6.2 Cass, New Zealand, earthquake of 24 November 1995: Reverse faulting in a strike-slip region, *N. Z. J. Geol. Geophys.*, *43*, 255–269.

Gonzalez, V. M. (2002), Magnetotelluric evidence for mid-crustal fluids in an active transpressive continental orogen, South Island, New Zealand, M.S. thesis, San Diego State University.

Haines, A. J. (1982), Calculating velocity fields across plate boundaries from observed shear rates, *Geophys. J. R. Astron. Soc.*, *68*, 203–209.

Haines, A. J., and W. E. Holt (1993), A procedure for obtaining the complete horizontal motions within zones of distributed deformation from the inversion of strain rate data, *J. Geophys. Res.*, *98*, 12,057–12,082.

Henrys, S. A., D. J. Woodward, D. Okaya, and J. Yu (2004), Mapping the Moho beneath the Southern Alps continent-continent collision, New Zealand, using wide-angle reflections, *Geophys. Res. Lett.*, *31*, L17602, doi:10.1029/2004GL020561.

Jackson, J. (2002), Strength of the continental lithosphere: Time to abandon the jelly sandwich?, *GSA Today*, September 2002, p. 4–10.

Kleffmann, S., F. Davey, A. Melhuish, D. Okaya, T. Stern, and the SIGHT Team, (1998), Crustal structure in the central South Island, New Zealand, from the Lake Pukaki seismic experiment, *N. Z. J. Geol. Geophys.*, *41*, 39–49.

Klosko, E. R., F. T. Wu, H. J., Anderson, D. Eberhart-Phillips, T. V. McEvilly, E. Audoine, M. K. Savage, and K. R. Gledhill (1999), Upper mantle anisotropy in the New Zealand region, *Geophys. Res. Lett.*, *26*, 1497–1500.

Leitner, B., D. Eberhart-Phillips, H. Anderson, and J. L. Nabelek (2001), A focused look at the Alpine Fault, New Zealand: Focal mechanisms and stress observations, *J. Geophys. Res.*, *106*, 2193–2220.

Little, T. A. (2004), Transpressive ductile flow and oblique ramping of lower crust in a two-sided orogen: Insight from quartz grain-shape fabrics near the Alpine fault, New Zealand, *Tectonics*, *23*, TC2013, doi:10.1029/2002TC001456.

Little, T. A., M. K. Savage, and B. Tikoff (2002), Relationship between crustal finitestrain and seismic anisotropy in the mantle, Pacific-Australia plate boundary zone, South Island, New Zealand, *Geophys. J. Int.*, *151*, 106–116.

Little, T. A., S. Cox, J. K. Vry, and G. Batt (2005), Variations in exhumation level and uplift rate along the oblique-slip Alpine Fault, central Southern Alps, New Zealand, *GSA Bull.*, *117*, 707–723; doi: 10.1130/B25500.

Long, D. T., S. C. Cox, S. Bannister, M. C. Gerstenberger, and D. Okaya (2003), Upper crustal structure beneath the Southern Alps and the Mackenzie Basin, New Zealand, derived from seismic reflection data, *N. Z. J. Geol. Geophys.*, *46*, 21–39.

McCaffrey, R. (2002), Crustal block rotations and plate coupling, in *Plate Boundary Zones*, *AGU Geodyn. Ser.*, vol. 30, edited by S. Stein and J. Freymueller, pp. 100–122, AGU, Washington, DC.

McCaffrey, R. (2005), Block kinematics of the Pacific-North America plate boundary in the southwestern United States from inversion of GPS, seismological and geologic data, *J. Geophys. Res.*, *110*, doi:10.1029/2004JB003307.

McCaffrey, R., M. D. Long, C. Goldfinger, P. C. Zwick, J. L. Nabelek, C. K. Johnson, and C. Smith (2000), Rotation and plate locking at the southern Cascadia subduction zone, *Geophys. Res. Lett.*, *27*, 3117–3120.

McClusky, S. C., Bjornstad, S. C., Hager, B. H., King, R. W., Meade, B. J., Miller, M. M., and Monastero, F. C. (2001), Present day kinematics of the eastern California Shear zone from a geodetically constrained block model, *Geophys. Res. Lett.*, *28*, 3369–3372.

Meade, B. J., and B. H. Hager (2005), Block models of crustal motion in southern California constrained by GPS measurements, *J. Geophys. Res.*, *110*, B03403, doi:10.1029/2004JB003209.

Molnar, P., H. J. Anderson, E. Audoine, D. Eberhart-Phillips, K. Gledhill, E. R. Klosko, T. V. McEvilly, D. Okaya, M. Savage, T. Stern, and F. T. Wu (1999), Continuous deformation versus faulting through the continental lithosphere of New Zealand, *Science*, *286*, 516–519.

Moore, M. (1999), Crustal deformation in the southern New Zealand region, Ph.D. thesis, Univ. of Oxford, Oxford, UK.

Moore, M., P. England, and B. Parsons, B. (2002), Relation between surface velocity field and shear wave splitting in the South Island of New Zealand. *J. Geophys. Res.*, *107*(B9), 2198, doi: 10.1029/2000JB000093.

Norris, R. J., and A. F. Cooper (2001), Late Quaternary slip rates and slip partitioning on the Alpine Fault, New Zealand, *J. Struc. Geol.*, *23*, 507–520.

Norris, R. J., and A. F. Cooper (2003), Very high strains recorded in mylonites along the Alpine Fault, New Zealand: Implications for the deep structure of plate boundary faults, *J. Struc. Geol.*, *25*, 2141–2157.

Norris, R. J., P. O. Koons, and A. F. Cooper (1990), The obliquely convergent plate boundary in the South Island of New Zealand: Implications for ancient collision zones, *J. Struc. Geol.*, *12*, 715–725.

Pearson, C., J. Beavan, D. J. Darby, G. H. Blick, and R. I. Walcott (1995), Strain distribution across the Australian-Pacific plate boundary in the central South Island, New Zealand, from 1992 GPS and earlier terrestrial observations, *J. Geophys. Res.*, *100*(B11), 22,071–22,081.

Pearson, C., D. Massonnet, N. Pourthie, and S. A. Israel (1999), SAR study of the Arthur's Pass earthquake of 18 June 1994: Implications for the tectonics of west-central Canterbury (abstract), *N. Z. Geophys. Soc. Met. Soc. N. Z. Joint Conf.*, Victoria Univ. of Wellington, Wellington, NZ, September 1–3, 1999.

Pearson, C., P. Denys, and K. Hodgkinson (2000), Geodetic constraints on the kinematics of the Alpine Fault in the southern South Island of New Zealand, using results from the Hawea-Haast GPS transect, *Geophys. Res. Lett.*, *27*(9), 1319–1322.

Pettinga, J. R., M. D. Yetton, R. J. Van Dissen, and G. L. Downes (2001), Earthquake source identification and characterization for the Canterbury region, South Island, New Zealand, *Bull. N. Z. Soc. Earthq. Eng.*, *34*, 282–317.

Prawirodirdjo, L., and Y. Bock (2004), Instantaneous global plate motion model from 12 years of continuous GPS observations: *J. Geophys. Res.*, *190*, doi:10.1029/2003JB002944.

Pysklywec, R. N., C. Beaumont, and P. Fullsack (2002), Lithospheric deformation during the early stages of continental collision: Numerical experiments and comparison with South Island, New Zealand, *J. Geophys. Res.*, *107*, doi:10.1029/2001JB000252.

Reyners, M. E., P. J. McGinty, S. C. Cox, I. M. Turnbull, T. O'Neill, K. R. Gledhill, G. T. Hancox, R. J. Beavan, D. W. Matheson, G. H. McVerry, W. J. Cousins, J. X. Zhao, H. A. Cowan, T. G. Caldwell, S. L. Bennie, and GeoNet team (2003), The Mw 7.2

Fiordland earthquake of August 21, 2003: Background and pre-liminary results. *Bull. N. Z. Soc. Earthq Eng.*, *36*(4): 233–248.

Robinson, R., and P. J. McGinty (2000), The enigma of the Arthur's Pass, New Zealand, earthquake—2. The aftershock distribution and its relation to regional and induced stress fields. *J. Geophys. Res.*, *105*(B7), 16,139–16,150.

Savage, J. C. (2000), Viscoelastic coupling model for the earthquake cycle driven from below, *J. Geophys. Res.*, *105*, 25,525–25,532.

Savage, J. C., and M. Lisowski (1998), Viscoelastic coupling model of the San Andreas fault along the big bend, southern California, *J. Geophys. Res.*, *103*(B4), 7281–7292.

Savage, M. K., K. M. Fischer, and C. E. Hall (2004), Strain model-ling, seismic anisotropy and coupling at strike-slip boundaries: Applications in New Zealand and the San Andreas Fault, in *Vertical Coupling and Decoupling in Lithosphere*, edited by J. Grocott, B. Tikoff, K. J. W. McCaffrey, and G. Taylor, *Geological Society of London, Special Publication*, *227*, 9–40.

Savage, M., A. Tommasi, S. Ellis, and J. Chéry (this volume), Modeling strain and anisotropy along the Alpine Fault, South Island, New Zealand.

Sella, G. F., T. H. Dixon, and A. L. Mao (2002), REVEL: A model for recent plate velocities from space geodesy, *J. Geophys. Res.*, *107*(B4), doi:10.1029/2000JB000033.

Spitzak, S., and C. DeMets (1996), Constraints on present-day plate motions south of 30°S from satellite altimetry, *Tectonophysics*, *253*, 167–208.

Stern, T. A., P. Molnar, D. Okaya, and D. Eberhart-Phillips (2000), Teleseismic P-wave delays and modes of shortening the man-tle lithosphere beneath South Island, New Zealand. *J. Geophys. Res.*, *105*, 21,615–21,631.

Stern, T., S. Kleffmann, M. Scherwath, D. Okaya, and S. Bannister (2001), Low seismic wave speeds and enhanced fluid pressures be-neath the Southern Alps of New Zealand, *Geology*, *29*, 679–682.

Stern T., D. Okaya, S. Kleffmann, M. Scherwath, S. Henrys, and F. Davey, Geophysical exploration and dynamics of the alpine fault zone, *this volume*.

Sutherland, R., et al. (this volume), Do great earthquakes occur on the Alpine fault in central South Island, New Zealand?

Upton, P., D. Craw, J. Zoe, and P. O. Koons (2004), Structure and late Cenozoic tectonics of the southern Two Thumb Range, mid Canterbury, New Zealand, *N. Z. J. Geol. Geophys.*, *47*,141–153.

van Avendonk, H. J. A., W. S. Holbrook, D. Okaya, J. K. Austin, F. Davey, and T. Stern (2004), Continental crust under compres-sion: A seismic refraction study of South Island Geophysical Transect I, South Island, New Zealand, *J. Geophys. Res.*, *109*, B06302, doi:10.1029/2003JB002790.

Van Dissen, R. J., A. G. Hull, and S. A. L. Read (1994), Timing of some large Holocene earthquakes on the Ostler Fault, New Zealand, in *Proceedings of the Eighth International Sympo-sium on Recent Crustal Movements (CRCM '93), Kobe, De-cember 6–11, 1993*, pp. 381–386, Geodetic Society of Japan, Japan.

Vergne, J., R. Cattin, R., and J. P. Avouac (2001), On the use of dislocations to model interseismic strain and stress build-up at intracontinental thrust faults, *Geophys. J. Int.*, *147*, 155–162.

Wallace, L. M., J. Beavan, R. McCaffrey, and D. J. Darby (2004), Subduction zone coupling and tectonic rotations in the North Island, New Zealand, *J. Geophys. Res.*, *109*, B12406, doi: 10.1029/2004JB003241.

Wallace, L. M., R. McCaffrey, J. Beavan, and S. Ellis (2005), Rapid microplate rotations and back-arc rifting at the transition between collision and subduction, *Geology*, *33*(11), 857–860.

Wallace, L. M., J. Beavan, R. McCaffrey, and K. Berryman (in press), Balancing the plate motion budget in the South Is-land, New Zealand, using GPS, geological and seismologi-cal data, *Geophys. J. Int.*, *168*, 332–352, doi:10.1111/j.1365-246X.2006.03183.x.

Walcott, R. I. (1979), Plate motion and shear strain rates in the vi-cinity of the Southern Alps, in *The Origin of the Southern Alps*, *R. Soc. N. Z. Bull.*, vol. 18, edited by R. I. Walcott and M. M. Cresswell, pp. 5–12, Royal Society of New Zealand, Thorndon, Wellington, NZ.

Walcott, R. I. (1998), Modes of oblique compression: Late Ceno-zoic tectonics of the South Island of New Zealand, *Rev. Geo-phys.*, *36*, 1–26.

Wang, K., and T. Dixon (2004), "Coupling" semantics and science in earthquake research, *Eos Trans. AGU*, *85*, 180.

Wannamaker, P. E., G. R. Jiracek, J. A. Stodt, T. G. Caldwell, V. M. Gonzalez, J. D. McKnight, and A. D. Porter (2002), Fluid generation and pathways beneath an active compressional orogen, the New Zealand Southern Alps, inferred from magne-totelluric data, *J. Geophys. Res.*, *107*(B6), 2117, doi:10.1029/2001JB000186.

Wellman, H. W. (1979), An uplift map for the South Island of New Zealand, and a model for uplift of the Southern Alps, in *The Origin of the Southern Alps*, *R. Soc. N. Z. Bull.*, vol. 18, edited by R. I. Walcott and M. M. Cresswell, pp. 13–20, Royal Society of New Zealand, Thorndon, Wellington, NZ.

Williams, S. D. P., Y. Bock, P. Fang, P. Jamason, R. Nikolaidis, L. Prawirodirdjo, M. Miller, and D. Johnson (2004), Error analysis of continuous GPS position time series, *J. Geophys. Res.*, *109*, B03412, doi:10.1029/2003/JB002741.

Seismic Anisotropy in South Island, New Zealand

Martha Kane Savage, Mathieu Duclos[1], and Katrina Marson-Pidgeon[2]

School of Geography, Environment and Earth Sciences, Victoria University of Wellington, New Zealand

Seismic anisotropy is one of the few tools for assessing the orientation of mantle deformation. We use it to study the plate boundary through New Zealand, which includes two oblique subduction zones of opposing polarity linked by the transpressional Alpine Fault. The mantle under South Island has strong, coherent deformation as evidenced by shear-wave splitting of *SKS* phases with fast polarizations oriented subparallel to the plate boundary, and with delay times averaging 1.7 s. *SKS* delay times are consistent through most of South Island and southern North Island, and fast polarizations are independent of distance from the plate boundary across the landmass. A fundamental change in anisotropy is nevertheless observed, which does not correspond in a simple way to changes between subduction and transform faulting. Instead, the change corresponds roughly with the northward narrowing of South Island (the "Waistline"). North of the Waistline, *SKS* splitting is parallel to the plate boundary, and the region of anisotropy is narrow (60–100 km wide) in the uppermost mantle as evidenced by *Pn* wave speeds and by shear wave splitting of local events. Lithospheric strain with dynamic recrystallization and sub-slab asthenospheric flow may both be contributing to the anisotropy. South of the Waistline, *SKS* splitting averages 20°, about 25° to the shear plane, and along the Waistline, *Pn* anisotropy extends about 200 km across the plate boundary. Lithospheric shearing without dynamic recrystallization, appropriate for smaller strain in this region, or asthenospheric flow directed at 20° or 200° could form the anisotropy.

1. INTRODUCTION

New Zealand's position as a continental mass located above a plate boundary provides a great natural laboratory to study processes associated with plate boundary deforma-

[1] Now at: Spectraseis, Switzerland.
[2] Now at: GNS Science, New Zeland.

A Continental Plate Boundary: Tectonics at South Island, New Zealand
Geophysical Monograph Series 175
Copyright 2007 by the American Geophysical Union.
10.1029/175GM06

tion beneath continents. Seismic anisotropy provides one of the few methods of determining orientational dependence of deformation in the mantle. The SAPSE/SIGHT project, described elsewhere in this volume [*Davey et al.*, this volume] provided the first widespread measurements of mantle anisotropy in New Zealand (NZ), and they remain an important base from which tectonic models are derived. These results have come to be considered the "type" model of anisotropy in strike-slip regions. Anisotropy may be caused by shear along the plate boundary, with continuous deformation through a wide zone on the continental landmass.

In this paper, we describe the principles behind seismic anisotropy and its relationship to deformation. Then we present the anisotropy measurements, which show strong, widespread and fairly consistent anisotropy throughout the region. Most of the measurements have been previously published, but we present new measurements in the Marlborough region of northern South Island. We discuss evidence for the presence of anisotropy in terms of its depth and lateral extent. Finally, we discuss the interpretations and plausible tectonic models, which range from some component being caused by processes at the core-mantle boundary, to asthenospheric flow, to control of anisotropy being caused by present or (our favoured model) past lithospheric shearing.

2. METHODS

Seismic anisotropy occurs when waves traveling or vibrating in one directtion travel faster than those in another. Causes of anisotropy in the Earth range from sedimentary layering of materials [*Backus*, 1962] to stress-related alignment of fluid-filled cracks in the crust [*Christensen*, 1966a; *Crampin*, 1994] to strain-aligned anisotropic minerals in the crust and mantle [*Christensen*, 1966a; 1966b; *Hess*, 1964].

Measurement of seismic anisotropy is relatively straightforward. Direct measurements of small samples of rocks can be made in pressure vessels containing samples of rocks cut in different orientations [*Birch*, 1960; *Okaya et al.*, 1995; *Godfrey et al.*, 2000]. Alternatively, measurements of orientations of individual grains of olivine and other minerals in ophiolites and mantle xenoliths can be averaged to determine seismic anisotropy [*Christensen*, 1984; *Mainprice*, 1990; *ben Ismail and Mainprice*, 1998]. Bulk anisotropy measurements of larger regions can be made using seismometers with controlled or earthquake sources. Those determined from the variation of wave speed with azimuth are called azimuthal anisotropy, and are often made with *Pn* waves [*Hess*, 1964; *Baldock and Stern*, 2005; *Galea*, 1993; *Smith and Ekstrom*, 1999; *Scherwath et al.*, 2002] or surface waves [*Forsyth*, 1975; *Laske*, 1998; *Montagner*, 2002; *Debayle*, 2005]. Polarization anisotropy, i.e., the difference in travel times of waves traveling with different polarizations, can be measured by shear-wave splitting or birefringence; when a shear wave enters an anisotropic medium, the component parallel to the fast polarization (ϕ) for that particular path, begins to lead the perpendicular (or near-perpendicular) component, leading to a separation in time (dt) between the waveforms) [e.g., *Byerly*, 1938; *Christensen*, 1966a; *Crampin*, 1981; *Klosko et al.*, 1999]. Polarization anisotropy also occurs with surface waves and is evidenced by differences between Love and Rayleigh wave speeds [*Anderson*, 1961] and coupling between Love and Rayleigh waves [*Kirkwood and Crampin*, 1981; *Yu and Park*, 1994; *Brisbourne et al.*, 1999].

Shear wave splitting in the mantle is most commonly measured using *SKS* phases, which travel as *P* waves in the liquid outer core [*Vinnik et al.*, 1984; *Silver and Chan*, 1988]. In the absence of anisotropy or strong lateral heterogeneities, these phases are expected to yield little energy on the transverse component, making diagnosis of anisotropy easy. Additionally, any measured splitting must have occurred between the core mantle boundary and the surface. However, teleseismic and local *S* phases can also be used to determine anisotropy. They have the advantage of allowing the possibility of measuring differing splitting for different earthquake depths, and hence paths. Also, they can separate effects of phase polarization, which depends on the focal mechanism for *S* phases, from back azimuth, which depends on earthquake-station geometry, and which is identical to the polarization for *SKS* phases.

The biggest problem in anisotropy research is how to interpret the measurements. While crustal anisotropy has often been considered to be caused by vertical, stress-aligned microcracks [*Nur and Simmons*, 1969; *Crampin*, 1994; *Miller and Savage*, 2001] mineral alignment may also be important [e.g., *Babuska and Cara*, 1991; *Zinke and Zoback*, 2000; *do Nascimento et al.*, 2004; *Balfour et al.*, 2005].

The interpretation of mantle anisotropy is likewise complex. Because olivine is the dominant mineral in the mantle and is highly anisotropic, it is well accepted that the fast polarization aligns with the dominant alignment of the olivine a-axes ([100]) [*Babuska and Cara*, 1991]. However, that alignment depends on strain history in complicated ways. Mineral preferred orientation in the mantle usually forms due to alignment of slip planes and orientations, respectively, with the shear plane and orientation during progressive rotational deformation [e.g., *Nicolas et al.*, 1973]. One simplified hypothesis is that the a-axes align with the maximum finite extension orientation, b-axes ([010]; slow) align with maximum compression, and c-axes ([001]; intermediate speed) align in the intermediate axis of the strain ellipse [*McKenzie*, 1979]. Small shear would therefore align the a-axes at 45° to the flow plane, while infinite shear would result in flow-parallel anisotropy. Such interpretations have been borne out in general by more elaborate models that follow the slip planes of individual mineral interactions in response to applied strain [*Wenk and Christie*, 1991; *Wenk et al.*, 1991; *Tommasi et al.*, 2000]. When large strains (greater than about 150%) occur, dynamic recrystallization via subgrain rotation enhances effects of the deformation, aligning the fast polarizations more rapidly to be parallel to the shear direction [*Zhang and Karato*, 1995; *Bystricky et al.*, 2000]. It was accepted practice for several years to interpret fast orientations of anisotropy as parallel or sub-parallel to the direction of shear in the lithosphere or flow in the asthenosphere [*Silver*, 1996; *Savage*, 1999].

Recent rock mechanics studies allowing deformation at higher pressure suggest that the relationship of flow to strain may change dramatically under different conditions. Some have suggested that there are five or more types of orientation depending on pressure and water content as well as strain rate [*Jung and Karato*, 2001; *Mainprice et al.*, 2005]. The presence of melt may also change the orientation of the fast axis [*Holtzman et al.*, 2003]. The wide-spread observation of trench-parallel fast polarizations of anisotropy in subduction regions were initially interpreted as caused by trench parallel flow [*Russo and Silver*, 1994]. However, recently some have been reinterpreted as caused by trench-perpendicular flow with such water or melt-modified fabric.

3. RESULTS

Figure 1 and Tables 1–3 present a summary of all the *SKS* and *Pn* anisotropy measurements published for New Zealand, along with some of the measurements from xenoliths and deep local earthquakes. Measurements from shallow local earthquakes and surface waves will be discussed, but are not as easily displayed. The *SKS* measurements include published results as well as new measurements in Marlborough.

The results show a clear signal of strong anisotropy in both *Pn* and splitting measurements. *Pn* anisotropy has been inferred up to $10 \pm 3\%$ [*Scherwath et al.*, 2002] (Table 2). Station average delay times from *SKS* phases range from 0.6 s (station JACA) [*Klosko et al.*, 1999] to 3.5 s (a single measurement at station WAG00) on South Island, averaging 1.7 ± 0.2 s, and are similar in southern North Island (Table 1; Figure 1). Fast polarizations on South Island average 38°, which is subparallel to the plate boundary. However, there are systematic changes in polarization over the island, which we discuss in more detail below. Any model of anisotropy must also explain the uniformity of SKS splitting measured above the stations in southern North Island, which occurs despite the paths to the stations having varying lengths through the mantle wedge and the sub-slab mantle (Figure 2)[*Marson-Pidgeon et al.*, 1999].

3.1 Location (Depth Extent) of Anisotropy

The depth extent of anisotropy is difficult to constrain from *SKS* phases alone, since the splitting could occur anywhere between the core mantle boundary and the surface. We therefore use a combination of methods to try to determine what part of the path is contributing most to the anisotropy.

3.1.1. Crustal anisotropy. Crustal anisotropy is often neglected in studies of shear wave splitting from *SKS* phases, since several studies suggested that it is unlikely to contribute to greater than a few tenths of a second of splitting [e.g.,

Barruol and Mainprice, 1993]. However, the similarity of *SKS* fast polarizations and crustal structure in many regions is often a concern and should be carefully evaluated. In particular, the schists outcropping in New Zealand, which are expected to be present at depth as well [e.g., *Suggate and Grindley*, 1972] have high anisotropy up to 17%, and as little as 10 km of such material could cause splitting up to 1.0 s [*Okaya et al.*, 1995; *Godfrey et al.*, 2000].

Surprisingly, although analysis of 3-component controlled source data from the SIGHT experiment yielded the expected plate-boundary-parallel fast polarization, small delay times with a maximum of 0.08 s were observed even over the most deformed part of the Alpine Fault region [*Pulford et al.*, 2003]. These small delay times could in part be a function of the difficulty of measuring splitting with high-frequency, controlled source data. So we examine other arguments for the contribution of the crust to the *SKS* splitting measurements.

SKS measurements do not change polarization in concert with the rotation of the outcropping schists in South Island Orocline [*Suggate and Grindley*, 1972]. This suggests that the crust may have only a small effect on the *SKS* measurements. Similarly, studies of crustal anisotropy from shallow local earthquakes in Marlborough [*Balfour et al.*, 2005] and southern North Island [*Gledhill*, 1991, 1993a] have shown rapid changes of fast polarization with station location, while both the fast polarization and delay times of *SKS* phases are consistent over most of the study region (Figure 1). This also suggests that the crust does not strongly affect the *SKS* measurements here.

Strong azimuthal variation of radial and transverse receiver functions in the Marlborough and Wellington regions have been interpreted in terms of crustal and upper mantle anisotropy [*Savage*, 1998; *Wilson et al.*, 2004]. However, the waveforms could be fit with small anisotropy of only 4-6% in the crust when the effect of dipping layers was taken into account [*Savage et al.*, 2007a]. For a 35-km thick crust, this yields only a few tenths of a second of splitting, again consistent with a small effect on *SKS* results.

3.1.2. Anisotropy in the subducting Pacific Plate. Perhaps the easiest locations to determine anisotropy as a function of depth ought to be subduction zones, because of the varying path lengths for earthquakes at different depths. Because the Alpine Fault collision zone is flanked by subduction zones, and because the *SKS* splitting above the subduction zone in southern North Island is very similar to that in South Island (Figure 1), we examine the depth distribution of anisotropy in this region first. The anisotropy measured above subduction zones must come from some combination of anisotropy in the upper plate, the subducting plate, the mantle wedge

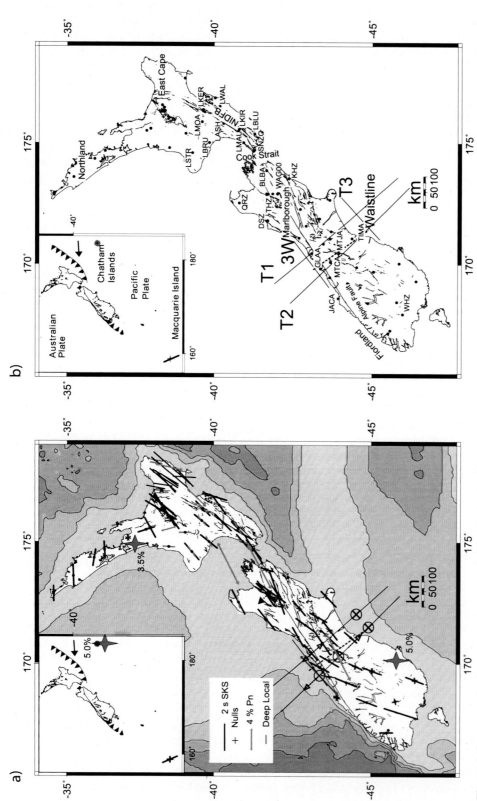

Figure 1. Summary of anisotropy measurements in New Zealand, with bathymetry in 1000 m contours. (a) Small filled circles: seismic stations on which SKS splitting has been attempted. [*Cochran*, 1999; *Marson-Pidgeon et al.*, 1999; *Klosko et al.*, 1999; *Audoine et al.*, 2000; *Audoine et al.*, 2004; *Duclos et al.*, 2005]. Bars: Positive measurements of station-averaged splitting. Orientation of the line represents average fast polarization. Black bars: *SKS* phases; length of the line is proportional to the average delay time as given in the scale. Gray bars: *S* phases from local earthquakes deeper than 100 km [*Audoine*, 2002]; length is arbitrary. Gray crosses: null measurements (in which no splitting was observed), with the possible fast orientations that could yield no splitting. Gray arrows: *Pn* anisotropy orientation, with length proportional to percent anisotropy. Circles enclose points measured between two lines [*Scherwath et al.*, 2002; *Baldock and Stern*, 2005; *Bourguignon et al.*, 2007]. Other measurements from *Smith and Ekstrom* [1999]. SIGHT lines T1 and T2 are marked, with dashed arrows delineating the width of the anisotropic region, as estimated by low *Pn* speeds [*Scherwath et al.*, 2002; *van Avendonk et al.*, 2004]. Dashed arrows in Marlborough represent the width of the proposed fault-parallel anisotropic region [*Audoine et al.*, 2000]. Stars show locations of xenoliths at which *S* wave % anisotropy was calculated [*Duclos et al.*, 2005]. Waistline is shown by the heavy dashed black line. (b) Locations and stations discussed in text and other figures are marked. Inset: Regional plate tectonic setting of NZ, and anisotropy measurements at nearby oceanic islands. Color version of this figure is on CDROM which accompanies this volume.

Table 1. Station averages of shear wave splitting of SKS and SKKS phases.

Lat	lon	Sta	φ	φ st. dev	dt	Dt st. dev.	# Events	# Nulls	Ref
177.6736	−39.0214	KNZ	46	4	2	0.3	2	3	Duclos
177.6736	−39.0214	KNZ	3	12	1.8	0.7	1	—	Duclos
177.1103	−38.2603	URZ	null	46	—	1			Duclos
176.8617	−40.0314	PWZ	15	8	2	0.4	2	3	Duclos
176.52	−40.188	LWAL	52	4	1.7	0.2	3	—	Marson-Pidgeon
176.38	−39.643	LKER	44	6	1.6	0.2	2	—	Marson-Pidgeon
175.88	−39.597	LMOA	38	5	1.5	0.1	5	—	Marson-Pidgeon
175.81	−40.309	LASH	44	5	1.8	0.2	2	—	Marson-Pidgeon
175.56	−40.809	LKIR	44	3	1.5	0.1	5	—	Marson-Pidgeon
175.5019	−37.7308	TOZ	−17	4	1.4	0.3	1	3	Duclos
175.1611	−37.1045	MKAZ	nulls	−19	−23	−75	—	3	Duclos
175.02	−40.967	LMAU	48	3	1.5	0.1	4	—	Marson-Pidgeon
175.02	−39.842	LBRU	37	7	1.7	0.3	2	—	Marson-Pidgeon
174.6723	−36.3003	MATA	86	8	1.3	0.4	2	7	Duclos
174.6	−39.273	LSTR	48	4	1.7	0.2	3	—	Marson-Pidgeon
174.3444	−35.9411	WCZ	−80	6	1.3	0.4	2	7	Duclos
173.5961	−35.2214	OUZ	88	9	1.7	0.3	2	6	Duclos
172.8647	−34.53	TIKO	89	10	1.2	0.5	2	4	Duclos
172.6522	−43.7078	MQZ	45	5	1.7	0.3	2	2	Duclos
171.8025	−41.7469	DSZ	48	4	1.64	0.18	1	3	Duclos
171.0539	−43.7192	RPZ	42	8	1.8	0.4	1	1	Duclos
167.9467	−45.8939	WHZ	17	7	2.7	0.4	2	4	Duclos
−37.87	175.59	WLZA	59	2	1.6	0.225	4	—	Klosko
−38.06	177.25	LWOO	10	7	1.5	0.2	3	—	Cochran
−38.24	176.31	RKIV	30	5	2.8	0.5	1	—	Audoine
−38.26	176.57	TAWV	37	5	2.4	0.4	1	—	Audoine
−38.3	177.35	LRED	8	5	1.5	0.1	3	—	Cochran
−38.3	177.55	LNOR	40	4	2.7	0.3	3	—	Cochran
−38.33	176.47	BTRV	38	14	2.5	0.8	1	—	Audoine
−38.36	176.336	HIHV	57	3	2.9	0.2	1	—	Audoine
−38.39	177.48	LMAH	47	5	2.3	0.2	1	—	Cochran
−38.39	177.55	LMNG	32	16	2.3	0.4	2	—	Cochran
−38.39	177.68	LTEK	29	17	2.3	0.4	2	—	Cochran
−38.42	176.23	PKMV	52	8	3.2	0.5	1	—	Audoine
−38.52	177.55	LENG	34	15	1.7	0.4	2	—	Cochran
−38.52	177.87	LSTN	39	20	3.6	0.9	1	—	Cochran
−38.61	178.05	LNOZ	40	5	3.7	0.4	1	—	Cochran
−40.82	172.53	QRZA	61	5	2.12	0.28	4	—	Klosko
−41	175.3	Tararua	28	5	1.5	0.4	8	4	Gledhill
−41.31	174.7	SNZO	50	16	2	0.6	11		*Marson-Pidgeon and Savage,* 2004

(continued on next page)

Table 1. (Continued)

Lat	lon	Sta	ϕ	ϕ st. dev	dt	Dt st. dev.	# Events	# Nulls	Ref
−41.71	173.87	BLBA	17	5	1.22	0.14	3	1	Klosko
−41.7102	172.8774	ALI00	31	11	2	0.4	4	—	This study
−41.7104	172.8892	ALIE2	54	3	2.3	0.2	1	—	This study
−41.7107	172.9074	ALIE5	43	3	2.1	0.2	1	—	This study
−41.74	171.8	DENA	49	9	2.1	0.28	3	—	Klosko
−41.76	172.9	TOPA	42	7	1.4	0.2	3	—	Klosko
−41.9083	172.9102	RSFG	34	4	2.6	0.2	1	—	This study
−42.0563	172.9007	WAGW5	57	3	0.9	0.1	1	—	This study
−42.0668	172.9308	WAG00	70	24	3.5	0.5	1	—	This study
−42.0711	172.9296	WAGS1	49	2	1.1	0.1	1	—	This study
−42.74	171.09	HOKA	59	4	1.68	0.27	2	—	Klosko
−42.78	172.27	LATA	49	6.5	1.52	0.16	5	2	Klosko
−42.97	171.58	ARPA	34	5	1.82	0.14	5	—	Klosko
−43.15	170.46	ABUA	48	17.5	1.6	0.39	3	—	Klosko
−43.39	172.88	SHEA	27	9	1.54	0.21	3	—	Klosko
−43.42	169.85	GLAA	26	7.5	1.66	0.24	4	—	Klosko
−43.51	170.85	EWZA	54	3.5	2.2	0.24	5	—	Klosko
−43.71	169.46	LAMA	9	10.5	1.14	0.19	2	—	Klosko
−43.73	170.09	MTCA	29	9	1.7	0.27	3	—	Klosko
−43.75	171.36	MAYA	51	10.5	1.82	0.27	3	—	Klosko
−43.95	−176.56	CHTA	x	x	x	X	—	5	Klosko
−43.97	168.6	JACA	24	11	0.64	0.15	—	6	Klosko
−43.98	170.46	MTJA	44	3.5	1.76	0.12	6	—	Klosko
−44.25	169.22	MAKA	21	12.5	1.76	0.36	3	—	Klosko
−44.53	169.88	BERA	24	7.5	1.12	0.11	3	2	Klosko
−44.87	168.4	GLEA	20	10	1.72	0.31	2	—	Klosko
−45.03	169.69	LUDA	20	4	1.46	0.1	10	4	Klosko
−45.53	167.27	DOTA/ COVA	0	22	1.04	0.5	3	2	Klosko
−46.29	169.31	CLIA	20	6	1.48	0.19	4	2	Klosko
−54.49	158.93	MACQ	−23	10.5	1.3	0.275	3	1	Klosko

Nulls represents the number of reported null measurements for the station. St. Dev. represents standard deviation. Reference key: Audoine: [*Audoine et al.*, 2004]; Cochran : [*Cochran*, 1999] ; Duclos: [*Duclos et al.*, 2005] ; Gledhill : [*Gledhill and Gubbins*, 1996] Klosko : [*Klosko et al.*, 1999] ; Marson-Pidgeon: [*Marson-Pidgeon et al.*, 1999] . Averages and standard deviations, or both, have been recalculated for references Cochran, Audoine, and Marson-Pidgeon and Savage. For Cochran, the recalculations leave out measurements with ϕ errors greater than 23°. For Marson-Pidgeon and Savage, Audoine, and Gledhill, averages and standard deviations were recalculated using circular statistics [*Mardia*, 1972] as discussed in *Gerst & Savage* [2004].

Table 2. Pn anisotropy measurements. Error estimates are as published; studies that determined anisotropy from only two lines have been given error estimates of 45°.

Lon	lat	φ(°)	φ error estimate (°)	Pn anisotropy (%)	% error estimate	Ref
174	−41.5	60	5	9	0.9	Smith & Ekstrom
174.088	−40.5	62	5	8.3	0.8	Smith & Ekstrom
176.763	−39.5	42	5	4.8	0.5	Smith & Ekstrom
175.532	−38.5	−14	5	2.7	0.3	Smith & Ekstrom
178.085	−38.5	15	10	5.0	1	Smith & Ekstrom
172.03	−44.55	—		0	3	Baldock & Stern
171.5	−44.92	45	45	6.5	2.5	Baldock & Stern
169.5	−43.42	45	45	10	3	Scherwath et al.
170.43	−44	45	45	10	3	Bourguignon et al.

between the two plates, and the sub slab asthenosphere. A puzzling constraint is that stations on southern North Island yield nearly identical orientations and delay times despite varying path lengths in the mantle wedge and sub slab asthenosphere (Figure 2; [*Marson-Pidgeon et al.*, 1999]). Below

we discuss the evidence for anisotropy in the different segments of the paths.

Several observations bear on the anisotropy in the subducting Pacific Plate. *P* residuals for earthquakes in the Tonga-Kermadec subduction zone recorded at three stations

Table 3. Station averages of shear-wave splitting polarizations from local events deeper than 100 km. References: Audoine (2004): [*Audoine et al.*, 2004]. Only the results for filter range 0.1 to 1.0 Hz were used. Audoine (2000): [*Audoine et al.*, 2000].

Lat	lon	sta	φ (°)	st. dev φ (°)	# events	Ref
−38.506	174.8	MOZ	4	39	33	Audoine (2004)
−39.047	175.39	OIZ	180	44	20	Audoine (2004)
−38.07	178.26	PUZ	7	29	14	Audoine (2004)
−37.87	175.6	WLZ	163	39	51	Audoine (2004)
−38.26	177.11	URZ	6	29	17	Audoine (2004)
−36.747	175.72	KUZ	164	29	27	Audoine (2004)
−40.82	172.53	QRZ	76	42	55	Audoine (2000)
−41.74	171.8	DSZ	85	42	21	Audoine (2000)
−41.76	172.9	THZ	22	13	66	Audoine (2000)
−42.42	173.54	KHZ	67	28	62	Audoine (2000)
−39.27	174.6	LSTR	54	11	7	Audoine (2000)
−39.6	175.88	LMOA	47	8	14	Audoine (2000)
−39.64	176.38	LKER	41	20	17	Audoine (2000)
−39.84	175.02	LBRU	60	21	13	Audoine (2000)
−40.19	176.52	LWAL	22	41	22	Audoine (2000)
−40.31	175.81	LASH	96	36	19	Audoine (2000)
−40.81	175.56	LKIR	58	11	20	Audoine (2000)
−40.97	175.02	LMAU	21	31	17	Audoine (2000)
−41.32	175.38	LBLU	51	11	17	Audoine (2000)
−41.31	174.7	SNZO	11	48	17	Audoine (2000)

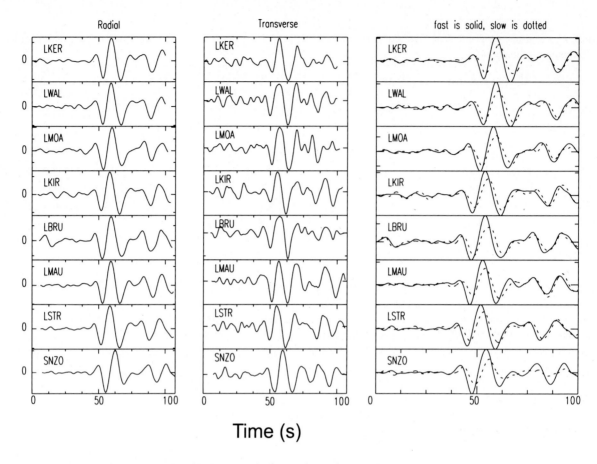

Figure 2. From *Marson-Pidgeon et al.* [1999], Figure 4. SKS waveforms for broadband stations in southern North Island recording a single event on Julian day 246, 1993. Radial (left) and transverse (middle) are shown separately, and (right) shows the components rotated into fast (solid) and slow (dotted) components. The events are lined up by distance from the earthquake and are plotted on the same absolute amplitude scale. Despite a large variation in the stations' distance from the plate boundary (Figure 1), the waveforms and splitting delay times are nearly identical at each station.

in the Pacific east and northeast of New Zealand (from latitudes −42° to −14°) yielded travel times that varied with azimuth [*Galea*, 1993]. This study found a fast orientation of 62° and a *P*-velocity anisotropy of 7%, which was attributed to the formation of mantle anisotropy with fast orientation parallel to the spreading direction as the oceanic plate was formed, similar to many other results in the Pacific Ocean [*Hess*, 1964] and specifically, the South Pacific [*Shearer and Orcutt*, 1986]. These values were later confirmed in a study of southern North Island, in which time-term analysis was used to determine *Pn*- anisotropy of 12–14% with a fast orientation of 65° [*Chadwick*, 1997]. Four *Pn* measurements in New Zealand were calculated from a global study, which used earthquake arrival times at a wide range of azimuths to determine anisotropy in caps near the station locations [*Smith and Ekstrom*, 1999]. Two of these measurements

were in Cook Strait between North and South Islands; these yield large anisotropy of 8–9% and confirm the 60° fast orientation (Figure 1).

So-called "quasi Love" phases, which occur when Rayleigh waves couple to Love waves in regions where anisotropy changes, have been observed at station SNZO near Wellington [*Yu and Park*, 1994]. They suggest a strong gradient in anisotropy, and have been modelled by 6% *P* wave anisotropy from the Moho to 210 km depth changing rapidly from NNE-SSW fast in the south Pacific to WNW-ESE about 1000 km northeast of Wellington. However, the forward modelling was not well constrained, and either deeper *S* velocity anisotropy, or a more gradual change in anisotropy were considered valid. It is difficult to observe changes near to the station using this method; therefore we do not consider it a strong constraint on the anisotropy orientation near SNZO and consider the 62° fast orientation from Galea, which is close to the shear wave splitting fast polarizations of NE/SW, to be more likely in the vicinity of SNZO. Anomalous surface-wave polarizations of Love waves recorded at SNZO and in the East Cape region of North Island also suggested plate-boundary parallel fast orientations of anisotropy within the subducting slab; they were modeled best with axes of symmetry plunging at 30° to the vertical [*Brisbourne et al.*, 1999].

A study of *S* waves from local earthquakes between 20-70 km depth beneath station SNZO provides the strongest constraint on *S*-wave anisotropy within the subducting slab [*Matcham et al.*, 2000]. Two populations of fast polarizations were recorded, with one set yielding NNE/SSW (29 ± 38°) fast and another E-W (86 ± 10 °) fast. Delay times increase with earthquake depth in both cases (Figure 3). The set of data with NE/SW fast polarizations, which we consider most representative of splitting in the region, yielded anisotropy of 4.4 ± 0.9% within the slab, based on the increase in delay time with path length. Smaller anisotropy with a NW/SE fast polarization was interpreted above the slab, based on the projected (negative) intercept of the delay times at the surface, as well as on *S*-wave splitting on a closely-spaced array deployed near SNZO, which examined shallow events [*Gledhill*, 1993b]. The measured anisotropy of 4.4% for *S* waves

compares well to the 6–9% anisotropy for *P* waves. Anisotropy determined via measuring fabric orientations of mantle xenoliths found that xenoliths with lowest strain yielded the smallest anisotropy, averaging about 8% for *P* waves and 6% for *S* waves (Figure 2 in *ben Ismail and Mainprice* [1998]). However, bulk measurements should almost always yield smaller anisotropy than laboratory measurements on small xenolith samples, because they will be averages of many samples which are unlikely to all be favorably oriented.

3.1.3. Sub-slab anisotropy. Below the slab, 1.4% anisotropy with fast orientation NE/SW was needed to explain the increase of delay times with depth for earthquakes to 250 km depth at an array located 50 km northeast of SNZO [*Gledhill and Stuart*, 1996] (Figure 3). There is a gap in measurements for earthquakes between 100 and 175 km depth, and it could be argued that, instead of a steady increase, the delay times jump by about 0.25 s somewhere between 100 and 200 km. Such a measurement would yield slightly smaller anisotropy of 1.2% averaged over 100 km. Similar anisotropy (1.2 ± 0.3%) was found at most of the rest of the stations in southern North Island (Figure 4) [*Audoine et al.*, 2000]. We think that the apparent isotropy determined from constant delay times with depth determined at some of the same stations [*Brisbourne et al.*, 1999] were caused by

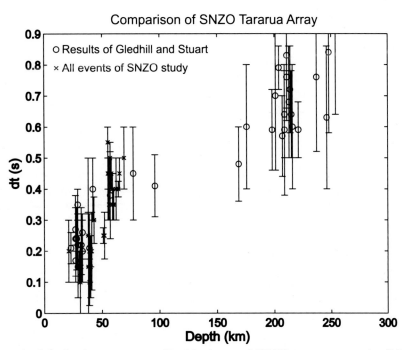

Figure 3. Delay time versus depth for local events measured by *Matcham et al.* [2000] at permanent station SNZO (black crosses) and by *Gledhill and Stuart* [1996] at stations in the Tararua ranges (blue circles). From *Matcham et al.* [2000].

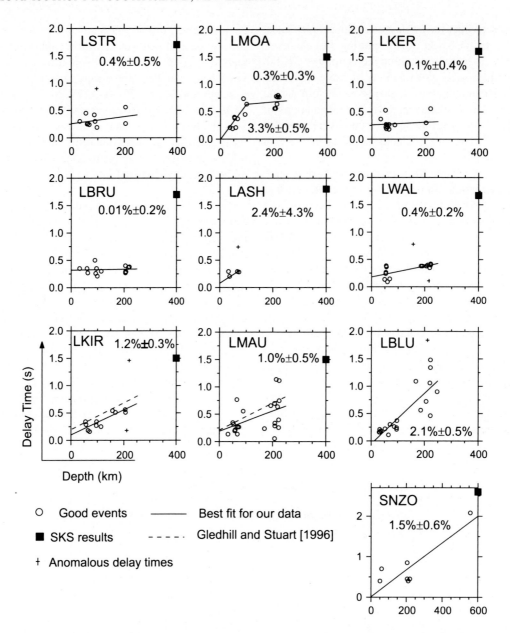

Figure 4. From *Audoine et al.* [2000]. Delay times versus earthquake depth for shear wave splitting measured on local S waves recorded at all the stations in the POMS II deployment [*Stuart et al.*, 1995]. The big squares on the right edge represent the results obtained using SKS phases [*Marson-Pidgeon et al.*, 1999]. The slopes in solid lines were determined by a least squares fit using good events only and are reported under the station name. The dashed slopes at stations LMAU and LKIR are from *Gledhill and Stuart* [1996].

the use of higher frequency phases. The waveforms may first be split by the mantle, and then resplit by the crust, which can be measured at high frequencies, but may be filtered out of low frequency records.

3.1.4. Two-layer models of anisotropy. Station SNZO in Wellington is the longest operating broadband seismic sta-

tion in New Zealand; it has been running since 1992. A detailed study revealed back azimuthal variations in fast polarizations and also strong variations in delay times [*Marson-Pidgeon and Savage*, 2004a]. These variations could not be explained by a single layer of anisotropy with orthorhombic or dipping symmetry axes. A double layer of anisotropy fit the fast polarizations but not the delay times, possibly indicating frequency dependence of anisotropy or lateral variations that could not be modelled. The two layer model that fit the fast polarizations best consisted of an upper layer with $\phi = 60°$, $dt = 0.4$ s and a lower layer of $\phi = 20°$, $dt = 1.0$ s. The delay times were poorly constrained, and the orientations could vary from 30° to 110° in the upper layer and −50° to 30° in the lower layer. The fast orientation in the upper layer is consistent with the idea that it represents the anisotropy within the subducting slab, discussed above, as measured by *Pn* anisotropy [*Galea*, 1993; *Chadwick*, 1997; *Smith and Ekstrom*, 1999]. It is also consistent with receiver function studies [*Savage*, 1998; *Savage et al.*, 2007a] We check the consistency of the measurements in an order of magnitude calculation: Using an intermediate *Pn* anisotropy of 9% [*Smith and Ekstrom*, 1999] (Table 2) in the region, and assuming a ratio of 1.4 between P-wave and S-wave anisotropy [*ben Ismail and Mainprice*, 1998], *S* anisotropy would be 6.4%. Using an S speed of 4 km/s and delay time of 0.4 s suggests a path length of only 25 km is needed for the upper layer of anisotropy. The top of the slab lies at only 20 km under SNZO [*Reading et al.*, 2001], so the upper layer of anisotropy need only extend to 45 km depth. The lower layer could represent anisotropy in the sub-slab asthenosphere. Assuming 1.2–1.4% anisotropy in the asthenosphere as determined above, and β_0 of 4.4 km/s, about 310–370 km of material is needed to explain 1.0 s of splitting, requiring that most of the mantle above the transition zone would need to be anisotropic.

In the first studies of SNZO, before good coverage was available, single layers of plate-boundary parallel fast polarizations were presented [*Vinnik et al.*, 1992]. Since most stations have only a few years of data, it is possible that further measurements will yield more complicated anisotropy throughout the rest of the region, as well. Indeed, some variations in parameters with back azimuth have been observed [*Klosko et al.*, 1999; *Hofmann*, 2002; *Marson-Pidgeon and Savage*, 2004a; *Greve et al.*, 2006], but were not easily interpreted due to the paucity of data.

3.1.5. Pn anisotropy away from the subducting slab. Away from subduction zones, other methods must be used to determine how anisotropy changes with depth. We infer anisotropy in the uppermost mantle lid from *Pn* measurements. In addition to the studies described above of *Pn* anisotropy in the subducting plate, four measurements were made using the controlled source lines from the SIGHT experiment and an earthquake refraction line, which crossed at nearly perpendicular locations (Figure 1; Table 2) [*Scherwath et al.*, 2002; *Baldock and Stern*, 2005; *Bourguignon et al.*, 2007].

Because only two azimuths were sampled in each case, the fast orientation of anisotropy cannot be determined. How-

ever, assuming that the fast orientation for *Pn* was the same as that from nearby splitting measurements allowed a percent anisotropy to be calculated. If *Pn* anisotropy were instead in the same ~60° orientation as that further north [*Galea*, 1993; *Smith and Ekstrom*, 1999], then the calculated anisotropy would have been higher. A value of 8% could fit within the error bars of measurements along the continental portion of the SIGHT T2 lines, while $0 \pm 2\%$ anisotropy was determined off the East coast along line T3 [*Baldock and Stern*, 2005] (Correction in preparation, personal communication) (Figure 1).

3.1.6. Estimates of anisotropy depth from lateral changes in splitting. Fresnel Zone arguments can be used in a few spots on South Island to determine the depth of changes in anisotropy. The change between the southern region, with fast polarizations that are 20°, and the central Alpine fault, with fast polarizations of 45°, occurs between stations separated by 100 km. Similarly, the change between small delay time at the southernmost station west of the fault (JACA), and larger delay times elsewhere, occurs in 100 km or less (Figure 1; [*Klosko et al.*, 1999; *Duclos et al.*, 2005]. If we require the Fresnel zones to be distinct, they should be no larger than 50 km in radius. If we allow some overlap, then a larger radius would be allowed. The average period of *SKS* waves for the SAPSE data is about 12 s (see Figure 4.5 in *Audoine* [2002]). The smallest Fresnel zone radius for this period is about 60 km at the surface [*Rümpker and Ryberg*, 2000]; thus the Fresnel zones should overlap somewhat at all depths. Calculations using another method [*Sheriff*, 1980] yield a 50 km radius Fresnel zone at a depth of about 100 km. If we allow a Fresnel zone of radius of 100 km, then the method of *Rümpker and Ryberg* [2000] yields a depth of 175 km, while that of Sheriff yields 350 km. This does not place a strong constraint on the depth of the variation, but suggests that it is likely to be in the lithosphere.

3.2 Lateral Extent of Anisotropy at Surface and at Depth

Strong, consistent anisotropy in South Island and southern North Island gives way, north of about 39°S, to more complicated anisotropy (Figure 1). Local splitting measurements in particular give widely varying fast polarizations, and delay times from *SKS* phases range up to 5 s or more [*Audoine et al.*, 2004; *Greve et al.*, 2006]. These changes are likely related to the back-arc spreading and volcanism in this region, and we do not discuss those results further, but refer interested readers to the references in Tables 1–3.

Splitting changes also at Chatham Island, where no splitting is observed despite good signals and a wide back-azimuthal coverage, and at MacQuarie Island, where a different fast polarization is observed (Figure 1).

Although the general pattern of *SKS* phase anisotropy is broadly consistent across the rest of NZ, there are variations that may be attributed to laterally varying deformation. Station JACA, which is the southernmost station on the Australian side of the Alpine Fault, has a much smaller delay time than the other stations (Figure 1). There is also a change in fast polarization from about 45°, i.e., plate-boundary parallel, in northern and central South Island, to about 20°, which is 25° counterclockwise from the plate boundary strike, in southern South Island. This change occurs at about the same position as SIGHT line T2, which is also the point

at which South Island narrows to the north. We call this line the "Waistline". In contrast to this N-S variation, *SKS* splitting measurements show no systematic variation of either φ or *dt* with distance from the Alpine fault However, in the upper crust, much of the deformation is localised along discrete faults, with about 450 km displacement along the Alpine fault [*Sutherland*, 1995]. Thus, the zone of deformation inferred from anisotropy is much wider in the mantle than the crust.

In the Marlborough region of northern South Island, variations of splitting fast polarizations with path for lo-

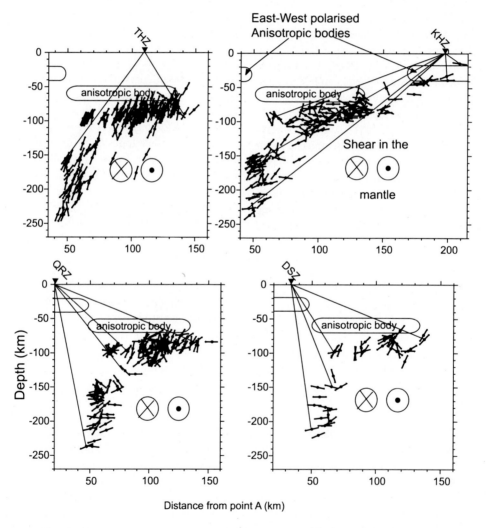

Figure 5. From *Audoine et al.* [2000]. Fast polarizations plotted at hypocenters of events, for stations THZ, QRZ, KHZ and DSZ, which all had 1-Hz seismometers, and which are shown in Figure 6. Orientation of bar represents horizontal orientation; e.g., apparent vertical bar represents N-S and apparent horizontal represents E-W polarizations. View is in cross section AB shown in Figure 6. The body in the middle of the plots represents the low-velocity zone first recognized by *Eberhart-Phillips and Reyners* [1997]. Phases that sample the anisotropic body all give the same fast polarization, 20°. Phases traveling west and east of the body are interpreted to measure anisotropy in the crust due to microcracks oriented in the main stress field orientation.

cal earthquakes were used to constrain the width and depth of a region of fault-parallel anisotropy to be about 75 km wide and between about 30-80 km depth. [*Audoine et al.*, 2000] (Figures 5, 6). Crustal anisotropy in the deforming region was inferred to be fault-parallel [*Audoine et al.*, 2000], which was supported by a later study of shallower, crustal earthquakes [*Balfour et al.*, 2005]. These orientations were also confirmed by an inversion for *P* anisotropy based on arrival times from local earthquakes [*Eberhart-Phillips and Henderson*, 2004]. Crustal anisotropy with an E-W fast orientation, parallel to the prevailing stress orientation, was inferred at least for stations east and west of the deforming region [*Audoine et al.*, 2000] (Figures 5, 6). Thus, in the

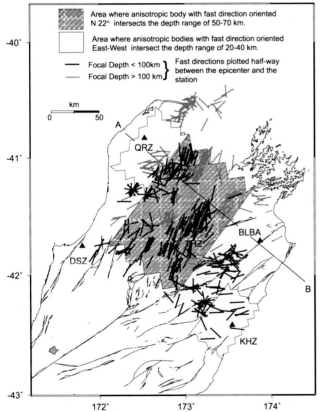

Figure 6. From *Audoine et al.* [2000]. Fast polarizations for all epicenter-station pairs in South Island, plotted halfway between the epicenter and the station. In the central Marlborough Fault System, around station THZ and further north, fast polarizations are mainly aligned at 22°. On the west side of the area, shallow events show a good agreement with results of the central Marlborough system whereas deep events give fast polarizations at 80°. On the east side of the area both shallow and deep events show less correlation than elsewhere. Patterned areas represent a map view of the anisotropic regions defined in Figure 5. AB is the line that defines the cross section presented in Figure 5.

Marlborough region, the zone of deformation appears to be widening with depth. This is typical of deformation from a strike-slip fault above a viscoelastic region, in which the bottom boundary is fixed, and has been suggested as a common mechanism for strike-slip margins in general [*Vauchez and Tommasi*, 2003], and South Island in particular [*Savage et al.*, 2004; *Duclos et al.*, 2005].

Interpretation of *Pn* measured on SIGHT Line T1 also suggests a narrower zone of anisotropy within the mantle lid, only 60 km wide, extending from 35 km west to 25 km east of the surface trace of the Alpine Fault [*van Avendonk et al.*, 2004]. This interpretation was based on slow *Pn* speeds of 7.8 km/s near the Alpine Fault, compared to published *Pn* speeds of 8.2 km/s along sight line 3W [*Melhuish et al.*, 2005]. The wave speeds determined in the van Avendonk study might represent a lower limit to the width of the low *Pn* speed/high anisotropy region, because it was calculated from a smoothed inversion, which will tend to return smaller variations than other methods. A careful study of *Pn* speeds at the intersection of lines T1 and T2 with T3 suggested that the anisotropy continues to the edge of the continent [*Baldock and Stern*, 2005], correction in preparation.

Pn measurements along line T2, just 50-60 km south of line T1 in central South Island suggest a much wider zone of deformation within the mantle lid. The highest *Pn* anisotropy measured in New Zealand was at the intersection of lines T2 and 3W; anisotropy was inferred to be 10 ± 3%, located within the Australian Plate, about 30 km west of the surface trace of the Alpine Fault (Figure 1; [*Scherwath et al.*, 2002]). This was measured from Vp = 7.68 ± 0.15 km/s along line 2, within 50 km of the intersection with line 3W, and 8.6 ± 0.14 km/s along line 3W, which matched that measured at station JACA for the same shots. A NE/SW oriented earthquake refraction line running nearly parallel but east of the Alpine Fault, yielded Pn speeds of 8.5 ± 0.2 km/s; compared with intersecting line T2, *Pn* anisotropy was determined as 7–13%, depending on the orientation chosen for the anisotropy [*Bourguignon et al.*, 2007.] *Pn* anisotropy is near zero at the intersection of lines T3 and T2 [*Baldock and Stern*, 2005](correction in preparation). A western limit to the mantle deformation of 100 km was determined from a change in *Pn* wave speed along SIGHT line T2, which was inferred to correspond to a change in anisotropy [*Scherwath et al.*, 2002]. Similar methods determine an eastern limit at the shoreline along line T2 [*Baldock and Stern*, 2005]. Station JACA in the SAPSE deployment yielded small splitting. Individual measurements from 6 events each yielded no splitting, but stacking of the events yielded an average of $\phi = 24°$, dt = 0.6 s [*Klosko et al.*, 1999]. This was the smallest value measured from *SKS* measurements on South Island. The smaller anisotropy at JACA may be due to its location closer to the edge of the continental mass. Station JACA is the one with the smallest *SKS* splitting, so that the largest *Pn* anisotropy is measured in nearly the same lateral position as the smallest *SKS* splitting. This suggests that the anisotropy in this region is larger in the mantle lid than at depth. An alternative possibility is that the anisotropy is changing not in its strength, but in its plunge; if the symmetry axis plunges, Pn anisotropy will be more strongly affected than splitting orientation [e.g., see diagrams in *ben Ismail and Mainprice*, 1998].

3.2.1. Change in fast SKS polarizations to south. North of the "Waistline," *SKS* splitting polarizations are plate-boundary parallel. South of the Waistline, polarizations are oriented at about 25° to the plate boundary. Along Line T2, the zone of low *Pn* speeds interpreted to be caused by anisotropy extends from 100 km west to 150 km east of the plate boundary, and the two western-most broadband stations GLAA and MTCA, which are closest to the plate boundary, yield fast polarizations at 20° to the plate boundary. The other two stations along line T2, MTJA and TIMA, are more like those further north, with plate-boundary parallel fast polarizations.

3.2.2. Fiordland. In the Fiordland region, i.e., southernmost New Zealand, the Australian plate is subducting obliquely under the Pacific plate. The fast polarizations of *SKS* phases in this region are nearly identical to that of the rest of the stations south of the Waistline. Local earthquakes in the Fiordland region showed interesting patterns of deformation that vary significantly from the consistent *SKS* polarizations, and may be related to patterns of asthenospheric flow in the region [*Gledhill and Savage*, 2002; *Duclos et al.*, 2005].

4. DISCUSSION: TECTONIC MODELS

Interpretation of anisotropy can be controversial. In the crust, the role of fluid-filled microcracks oriented with the present stress field vs. the role of aligned minerals is debated [*Crampin*, 1994; *do Nascimento et al.*, 2004]. Similarly, the contribution of present or past lithospheric strain versus the contribution of present asthenospheric flow is debated for *SKS* splitting [*Vinnik et al.*, 1992; *Silver*, 1996]. Measurements in New Zealand bear on all these possibilities.

4.1. Crustal Anisotropy—Minerals or Stress?

Shear-wave splitting from shallow local earthquakes in the Marlborough and Wellington regions show a variety of fast polarizations, with a tendency for them to be fault-parallel despite consistent stress orientations that are aligned at high angles to the fault planes; this has been interpreted to be caused by near-fault structure and mineral

alignment [*Gledhill*, 1993b; *Balfour et al.*, 2005]. On the other hand, crustal anisotropy with an E-W fast orientation, parallel to the prevailing stress orientation, was inferred for stations east and west of the deforming region [*Audoine et al.*, 2000; *Eberhart-Phillips and Henderson*, 2004] (Figure 6). Also, changing anisotropy with changing volcanic activity on Mt. Ruapehu on North Island occurs too rapidly for reorientation of minerals, and is most likely explained by stress changes reorienting cracks and microcracks [*Miller and Savage*, 2001; *Gerst and Savage*, 2004]. Therefore, in New Zealand, both stress-controlled microcracks and structure-controlled mineral orientations affect the anisotropy.

4.2. Mantle Anisotropy

The strong consistency of *SKS* splitting results across most of New Zealand is surprising, given the wide range of tectonic settings in the region. Three different tectonic boundaries exist at present: oblique subduction of the Pacific Plate beneath the Australian plate under North Island and northern South Island (Marlborough region), transpression across the Alpine Fault, and oblique subduction of the Australian plate under the Pacific plate beneath the Fiordland region in southern South Island. In other regions with changing plate boundaries, fast polarizations have rotated from slab-parallel to slab-perpendicular in such a manner that they seemed to follow flow around slab edges [*Peyton et al.*, 2001]. However, in our study region, most of the fast polarizations are parallel or subparallel to the plate boundary, and the delay times are nearly constant (Figures 1, 2). These attributes suggest a common mechanism for anisotropy across the entire region. One possibility would be that very deep anisotropy, for instance, in the D'' layer, controlled all the measurements, but the back azimuthal and station-to-station variations observed still would require some shallower anisotropy [*Hofmann and Savage*, 2006]. Interpretations of the anisotropy have usually differed between whether studies were focused on North Island, where asthenospheric flow was usually invoked [e.g., *Gledhill and Gubbins*, 1996; *Gledhill and Stuart*, 1996; *Marson-Pidgeon et al.*, 1999], or South Island, where present or past lithospheric strain is inferred [e.g., *Klosko et al.*, 1999; *Molnar et al.*, 1999; *Duclos et al.*, 2005].

3.3. Asthenospheric Flow

Trench-parallel fast polarizations, and the increase in splitting with depth for local earthquakes with paths that traveled mainly beneath the slab, were interpreted to be caused by trench-parallel flow beneath the slab [*Gledhill and Stuart*,

1996]. Nearly identical traces were observed on *SKS* phases for stations throughout southern North Island, independent of distance from the trench, and therefore path length through the asthenosphere [*Marson-Pidgeon et al.*, 1999] (Figures 1; 2). This consistency was used to argue that the flow continued above the slab as well [*Marson-Pidgeon et al.*, 1999]. Because the anisotropy beneath the slab occurs where little water or melt is likely, and anisotropy above the slab is in a region with low heat-flow, no volcanism, and none of the low-velocity, low-attenuation, bodies usually associated with dewatering of a slab, it seems unlikely that mechanisms of "b-fabric" [*Jung and Karato*, 2001] or melt-induced anisotropy [*Holtzman et al.*, 2003], could be causing trench-parallel fast polarizations with flow perpendicular to the slab. Therefore, if asthenospheric flow is causing the olivine alignment under North Island, it's possible that the (20° or 200°-directed) flow inferred from the lower layer in the two-layer modeling [*Marson-Pidgeon and Savage,* 2004a], is continuing beneath South Island as well. The change in average fast polarization from trench-parallel on stations north of the Waistline to 20° on stations to the south, could be caused by the same lower layer in both cases, with an upper layer disappearing in southern South Island.

If the anisotropy in North Island is caused by asthenospheric flow, then the lack of observed splitting around the presumed slab edge in Marlborough could suggest that the slab abuts the cold, transpressional root that has been inferred beneath central South Island from *P* residuals and from gravity measurements [*Stern et al.*, 2000]. *Reyners and Robertson* [2004] have also suggested the slab may continue, on the basis of focal mechanisms within the Benioff zone that do not change with distance from the inferred slab edge. A large *SKS* delay time (3.0 ± 0.5 s) for station WHZ (compared to the average of 1.7 s), located above the eastern edge of the Fiordland block, near the right-lateral east Fiordland fault, was interpreted as being caused by a combination of shear on the fault with the general mantle shear due to the plate boundary deformation in the region, and to present trench-parallel asthenospheric flow around the subducting slab [*Duclos et al.*, 2005].

4.4. Lithospheric Deformation

However, other arguments suggest lithospheric shear may be more important than asthenospheric flow. Delay times on local *S* phases are notoriously difficult to measure, and are quite scattered (e.g., see Figures 3, 4). The stations with the strongest evidence for increasing delay times with earthquake depth were those within the North Island Dextral Fault Belt (NIDFB) [*Beanland*, 1995] (SNZO, LMAU, LBLU and LKIR, LASH and LMOA). Stations east and west of the NIDFB showed little dt increase with earthquake depth. This suggests that the local earthquakes may be "seeing" different properties than the *SKS* phases with their longer periods. Studies of local [*Gledhill and Stuart,* 1996] and ScS phases [*Marson-Pidgeon and Savage*, 2004b] on the closely-spaced Tararua array also showed stronger variation than the *SKS* phases had shown, suggesting lateral variations that are measurable at short periods (up to 4 s) but indistinguishable at the (10–16 s) periods of the *SKS* phases. The apparent increase in delay time in the asthenosphere beneath North Island, measured with local events traveling through the asthenosphere, could stem from frequency dependence of the delay times, in which longer period waves return larger splitting [*Marson-Pidgeon and Savage*, 1997]. Deeper earthquakes tend to have longer periods because only the bigger ones are well-recorded. Therefore an increase in splitting with increasing period would result in an apparent increase in splitting with earthquake depth. If this is the case, then the anisotropy could be fit within the upper plate and the slab, and lithospheric deformation associated with shearing, discussed below, could be causing the anisotropy in southern North Island.

Pn measurements yield similar anisotropy of around 8% everywhere from southern North Island and the Pacific plate northeast of North Island, through to South Island (Figure 1); thus one interpretation could be that the anisotropy is formed in the same manner, e.g., that it comes from the formation of the Pacific Plate and is not related to the anisotropy due to present deformation. Two problems with this interpretation are that (1) the high *Pn* anisotropy measured west of the Alpine Fault in central South Island is presumably beneath the Australian plate [*Scherwath et al.*, 2002], and (2) the lack of strong anisotropy measured to the east on SIGHT lines T1 and T3 [*Baldock and Stern*, 2005] argues against pervasive anisotropy.

A global study of *Pn* anisotropy yielded a maximum of 9.6% [*Smith and Ekstrom*, 1999]. Therefore, the *Pn* anisotropy measured in New Zealand is close to the highest expected values, and should correspond to the largest strains. Thus, as argued below and in *Savage et al.* [2007b], it seems likely that the anisotropy was formed in a strong shearing environment in the past.

The nearly plate-boundary parallel fast orientation of anisotropy on South Island has been used to argue that anisotropy is caused by cumulative shear strain in the lithosphere from ca 45 Ma to the present [*Klosko et al.*, 1999; *Little et al.*, 2002; *Molnar et al.*, 1999; *Scherwath et al.*, 2002; *Baldock and Stern*, 2005]. *SKS* splitting measurements show little variation of either ϕ or dt with distance from the Alpine fault, except for station JACA, and the changes at the distant stations on Campbell Island and MacQuarie Island (Figure 1). This suggests widespread deformation at the depths at which *SKS* waves are most sensitive. Simple arguments relating the rotation of olivine fast orientations to the orientation of maximum finite stretching, and assuming a 400 km wide deformation region, require distributed deformation throughout the lithosphere and allow anisotropy to be purely strain-controlled in southern South Island [*Molnar et al.*, 1999]. This calculation was refined when a more quantitative limit to the width of deformation was determined from *Pn* measurements [*Baldock and Stern*, 2005]. The smaller width of the shearing region at shallow depths may represent the spreading out of deformation with depth expected in wrench-fault type fabric in which motion is driven from the upper plate, and a bottom region is fixed at great depth [*Vauchez and Tommasi*, 2003; *Savage et al.*, 2004; *Duclos et al.*, 2005]. Some enhancement of rotation in the northern regions is needed to explain the plate-boundary parallel fast polarizations, either through dynamic recrystallization [*Zhang and Karato*, 1995] or pure shear with mantle thickening [*Little et al.*, 2002]. Coincidence of the splitting fast polarizations with the orientation of structurally-determined finite strain in the crust was used to argue that deformation is coherent between the crust and mantle, suggesting strong coupling between them [*Little et al.*, 2002]. It is important to remember that shear strain can only explain anisotropy between 45° to the shear plane, and parallel to the shear plane, i.e., between 0° and 45° for the Alpine Fault's strike of 45°. Therefore, the 60° fast orientations measured with *Pn* near Cook Strait could not be caused by plate boundary shear alone.

Numerical modeling has attempted to quantitatively test these hypotheses. Using the present velocity field, GPS measurements [*Beavan et al.*, 1999] can be fit with either discrete faults reaching through the lithosphere, or with continuous lithospheric deformation [*Moore et al.*, 2002]. However, to fit the shear-wave splitting, continuous lithospheric deformation is required. The fast polarizations were matched well in both the southern and northern regions by deformation of a continuous lithospheric sheet if plate motion from the last 6.5 Ma as determined by *Walcott* [1998] were imposed (Figure 7). However, the small strains in the southern regions should have produced smaller delay times if normal relations between delay times and deformation are used [*ben Ismail and Mainprice*, 1998]. A systematic mismatch in central South Island, in which the fast polarizations are closer to fault-parallel than the calculated finite strain ellipses, may also require some dynamic recrystallisation.

Geodynamic modeling assuming constant or layered viscosity could match the fast polarizations if strain equivalent to the 45 Ma of deformation occurred in central South Island, and if smaller strain occurred in southern South Island [*Savage et al.*, 2004]. Geodynamic modeling including con-

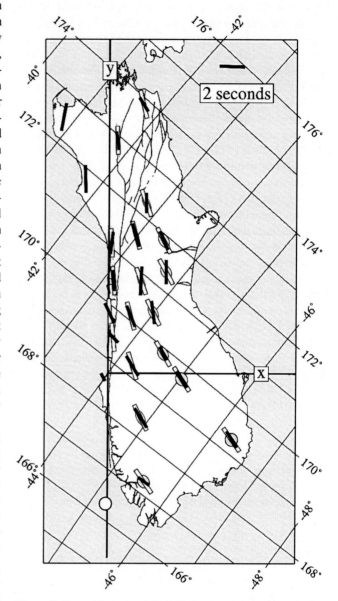

Figure 7. From *Moore et al.* [2002]. Observed and calculated fast polarizations for shear waves recorded in South Island, New Zealand. Black symbols show polarizations observed by *Klosko et al.* [1999], with length of symbol proportional to the delay time. Ellipses represent the finite strain calculated from a model velocity field based on the present-day field of velocity in central South Island. Fast polarizations calculated from the finite strain are shown by white bars, parallel to the major axis of the ellipse. Thick lines show the coordinate system used by *Moore et al.* [2002] for the strain calculation.

trasting rheologies in oceanic and continental lithosphere also suggests that weak continental lithosphere may spread out the mantle deformation across the entire continental region [Savage et al., 2007b]. Station JACA in the SAPSE deployment yielded little splitting. This station is close to the Alpine Fault, but on its western side. It seems unlikely that vertical variation in anisotropy could be causing waveform complexity, since the Pn and SKS directions are similar (Figure 1.) One possibility is that a lateral boundary of anisotropy is close to this station. It is within about 50 km of the transition from oceanic bathymetry to the continental material of the Lord Howe Rise; it is likely that the station is experiencing the effect of the strain preferentially occurring beneath the continental material. Thus, smaller strain in the region, or perhaps some interference due to nearby lateral variations in anisotropy, could be causing the smaller splitting.

4.5. Past vs. Present Deformation

Recent geodynamic modeling of South Island deformation, in which temperature-dependent viscosity is included [Savage et al., 2007b], suggests that the cold root present now under South Island should inhibit anisotropy formation. This implies that the anisotropy observed there may come from the past lithospheric shearing that occurred prior to the convergence that caused the root. It is not clear if such shearing continues into North Island, but it seems likely; the North Island Dextral Fault belt is a current feature that extends across much of the island [Beanland, 1995], and past shear is observed all along the fault, including the aeromagnetic anomalies in Northland [Eccles et al., 2005]. We suggest, therefore, that the splitting observed on the SKS phases in southern North Island and South Island all come from the same mechanism: shearing in the lithosphere due to the strike-slip deformation that took place prior to 10 Ma. However, an alternative explanation, that present deformation through trench-parallel asthenospheric flow continues all through the region, cannot be ruled out.

The Waistline (line T2) appears to cross near a fundamental, and abrupt, change in the mantle. North of the line, SKS splitting measurements are plate-boundary parallel, and anisotropy measured from Pn and local earthquakes suggests a much narrower zone of anisotropy, of 60–75 km width, than the splitting from SKS phases, which are caused by anisotropy along the entire path from the core-mantle boundary to the surface. South of the Waistline, SKS measurements are oriented at about 25° to the plate boundary. Along Line T2, the zone of low Pn speeds interpreted to be caused by anisotropy extends from 100 km west to 150 km east of the plate boundary. Other changing properties south of the Waistline are: Deformation in the crust gets wider in this region [Norris and Cooper, 2001]; the crustal root beneath the Southern Alps thickens; the Bouguer gravity anomaly gets more positive [Bourguignon et al., 2007]; earthquakes are more broadly distributed [Leitner et al., 2001], and the maximum shear strain rate is smaller [Beavan and Haines, 2001]. The buttressing effect of the Fiordland block and/or the mantle east of Fiordland could be contributing to all these properties by providing an "indentor" pushing against the full motion of the Pacific Plate [Bourguignon et al., 2007]. Alternatively, the region south of the Waistline may correspond to a widening of the region of shear to the south, with accompanying smaller average strain [Baldock and Stern, 2005].

5. SUMMARY

Between Latitude 39°S and the Waistline (coincident with SIGHT line T2), seismic anisotropy beneath New Zealand as determined from station averages of SKS phases is consistent, with plate-boundary parallel anisotropy averaging ϕ 45° and dt of 1.7 s, and extending throughout all the stations on the continental landmass, for a width of at least 200 km. South of the Waistline, station JACA, located just west of the Alpine Fault, has little splitting. Other stations south of the Waistline have similar delay times to those in central NZ, but the fast polarizations are rotated approximately 20° counterclockwise, oblique to the plate boundary. Lithospheric shear and trench-parallel flow may both be combining to form these features.

Local earthquakes and Pn anisotropy show more complicated patterns. North of the Waistline, the zone of plate-boundary parallel anisotropic fast orientations, and hence the interpreted zone of shear, appears to widen with depth, so that the zone within the lithospheric lid is only 60–75 km wide. South of the Waistline, shear in the mantle lid determined from Pn anisotropy is wide (200 km), extending further west than the SKS measurements require. The zone has tentatively been attributed to deformation getting weaker but spreading more broadly in this region.

Crustal anisotropy appears to be caused by mineral orientation in some regions and stress-aligned microcracks in others. It does not appear to be affecting the SKS measurements.

Acknowledgements. N. Koehler performed some of the preliminary measurements of the Marlborough data. Discussions with K. Gledhill and E. Audoine have been particularly helpful. Vadim Levin and an anonymous reviewer provided insightful comment. The thinking behind this paper was spurred by two grants from the NZ Marsden Fund, and grants from the U.S. National Science Foundation and NZ Foundation for Research, Science and Technology.

REFERENCES

Anderson, D. L. (1961), Elastic wave propagation in layered anisotropic media, *J. Geophys. Res.*, *66*, 2953–2963.

Audoine, E. (2002), Upper mantle and crustal seismic anisotropy across the Pacific-Australian plate boundary, New Zealand, PhD thesis, 219 pp, Victoria University of Wellington, Wellington.

Audoine, E., M. K. Savage, and K. Gledhill (2000), Seismic anisotropy from local earthquakes in the transition region from a subduction to a strike-slip plate boundary, New Zealand, *J. Geophys. Res.*, *105*, 8013–8033.

Audoine, E. L., M. K. Savage, and K. R. Gledhill (2004), Anisotropic structure under a back-arc spreading region, the Taupo Volcanic Zone, New Zealand, *J. Geophys. Res.*, *109*, doi:10.1029/2003JB002932 (002922 pp.).

Babuska, V., and M. Cara (1991), *Seismic Anisotropy in the Earth*, 217 pp., Kluwer Academic Publishers, Dordrecht/Boston/London.

Backus, G. E. (1962), Long-wave elastic anisotropy produced by horizontal layering, *J. Geophys. Res.*, *67*, 4427–4440.

Baldock, G., and T. Stern (2005), Width of mantle deformation across a continental transform: Evidence from upper mantle (Pn) seismic anisotropy measurements, *Geology*, *33*, 741–744.

Balfour, N. J., M. K. Savage, and J. Townend (2005), Stress and crustal anisotropy in Marlborough, New Zealand: Evidence for low fault strength and structure-controlled anisotropy., *Geophys. J. Int.*, *163*, 1073–1086, doi:1010.1111/j.1365-1246X.2005.02783.x.

Barruol, G., and D. Mainprice (1993), A quantitative evaluation of the contribution of crustal rocks to the shear-wave splitting of teleseismic *SKS* waves, *Phys. Earth Planet. Inter.*, *78*, 281–300.

Beanland, S. (1995), The North Island Dextral Fault Belt, Hikurangi subduction margin, New Zealand, PhD thesis, 234 pp, Victoria University of Wellington, Wellington.

Beavan, J., and A. J. Haines (2001), Contemporary horoizontal velocity and strain fields of the Pacific-Australian plate boundary zone through New Zealand, *J. Geophys. Res.*, *106*, 741–770; doi:710.1029/2000JB900302.

Beavan, J., M. Moore, C. Pearson, M. Henderson, B. Parsons, S. Bourne, P. England, D. Walcott, G. Blick, D. Darby, and K. Hodgkinson (1999), Crustal deformation during 1994-1998 due to oblique continental collision in the central Southern Alps, New Zealand, and implications for seismic potential of the Alpine Fault, *J. Geophys. Res.*, *104*, 25,233–25,255.

ben Ismail, W., and D. Mainprice (1998), An olivine fabric database: an overview of upper mantle fabrics and seismic anisotropy, *Tectonophysics*, *296*, 145–157.

Birch, F. (1960), The velocity of compressional waves in rocks to 10 kilobars, Part 1, *J. Geophys. Res.*, *65*, 1083–1102.

Bourguignon, S., T. A. Stern, and M.K. Savage (2007), Crust and mantle thickening beneath the southern portion of the Southern Alps, New Zealand, *Geophys. J. Int.*, *168*(2), (Feb.) pp. 681–690, doi:10.1111/j.1365-246X.2006.03208.x.

Brisbourne, A. G., G. Stuart, and J. M. Kendall (1999), Anisotropic structure of the Hikurangi subduction zone, New Zealand—integrated interpretation of surface-wave and body-wave observations, *Geophys. J. Int.*, *137*, 214–230.

Byerly, P. (1938), The earthquake of July 6, 1934: Amplitudes and first motion, *Bull. Seismol. Soc. Am.*, *28*, 1–13.

Bystricky, M., K. Kunze, L. Burlini, and J.-P. Burg (2000), High shear strain of olivine aggregates: rheological and seismic consequences, *Science*, *290*, 1564–1567.

Chadwick, M. (1997), Hikurangi Margin Seismic Experiment, PhD thesis, 230 pp, Victoria University of Wellington, Wellington.

Christensen, N. I. (1966a), Shear wave velocities in metamorphic rocks at pressures to 10 kilobars, *J. Geophys. Res.*, *71*, 3549–3556.

Christensen, N. I. (1966b), Elasticity of ultrabasic rocks, *J. Geophys. Res.*, *71*, 5921–5931.

Christensen, N. I. (1984), The magnitude, symmetry and origin of upper mantle anisotropy based on fabric analyses of ultramafic tectonites, *Geophys. J. R. Ast. Soc*, *76*, 89–111.

Cochran, E. S. (1999), SKS shear-wave splitting observations in the mantle beneath East Cape, New Zealand, 31 pp, Univ. of Leeds, Leeds.

Crampin, S. (1981), A review of wave motion in anisotropic and cracked elastic media, *Wave Motion*, *3*, 343–391.

Crampin, S. (1994), The fracture criticality of crustal rocks, *Geophys. J. Int.*, *118*, 428–438.

Davey, F. J., D. Eberhart-Phillips, M. D. Kohler, S. Bannister, G. Caldwell, S. Henrys, M. Scherwath, T. Stern, and H. van Avendonk (this volume), Geophysical structure of the Southern Alps orogen, South Island, New Zealand.

Debayle, E., Kennett, B., & Priestley, K. (2005), Global azimuthal seismic anisotropy and the unique plate-motion deformation of Australia, *Nature*, *433*, 509–512.

do Nascimento, A. F., F. H. R. Bezerra, and M. K. Takeya (2004), Ductile Precambrian fabric control of seismic anisotropy in the Acu Dam area, northeastern Brazil, *J. Geophys. Res.*, *109*, doi:10.1029/2004JB003120.002004.

Duclos, M., M. K. Savage, A. Tommasi, and K. R. Gledhill (2005), Mantle Tectonics beneath New Zealand Inferred from SKS Splitting and Petrophysics, *Geophys. J. Int.*, *163*, 760–774, doi:710.1111/j.1365-1246X.2005.02793.x.

Eberhart-Phillips, D., and C. M. Henderson (2004), Including anisotropy in 3-D velocity inversion and application to Marlborough, New Zealand, *Geophys. J. Int.*, *156*, 237–254.

Eberhart-Phillips, D., and M. Reyners (1997), Continental subduction and three-dimesional crustal structure: the northern South Island, New Zealand., *J. Geophys. Res.*, *102*, 11,843–11,861.

Eccles, J. D., J. Cassidy, C. A. Locke, and K. B. Spoerli (2005), Aeromagnetic imaging of the Dun Mountain ophiolite belt in northern New Zealand; insight into the fine structure of a major SW Pacific terrane suture, *J. Geological Soc. London*, *162*, 723–735.

Forsyth, D. W. (1975), The early structural evolution and anisotropy of the oceanic upper mantle, *Geophys. J. R. Ast. Soc*, *43*, 103–162.

Galea, P. (1993), Upper mantle anisotropy in the S. W. Pacific from earthquake travel-time analysis, *Phys. Earth Planet. Inter.*, *76*, 229–239.

Gerst, A., and M. K. Savage (2004), Seismic anisotropy beneath Ruapehu Volcano: A possible eruption forecasting tool, *Science*, *306*, 1543–1547.

Gledhill, K. R. (1991), Evidence for shallow and pervasive seismic anisotropy in the Wellington region, New Zealand, *J. Geophys. Res.*, *96*, 21,503–21,516.

Gledhill, K. R. (1993a), Shear waves recorded on close-spaced seismographs: I. Shear-wave splitting results, *Can. J. Expl. Geophys.*, *29*, 285–298.

Gledhill, K. R. (1993b), Shear waves recorded on close-spaced Seismographs: II. The complex anisotropic structure of the Wellington Peninsula, New Zealand, *Can. J. Expl. Geophys.*, *29*, 299–314.

Gledhill, K. R., and D. Gubbins (1996), *SKS* splitting and the seismic anisotropy of the mantle beneath the Hikurangi subduction zone, New w Zealand, *Phys. Earth Planet. Inter.*, *95*, 227–236.

Gledhill, K. R., and M. K. Savage (2002), Seismic anisotropy above the Fiordland subduction zone, *New Zealand Seismol. Res. Lett.*, *73*, 222.

Gledhill, K. R., and G. Stuart (1996), Seismic anisotropy in the fore-arc region of the Hikurangi subduction zone, New Zealand, *Phys. Earth Planet. Inter.*, *95*, 211–225.

Godfrey, N. J., N. I. Christensen, and D. A. Okaya (2000), Anisotropy of schists: contributions of crustal anisotropy to active-source seismic experiments and shear-wave splitting observations, *J. Geophys. Res.*, *105*, 27,991–28,007.

Greve, S., S. D. Hofmann, and M. K. Savage (2006), Strong variations in seismic anisotropy across the Hikurangi subduction zone, North Island, New Zealand., *Tectonophysics*, submitted.

Hess, H. (1964), Seismic anisotropy of the uppermost mantle under oceans, *Nature*, *203*, 629–631.

Hofmann, S. D. (2002), Seismic anisotropy in the crust and mantle: A study at the western edge of the Central Volcanic Region, New Zealand, MSc. thesis, 135 pp, Victoria University of Wellington, Wellington.

Holtzman, B. K., D. L. Kohlstedt, M. E. Zimmerman, F. Heidelbach, T. Hiraga, and J. Hustoft (2003), Melt segregation and strain partitioning: Implications for seismic anisotropy and mantle flow, *Science*, *301*, 1227–1230.

Jung, H., and S.-I. Karato (2001), Water-induced fabric transitions in olivine, *Science*, *293*, 1460–1463.

Kirkwood, S. C., and S. Crampin (1981), Surface-wave propagation in an ocean basin with an anisotropic upper mantle: observation of polarization anomalies, *Geophys. J. R. Ast. Soc*, *64*, 487–497.

Klosko, E. R., F. T. Wu, H. J. Anderson, D. Eberhardt-Phillips, T. V. McEvilly, E. Audoine, M. K. Savage, and K. R. Gledhill (1999), Upper mantle anisotropy in the New Zealand region, *Geophys. Res. Lett.*, *26*, 1497–1500.

Laske, G., & Masters, G. (1998), Surface-wave polarization data and global anisotropic structure, *Geophys. J. Int.*, *132*, 508–520.

Leitner, B., D. Eberhart-Phillips, H. Anderson, and J. L. Nableck (2001), A focused look at the Alpine fault, New Zealand: Seismicity, focal mechanisms, and stress observations, *J. Geophys. Res.*, *106*, 2193–2220.

Little, T. A., M. K. Savage, and B. Tikoff (2002), Relationship between crustal finite strain and seismic anisotropy in the mantle, Pacific-Australia plate boundary zone, South Island, New Zealand, *Geophys. J. Int.*, *151*, 106–116.

Mainprice, D. (1990), A FORTRAN program to calculate seismic anisotropy from lattice preferred orientation of minerals, *Comput. Geosci.*, *16*, 385–393.

Mainprice, D., A. Tommasi, H. Couvy, P. Cordier, and D. J. Frost (2005), Pressure sensitivity of olivine slip systems and seismic anisotropy of Earth's upper mantle, *Nature*, *433*, 731–733, doi:710.1038/nature03266.

Mardia, K. V. (1972), *Statistics of Directional Data*, Academic Press, New York and London.

Marson-Pidgeon, K., and M. Savage (1997), Frequency-dependent anisotropy in Wellington, New Zealand, *Geophys. Res. Letters*, *24*, 3297–3300.

Marson-Pidgeon, K., M. K. Savage, K. R. Gledhill, and G. Stuart (1999), Seismic anisotropy beneath the lower half of the North Island, New Zealand, *J. Geophys. Res.*, *104*, 20,277–220,286.

Marson-Pidgeon, K. A., and M. K. Savage (2004a), Modelling shear wave splitting observations from Wellington, New Zealand, *Geophys. J. Int.*, *157*, 853–864, doi:810.1111/j.1365-1246X.2004.02274.x.

Marson-Pidgeon, K. A., and M. K. Savage (2004b), Shear-wave splitting variations across an array in the lower North Island, New Zealand., *Geophys. Res. Lett.*, *31*, doi:10.1029/2004GL021190.

Matcham, I., M. K. Savage, and K. R. Gledhill (2000), Distribution of seismic anisotropy in the subduction zone beneath the Wellington region, New Zealand, *Geophys. J. Int.*, *140*, 1–10.

McKenzie, D. (1979), Finite deformation during fluid flow, *Geophys. J. R. astr. Soc.*, *58*, 689–715.

Melhuish, A., W. S. Holbrook, F. Davey, D. A. Okaya, and T. Stern (2005), Crustal and upper mantle seismic structure of the Australian Plate, South Island, New Zealand, *Tectonophysics*, *395*, 113–135.

Miller, V. L., and M. K. Savage (2001), Changes in seismic anisotropy after a volcanic eruption: Evidence from Mount Ruapehu, *Science*, *293*, 2231–2233.

Molnar, P., H. Anderson, E. Audoine, D. Eberhart-Philips, K. Gledhill, E. Klosko, T. McEvilly, D. Okaya, M. Savage, T. Stern, and F. Wu (1999), Continuous deformation versus faulting through the continental lithosphere of New Zealand, *Science*, *286*, 516–519.

Montagner, J. P. (2002), Upper mantle low anisotropy channels below the Pacific Plate, *Earth and Planet. Science Lett.*, *202*, 263–274.

Moore, M., P. England, and B. Parsons (2002), Relation between surface velocity field and shear-wave splitting in the South Island of New Zealand, *J. Geophys. Res.*, *107*, 2198, doi:2110.1029/2000JB000093.

Nicolas, A., F. Boudier, and A. M. Boullier (1973), Mechanisms of flow in naturally and experimentally deformed peridotites, *Am. J. Science*, *273*, 853–876.

Norris, R. J., and A. F. Cooper (2001), Late Quaternary slip rates and slip partitioning on the Alpine Fault, New Zealand *J. Structural Geol.*, *23*, 507–520.

Nur, A., and G. Simmons (1969), Stress-induced velocity anisotropy in rock: an experimental study, *J. Geophys. Res.*, *74*, 6667–6674.

Okaya, D., N. Christensen, D. Stanley, and T. Stern (1995), Crustal anisotropy in the vicinity of the Alpine Fault Zone , South Island New Zealand., *N. Z. J. Geol. & Geophys.*, *38*, 579–583.

Peyton, V., J. Lees, E. Gordeev, A. Ozerov, V. Levin, J. Park, and M. Brandon (2001), Mantle flow at a slab edge: Seismic anisotropy in the Kamchatka region, *Geophys. Res. Lett.*, *28*, 379–382.

Pulford, A., M. K. Savage, and T. Stern (2003), Absent anisotropy; the paradox of the Southern Alps orogen, *Geophys. Res. Lett.*, *30*, 2051.

Reading, A. M., D. Gubbins, and W. J. Mao (2001), A multiphase seismic investigation of the shallow subduction zone, southern North Island, New Zealand, *Geophys. J. Int.*, *147*, 215–226.

Reyners, M., and E. d. J. Robertson (2004), Intermediate depth earthquakes beneath Nelson, New Zealand, and the southwestern termination of the subducted Pacific Plate *Geophys. Res. Lett.*, *31*, 4.

Rümpker, G., and T. Ryberg (2000), New 'Fresnel-zone' estimates for shear-wave splitting observations from finite-difference modeling, *Geophys. Res. Letters*, *27*, 2005–2008.

Russo, R. M., and P. G. Silver (1994), Trench-parallel flow beneath the Nazca Plate from seismic anisotropy, *Science*, *263*, 1105–1111.

Savage, M. K. (1998), Lower Crustal anisotropy or dipping layers? Effects on receiver functions and a case study in New Zealand, *J. Geophys. Res.*, *103*, 15,069–15,087.

Savage, M. K. (1999), Seismic anisotropy and mantle deformation: what have we learned from shear wave splitting?, *Rev. Geophys.*, *37*, 65–106.

Savage, M. K., K. M. Fischer, and C. E. Hall (2004), Strain modelling, seismic anisotropy and coupling at strike-slip boundaries: Applications in New Zealand and the San Andreas Fault, in *Vertical Coupling and Decoupling in the Lithosphere*, edited by J. Grocott, Tikoff, B., McCaffrey, K. J. W. & Taylor, G., pp. 9–40, Geological Society of London, Special Publication, London.

Savage, M. K., J. Park, and H. Todd (2007a), Velocity and anisotropy structure at the Hikurangi subduction margin, New Zealand from receiver functions, *Geophys. J. Int.*, *168.*, 1034–1050, doi: 1010.1111/j.1365-1246X.2006.03086.x, .

Savage, M. K., A. Tommasi, S. Ellis, and J. Chery (2007b), Modeling strain and anisotropy along the Alpine Fault, South Island, New Zealand, *this volume.*

Scherwath, M., T. Stern, A. Melhuish, and P. Molnar (2002), Pn anisotropy and distributed upper mantle deformation associated with a continental transform fault, *Geophys. Res. Lett.*, *29*, 16-11 – 16-14, doi:10.1029/2001GL014179.

Shearer, P. M., and J. A. Orcutt (1986), Compressional and shear wave anisotropy in the oceanic lithosphere—The Ngendei seismic refraction experiment, *Geophys. J. R. Ast. Soc*, *87*, 967–1003.

Sheriff, R. E. (1980), Nomogram for Fresnel-zone calculation, *Geophysics*, *45*, 968–972.

Silver, P. G. (1996), Seismic anisotropy beneath the continents: probing the depths of geology, *Ann. Rev. Earth Planet. Sci.*, *24*, 385–432.

Silver, P. G., and W. W. Chan (1988), Implications for continental structure and evolution from seismic anisotropy, *Nature*, *335*, 34–49.

Smith, G. P., and G. Ekstrom (1999), A global study of Pn anisotropy beneath continents, *J. Geophys. Res.*, *104*, 963–980.

Stern, T. A., P. Molnar, D. Okaya, and D. Eberhart-Phillips (2000), Teleseismic *P* wave delays and modes of shortening the mantle lithosphere beneath South Island, New Zealand, *J. Geophys. Res.*, *105*, 21,615–21,631.

Stuart, G., D. Francis, D. Gubbins, and G. Smith. (1995), Tararua broadband array, North Island, New Zealand, *Bull. Seismol. Soc. Am.*, *85*, 325–333.

Suggate, R. P., and G. W. Grindley (1972), Geological map of New Zealand, 1:250,000 N.Z. Geol. Surv., Dep. Sci. and Ind. Res., Wellington, New Zealand (NZL), Wellington.

Sutherland, R. (1995), The Australia-Pacific boundary and Cenozoic plate motions in the SW Pacific: Some constraints from Geosat data, *Tectonics*, *14*, 819–831.

Tommasi, A., D. Mainprice, G. Canova, and Y. Chastel (2000), Viscoplastic self-consistent and equilibrium-based modeling of olivine lattice preferred orientations: Implications for the upper mantle seismic anisotropy, *J. Geophys. Res.*, *105*, 7893–7908.

van Avendonk, H. J., W. S. Holbrook, D. A. Okaya, J. K. Austin, F. Davey, and T. Stern (2004), Continental crust under compression: A seismic refraction study of South Island Geophysical Transect I, South Island, New Zealand, *J. Geophys. Res.*, *109*, doi:10.1029/2003JB002790 (002716 pp.).

Vauchez, A., and A. Tommasi (2003), Wrench faults down to the asthenosphere: Geological and geophysical evidence and thermo-mechanical effects, in *Intraplate Strike-Slip Deformation Belts*, edited by F. Storti, R. E. Holdsworth and F. Salvini, pp. 15–34, Geological Society of London, London.

Vinnik, L. P., G. L. Kosarev, and L. I. Makeyeva (1984), Anizotropiya litosfery po nablyudeniyam voln SKS and SKKS, *Dokl. Akad. Nauk USSR 278*, 1335–1339.

Vinnik, L. P., L. I. Makeeva, A. Milev, and A. I. Usenko (1992), Global patterns of azimuthal anisotropy and deformations in the continental mantle, *Geophys. J. Int*, *111*, 433–447.

Walcott, R. I. (1998), Modes of oblique compression: late Cenozoic tectonics of the South Island of New Zealand., *Rev. Geophys.*, *36*, 1–26.

Wenk, H.-R., K. Bennett, G. R. Canova, and A. Molinari (1991), Modelling plastic deformation of peridotite with the self-consistent theory, *J. Geophys. Res.*, *96*, 8337–8349.

Wenk, H.-R., and J. M. Christie (1991), Review Paper: Comments on the interpretation of deformation textures in rocks, *J. Structural Geol.*, *13*, 1091–1110.

Wilson, C. K., C. H. Jones, P. Molnar, A. F. Sheehan, and O. S. Boyd (2004), Distributed deformation in the lower crust and upper mantle beneath a continental strike-slip fault zone; Marlborough fault system, South Island, New Zealand, *Geology*, *32*, 837–840.

Yu, Y., and J. Park (1994), Hunting for azimuthal anisotropy beneath the Pacific Ocean region, *J. Geophys. Res.*, *99*, 15,399–15,421.

Zhang, S., and S.-I. Karato (1995), Lattice preferred orientation of olivine aggregates deformed in simple shear, *Nature*, *375*, 774–777.

Zinke, J. C., and M. D. Zoback (2000), Structure-related and stress-induced shear-wave velocity anisotropy: observations from microearthquakes near the Calaveras Fault in central California, *Bull. Seismol. Soc. Am.*, *90*, 1305–1312.

Crustal Thickness and Pn Anisotropy Beneath the Southern Alps Oblique Collision, New Zealand

S. Bourguignon, M. K. Savage, and T. Stern

Institute of Geophysics, School of Geography, Environment and Earth Sciences,
Victoria University of Wellington, New Zealand

Over-thickened crust and fast, anisotropic mantle material are interpreted beneath South Island, New Zealand, from an earthquake refraction study along the Southern Alps foothills. An 8.54 ± 0.20 km/s Pn speed is estimated along the N60°E striking refraction profile and a maximum crustal thickness of 48 ± 4 km is inferred near Wanaka township, at the southern end of the profile. The crustal thickness represents an 18 km thick crustal root relative to a 30 km coastal average. Thus, the root is 2–3 times thicker than expected for Airy isostatic compensation of the mean ~1000 m Southern Alps topographic load. This suggests that the underlying mantle plays an active role in depressing topography. Comparison of the 8.54 ± 0.20 km/s Pn-speed estimate with cross profiles suggests anisotropy arising from finite strain of the mantle lid rocks. The Pn anisotropy is estimated near Lake Tekapo, at the northern end of the profile, to be a minimum of $6.5 \pm 3.5\%$. We predict a maximum Pn anisotropy of 7–13% and an average isotropic Pn speed of ~8.3 km/s by adopting the fast polarization orientation from previous SKS splitting measurements done at the profile intersection. The Pn speed of 8.3 km/s is consistent with previous studies showing high average Pn speeds below the southern half of South Island and the presence of cold, dense mantle lithosphere.

1. INTRODUCTION

A long standing question on collision zones is how the upper mantle accommodates shortening. Two end-member models, intra-continental subduction [*Wellman*, 1979; *Beaumont*, 1996] and continuous thickening [*Molnar et al.*, 1999; *Stern et al.*, 2000] have been proposed for the Southern Alps of South Island, New Zealand. These two modes of deformation are analogous to the simple shear [*Wernicke*, 1985] and pure shear [e.g., *McKenzie*, 1978] models of extension, respectively. In the subduction-type end-member, deformation by simple shear is localized in a narrow and obliquely-

dipping shear zone at the slab top interface. In the continuous thickening end-member, shortening is accommodated by distributed pure shear. While both end-member models involve cold temperature contrasts in the mantle that produce faster wave speeds, the distribution of deformation and that of seismic anisotropy, i.e., localized on a narrow discontinuity vs. widespread, may help discriminate between them.

Continuous and distributed thickening has been suggested based on teleseismic traveltime residuals [*Kohler and Eberhart-Phillips*, 2002; *Stern et al.*, 2000] showing a symmetric pattern. However, the simple shear model is preferred by a number of numerical models that intend to fit the GPS velocity field across the Southern Alps [*Beavan et al.*, 1999; *Ellis et al.*, 2006; *Liu and Bird*, 2006]. Numerical investigations of the development of continental collision [*Pysklywec et al.*, 2002] have shown that both modes of shortening may be combined depending on the thermal structure and the convergence rate.

A Continental Plate Boundary: Tectonics at South Island, New Zealand
Geophysical Monograph Series 175
Copyright 2007 by the American Geophysical Union.
10.1029/175GM07

Determining Pn speeds and their azimuthal variation with respect to the orientation of the plate boundary can provide insight on physical conditions and deformation experienced in the uppermost mantle. Pn anisotropy combined with SKS-splitting measurements [*Savage*, 1999; *Savage et al.*,

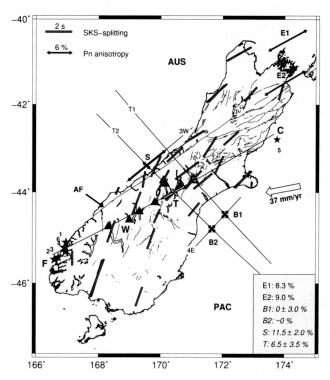

Figure 1. Location of Fiordland-Cheviot refraction study within South Island. AUS and PAC denote Australian and Pacific plates, respectively, AF is the Alpine Fault and light lines denote additional faults. Open arrow indicates the Australian-Pacific relative plate motion [*DeMets*, 1994]. Locations cited in this study are: Cheviot (C), Tekapo (T), Wanaka (W) and Fiordland (F). The Fiordland-Cheviot refraction profile is represented by line F–C. Earthquake sources (stars) numbered 1–6 are described in Table 1. Seismic data were collected by our deployed seismographs (black triangles) and the Geonet permanent station RPZ (open triangle). SIGHT seismic profiles are T1, T2, 3W and 4E [*Okaya et al.*, 2002]. SKS-splitting fast polarization orientations are light gray bars (lengths proportional to delay times; [*Klosko et al.*, 1999; *Duclos et al.*, 2005]). Single double arrows labeled E1 and E2 indicate Pn fast propagation azimuths from a global earthquake study [*Smith and Ekström*,1999]. Crossing long and small arrows denote apparent fast and slow Pn propagation orientations, respectively, measured along intersecting profiles. Apparent measurements are located at S [*Scherwath et al.*, 2002], B1 and B2 [*Baldock and Stern*, 2005, correction in prep.] and T (this study). Double arrows oriented along the absolute and apparent fast azimuths scale with the amount of anisotropy. The anisotropy percentage, i.e., absolute or apparent (italic), is indicated in the lower right corner [after *Bourguignon et al.*, 2007].

this volume] can help quantify anisotropy and constrain the depth and lateral extent of upper mantle deformation.

We used the M_w 7.2 Fiordland (21st of August 2003) aftershocks [*Reyners et al.*, 2003] and a M_L 4.1 event offshore Cheviot on the east coast of South Island (Tab. 1) to determine the Pn speed along a line, herein called Fiordland-Cheviot profile (Figure 1), and infer the azimuthal anisotropy in the mantle lid below the Southern Alps. The thickness of the crust and that of the Southern Alps crustal root were estimated near Wanaka and compared with that expected for Airy isostatic compensation of the Southern Alps topographic load.

2. REFRACTION ANALYSIS

Shortly after the Fiordland mainshock, we deployed an array of seven short-period seismographs in alignment with the Rata Peak (RPZ) Geonet broad-band permanent station and along the eastern foothills of the Southern Alps. The resulting Fiordland-Cheviot profile was oriented N60°E (Figure 1), i.e., 5° and ~15°clockwise from the Alpine Fault and the trend of the Bouguer gravity anomaly, respectively. Five aftershocks of $M_L \geq 5$ were recorded at the SW end of the profile line in Fiordland along with an additional M_L 4.1 earthquake off the coast of Cheviot (Tab. 1, Figure 1) that occurred at the NE end of the profile. Sources at the two ends of the profile and epicentral distances as large as 490 km enabled a simple one-layer refraction travel-time analysis along the root of the Southern Alps. An 8.21 ± 0.27 km/s apparent Pn speed was measured from the offshore Cheviot event, while an average apparent Pn-speed of 8.92 ± 0.18 km/s was inferred from Fiordland aftershocks (error bars are 95% confidence intervals) (Table 1). Single measurements were weighted with the inverse of their standard deviation (Figure 2). A Pn speed of 8.54 ± 0.20 km/s and an apparent dip of 2.5 ± 1.3° SW were calculated by assuming a uniform dipping Moho and by taking an average crustal wave-speed of 6.10-6.23 km/s [*Scherwath et al.*, 2003; *van Avendonk et al.*, 2004]. The Pn-speed of 8.54 ± 0.20 km/s is slightly greater than that inferred in two previous studies using regional earthquakes: 8.3 ± 0.1 km/s [*Haines*, 1979] and 8.4 km/s [*Smith and Davey*, 1984]. The ~2.5° SW dip is much smaller than a ~8°SW apparent dip calculated independently from the Moho depths at intersections of the Fiordland-Cheviot profile with SIGHT T1 [*van Avendonk et al.*, 2004] and T2 [*Scherwath et al.*, 2003]. The discrepancy may result from the assumed uniform dipping Moho. In addition, the use of different inversion methods and, in particular, different smoothing on boundaries may be a reason for the large variation of crustal thickness between SIGHT T1 and T2 [*Van Avendonk et al.*, 2004]. Comparison of this experiment with others of this type in the Sierra Nevada [*Jones et al.*, 1994; *Savage et al.*, 1994; *Ruppert*, 1998;

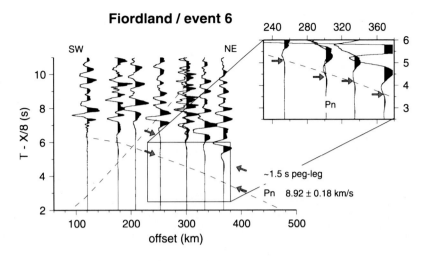

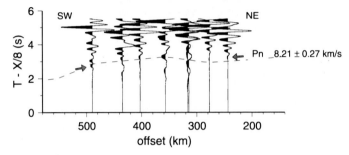

Figure 2. Top: Arrivals from the Fiordland aftershock are bandpass filtered at cut-off and corner frequencies of 0.5-1-5-10 Hz. First-break Pn are indicated by the bottom pair of arrows and single arrows in blow-up on the right, and predicted Pg and Pn travel-time curves by dashed curves (see model Figure 3c). Pn arrivals are followed ca. 1.5 s later by arrivals (~1.5-s peg-leg indicated with top pair of arrows) with much larger amplitude. These second arrivals have the same apparent wave speed as the Pn and are interpreted as an internal reflection ~5 km near the source. The Pn-speed estimate and corresponding 95% confidence interval (right-hand side of graph) is the mean of single regression slopes weighted with their respective standard deviations. Bottom: Arrivals from the ML 4.1 offshore Cheviot event are bandpass filtered at cut-off and corner frequencies of 0.5-1-3-5 Hz. Note the offset axis is in the opposite direction to that of the top figure. First-break Pn are indicated by the pair of arrows and the predicted Pn travel-time curve by a dashed curve (see model Figure 3c). The Pn-speed estimate is the result of a single linear regression and is given with corresponding 95% confidence interval (right-hand side of graph). In both graphs the trace of the third station from the left was shifted by 3.5 s to correct a timing error. However, the pick wasn't included in Pn-speed calculations because of uncertainty in the timing error [after *Bourguignon et al.*, 2007].

Louie et al., 2004] suggests that two factors may have made the observation of Pn along the crustal root possible. (1) The Moho is smoothly dipping along the profile line, as a result of both the obliquity of the profile relative to the Southern Alps crustal root and gentle thickening of the root in the east. (2) Fast wave speeds within the cold mantle lithosphere may contribute to efficient refraction of seismic waves along the Moho boundary.

3. CRUSTAL THICKNESS

Previous crustal studies in South Island [*Davey and Broadbent*, 1980; *Reyners et al.*, 1993; *Eberhart-Phillips*

and Reyners, 2001; *Eberhart-Phillips and Bannister*, 2002; *Scherwath et al.*, 2003; *van Avendonk et al.*, 2003] were integrated into a crustal model taken along our refraction profile (Figure 3c). In addition, the above measurements were used to constrain the dip of the Moho to 2.5°SW, to the southwest of the intersection of the Fiordland-Cheviot profile with SIGHT T2. The uppermost mantle wave speed was set to 8.54 km/s. Ray tracing [*Luetgert*, 1992] on this 2-D crustal model (Figure 3c) shows that rays propagating from Fiordland and offshore Cheviot constrain a ~150 km Moho portion extending from Wanaka (southern SI) to the intersection of SIGHT T2 with the Fiordland-Cheviot profile, i.e., west of Tekapo (central SI). Along this profile,

Table 1. Events 1–4 and 6 are the Fiordland aftershocks located using data collected by a temporary seismograph deployment [*Reyners et al.*, 2003] and a 1-D velocity model for Fiordland [*Eberhart-Phillips and Reyners*, 2001]. Event 5 is the offshore Cheviot event located by GeoNet using the standard 1-D model for New Zealand [*Maunder*, 1999].

Event		Origin time		Location			Magnitude M_L
nb	ID	Date	Time	Lat.	Long.	Depth (km)	
1	2105255	2003/08/25	03:36:30.26	−45.111	166.964	20.5	5.0
2	2106280	2003/08/26	23:56:26.57	−45.486	166.596	19.7	5.5
3	2106314	2003/08/27	01:29:40.41	−45.442	166.716	22.7	5.1
4	2106319	2003/08/27	01:42:54.12	−45.314	166.945	24.4	5.6
5	2106361	2003/08/27	03:39:37.23	−42.802	173.758	35	4.1
6	2110611	2003/09/04	08:40:44.25	−45.224	166.921	22.7	6.1

the Moho depth is maximum near Wanaka, located at the southwestern tip of the constrained zone, and estimated to 48 ± 4 km (Figure 3c). This crustal thickness represents an 18 ± 4 km thick crustal root beneath the Southern Alps relative to a coastal average of 30 km crustal thickness in South Island [*Godfrey et al.*, 2001; *Melhuish et al.*, 2005]. Hence, near Wanaka the crustal root is estimated to be 4 km thicker than near Mount Cook [*Scherwath et al.*, 2003] and suggests thickening of the Southern Alps crustal root from the NE to the SW. In contrast, mean elevations decrease from ~1500 m near Mt Cook to ~1000 m near Wanaka (Figure 3b). Thus, topography and crustal root thickness have an inverse relationship in the Southern Alps region. For 1000 m elevation and an Airy root with a density contrast of −300 *kg/m³* [*Bourguignon et al.*, 2007] or −400 *kg/m³* [*Scherwath*, 2002] with the surrounding mantle, the crustal root should be 9 or 6 km thick, respectively, i.e., less than half the ca. 18 km inferred near Wanaka. A similar conclusion was drawn further north where *Stern et al.* [2000] deduced the presence of a dense mantle body beneath the central Southern Alps. Modeling of the crustal structure along SIGHT T2 with an assumed −450 *kg/m³* density contrast between crustal root and mantle predicts a Bouguer gravity anomaly far more negative than observed [*Reilly and Whiteford*, 1979], indicating a ca. 10 km excess of crustal root thickness (relative to that expected from Airy isostasy). *Stern et al.* [2000] attributed this excess of crustal thickness to the downward pull of a cold, and therefore dense, lithospheric root (Figure 3d).

Intermediate depth seismicity beneath Fiordland indicates steepening of the Australian slab in proximity to the Southern Alps collision zone [*Reyners et al.*, 2002, also see Figure 3d]. It has been suggested that the Australian slab may act as a backstop that converges at ~26 mm/yr, i.e., 3/4 of the full plate speed, with the Pacific lithosphere and contributes to thicken

the Pacific mantle lithosphere from the southwest (Figure 3d; *Malservisi et al.*, 2003; *Bourguignon et al.*, 2007].

4. PN ANISOTROPY

We combine the Pn speed from this study's earthquake refraction and that from SIGHT T2's seismic line [*Scherwath et al.*, 2003] to infer the Pn anisotropy at these two profile intersections. The first terms of the Taylor expansion of the azimuth-dependent wave speed, $\alpha(\varphi) = om$, $\alpha_0 + C\cos(2\varphi) + D\sin(2\varphi)$ [*Smith and Ekström*, 1999] are employed to rotate the intersecting profiles into fast and slow orientations and infer a maximum anisotropy. Three equations are found, which enable us to solve for the three unknown parameters, α_0, the average Pn-speed and both constants C and D, simultaneously. Two equations are determined by substituting $\alpha(\varphi)$ and φ for the known Pn-speed values and azimuths of the two respective intersecting profiles. A third equation, $\frac{d\alpha(\varphi)}{d\varphi}\Big|_{\varphi=\Phi}=0$, is found by assuming $\alpha(\varphi)$ is maximum for the fast propagation azimuth, Φ, from a nearby SKS-splitting measurement.

Taking our result of 8.54 ± 0.20 km/s and a Pn speed of 8.0 ± 0.2 km/s on the nearly perpendicular profile SIGHT T2 [*Scherwath et al.*, 2003], implies 6.5 ± 3.5% apparent anisotropy (T in Figure 1). Assuming the fast orientation to be that of the nearby SKS fast polarization orientation, we can calculate the maximum anisotropy. However, the nearby fast-polarization measurement [*Klosko et al.*, 1999] is located at the transition between two domains of anisotropy and this must be an average resulting from the overlap of Fresnel zones over the two domains. In central South Island, SKS fast polarization orientations, Φ, are sub-parallel to the Alpine Fault, while in southern South Island these are consistently oblique to the Alpine fault, i.e., the orientation of shear

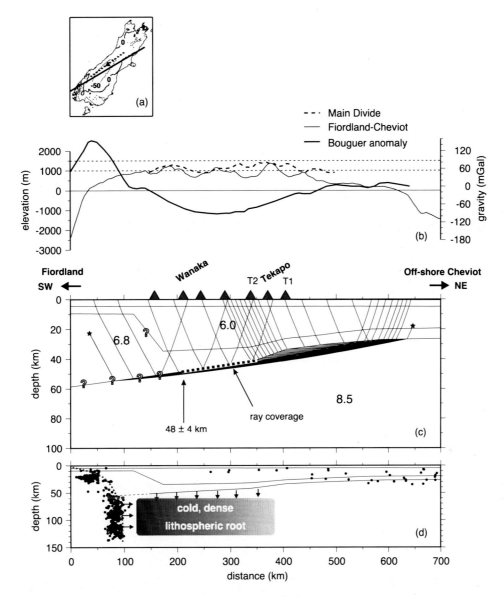

Figure 3. *(a)*: Bouguer gravity anomaly [*Reilly and Whiteford*, 1979] in 50 mGal contours is supplemented with locations of Mt Cook (triangle) and of the main divide (dashed line) and Fiordland-Cheviot (solid line) profiles of the graph below. *(b)*: Mean topography in a 10 km wide swath along the Fiordland-Cheviot profile (thin curve) and the Main Divide (thick dashed curve) and Bouguer anomaly (thick curve; [*Reilly and Whiteford*, 1979]) along the Fiordland-Cheviot profile (solid line in Figure 3a). *(c)*: 2-D velocity model based on this study's Pn speed and Moho dip estimates and on results from: *Davey and Broadbent* [1980], *Reyners et al.* [1993], *Eberhart-Phillips and Reyners* [2001], *Eberhart-Phillips and Bannister* [2002], *Scherwath et al.* [2003] and *van Avendonk et al.* [2003]. Triangles denote deployed instruments, T1/T2 indicate intersections with SIGHT previous crustal studies, ray-tracing is for events 5 and 6 (predicted travel-time curves in Figure 2), thick dashed line indicates the constrained portion of the Moho that extends from Wanaka to Tekapo, question marks denote unconstrained interfaces [after *Bourguignon et al.*, 2007]. *(d)*: 2-D model of the Fiordland-Cheviot profile (no vertical exaggeration) with seismicity within a 10 km wide swath and interpretation.

(Figure 1) [*Klosko et al.*, 1999; *Molnar et al.*, 1999]. Therefore, two possible fast orientations need to be considered. Taking central South Island Φ of $56 \pm 2°$ implies a maximum Pn anisotropy δP of 7 ± 3.5 % and an average Pn speed α_0 of 8.25 ± 0.24 km/s while taking southern South Island mean Φ of $21 \pm 1°$ implies a δP of $13.3 \pm 3.5\%$ and an α_0 of 8.42 ± 0.28 km/s. A mean α_0 of 8.3 ± 0.3 km/s fits both results and is consistent with Haines' average 8.3 ± 0.1 km/s for southern South Island (1979) and wave-speed perturbations of ~2 % [relative to the IASP91 Earth's model; *Kennett and Engdahl*, 1991] imaged by inversion of teleseismic travel times [*Kohler and Eberhart-Phillips*, 2002]. The 8.3 km/s Pn speed is ~3% more than the 8.1 km/s world-wide average [*Kennett and Engdahl*, 1991] and suggests cold and dense upper mantle material.

We replicate the analysis further north at the intersection with SIGHT T1 and make the important assumption that the Pn speed northeast of the constrained portion of the Cheviot-Fiordland refraction profile is 8.54 ± 0.20 km/s as well. The Pn speed on the intersecting SIGHT T1 line is 7.9–8.0 km/s [*van Avendonk et al.*, 2003], i.e., slightly less than along SIGHT T2. Here the azimuths of SIGHT T1 and the Cheviot-Fiordland profile almost align with the slow and fast orientations of wave propagation. Taking a 7.9 ± 0.2 km/s Pn speed along SIGHT T1, an 8.54 ± 0.20 km/s along our refraction profile and a Φ of $56 \pm 2°$, results in a 8.1 ± 3.5 % Pn anisotropy and an average Pn speed α_0 of 8.21 ± 0.24 km/s.

If the 7% or 13% anisotropy at the intersection with SIGHT T2 is constant throughout the mantle lid, then an anisotropic layer of about 100 km or 50 km thickness, respectively, would account for the observed SKS-delay time of 1.76 s [*Klosko et al.*, 1999] (assuming a P-to S-anisotropy ratio of 1.4 and a 4.7 km/s average S-wave speed in the uppermost mantle).

Dynamic slip alone can't explain in situ anisotropy greater than a theoretical maximum of 10% [*Ribe*, 1992] as calculated for southern South Island, but requires additional dynamic re-crystallization by subgrain rotation and grain-boundary migration [*Nicolas et al.*, 1973; *Karato*, 1988], additional pure shear or infinite strain. All these processes have the effect of rotating fast propagation orientations parallel to the shear orientation, i.e., reducing the obliquity of fast orientations to that of shear. However, SKS fast polarizations of southern South Island are ~28° oblique to the Alpine Fault and the shear orientation. Hence, a $13.3 \pm 3.5\%$ Pn anisotropy as calculated for southern South Island Φ seems incompatible with the obliquity of fast polarization orientations from SKS splitting to that of shear. A $7 \pm 3.5\%$ anisotropy, calculated for central South Island, is a more reasonable result. However, there fast polarizations are oriented parallel to the shear orientation. An amount of anisotropy intermediate to

7% and 13% or rotation of material independent of strain would resolve this paradox.

4.1. Comparison With Previous Pn-Anisotropy Measurements in South Island

Three other Pn-anisotropy measurements were made on crossing refraction lines. The Pn anisotropy is $11.5 \pm 2.0\%$ on the Australian side, 30 km west of the surface trace of the Alpine Fault (S in Figure 1) [*Scherwath et al.*, 2002]. If we take the dip of the Alpine Fault as 40°SE [*Kleffmann et al.*, 1998], then at the Moho, the measurement on the Australian side is at a similar distance to the fault as our measurement of 7–13% anisotropy on the Pacific side. Offshore 230 km east of the Alpine Fault, two null Pn anisotropy measurements on crossing lines SIGHT T1 and T3 and SIGHT T2 and T3 (B1 and B2 in Figure 1) show that upper mantle anisotropy does not extend 50 km east of South Island [*Baldock and Stern*, 2005; correction in prep.]. The Pn speed is 8.1 ± 0.1 km/s in both transect azimuths and can be assumed as the isotropic Pn speed. Beneath the Canterbury plains (east of the Southern Alps) into the offshore, however, northwest-southeast raypaths define a broad region of 7.8 ± 0.1 km/s Pn speed [*Baldock*, 2004]. Assuming 7.8 km/s and 8.1 km/s are the minimum and isotropic Pn speeds, respectively, the Pn anisotropy beneath the Canterbury plains is $7.5 \pm 3.0\%$ [*Baldock and Stern*, 2005; correction in prep.].

These measurements suggest that the Pn anisotropy is strong up to ~70–80 km distance from the Alpine Fault at depth. *Scherwath et al.* [2002] noted that the $11.5 \pm 2.0\%$ Pn anisotropy (S in Figure 1) is slightly greater than the theoretical maximum of ~10% for strain-induced anisotropy [*Ribe*, 1992]. They suggested dynamic recrystallization, some pure shear component and/or infinite strain as possible mechanisms to explain the high observed anisotropy. In the east of South Island, the Pn anisotropy is less strong, and possibly extends as far as the east coast, ~150 km east from the Alpine Fault at depth. Modeling of *Ellis et al.* [2006], in contrast, predicts a width of ~100 km maximum. *Duclos et al.* [2005] interpret constant SKS-splitting delay times but decreasing Pn anisotropy with distance to the Alpine Fault as the result of widening of a continuously distributed zone of mantle deformation with depth.

5. CONCLUSIONS

A Pn speed of 8.54 ± 0.20 km/s and a Moho apparent dip of $2.5 \pm 1.3°$SW are determined from an earthquake refraction travel-time analysis along the Southern Alps crustal root. The profile line is oriented N60°E, ~N5°E from the Alpine Fault.

We estimate a 48 ± 4 km crustal thickness near Wanaka, which is a ca. 18 km thick crustal root (relative to a coastal average of 30 km). Here the root is at least twice as thick as expected for Airy isostatic compensation of the Southern Alps topographic load.

The relatively high wave speed of 8.54 ± 0.20 km/s is interpreted to be the result of both anisotropy in the mantle lid and a relatively high average Pn speed of ca. 8.3 km/s below southern South Island. The Pn anisotropy is 7–13% 80 km east of the Alpine Fault. Pn anisotropy values across South Island confine the deformation to a maximum ~100 km thick layer in the mantle lid assuming anisotropy is constant with depth. An average 8.3 km/s P-wave speed is interpreted as mantle lithosphere colder and, hence, denser and with higher P-wave speeds than surrounding mantle rocks. This colder zone results from the downward deflection of isotherms and acts as an effective load at the base of the crust.

Acknowledgments. We thank David Okaya, John Louie and an anonymous reviewer for their suggested improvements on the manuscript; GeoNet (www.geonet.org.nz) for providing data of the RPZ permament broad-band station. This project was funded with a grant from the New Zealand Public Good Science and Technology Fund and a VUW scholarship. Instruments used were paid for by a grant from the Lotteries Commission and the Planet Earth Fund.

REFERENCES

Baldock, G. (2004), High resolution crustal and upper mantle structure adjacent to a continental transform, South Island, New Zealand, *M.S. thesis*, Victoria University of Wellington, Wellington, New Zealand.

Baldock, G. and T. Stern (2005), Width of mantle deformation across a continental transform: evidence from upper mantle (Pn) seismic anisotropy measurements, New Zealand, *Geology*, *33*(9), 741–744.

Beaumont, C., P. Kamp, J. Hamilton, P. Fullsack (1996), The continental collision zone, South Island, New Zealand: Comparison of geodynamical models and observations, *J. Geophys. Res.*, *101*(B2), 3333–3359.

Beavan, J., M. Moore, C. Pearson, M. Henderson, B. Parsons, S. Bourne, P. England, D. Walcott, G. Blick, D. Darby and K. Hodgkinson (1999), Crustal deformation during 1994–1998 due to oblique continental collision in the central Southern Alps, New Zealand, and implications for seismic potential of the Alpine Fault, *J. Geophys. Res.*, *104*(B11), 25,233–25,255.

Bourguignon, S., T. A. Stern and M. K. Savage (2007), Crust and mantle thickening beneath the southern portion of the Southern Alps, New Zealand, *Geophys. J. Int.*, *168*(2), 681–690.

Davey, F. J. and M. Broadbent (1980), Seismic refraction measurements in Fiordland southwest New Zealand, *New Zealand J. Geol. Geophys.*, *23*, 395–406.

DeMets, C., R. G.Gordon, D. F. Argus and S. Stein (1994), Effect of recent revisions to the geomagnetic reversal time scale on estimates of current plate motions, *Geophys. Res. Lett.*, *21*, 2191–2194.

Duclos, M., M. K. Savage, A. Tommasi and K. R. Gledhill (2005), Mantle tectonics beneath New Zealand inferred from SKS and petrophysics, *Geophys. J. Int.*, *162*, 1–15.

Eberhart-Phillips, D. and S. Bannister (2002), Three-dimensional crustal structure in the Southern Alps region of New Zealand from inversion of local earthquake and active source data, *J. Geophys. Res.*, *107*(B10), 2262, doi:10.1029/2001JB000182.

Eberhart-Phillips, D. and M. Reyners (2001), A complex, young subduction zone imaged by three-dimensional seismic velocity, Fiordland, New Zealand, *Geophys. J. Int.*, *146*, 731–746.

Ellis, S, J. Beavan and D. Eberhart-Phillips (2006), Bounds on the width of mantle lithosphere flow derived from surface geodetic measurements: application to the central Southern Alps, New Zealand, *Geophys. J. Int.*, *166*(1), 403–417.

Godfrey, N. J., F. Davey, T. A. Stern and D. Okaya (2001), Crustal structure and thermal anomalies of the Dunedin Region, South Island, New Zealand, *J. Geophys. Res.*, *106*(B12), 30,835–30,848.

Haines, A. J. (1979), Seismic wave velocities in the uppermost mantle beneath New Zealand, *NZ J. Geol. Geophys.*, *22*(2), 245–257.

Jones, C. H., H. Kanamori and S. W. Roecker (1994), Missing roots and mantle "drips": regional Pn and teleseismic arrival times in the southern Sierra Nevada, *J. Geophys. Res.*, *99*(B3), 4567–4601.

Karato, S. (1988), The role of recrystallization in the preferred orientation of olivine, *Phys. Earth Planet. Inter.*, *51*, 107–122.

Kennett, B. L. N. and E. R. Engdahl (1991), Traveltimes for global earthquake location and phase identification, *Geophys. J. Int.*, *105*, 429–465.

Kleffmann, S., F. Davey, A. Melhuish, D. Okaya and T. Stern (1998) Crustal structure in central South Island from the Lake Pukaki seismic experiment, *NZ J. Geol. Geophys.*, *41*, 39–49.

Klosko, E. R., F. T. Wu, H. J. Anderson, D. Eberhart-Phillips, T. V. McEvilly, E. Audoine, M. K. Savage and K. R. Gledhill (1999), Upper Mantle Anisotropy in the New Zealand Region, *Geophys. Res. Lett.*, *26*(10), 1497–1500.

Kohler, M. and D. Eberhart-Phillips (2002), Three-dimensional lithospheric structure below the New Zealand Southern Alps, *J. Geophys. Res.*, *107*(B10), 2225, doi:10.1029/2001JB000182.

Liu, Z. and P. Bird (2006), Two-dimensional and three-dimensional finite element modelling of mantle processes beneath the central South Island, New Zealand, *Geophys. J. Int.*, *165*(3), 1003–1028.

Louie, J. N., W. Thelen, S. B. Smith, J. B. Scott, M. Clark and S. Pullammanappallil (2004), The northern Walker Lane refraction experiment: Pn arrivals and the northern Sierra Nevada root, *Tectonoph.*, *388*, 253–269.

Luetgert, J. H. (1992), MacRay-interactive two-dimensional seismic raytracing for the Macintosh, *Open File Rep. 92–0356*, 45 pp., U.S. Geological Survey, U.S.A.

McKenzie, D. (1978), Some remarks on the development of sedimentary basins, *Earth Planet. Sci. Lett.*, *40*(1), 25–32.

Melhuish, A., W. S. Holbrook, F. Davey, D. Okaya and T. A. Stern (2005), Crustal and upper mantle seismic structure of the Australian plate, South Island, New Zealand, *Tectonoph.*, *395*, 113–135.

Malservisi, R., K. P. Furlong and H. Anderson (2003), Dynamic uplift in a transpressional regime: numerical model of the subduction area in Fiordland, New Zealand, *Earth Planet. Sci. Lett.*, *206*, 349–364.

Maunder, D. E. (ed.) (2001), New Zealand Seismological Report 1999, *Seismological Observatory Bulletin E-1822001/7*, 156 pp., Institute of Geological & Nuclear Science Report, Wellington, New Zealand.

Molnar, P., H. J. Anderson, E. Audoine, D. Eberhart-Phillips, K. R. Gledhill, E. R. Klosko, T. V. McEvilly, D. Okaya, M. K. Savage, T. Stern and F. T. Wu (1998), Continuous Deformation Versus Faulting Through the Continental Lithosphere of New Zealand, *Geophys. J. Int.*, *133*, 568–584.

Nicolas, A., F. Boudier and A. M. Boullier (1973), Mechanisms of flow in naturally and experimentally deformed peridotites, *Am. J. Sci.*, *273*, 853–876.

Okaya, D., S. Henry and T. Stern (2002), Double-sided onshore-offshore seismic imaging of a plate boundary: "super-gathers" across South Island, New Zealand, *Tectonophys.*, *355*, 247– 263.

Pysklywec, R. N., C. Beaumont and P. Fullsack (2002), Lithospheric deformation during the early stage of continental collision: Numerical experiments and comparison with South Island, New Zealand, *J. Geophys. Res.*, *107*(B7), doi:10.1029/ 2001JB000252.

Reilly, W. I. and C. M. Whiteford (1979), Gravity Map of New Zealand, 1:1,000,000, Bouguer Anomaly, Dept. of Scientific and Industrial Research, Wellington, New Zealand.

Reyners, M. and H. Cowan (1993), The transition from subduction to continental collision: crustal structure in the North Canterbury region, New Zealand, *Geophys. J. Int.*, *115*, 1124–1136.

Reyners, M., R. Robinson, A. Pancha, P. McGinty (2002), Stresses and strains in a twisted subduction zone — Fiordland, New Zealand, *Geophys. J. Int.*, *148*, 637–648.

Reyners, M. E., P. J. McGinty, S. C. Cox, I. M. Turnbull, T. O'Neill, K. R. Gledhill, G. T. Han-cox, R. J. Beavan, D. W. Matheson, G. H. McVerry, W. J. Cousins, J. X. Zhao, H. A. Cowan, T. G. Caldwell, S. L. Bennie and GeoNet team (2003), The M_W 7.2 Fiordland earthquake of August 21, 2003: background and preliminary results, *Bulletin of the New Zealand Society for Earthquake Engineering*, *36*(4), 233–248.

Ribe, N. M. (1992), On the relation between seismic anisotropy and finite strain, *J. Geophys. Res.*, *97*(B6), 8737–8747.

Ruppert, S., M. M. Fliedner and G. Zandt (1998), Thin crust and active upper mantle beneath the Southern Sierra Nevada in the western United States, *Tectonophys.*, *286*, 237–252.

Savage, M. K. (1999), Seismic anisotropy and mantle deformation: what have we learned from shear wave splitting?, *Rev. of Geophys.*, *37*(1), 65–106.

Savage, M. K., L. Li, J. P. Eaton, C. H. Jones and J. N. Brune (1994), Earthquake refraction profiles of the root of the Sierra Nevada, *Tectonics*, *13*(4), 803–817.

Scherwath, M., T. Stern, A. Melhuish and P. Molnar (2002), Pn anisotropy and distributed upper mantle deformation associated with a continental transform fault, *Geophys. Res. Lett.*, *29*(8), doi:10.1029/2001GL014179.

Scherwath, M., T. Stern, F. Davey, D. Okaya, W.S. Holbrook, R. Davies and S. Kleffmann (2003), Lithospheric structure across oblique continental collision in New Zealand from wide-angle P-wave modelling, *J. Geophys. Res.*, *108*(B12), 2566, doi:10.1029/ 2002JB002286.

Smith, E. and F. J. Davey (1984), Joint-hypocentre determination of intermediate depth earthquakes in Fiordland, New Zealand, *Tectonophys.*, *104*, 127–144.

Smith, G. P. and G. Ekström (1999), A global study of Pn anisotropy beneath continents, *J. Geophys. Res.*, *104*, 963–980.

Stern, T., P. Molnar, D. Okaya and D. Eberhart-Philips (2000), Teleseismic P wave delays and modes of shortening the mantle lithosphere beneath South Island, New Zealand, *J. Geophys. Res.*, *105*, 21,615–21,631.

van Avendonk, H. J. A., W. S. Holbrook, D. Okaya, J. Austin, F. Davey and T. Stern (2004), Continental crust under compression: A seismic refraction study of SIGHT Transect I, South Island, New Zealand, *J. Geophys. Res.*, *109*, B06302, doi:10.1029/ 2003JB002790.

Wellman, H. W. (1979), An uplift map for the South Island of New Zealand, and a model for uplift of the Southern Alps, *Bull. Roy. Soc. NZ*, *18*, 13–20.

Wernicke, B. (1985), Uniform-sense normal simple shear of the continental lithosphere, *Can. J. Earth Sci.*, *22*, 108–125.

Compressional and Shear Wave Velocities in South Island, New Zealand Rocks and Their Application to the Interpretation of Seismological Models of the New Zealand Crust

Nikolas I. Christensen

Department of Geology and Geophysics, University of Wisconsin-Madison, Madison, Wisconsin
and Department of Earth and Ocean Sciences, University of British Columbia, Vancouver, Canada

David A. Okaya

Department of Earth Sciences, University of Southern California, Los Angeles, California

The seismic properties of a suite of metamorphic and igneous rocks from South Island, New Zealand have been investigated using laboratory velocity measurements as a function of confining pressures up to 1000 MPa. Representative samples for the velocity measurements were collected from the Caples, Aspiring, Torlesse, Buller, and Takaka Terranes. Lithologies studied include quartzofeldspathic schists, amphibolites, greenschists, slates, argillites, graywackes, gabbros and granites. Seismic velocities, in general, increase with increasing density. Whole rock geochemical analyses show velocities increase with decreasing SiO_2 contents. Compressional wave velocities increase with increasing Mg numbers for Haast Schist and Torlesse samples, however velocities of the Haast schist rocks do not show any correlation with metamorphic grade. Haast schists are highly anisotropic due to preferred mineral orientations, whereas the Torlesse greywackes are isotropic. Symmetry of the anisotropic rocks varies from axial to orthorhombic, with fast compressional wave velocities parallel to lineations. Shear wave splitting is only significant for propagation directions at high angles to foliation normals; minimal production at other angles can mistakenly be interpreted as due to isotropic media. A refraction experiment shot in orthogonal directions across the Haast Schist shows 6% compressional wave anisotropy, in good agreement with the laboratory measurements. We conclude that seismic anisotropy is a pervasive feature throughout much of the South Island crust.

INTRODUCTION

For interpreting structures modeled by seismic experiments, which are based on P-wave and S-wave velocities, a systematic data base of velocities for appropriate rock types are required with a precision at least comparable with that of seismological measurements. In particular, for continental crustal studies we require velocity measurements under conditions approaching 30 to 50 km depths. Of fundamental importance

A Continental Plate Boundary: Tectonics at South Island, New Zealand
Geophysical Monograph Series 175
10.1029/175GM08

is the role of pressure which increases with depth and reaches about 1000 MPa at 35 km. In this paper we present a database of compressional and shear wave velocities for a wide variety of South Island, New Zealand rocks at hydrostatic confining pressures to 1000 MPa. The measurements were part of the joint USA-New Zealand geophysical study of the continent-continent transform plate boundary in South Island, New Zealand, which separates the Australian and Pacific plates. In the following sections emphasis is placed on the correlation of seismic velocities with various physical and chemical parameters, with the goal of interpreting field studies of seismic velocities of South Island, New Zealand.

SAMPLE LOCALITIES

The physical property measurements encompass rocks from several tectonostratigraphic terranes of South Island New Zealand. Based on the classifications of *Bishop et al.* [1985] and *Frost and Coombs* [1989] these include samples from the Buller and Takaka terranes of the Western Province and the Murihiku, Caples, Aspiring and older Torlesse terranes of the Eastern Province (Plate 1). These papers provide detailed information on the ages, lithologies, and tectonic settings of the various terranes; additional information is available in *Cox and Sutherland* [this volume]. Much of the following discussion focuses on seismic properties of rocks from the Caples, Aspiring and Torlesse terranes, since the majority of the samples were collected from these terranes and recent seismic studies have targeted these regions. The rocks are primarily quartzofeldspathic schists, greywackes, argillites and mafic volcanics, which have been subjected to zeolite, greenschist and higher grade metamorphism (Table 1).

The rocks selected for this study were collected during the South Island Geophysical Transect [e.g., *Davey et al.*, this volume] from several major South Island localities representing Alpine and Otago schists, the Alpine fault zone, Torlesse greywacke, Tuhua granite, Fiordland gabbro and Murihiku greywacke and argillite (Plate 1). All samples were collected in place from outcrops, except the jadestone (A-26) and garnet amphibolite (A-79) which were collected as float.

EXPERIMENTAL PROCEDURES

The new measurements are based on a pulse transmission technique described by *Birch* [1960]. To investigate anisotropy, three cores, 2.54 cm in diameter and 5–8 cm in length, were cut with mutually perpendicular axes from all specimens. Using specially designed sample holders and protractors, the cores were taken to within a few degrees of the desired directions. A-cores were taken normal to foliations and either B or C-cores were cut parallel to lineations, when

present. Velocities were measured in all samples at room temperature with both rising and descending pressure at 20 MPa intervals for confining pressures between 0 and 100 MPa, at 50 MPa intervals between 100 and 200 MPa, and at 100 MPa intervals between 200 and 1000 MPa. Compressional and shear waves were generated across each sample by means of lead-zirconate-titanate (PZT) and AC-cut quartz transducers, respectively, having resonant frequencies of 2 MHz. Vibration directions for the shear wave A-core velocity measurements were oriented parallel to the B-core axes. For foliated rocks, B-core vibration directions were parallel to foliations and C-core vibration directions were normal to foliations. Thus for transversely isotropic samples, comparisons of B and C-core shear wave velocities are a measure of shear wave splitting for propagation within the foliation. Compressional and shear wave velocity data, together with V_p/V_s and Poisson's ratios calculated from mean velocities are presented in Table 2 for ten pressures up to 1000 MPa. The cumulative error limits for compressional and shear wave velocities are estimated to be 0.5% and 1.0% respectively. Confining pressures, measured by means of a manganin coil exposed directly to the pressure medium, are accurate to within 1%. The reported densities are bulk densities calculated from the weights and dimensions of the cylindrical samples.

Pressure Induced Crack Closure

To be useful in the interpretation of crustal seismic investigations laboratory measured velocities need to be obtained at appropriate elevated confining pressures. At atmospheric pressure most igneous and metamorphic rocks contain cracks characterized by very low aspect-ratios (minimum/maximum dimensions). These cracks are usually attributed to cooling and decompression accompanying their uplift to the Earths surface [e.g., *Birch,* 1960]. It has long been recognized that these cracks significantly lower velocities and many theoretical studies of the elasticity of cracked solids have focused on the influence of low aspect-ratio cracks on velocities and their changes with applied stress [e.g., *O'Connell and Budiansky,* 1974].

As expected the measured wave velocities in Table 2 are low and extremely pressure sensitive for pressures below approximately 100 to 200 MPa. At higher pressures the grain boundary cracks are closed and the velocities are related to the elastic properties of the constituent minerals of the rocks. Thus most interpretations of field measured velocities at significant crustal depths require laboratory measurements at elevated confining pressures. Likewise correlations of rock velocities with mineralogy require measurements at pressures above those of crack closure.

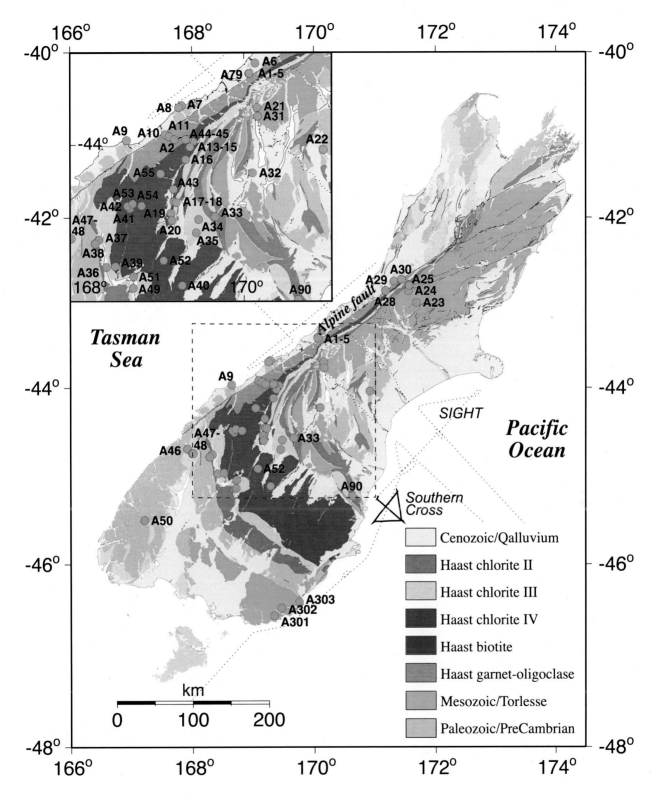

Plate 1. Terrane map of South Island showing locations of rocks collected for velocity measurements. Circles denote sample locations; inset map provides names of samples within dashed box. See Tables 1–3 for sample information. Dotted lines denote SIGHT marine ship tracks. Location of Southern Cross experiment shown (for Plate 2).

Table 1. Rock samples used in this study. Locations shown in Plate 1. Digital version of this table is on the CDROM which accompanies this volume.

Sample	Rock	Zone	Terrane	density (kg/m^3)	Longitude (E)	Latitude (S)	Location
A-1	Garnet schist	garnet zone	Alpine Schist	2723	170°03′31″	43°25′03″	Fox Glacier
A-2	Mica schist	garnet zone	Alpine Schist	2765	170°03′31″	43°25′06″	Fox Glacier
A-3	Garnet schist	garnet zone	Alpine Schist	2735	170°03′33″	43°25′06″	Fox Glacier
A-4	Garnet schist	garnet zone	Alpine Schist	2668	170°03′40″	43°25′10″	Fox Glacier
A-5	Garnet schist	garnet zone	Alpine Schist	2656	170°03′40″	43°25′10″	Fox Glacier
A-6	Phyllite		Greenland Group	2713	170°07′42″	43°20′00″	Omoeroa R.
A-7	Phyllite		Greenland Group	2640	169°17′10″	43°40′45″	Lake Moeraki
A-8	Phyllite		Greenland Group	2642	169°15′00″	43°41′50″	Whakapohai R.
A-9	Phyllite		Greenland Group	2644	168°39′00″	43°57′45″	Jackson Bay
A-10	Schist	garnet zone	Alpine Schist	2700	169°05′40″	43°54′58″	Big Bluff
A-11	Schist	garnet zone	Alpine Schist	2694	169°08′30″	43°55′53″	Thomas Bluff
A-12	Schist	garnet zone	Alpine Schist	2713	169°12′10″	43°57′20″	Halfway Bluff
A-13	Schist	biotite zone	Alpine Schist	2712	169°22′00″	43°57′40″	Evans Creek
A-14	Greenschist	chlorite zone IV	Alpine Schist	2681	169°25′07″	43°58′15″	Clarke Bluff
A-15	Greenschist	chlorite zone IV	Alpine Schist	2656	169°22′53″	44°00′48″	Mather Creek
A-16	Greenschist	chlorite zone IV	Alpine Schist	2714	169°19′55″	44°06′55″	Fish R.
A-17	Greenschist	chlorite zone IV	Otago Schist	3002	169°11′40″	44°27′15″	Lake Hawea
A-18	Greenschist	chlorite zone III	Otago Schist	2726	169°13′15″	44°27′45″	Lake Hawea
A-19	Greenschist (Greywacke)	chlorite zone II	Otago Schist	2735	169°09′45″	44°32′50″	Lake Hawea
A-20	Greenschist (Greywacke)	chlorite zone II	Otago Schist	2708	169°10′20″	44°36′20″	Lake Hawea
A-21	Torlesse greywacke		Rakaia Terrane	2737	170°09′10″	43°41′50″	Wakefield Falls
A-22	Torlesse greywacke		Rakaia Terrane	2690	170°54′35″	44°02′20″	Allandale
A-23	Torlesse greywacke		Rakaia Terrane	2688	171°41′10″	43°00′20″	Corner Knob
A-24	Torlesse greywacke		Rakaia Terrane	2654	171°33′06″	42°52′08″	Windy Pt
A-25	Torlesse greywacke		Rakaia Terrane	2706	171°34′04″	42°48′11″	Kellys Creek
A-26	Jadestone		Alpine Schist	2971			Hokitika
A-28	Tuhua Granite		Tuhua Granite	2631	171°10′15″	42°52′00″	Mt Tuhua
A-29	Tuhua Granite		Tuhua Granite	2650	171°18′41″	42°45′25″	Fizgerald Ck
A-30	Alpine fault mylonite		Alpine Schist	2677	171°26′18″	42°44′00″	Rocky Pt
A-3 IB	Torlesse greywacke		Rakaia Terrane	2705	170°09′40″	43°46′00″	Lake Pukaki
A-32B	Torlesse greywacke		Rakaia Terrane	2756	170°06′00″	44°13′30″	Twizel Power House
A-33	Greenschist	chlorite zone II	Otago Schist	2624	169°39′50″	44°34′30″	Lindis Pass
A-34	Greenschist	chlorite zone III	Otago Schist	2701	169°28′40″	44°35′48″	Dip Creek

Table 1. (Continued)

Sample	Rock	Zone	Terrane	density (kg/m^3)	Longitude (E)	Latitude (S)	Location
A-35	Greenschist	chlorite zone III	Otago Schist	2741	169°27'00″	44°42'20″	N Lindis Valley
A-36	Greenschist	chlorite zone III	Otago Schist	2783	169°25'00″	44°58'45″	White Pt
A-37	Greenschist	chlorite zone III	Otago Schist	2742	169°19'50″	44°45'40″	Glenorchy
A-38	Greenschist (Greywacke)	chlorite zone II	Otago Schist	2749	168°17'10″	44°47'30″	S Kinloch
A-39	Greenschist	chlorite zone II	Otago Schist	2762	168°31'10″	44°58'18″	Lake Dispute
A-40	Biotite schist	chlorite zone IV	Otago Schist	2649	169°17'12″	45°07'21″	Waikeriker
A-41	Epidote amphibolite		Otago Schist	2955	168°41'07″	44°30'30″	Brides Veil Str
A-42	Greenschist	chlorite zone IV	Otago Schist	2659	168°39'37″	44°29'25″	Aspiring Cascade Huts
A-43	Greenschist	chlorite zone IV	Otago Schist	2746	169°11'30″	44°21'15″	Lake Wanaka
A-44	Biotite schist	biotite zone	Otago Schist	2715	169°21'08″	43°57'10″	Douglas Bluff
A-45	Biotite schist	biotite zone	Otago Schist	2700	169°19'50″	43°57'20″	Douglas Crk
A-46	Hornblende gabbro		Darran	2896	167°55'15″	44°42'01″	Homer Tunnel
A-47	Hornblende gabbro		Mackay Intrusives	3032	168°00'00″	44°45'00″	Hollylord R
A-48	Gabbro		Mackay Intrusives	2857	168°00'50″	44°45'05″	Falls Creek
A-49	Greenschist	chlorite zone IV	Otago Schist	2768	168°43'15″	45°08'40″	Devils Staircase
A-50	Felsic gneiss			2679	167°12'45″	45°30'27″	Pahiri Peak
A-51	Greenschist	chlorite zone III	Otago Schist	2732	168°43'35″	45°03'10″	Wye Creek
A-52	Epidote amphibolite		Otago Schist	2968	169°04'35″	44°55'25″	Kawarau Gorge
A-53	Epidote amphibolite		Otago Schist	3030	168°43'10″	44°28'16″	Rob Roy Str
A-54	Amphibolite		Otago Schist	2819	168°49'17″	44°29'17″	Glenlinnan Str
A-55	Biotite schist	biotite zone	Otago Schist	2641	169°02'30″	44°13'52″	Kerin Forks
A-79	Garnet amphibolite		Alpine Schist	2911	170°03'42″	43°25'10″	Fox Glacier
A-90	Greenschist	chlorite zone II	Otago Schist	2512	170°37'20″	46°14'12″	
A-301	Greywacke		Murihiku	2710	169°45'00″	46°34'10″	
A-302	Greywacke		Murihiku	2509	169°27'35″	46°29'15″	
A-303	Argillite		Murihiku	2596	169°20'30″	46°25'10″	

Table 2. Velocity measurements

#		core	Pressure (MPa)									
			20	40	60	80	100	200	400	600	800	1000
A-1	Vp	A	3.855	4.430	4.783	5.011	5.163	5.472	5.651	5.750	5.823	5.879
	Vp	B	5.419	5.661	5.826	5.947	6.038	6.268	6.405	6.468	6.511	6.545
	Vp	C	5.499	5.769	5.642	6.060	6.143	6.339	6.466	6.534	6.583	6.621
	Vp	mean	4.924	5.287	5.417	5.672	5.781	6.026	6.174	6.251	6.306	6.349
	Vs	A	2.608	2.843	2.997	3.103	3.177	3.338	3.418	3.457	3.484	3.506
	Vs	B	3.372	3.524	3.627	3.699	3.750	3.862	3.911	3.931	3.944	3.955
	Vs	C	2.623	2.800	2.903	3.101	3.009	3.101	3.168	3.207	3.235	3.257
	Vs	mean	2.868	3.055	3.175	3.301	3.312	3.434	3.499	3.531	3.554	3.572
	Vp/Vs	mean	1.717	1.730	1.706	1.719	1.746	1.755	1.765	1.770	1.774	1.777
	σ	mean	0.243	0.249	0.238	0.244	0.256	0.260	0.263	0.266	0.267	0.268
A-2	Vp	A	3.790	4.377	4.721	4.930	5.063	5.312	5.458	5.541	5.600	5.648
	Vp	B	4.997	5.443	5.718	5.895	6.011	6.239	6.354	6.415	6.459	6.493
	Vp	C	4.559	5.088	5.433	5.667	5.828	6.156	6.300	6.366	6.744	6.451
	Vp	mean	4.449	4.969	5.291	5.497	5.634	5.902	6.037	6.107	6.268	6.197
	Vs	A	2.608	2.784	2.882	2.911	2.980	3.071	3.148	3.194	3.228	3.254
	Vs	B	3.155	3.464	3.657	3.780	3.856	4.002	4.054	4.078	4.096	4.109
	Vs	C	2.607	2.862	3.006	3.105	3.169	3.288	3.334	3.357	3.373	3.385
	Vs	mean	2.790	3.037	3.182	3.265	3.335	3.453	3.512	3.543	3.565	3.583
	Vp/Vs	mean	1.595	1.636	1.663	1.684	1.689	1.709	1.719	1.724	1.758	1.730
	σ	mean	0.176	0.202	0.217	0.227	0.230	0.240	0.244	0.246	0.261	0.249
A-3	Vp	A	4.409	4.738	4.949	5.094	5.199	5.449	5.622	5.717	5.785	5.839
	Vp	B	5.216	5.533	5.735	5.873	5.971	6.197	6.341	6.418	6.474	6.517
	Vp	C	5.427	5.674	5.836	5.952	6.038	6.256	6.404	6.481	6.563	6.579
	Vp	mean	5.017	5.315	5.507	5.640	5.736	5.967	6.122	6.205	6.274	6.312
	Vs	A	2.889	3.023	3.114	3.179	3.228	3.351	3.424	3.459	3.483	3.503
	Vs	B	3.324	3.457	3.550	3.617	3.666	3.778	3.824	3.841	3.852	3.860
	Vs	C	2.517	2.776	2.935	3.036	3.102	3.226	3.286	3.318	3.341	3.359
	Vs	mean	2.910	3.086	3.200	3.277	3.332	3.451	3.512	3.539	3.559	3.574
	Vp/Vs	mean	1.724	1.723	1.721	1.721	1.721	1.729	1.743	1.753	1.763	1.766
	σ	mean	0.247	0.246	0.245	0.245	0.245	0.249	0.255	0.259	0.263	0.264
A-4	Vp	A	3.551	4.311	4.810	5.147	5.377	5.838	6.025	6.110	6.172	6.220
	Vp	B	4.480	4.966	5.301	5.539	5.711	6.089	6.245	6.304	6.344	6.376
	Vp	C	4.133	4.756	5.173	5.460	5.661	6.072	6.232	6.297	6.344	6.380
	Vp	mean	4.054	4.677	5.095	5.382	5.583	6.000	6.168	6.237	6.286	6.325
	Vs	A	2.412	2.801	3.058	3.233	3.353	3.587	3.666	3.697	3.719	3.735
	Vs	B	2.692	3.018	3.239	3.390	3.494	3.693	3.744	3.757	3.765	3.772
	Vs	C	2.353	2.748	3.004	3.172	3.285	3.490	3.552	3.578	3.596	3.609
	Vs	mean	2.486	2.856	3.100	3.265	3.377	3.590	3.654	3.677	3.693	3.706
	Vp/Vs	mean	1.631	1.638	1.643	1.648	1.653	1.671	1.688	1.696	1.702	1.707
	σ	mean	0.199	0.203	0.206	0.209	0.211	0.221	0.230	0.234	0.236	0.239

Table 2. (Continued)

#		core	Pressure (MPa)									
			20	40	60	80	100	200	400	600	800	1000
A-5	Vp	A	4.104	4.433	4.643	4.793	4.905	5.207	5.454	5.596	5.699	5.780
	Vp	B	5.382	5.600	5.740	5.835	5.901	6.045	6.123	6.163	6.191	6.213
	Vp	C	5.665	5.979	6.148	6.243	6.300	6.041	6.476	6.517	6.546	6.569
	Vp	mean	5.050	5.337	5.510	5.624	5.702	5.764	6.018	6.092	6.145	6.188
	Vs	A	2.708	2.866	2.977	3.061	3.127	3.303	3.409	3.452	3.481	3.503
	Vs	B	3.027	3.187	3.303	3.390	3.457	3.620	3.691	3.710	3.722	3.731
	Vs	C	1.974	2.125	2.232	2.317	2.385	2.840	2.752	2.817	2.859	2.890
	Vs	mean	2.569	2.726	2.837	2.923	2.990	3.254	3.284	3.326	3.354	3.375
	Vp/Vs	mean	1.965	1.958	1.942	1.924	1.907	1.771	1.832	1.831	1.832	1.833
	σ	mean	0.325	0.324	0.320	0.315	0.310	0.266	0.288	0.288	0.288	0.288
A-6	Vp	A	5.657	5.776	5.847	5.894	5.927	6.017	6.097	6.143	6.177	6.203
	Vp	B	5.440	5.581	5.668	5.729	5.773	5.893	5.994	6.052	6.093	6.125
	Vp	C	5.323	5.476	5.558	5.609	5.645	5.742	5.834	5.889	5.929	5.959
	Vp	mean	5.473	5.611	5.691	5.744	5.782	5.884	5.975	6.028	6.066	6.096
	Vs	A	3.668	3.734	3.776	3.805	3.826	3.877	3.914	3.935	3.949	3.960
	Vs	B	3.345	3.409	3.453	3.484	3.507	3.563	3.593	3.607	3.617	3.624
	Vs	C	3.448	3.498	3.533	3.559	3.578	3.628	3.657	3.668	3.676	3.682
	Vs	mean	3.487	3.547	3.587	3.616	3.637	3.689	3.721	3.737	3.747	3.755
	Vp/Vs	mean	1.570	1.582	1.586	1.588	1.590	1.595	1.606	1.613	1.619	1.623
	σ	mean	0.158	0.167	0.170	0.172	0.173	0.176	0.183	0.188	0.191	0.194
A-7	Vp	A	3.622	4.167	4.496	4.706	4.847	5.150	5.359	5.481	5.570	5.641
	Vp	B	5.784	5.903	5.983	6.042	6.088	6.213	6.299	6.339	6.367	6.389
	Vp	C	5.387	5.558	5.668	5.747	5.807	5.967	6.088	6.153	6.199	6.235
	Vp	mean	4.931	5.209	5.382	5.499	5.581	5.777	5.915	5.991	6.046	6.088
	Vs	A	2.837	2.968	3.054	3.118	3.168	3.311	3.421	3.478	3.517	3.548
	Vs	B	3.628	3.690	3.732	3.762	3.786	3.845	3.881	3.897	3.908	3.917
	Vs	C	2.817	2.935	3.015	3.074	3.120	3.247	3.332	3.372	3.400	3.421
	Vs	mean	3.094	3.197	3.267	3.318	3.358	3.467	3.545	3.582	3.608	3.629
	Vp/Vs	mean	1.594	1.629	1.648	1.657	1.662	1.666	1.669	1.672	1.675	1.678
	σ	mean	0.175	0.198	0.208	0.214	0.216	0.218	0.220	0.222	0.223	0.224
A-8	Vp	A	3.070	3.350	3.560	6.729	3.871	4.327	4.686	4.819	4.891	4.942
	Vp	B	5.929	6.044	6.129	6.196	6.249	6.396	6.479	6.505	6.519	6.530
	Vp	C	6.227	6.351	6.430	6.484	6.522	6.607	6.658	6.685	6.704	6.718
	Vp	mean	5.075	5.248	5.373	6.470	5.547	5.777	5.941	6.003	6.038	6.064
	Vs	A	1.945	2.079	2.167	2.235	2.289	2.470	2.651	2.748	2.814	2.864
	Vs	B	3.083	3.198	3.278	3.341	3.392	3.560	3.706	3.770	3.809	3.838
	Vs	C	1.755	1.896	1.993	2.068	2.130	2.336	2.534	2.632	2.694	2.741
	Vs	mean	2.261	2.391	2.479	2.548	2.604	2.789	2.964	3.050	3.106	3.148
	Vp/Vs	mean	2.245	2.195	2.167	2.539	2.130	2.071	2.005	1.968	1.944	1.926
	σ	mean	0.376	0.369	0.365	0.408	0.359	0.348	0.334	0.326	0.320	0.316

Table 2. (Continued)

#		core	Pressure (MPa)									
			20	40	60	80	100	200	400	600	800	1000
A-9	Vp	A	4.296	4.758	5.042	5.224	5.345	5.586	5.666	5.788	5.839	5.879
	Vp	B	4.772	5.064	5.259	5.396	5.495	5.724	5.849	5.909	5.952	5.985
	Vp	C	5.271	5.539	5.697	5.796	5.861	5.993	6.079	6.127	6.162	6.189
	Vp	mean	4.780	5.120	5.333	5.472	5.567	5.768	5.865	5.941	5.984	6.017
	Vs	A	2.959	3.161	3.281	3.356	3.406	3.507	3.571	3.608	3.634	3.655
	Vs	B	3.198	3.334	3.425	3.488	3.533	3.633	3.681	3.703	3.718	3.730
	Vs	C	3.122	3.210	3.269	3.312	3.344	3.420	3.459	3.476	3.487	3.496
	Vs	mean	3.093	3.235	3.325	3.385	3.428	3.520	3.570	3.596	3.613	3.627
	Vp/Vs	mean	1.545	1.583	1.604	1.616	1.624	1.639	1.643	1.652	1.656	1.659
	σ	mean	0.140	0.168	0.182	0.190	0.195	0.203	0.206	0.211	0.213	0.215
A-10	Vp	A	3.999	4.416	4.707	4.922	5.087	5.511	5.763	5.874	5.952	6.014
	Vp	B	4.284	4.757	5.073	5.294	5.452	5.810	5.998	6.088	6.152	6.203
	Vp	C	4.620	5.124	5.448	5.668	5.820	6.146	6.319	6.408	6.472	6.522
	Vp	mean	4.301	4.766	5.076	5.295	5.453	5.822	6.027	6.123	6.192	6.246
	Vs	A	2.472	2.730	2.907	3.034	3.126	3.338	3.440	3.483	3.513	3.536
	Vs	B	2.880	3.127	3.288	3.398	3.474	3.640	3.724	3.765	3.794	3.817
	Vs	C	2.378	2.634	2.807	2.930	3.018	3.218	3.314	3.356	3.387	3.410
	Vs	mean	2.577	2.830	3.001	3.120	3.206	3.399	3.493	3.535	3.565	3.588
	Vp/Vs	mean	1.669	1.684	1.692	1.697	1.701	1.713	1.726	1.732	1.737	1.741
	σ	mean	0.220	0.228	0.231	0.234	0.236	0.242	0.247	0.250	0.252	0.254
A-11	Vp	A	2.841	3.708	4.281	4.670	4.939	5.486	5.723	5.836	5.918	5.984
	Vp	B	4.887	5.347	5.656	5.871	6.024	6.351	6.494	6.554	6.597	6.631
	Vp	C	4.328	4.885	5.270	5.521	5.707	6.105	6.279	6.353	6.406	6.447
	Vp	mean	4.019	4.646	5.069	5.354	5.556	5.981	6.165	6.248	6.307	6.354
	Vs	A	2.203	2.559	2.787	2.938	3.039	3.237	3.322	3.364	3.393	3.416
	Vs	B	3.059	3.298	3.455	3.561	3.634	3.785	3.850	3.880	3.901	3.917
	Vs	C	2.202	2.548	2.768	2.913	3.010	3.197	3.278	3.318	3.347	3.370
	Vs	mean	2.488	2.802	3.003	3.137	3.228	3.406	3.483	3.521	3.547	3.568
	Vp/Vs	mean	1.615	1.659	1.688	1.707	1.721	1.756	1.770	1.775	1.778	1.781
	σ	mean	0.189	0.214	0.230	0.239	0.245	0.260	0.266	0.267	0.269	0.270
A-12	Vp	A	3.324	4.044	4.518	4.841	5.066	5.536	5.757	5.865	5.943	6.005
	Vp	B	4.008	4.624	5.040	5.330	5.537	5.986	6.192	6.283	6.348	6.399
	Vp	C	4.704	5.189	5.521	5.756	5.926	6.303	6.469	6.537	6.584	6.621
	Vp	mean	4.012	4.619	5.026	5.309	5.509	5.942	6.140	6.228	6.292	6.342
	Vs	A	2.271	2.588	2.804	2.955	3.062	3.289	3.379	3.415	3.440	3.460
	Vs	B	2.768	3.039	3.228	3.366	3.469	3.715	3.834	3.881	3.913	3.938
	Vs	C	1.870	2.287	2.545	2.709	2.815	3.002	3.074	3.109	3.134	3.154
	Vs	mean	2.303	2.638	2.859	3.010	3.115	3.335	3.429	3.468	3.496	3.517
	Vp/Vs	mean	1.742	1.751	1.758	1.764	1.768	1.781	1.790	1.796	1.800	1.803
	σ	mean	0.254	0.258	0.261	0.263	0.265	0.270	0.273	0.275	0.277	0.278

Table 2. (Continued)

#		core	\multicolumn{10}{c}{Pressure (MPa)}									
			20	40	60	80	100	200	400	600	800	1000
A-13	Vp	A	2.931	3.840	4.431	4.825	5.090	5.596	5.784	5.869	5.931	5.980
	Vp	B	5.496	6.802	6.368	6.571	6.700	6.918	6.997	7.035	7.063	7.084
	Vp	C	4.786	5.258	5.583	5.815	5.983	6.355	6.512	6.572	6.614	6.646
	Vp	mean	4.404	5.300	5.461	5.737	5.924	6.289	6.431	6.492	6.536	6.570
	Vs	A	2.280	2.624	2.840	2.980	3.073	3.253	3.337	3.380	3.411	3.436
	Vs	B	2.848	3.254	3.499	3.650	3.746	3.909	3.972	4.005	4.028	4.046
	Vs	C	2.112	2.524	2.789	2.962	3.077	3.286	3.353	3.382	3.402	3.417
	Vs	mean	2.414	2.801	3.043	3.198	3.299	3.482	3.554	3.589	3.614	3.633
	Vp/Vs	mean	1.825	1.892	1.795	1.794	1.796	1.806	1.809	1.809	1.809	1.809
	σ	mean	0.285	0.306	0.275	0.275	0.275	0.279	0.280	0.280	0.280	0.280
A-14	Vp	A	2.771	3.666	4.279	4.709	5.013	5.656	5.910	6.012	6.085	6.144
	Vp	B	3.549	4.298	4.806	5.161	5.412	5.940	6.151	6.237	6.298	6.375
	Vp	C	4.176	4.785	5.204	5.500	5.712	6.174	6.361	6.433	6.483	6.522
	Vp	mean	3.499	4.250	4.763	5.123	5.379	5.923	6.141	6.227	6.288	6.347
	Vs	A	2.174	2.653	2.959	3.158	3.290	3.528	3.607	3.642	3.666	3.685
	Vs	B	2.577	2.944	3.180	3.336	3.441	3.638	3.710	3.743	3.766	3.783
	Vs	C	1.918	2.482	2.844	3.082	3.239	3.526	3.621	3.663	3.693	3.716
	Vs	mean	2.223	2.693	2.994	3.192	3.323	3.564	3.646	3.682	3.708	3.728
	Vp/Vs	mean	1.574	1.578	1.591	1.605	1.619	1.662	1.684	1.691	1.696	1.702
	σ	mean	0.161	0.165	0.173	0.183	0.191	0.216	0.228	0.231	0.233	0.237
A-15	Vp	A	2.904	3.577	4.066	4.439	4.728	5.495	5.936	6.112	6.233	6.329
	Vp	B	4.963	5.222	5.401	5.538	5.646	5.963	6.193	6.294	6.362	6.414
	Vp	C	4.309	4.836	5.204	5.474	5.675	6.163	6.411	6.513	6.584	6.639
	Vp	mean	4.059	4.545	4.891	5.150	5.350	5.874	6.180	6.306	6.393	6.461
	Vs	A	2.053	2.420	2.674	2.857	2.993	3.314	3.480	3.554	3.606	3.647
	Vs	B	3.062	3.224	3.341	3.430	3.499	3.685	3.791	3.829	3.845	3.874
	Vs	C	1.932	2.345	2.624	2.821	2.961	3.268	3.410	3.474	3.520	3.557
	Vs	mean	2.349	2.663	2.880	3.036	3.151	3.422	3.560	3.619	3.657	3.693
	Vp/Vs	mean	1.728	1.707	1.698	1.696	1.698	1.716	1.736	1.743	1.748	1.750
	σ	mean	0.248	0.239	0.235	0.234	0.234	0.243	0.252	0.254	0.257	0.257
A-16	Vp	A	2.939	3.541	3.961	4.270	4.504	5.102	5.474	5.656	5.787	5.892
	Vp	B	5.800	6.023	6.178	6.295	6.386	6.632	6.786	6.850	6.893	6.927
	Vp	C	5.441	5.755	5.962	6.107	6.212	6.453	6.588	6.654	6.701	6.737
	Vp	mean	4.727	5.106	5.367	5.557	5.700	6.062	6.283	6.386	6.460	6.519
	Vs	A	2.195	2.425	2.580	2.691	2.773	2.976	3.101	3.164	3.208	3.244
	Vs	B	3.523	3.674	3.778	3.851	3.903	4.014	4.054	4.068	4.078	4.085
	Vs	C	2.364	2.581	2.727	2.819	2.903	3.068	3.147	3.182	3.207	3.223
	Vs	mean	2.694	2.893	3.028	3.120	3.193	3.353	3.434	3.471	3.498	3.517
	Vp/Vs	mean	1.755	1.765	1.772	1.781	1.785	1.808	1.829	1.840	1.847	1.853
	σ	mean	0.259	0.264	0.267	0.270	0.271	0.280	0.287	0.290	0.293	0.295

Table 2. (Continued)

#		core	\multicolumn{10}{c}{Pressure (MPa)}									
			20	40	60	80	100	200	400	600	800	1000
A-17	Vp	A	3.309	4.588	5.052	5.335	5.513	5.828	5.983	6.069	6.131	6.180
	Vp	B	5.000	5.402	5.673	5.864	6.003	6.316	6.471	6.541	6.590	6.629
	Vp	C	5.517	5.799	5.987	6.121	6.220	6.457	6.593	6.657	6.702	6.738
	Vp	mean	4.609	5.263	5.571	5.774	5.912	6.200	6.349	6.422	6.474	6.515
	Vs	A	2.747	2.973	3.115	3.209	3.271	3.392	3.446	3.472	3.491	3.506
	Vs	B	3.292	2.463	3.575	3.622	3.705	3.815	3.862	3.883	3.898	3.909
	Vs	C	2.551	2.793	2.950	3.056	3.130	3.286	3.366	3.407	3.436	3.458
	Vs	mean	2.863	2.743	3.214	3.296	3.369	3.498	3.558	3.587	3.608	3.624
	Vp/Vs	mean	1.610	1.919	1.734	1.752	1.755	1.772	1.784	1.790	1.794	1.798
	σ	mean	0.186	0.314	0.251	0.258	0.260	0.267	0.271	0.273	0.275	0.276
A-18	Vp	A	3.643	4.053	4.344	4.569	4.749	5.283	5.675	5.850	5.967	6.059
	Vp	B	5.430	5.674	5.846	5.980	6.087	6.408	6.640	6.733	6.793	6.838
	Vp	C	5.202	5.449	5.319	5.749	5.854	6.163	6.395	6.498	6.566	6.619
	Vp	mean	4.758	5.059	5.170	5.433	5.563	5.952	6.237	6.360	6.442	6.505
	Vs	A	2.939	2.587	2.716	2.814	2.892	3.113	3.274	3.350	3.403	3.444
	Vs	B	3.317	3.434	3.520	3.587	3.640	3.787	3.872	3.900	3.918	3.931
	Vs	C	2.351	2.530	2.659	2.759	2.839	3.078	3.246	3.314	3.358	3.391
	Vs	mean	2.869	2.850	2.965	3.053	3.124	3.326	3.464	3.522	3.559	3.589
	Vp/Vs	mean	1.659	1.775	1.744	1.779	1.781	1.789	1.800	1.806	1.810	1.813
	σ	mean	0.214	0.267	0.255	0.269	0.270	0.273	0.277	0.279	0.280	0.281
A-19	Vp	A	4.250	4.722	5.038	5.261	5.423	5.801	6.014	6.119	6.194	6.253
	Vp	B	5.802	6.026	6.169	6.267	6.337	6.500	6.607	6.664	6.704	6.736
	Vp	C	4.950	5.274	5.487	5.638	5.751	6.034	6.227	6.329	6.401	6.458
	Vp	mean	5.001	5.341	5.565	5.722	5.837	6.112	6.283	6.370	6.433	6.482
	Vs	A	2.979	3.152	3.265	3.342	3.397	3.520	3.589	3.624	3.650	3.669
	Vs	B	3.423	3.517	3.583	3.632	3.669	3.761	3.810	3.829	3.842	3.852
	Vs	C	3.191	3.300	3.363	3.400	3.423	3.462	3.482	3.492	3.500	3.505
	Vs	mean	3.198	3.323	3.404	3.458	3.496	3.581	3.627	3.648	3.664	3.675
	Vp/Vs	mean	1.564	1.607	1.635	1.655	1.669	1.707	1.732	1.746	1.756	1.764
	σ	mean	0.154	0.184	0.201	0.212	0.220	0.239	0.250	0.256	0.260	0.263
A-20	Vp	A	4.827	5.182	5.360	5.460	5.521	5.659	5.782	5.854	5.907	5.948
	Vp	B	6.302	6.425	6.495	6.539	6.569	6.643	6.708	6.745	6.772	6.793
	Vp	C	5.244	5.528	5.702	5.815	5.891	6.056	6.160	6.217	6.259	6.291
	Vp	mean	5.457	5.712	5.852	5.938	5.994	6.119	6.216	6.272	6.313	6.344
	Vs	A	3.237	3.298	3.338	3.368	3.391	3.460	3.514	3.541	3.559	3.574
	Vs	B	3.656	3.723	3.760	3.782	3.796	3.826	3.850	3.864	3.873	3.881
	Vs	C	3.493	3.529	3.550	3.566	3.577	3.609	3.635	3.650	3.660	3.669
	Vs	mean	3.462	3.517	3.549	3.572	3.588	3.632	3.666	3.685	3.697	3.708
	Vp/Vs	mean	1.576	1.624	1.649	1.662	1.670	1.685	1.696	1.702	1.707	1.711
	σ	mean	0.163	0.195	0.209	0.216	0.221	0.228	0.233	0.236	0.239	0.241

Table 2. (Continued)

#		core	\multicolumn{10}{c}{Pressure (MPa)}									
			20	40	60	80	100	200	400	600	800	1000
A-21	Vp	A	5.874	6.000	6.068	6.111	6.140	6.214	6.284	6.324	6.353	6.376
	Vp	B	5.859	5.952	6.007	6.045	6.073	6.152	6.225	6.268	6.298	6.322
	Vp	C	5.890	5.991	6.049	6.087	6.115	6.189	6.259	6.300	6.329	6.352
	Vp	mean	5.874	5.981	6.041	6.081	6.109	6.185	6.256	6.297	6.327	6.350
	Vs	A	3.542	3.576	3.597	3.612	3.623	3.654	3.680	3.695	3.706	3.714
	Vs	B	3.541	3.573	3.593	3.607	3.619	3.650	3.676	3.689	3.699	3.706
	Vs	C	3.548	3.580	3.602	3.617	3.629	3.660	3.684	3.697	3.705	3.712
	Vs	mean	3.544	3.576	3.597	3.612	3.623	3.655	3.680	3.694	3.703	3.711
	Vp/Vs	mean	1.658	1.672	1.680	1.684	1.686	1.692	1.700	1.705	1.708	1.711
	σ	mean	0.214	0.222	0.225	0.227	0.229	0.232	0.235	0.238	0.239	0.241
A-22	Vp	A	5.838	5.895	5.925	5.945	5.961	6.013	6.078	6.122	6.154	6.179
	Vp	B	5.913	5.969	6.001	6.024	6.042	6.099	6.161	6.199	6.226	6.247
	Vp	C	5.978	6.032	6.063	6.085	6.102	6.158	6.216	6.250	6.275	6.294
	Vp	mean	5.910	5.965	5.996	6.018	6.035	6.090	6.152	6.190	6.218	6.240
	Vs	A	3.551	3.571	3.583	3.592	3.598	3.619	3.641	3.654	3.663	3.670
	Vs	B	3.507	3.539	3.558	3.570	3.580	3.604	3.625	3.637	3.646	3.653
	Vs	C	3.513	3.536	3.550	3.559	3.567	3.590	3.615	3.630	3.640	3.649
	Vs	mean	3.524	3.549	3.564	3.574	3.581	3.604	3.627	3.640	3.650	3.657
	Vp/Vs	mean	1.677	1.681	1.683	1.684	1.685	1.690	1.696	1.700	1.704	1.706
	σ	mean	0.224	0.226	0.227	0.228	0.228	0.230	0.234	0.236	0.237	0.238
A-23	Vp	A	5.430	5.291	5.826	5.902	5.948	6.045	6.125	6.172	6.205	6.232
	Vp	B	5.645	5.783	5.865	5.920	5.958	6.056	6.138	6.185	6.219	6.245
	Vp	C	5.443	5.640	5.758	5.835	5.887	6.011	6.106	6.160	6.198	6.228
	Vp	mean	5.506	5.571	5.816	5.885	5.931	6.038	6.123	6.172	6.208	6.235
	Vs	A	3.431	3.488	3.524	3.549	3.566	3.607	3.632	3.640	3.655	3.663
	Vs	B	2.386	3.425	3.452	3.472	3.488	3.534	3.564	3.576	3.584	3.590
	Vs	C	3.383	3.437	3.473	3.500	3.520	3.572	3.603	3.617	3.627	3.634
	Vs	mean	3.067	3.450	3.483	3.507	3.525	3.571	3.600	3.611	3.622	3.629
	Vp/Vs	mean	1.795	1.615	1.670	1.678	1.683	1.691	1.701	1.709	1.714	1.718
	σ	mean	0.275	0.189	0.220	0.225	0.227	0.231	0.236	0.240	0.242	0.244
A-24	Vp	A	5.275	5.528	5.696	5.813	5.896	6.086	6.187	6.235	6.269	6.295
	Vp	B	5.111	5.437	5.644	5.781	5.873	6.064	6.163	6.214	6.251	6.279
	Vp	C	5.168	5.428	5.604	5.731	5.823	6.043	6.160	6.213	6.250	6.278
	Vp	mean	5.184	5.465	5.648	5.775	5.864	6.064	6.170	6.221	6.256	6.284
	Vs	A	3.251	3.374	3.453	3.507	3.544	3.622	3.661	3.679	3.692	3.702
	Vs	B	3.192	3.316	3.399	3.456	3.497	3.585	3.623	3.640	3.651	3.660
	Vs	C	3.203	3.320	3.400	3.457	3.498	3.588	3.625	3.639	3.649	3.656
	Vs	mean	3.216	3.337	3.417	3.473	3.513	3.598	3.636	3.653	3.664	3.673
	Vp/Vs	mean	1.612	1.638	1.653	1.663	1.669	1.685	1.697	1.703	1.708	1.711
	σ	mean	0.187	0.203	0.211	0.217	0.220	0.228	0.234	0.237	0.239	0.241

Table 2. (Continued)

#		core	\multicolumn{10}{c}{Pressure (MPa)}									
			20	40	60	80	100	200	400	600	800	1000
A-25	Vp	A	5.663	5.802	5.887	5.945	5.987	6.093	6.176	6.223	6.257	6.283
	Vp	B	5.475	5.690	5.816	5.895	5.947	6.067	6.159	6.212	6.250	6.279
	Vp	C	5.416	5.649	5.775	5.850	5.898	6.010	6.106	6.162	6.203	6.235
	Vp	mean	5.518	5.714	5.826	5.896	5.944	6.056	6.147	6.199	6.237	6.266
	Vs	A	3.415	3.476	3.516	3.543	3.562	3.608	3.637	3.652	3.663	3.671
	Vs	B	3.429	3.473	3.502	3.522	3.537	3.576	3.605	3.621	3.632	3.641
	Vs	C	3.339	3.379	3.403	3.422	3.429	3.485	3.532	3.556	3.573	3.585
	Vs	mean	3.395	3.443	3.474	3.495	3.509	3.556	3.591	3.610	3.623	3.632
	Vp/Vs	mean	1.626	1.660	1.677	1.687	1.694	1.703	1.712	1.717	1.722	1.725
	σ	mean	0.196	0.215	0.224	0.229	0.232	0.237	0.241	0.243	0.245	0.247
A-26	Vp	A	6.542	6.693	6.721	6.760	6.787	6.856	6.920	6.957	6.984	7.005
	Vp	B	7.087	7.205	7.268	7.305	7.328	7.378	7.419	7.442	7.459	7.472
	Vp	C	6.863	6.992	7.056	7.093	7.116	7.173	7.226	7.257	7.279	7.296
	Vp	mean	6.831	6.963	7.015	7.053	7.077	7.136	7.188	7.219	7.241	7.258
	Vs	A	3.881	3.920	3.944	3.959	3.970	3.996	4.012	4.021	4.028	4.033
	Vs	B	4.004	4.042	4.069	4.090	4.105	4.147	4.171	4.179	4.184	4.189
	Vs	C	3.923	3.960	3.981	3.996	4.008	4.051	4.102	4.133	4.155	4.172
	Vs	mean	3.936	3.974	3.998	4.015	4.028	4.065	4.095	4.111	4.122	4.131
	Vp/Vs	mean	1.735	1.752	1.755	1.756	1.757	1.756	1.755	1.756	1.756	1.757
	σ	mean	0.251	0.258	0.259	0.260	0.260	0.260	0.260	0.260	0.260	0.260
A-28	Vp	A	5.242	5.594	5.813	5.954	6.047	6.232	6.326	6.375	6.410	6.437
	Vp	B	5.018	5.443	5.712	5.888	6.006	6.235	6.338	6.389	6.426	6.454
	Vp	C	4.966	5.284	5.499	5.652	5.763	6.018	6.145	6.202	6.241	6.272
	Vp	mean	5.075	5.440	5.675	5.831	5.939	6.162	6.270	6.322	6.359	6.388
	Vs	A	3.083	3.282	3.408	3.489	3.542	3.636	3.666	3.678	3.687	3.694
	Vs	B	2.950	3.193	3.342	3.437	3.500	3.624	3.692	3.729	3.755	3.776
	Vs	C	2.892	3.124	3.278	3.382	3.452	3.585	3.622	3.635	3.644	3.650
	Vs	mean	2.975	3.200	3.343	3.436	3.498	3.615	3.660	3.681	3.695	3.707
	Vp/Vs	mean	1.706	1.700	1.698	1.697	1.698	1.704	1.713	1.718	1.721	1.723
	σ	mean	0.238	0.236	0.234	0.234	0.234	0.238	0.242	0.244	0.245	0.246
A-29	Vp	A	4.373	4.934	5.304	5.557	5.733	6.100	6.261	6.334	6.386	6.427
	Vp	B	4.440	5.062	5.450	5.699	5.864	6.175	6.315	6.387	6.439	6.479
	Vp	C	4.932	5.378	5.664	5.854	5.984	6.250	6.381	6.447	6.493	6.530
	Vp	mean	4.582	5.125	5.473	5.704	5.860	6.175	6.319	6.389	6.440	6.479
	Vs	A	2.650	3.036	3.276	3.427	3.524	3.686	3.732	3.751	3.764	3.775
	Vs	B	2.887	3.125	3.281	3.385	3.456	3.591	3.636	3.654	3.667	3.676
	Vs	C	2.853	3.127	3.309	3.432	3.517	3.685	3.746	3.771	3.788	3.802
	Vs	mean	2.797	3.096	3.289	3.415	3.499	3.654	3.705	3.725	3.740	3.751
	Vp/Vs	mean	1.638	1.655	1.664	1.670	1.675	1.690	1.706	1.715	1.722	1.727
	σ	mean	0.203	0.213	0.217	0.221	0.223	0.231	0.238	0.242	0.246	0.248

Table 2. (Continued)

#		core	20	40	60	80	100	200	400	600	800	1000
A-30	Vp	A	4.595	5.023	5.303	5.492	5.621	5.883	5.992	6.041	6.076	6.103
	Vp	B	4.910	5.309	5.561	5.727	5.839	6.066	6.179	6.236	6.278	6.310
	Vp	C	5.358	5.624	5.803	5.929	6.020	6.225	6.325	6.369	6.401	6.425
	Vp	mean	4.954	5.319	5.556	5.716	5.827	6.058	6.165	6.216	6.251	6.279
	Vs	A	3.019	3.176	3.280	3.351	3.401	3.506	3.555	3.578	3.594	3.606
	Vs	B	3.303	3.451	3.546	3.610	3.652	3.735	3.766	3.780	3.789	3.797
	Vs	C	2.944	3.154	3.285	3.369	3.425	3.531	3.580	3.605	3.622	3.636
	Vs	mean	3.089	3.260	3.370	3.443	3.493	3.591	3.634	3.654	3.668	3.680
	Vp/Vs	mean	1.604	1.631	1.648	1.660	1.668	1.687	1.697	1.701	1.704	1.706
	σ	mean	0.182	0.199	0.209	0.215	0.220	0.229	0.234	0.236	0.237	0.238
A-31b	Vp	A	5.710	5.892	6.002	6.073	6.122	6.231	6.306	6.349	6.379	6.403
	Vp	B	5.669	5.829	5.929	5.996	6.044	6.160	6.241	6.286	6.318	6.343
	Vp	C	5.349	5.614	5.769	5.865	5.927	6.058	6.148	6.200	6.236	6.264
	Vp	mean	5.576	5.778	5.900	5.978	6.031	6.149	6.232	6.278	6.311	6.337
	Vs	A	3.490	3.554	3.596	3.625	3.646	3.696	3.727	3.742	3.752	3.761
	Vs	B	3.497	3.552	3.589	3.614	3.632	3.672	3.691	3.699	3.705	3.710
	Vs	C	3.398	3.456	3.493	3.519	3.537	3.582	3.610	3.625	3.635	3.643
	Vs	mean	3.462	3.521	3.559	3.586	3.605	3.650	3.676	3.689	3.698	3.705
	Vp/Vs	mean	1.611	1.641	1.658	1.667	1.673	1.685	1.695	1.702	1.707	1.711
	σ	mean	0.186	0.205	0.214	0.219	0.222	0.228	0.233	0.236	0.239	0.241
A-32b	Vp	A	5.327	5.465	5.556	5.627	5.684	5.877	6.064	6.156	6.213	6.254
	Vp	B	5.515	5.631	5.705	5.761	5.807	5.962	6.125	6.213	6.270	6.310
	Vp	C	5.519	5.641	5.720	5.781	5.831	5.997	6.164	6.248	6.301	6.338
	Vp	mean	5.454	5.579	5.661	5.723	5.774	5.946	6.118	6.206	6.261	6.301
	Vs	A	3.220	3.286	3.332	3.368	3.397	3.484	3.546	3.570	3.584	3.595
	Vs	B	3.383	3.444	3.483	3.511	3.532	3.587	3.627	3.648	3.663	3.675
	Vs	C	3.378	3.442	3.483	3.512	3.533	3.583	3.610	3.624	3.633	3.640
	Vs	mean	3.327	3.391	3.433	3.464	3.487	3.551	3.594	3.614	3.627	3.637
	Vp/Vs	mean	1.639	1.645	1.649	1.652	1.656	1.674	1.702	1.717	1.726	1.732
	σ	mean	0.204	0.207	0.209	0.211	0.213	0.223	0.236	0.243	0.248	0.250
A-33	Vp	A	3.404	3.900	4.237	4.500	4.711	5.330	5.768	5.954	6.079	6.176
	Vp	B	4.251	4.623	4.892	5.102	5.271	5.760	6.084	6.207	6.286	6.345
	Vp	C	4.578	4.988	5.267	5.470	5.622	5.996	6.213	6.313	6.384	6.439
	Vp	mean	4.078	4.504	4.799	5.024	5.201	5.695	6.022	6.158	6.249	6.320
	Vs	A	2.100	2.439	2.566	2.845	2.974	3.292	3.475	3.559	3.620	3.664
	Vs	B	2.699	2.980	3.172	3.310	3.412	3.656	3.788	3.848	3.890	3.923
	Vs	C	2.615	2.805	2.937	3.035	3.110	3.309	3.430	3.482	3.519	3.547
	Vs	mean	2.471	2.741	2.892	3.063	3.165	3.419	3.564	3.630	3.676	3.711
	Vp/Vs	mean	1.650	1.643	1.660	1.640	1.643	1.666	1.690	1.697	1.700	1.703
	σ	mean	0.210	0.206	0.215	0.204	0.206	0.218	0.230	0.234	0.235	0.237

Pressure (MPa)

Table 2. (Continued)

#		core	Pressure (MPa)									
			20	40	60	80	100	200	400	600	800	1000
A-34	Vp	A	4.049	4.483	4.763	4.955	5.093	5.419	5.636	5.756	5.843	5.911
	Vp	B	5.682	5.939	6.109	6.229	6.316	6.524	6.647	6.707	6.750	6.783
	Vp	C	4.625	5.012	5.268	5.448	5.578	5.881	6.056	6.143	6.206	6.255
	Vp	mean	4.786	5.145	5.380	5.544	5.662	5.941	6.113	6.202	6.266	6.316
	Vs	A	2.747	2.921	3.034	3.114	3.171	3.306	3.387	3.428	3.457	3.480
	Vs	B	3.258	3.446	3.566	3.647	3.702	3.818	3.878	3.908	3.929	3.946
	Vs	C	3.369	3.491	3.576	3.639	3.685	3.797	3.852	3.874	3.888	3.899
	Vs	mean	3.125	3.286	3.392	3.466	3.519	3.640	3.706	3.736	3.758	3.775
	Vp/Vs	mean	1.531	1.566	1.586	1.599	1.609	1.632	1.650	1.660	1.667	1.673
	σ	mean	0.128	0.156	0.170	0.179	0.185	0.199	0.210	0.215	0.219	0.222
A-35	Vp	A	4.564	4.843	5.020	5.145	5.239	5.488	5.688	5.802	5.884	5.949
	Vp	B	5.396	5.583	5.710	5.801	5.870	6.065	6.218	6.299	6.356	6.401
	Vp	C	6.087	6.265	6.376	6.454	6.509	6.655	6.768	6.831	6.876	6.912
	Vp	mean	5.349	5.564	5.702	5.800	5.873	6.069	6.225	6.311	6.372	6.420
	Vs	A	2.910	3.026	3.100	3.153	3.193	3.298	3.377	3.420	3.451	3.475
	Vs	B	3.615	3.676	3.716	3.745	3.769	3.834	3.882	3.905	3.922	3.934
	Vs	C	3.037	3.141	3.211	3.232	3.300	3.399	3.461	3.491	3.512	3.528
	Vs	mean	3.187	3.281	3.342	3.377	3.421	3.510	3.573	3.605	3.628	3.646
	Vp/Vs	mean	1.678	1.696	1.706	1.718	1.717	1.729	1.742	1.750	1.756	1.761
	σ	mean	0.225	0.233	0.238	0.244	0.243	0.249	0.254	0.258	0.260	0.262
A-36	Vp	A	4.122	4.677	5.042	5.294	5.472	5.859	6.055	6.150	6.219	6.273
	Vp	B	5.946	6.236	6.400	6.510	6.588	6.762	6.872	6.930	6.972	7.005
	Vp	C	5.883	6.185	6.374	6.499	6.584	6.758	6.849	6.897	6.931	6.957
	Vp	mean	5.317	5.699	5.939	6.101	6.215	6.460	6.592	6.659	6.707	6.745
	Vs	A	2.753	2.972	3.108	3.198	3.260	3.393	3.475	3.520	3.552	3.577
	Vs	B	3.743	3.843	3.908	3.951	3.980	4.042	4.074	4.094	4.102	4.111
	Vs	C	3.165	3.294	3.386	3.453	3.502	3.611	3.649	3.659	3.666	3.671
	Vs	mean	3.220	3.370	3.467	3.534	3.581	3.682	3.733	3.758	3.773	3.786
	Vp/Vs	mean	1.651	1.691	1.713	1.726	1.736	1.754	1.766	1.772	1.778	1.781
	σ	mean	0.210	0.231	0.241	0.248	0.252	0.259	0.264	0.266	0.269	0.270
A-37	Vp	A	4.197	4.709	5.028	5.241	5.389	5.730	5.971	6.109	6.210	6.289
	Vp	B	5.608	5.953	6.190	6.359	6.482	6.769	6.908	6.967	7.008	7.040
	Vp	C	5.368	5.801	6.086	6.282	6.421	6.721	6.870	6.941	6.991	7.031
	Vp	mean	5.058	5.488	5.768	5.961	6.097	6.406	6.583	6.672	6.736	6.787
	Vs	A	2.628	2.795	2.902	2.975	3.028	3.158	3.252	3.305	3.343	3.373
	Vs	B	3.530	3.708	3.826	3.907	3.964	4.083	4.135	4.158	4.173	4.186
	Vs	C	2.438	2.636	2.770	2.866	2.936	3.099	3.185	3.224	3.251	3.273
	Vs	mean	2.865	3.047	3.166	3.249	3.309	3.447	3.524	3.562	3.589	3.610
	Vp/Vs	mean	1.765	1.801	1.822	1.834	1.843	1.859	1.868	1.873	1.877	1.880
	σ	mean	0.264	0.277	0.284	0.289	0.291	0.296	0.299	0.301	0.302	0.303

Table 2. (Continued)

#		core	Pressure (MPa)									
			20	40	60	80	100	200	400	600	800	1000
A-38	Vp	A	4.760	5.064	5.239	5.350	5.423	5.592	5.727	5.806	5.862	5.907
	Vp	B	5.982	6.080	6.141	6.185	6.219	6.321	6.414	6.467	6.505	6.534
	Vp	C	6.120	6.285	6.372	6.424	6.457	6.540	6.614	6.657	6.688	6.712
	Vp	mean	5.620	5.810	5.917	5.986	6.033	6.151	6.252	6.310	6.352	6.385
	Vs	A	3.043	3.125	3.180	3.220	3.251	3.332	3.388	3.416	3.436	3.451
	Vs	B	3.782	3.814	3.833	3.846	3.856	3.887	3.918	3.937	3.950	3.960
	Vs	C	3.642	3.744	3.780	3.798	3.810	3.846	3.883	3.904	3.920	3.932
	Vs	mean	3.489	3.561	3.598	3.621	3.639	3.688	3.730	3.752	3.769	3.781
	Vp/Vs	mean	1.611	1.631	1.645	1.653	1.658	1.668	1.676	1.682	1.685	1.689
	σ	mean	0.187	0.199	0.207	0.211	0.214	0.219	0.224	0.226	0.228	0.230
A-39	Vp	A	3.740	4.153	4.441	4.659	4.828	5.298	5.619	5.768	5.873	5.955
	Vp	B	5.250	5.551	5.753	5.901	6.012	6.306	6.505	6.604	6.675	6.730
	Vp	C	5.653	5.889	6.051	6.172	6.265	6.517	6.679	6.751	6.801	6.839
	Vp	mean	4.881	5.198	5.415	5.577	5.702	6.040	6.268	6.375	6.449	6.508
	Vs	A	2.543	2.707	2.820	2.904	2.968	3.140	3.247	3.295	3.328	3.354
	Vs	B	3.389	3.550	3.655	3.728	3.781	3.905	3.979	4.016	4.042	4.063
	Vs	C	2.647	2.775	2.865	2.934	2.989	3.145	3.250	3.293	3.321	3.343
	Vs	mean	2.860	3.011	3.113	3.189	3.246	3.397	3.492	3.535	3.564	3.587
	Vp/Vs	mean	1.707	1.727	1.739	1.749	1.757	1.778	1.795	1.803	1.810	1.815
	σ	mean	0.239	0.248	0.253	0.257	0.260	0.269	0.275	0.278	0.280	0.282
A-40	Vp	A	2.583	3.231	3.699	4.054	4.332	5.083	5.561	5.778	5.933	6.057
	Vp	B	4.151	4.644	4.984	5.234	5.423	5.911	6.213	6.355	6.455	6.535
	Vp	C	5.445	6.059	6.367	6.533	6.629	6.818	6.967	7.055	7.119	7.169
	Vp	mean	4.060	4.645	5.016	5.274	5.461	5.937	6.247	6.396	6.502	6.587
	Vs	A	2.064	2.351	2.555	2.708	2.826	3.135	3.316	3.391	3.443	3.484
	Vs	B	2.994	3.172	3.303	3.405	3.486	3.703	3.819	3.852	3.872	3.887
	Vs	C	3.145	3.335	3.452	3.530	3.584	3.706	3.790	3.837	3.871	3.898
	Vs	mean	2.734	2.953	3.104	3.215	3.299	3.515	3.641	3.694	3.729	3.757
	Vp/Vs	mean	1.485	1.573	1.616	1.641	1.656	1.689	1.715	1.732	1.744	1.753
	σ	mean	0.085	0.161	0.190	0.204	0.213	0.230	0.243	0.250	0.255	0.259
A-41	Vp	A	2.953	3.753	4.330	4.763	5.092	5.912	6.324	6.478	6.585	6.670
	Vp	B	5.087	5.538	5.865	6.119	6.324	6.918	7.315	7.467	7.564	7.638
	Vp	C	3.608	4.269	4.747	5.112	5.397	6.157	6.600	6.775	6.893	6.987
	Vp	mean	3.883	4.520	4.981	5.332	5.604	6.329	6.746	6.907	7.014	7.098
	Vs	A	2.289	2.630	2.867	3.037	3.163	3.461	3.613	3.379	3.726	3.763
	Vs	B	2.702	2.991	3.195	3.346	3.462	3.757	3.922	3.991	4.038	4.075
	Vs	C	2.305	2.604	2.816	2.973	3.092	3.384	3.528	3.582	3.619	3.648
	Vs	mean	2.432	2.742	2.959	3.119	3.239	3.534	3.688	3.651	3.794	3.828
	Vp/Vs	mean	1.597	1.649	1.683	1.709	1.730	1.791	1.829	1.892	1.849	1.854
	σ	mean	0.177	0.209	0.227	0.240	0.249	0.273	0.287	0.306	0.293	0.295

Table 2. (Continued)

#		core	Pressure (MPa)									
			20	40	60	80	100	200	400	600	800	1000
A-42	Vp	A	3.386	4.281	4.806	5.122	5.316	5.650	5.807	5.895	5.959	6.009
	Vp	B	4.408	5.087	5.472	5.699	5.837	6.083	6.222	6.301	6.359	6.404
	Vp	C	5.232	5.722	6.000	6.167	6.271	6.469	6.596	6.669	6.722	6.763
	Vp	mean	4.342	5.030	5.426	5.662	5.808	6.067	6.208	6.288	6.346	6.392
	Vs	A	2.781	2.951	3.071	3.159	3.223	3.375	3.440	3.463	3.478	3.490
	Vs	B	3.222	3.431	3.550	3.620	3.664	3.746	3.795	3.824	3.844	3.849
	Vs	C	2.877	3.057	3.161	3.228	3.273	3.384	3.479	3.535	3.576	3.608
	Vs	mean	2.960	3.147	3.261	3.336	3.387	3.502	3.572	3.607	3.633	3.649
	Vp/Vs	mean	1.467	1.599	1.664	1.698	1.715	1.733	1.738	1.743	1.747	1.752
	σ	mean	0.066	0.179	0.217	0.234	0.242	0.250	0.253	0.255	0.256	0.258
A-43	Vp	A	3.238	4.198	4.805	5.197	5.454	5.919	6.080	6.171	6.230	6.277
	Vp	B	4.793	5.322	5.679	5.928	6.105	6.479	6.632	6.695	6.739	6.773
	Vp	C	5.440	5.864	6.148	6.345	6.484	6.779	6.904	6.958	6.996	7.025
	Vp	mean	4.490	5.128	5.544	5.824	6.014	6.392	6.539	6.608	6.655	6.692
	Vs	A	2.497	2.816	3.019	3.153	3.243	3.416	3.486	3.520	3.544	3.563
	Vs	B	3.150	3.395	3.554	3.662	3.736	3.885	3.949	3.978	3.999	4.015
	Vs	C	2.754	3.003	3.177	3.293	3.377	3.550	3.608	3.626	3.639	3.649
	Vs	mean	2.800	3.071	3.250	3.369	3.452	3.617	3.681	3.708	3.727	3.742
	Vp/Vs	mean	1.604	1.670	1.706	1.728	1.742	1.767	1.776	1.782	1.785	1.788
	σ	mean	0.182	0.220	0.238	0.248	0.254	0.265	0.268	0.270	0.271	0.272
A-44	Vp	A	2.467	3.446	4.098	4.543	4.850	5.476	5.743	5.871	5.965	6.041
	Vp	B	4.590	5.200	5.611	5.899	6.104	6.550	6.751	6.837	6.899	6.947
	Vp	C	4.996	5.507	5.850	6.087	6.254	6.610	6.765	6.832	6.879	6.916
	Vp	mean	4.018	4.718	5.186	5.510	5.736	6.212	6.420	6.513	6.581	6.635
	Vs	A	2.015	2.397	2.649	2.821	2.941	3.189	3.299	3.350	3.387	3.416
	Vs	B	2.974	3.320	3.542	3.689	3.788	3.978	4.053	4.087	4.111	4.130
	Vs	C	2.297	2.580	2.770	2.900	2.990	3.168	3.228	3.250	3.266	3.278
	Vs	mean	2.429	2.766	2.987	3.137	3.240	3.445	3.527	3.563	3.588	3.608
	Vp/Vs	mean	1.654	1.706	1.736	1.756	1.771	1.803	1.820	1.828	1.834	1.839
	σ	mean	0.212	0.238	0.252	0.260	0.266	0.278	0.284	0.287	0.289	0.290
A-45	Vp	A	3.629	4.277	4.683	4.950	5.130	5.507	5.727	5.849	5.937	6.008
	Vp	B	4.560	5.128	5.516	5.791	5.991	6.443	6.656	6.748	6.813	6.864
	Vp	C	4.951	5.369	5.671	5.898	6.074	6.521	6.747	6.823	6.872	6.911
	Vp	mean	4.380	4.925	5.290	5.547	5.732	6.157	6.377	6.473	6.541	6.594
	Vs	A	2.226	2.381	2.491	2.574	2.640	2.819	2.931	2.975	3.005	3.028
	Vs	B	2.841	3.179	3.406	3.564	3.675	6.907	3.998	4.035	4.061	4.081
	Vs	C	2.072	2.420	2.654	2.814	2.926	3.153	3.239	3.274	3.299	3.319
	Vs	mean	2.380	2.660	2.850	2.984	3.080	4.293	3.389	3.428	3.455	3.476
	Vp/Vs	mean	1.841	1.851	1.856	1.859	1.861	1.434	1.881	1.888	1.893	1.897
	σ	mean	0.291	0.294	0.295	0.296	0.297	0.027	0.303	0.305	0.307	0.308

Table 2. (Continued)

#		core	Pressure (MPa)									
			20	40	60	80	100	200	400	600	800	1000
A-46	Vp	A	3.623	4.639	5.313	5.770	6.084	6.703	6.932	7.028	7.098	7.153
	Vp	B	3.787	4.759	5.407	5.845	6.152	6.753	6.966	7.051	7.112	7.160
	Vp	C	4.509	5.294	5.802	6.140	6.366	6.801	6.966	7.040	7.093	7.135
	Vp	mean	3.973	4.898	5.508	5.918	6.201	6.753	6.955	7.040	7.101	7.149
	Vs	A	2.505	2.957	3.252	3.450	3.583	3.841	3.933	3.972	4.000	4.021
	Vs	B	2.477	2.979	3.296	3.499	3.630	3.858	3.925	3.953	3.973	3.989
	Vs	C	2.720	3.226	3.485	3.621	3.695	3.800	3.849	3.877	3.898	3.914
	Vs	mean	2.567	3.054	3.344	3.523	3.636	3.833	3.902	3.934	3.957	3.975
	Vp/Vs	mean	1.548	1.604	1.647	1.680	1.705	1.762	1.782	1.789	1.795	1.799
	σ	mean	0.142	0.182	0.208	0.226	0.238	0.262	0.270	0.273	0.275	0.276
A-47	Vp	A	4.243	5.090	5.667	6.066	6.345	6.906	7.084	7.142	7.182	7.214
	Vp	B	3.936	4.488	5.535	5.985	6.300	6.931	7.127	7.188	7.231	7.264
	Vp	C	3.795	4.718	5.361	5.817	6.143	6.832	7.006	7.135	7.183	7.221
	Vp	mean	3.991	4.765	5.521	5.956	6.262	6.890	7.073	7.155	7.199	7.233
	Vs	A	2.683	3.118	3.396	3.577	3.695	3.904	3.964	3.988	4.005	4.018
	Vs	B	2.533	3.013	3.320	3.521	3.653	3.889	3.959	3.987	4.007	4.022
	Vs	C	2.667	3.046	3.303	3.482	3.607	3.861	3.944	3.971	3.990	4.005
	Vs	mean	2.628	3.059	3.340	3.526	3.652	3.885	3.956	3.982	4.001	4.015
	Vp/Vs	mean	1.519	1.558	1.653	1.689	1.715	1.774	1.788	1.797	1.799	1.801
	σ	mean	0.117	0.150	0.211	0.230	0.242	0.267	0.272	0.276	0.277	0.277
A-48	Vp	A	4.167	4.900	5.401	5.753	6.002	6.524	6.713	6.780	6.826	6.863
	Vp	B	3.770	4.612	5.184	5.581	5.861	6.437	6.476	6.727	6.782	6.826
	Vp	C	4.459	5.119	5.558	5.586	6.064	6.483	6.644	6.711	6.759	6.796
	Vp	mean	4.132	4.877	5.381	5.640	5.976	6.481	6.611	6.739	6.789	6.829
	Vs	A	2.490	2.939	3.221	3.402	3.520	3.730	3.803	3.836	3.860	3.879
	Vs	B	2.573	2.917	3.150	3.310	3.422	3.645	3.716	3.739	3.754	3.767
	Vs	C	2.742	3.064	3.272	3.408	3.498	3.662	3.713	3.733	3.748	3.759
	Vs	mean	2.602	2.973	3.214	3.373	3.480	3.679	3.744	3.770	3.788	3.802
	Vp/Vs	mean	1.588	1.640	1.674	1.672	1.717	1.762	1.766	1.788	1.793	1.796
	σ	mean	0.172	0.204	0.223	0.221	0.243	0.262	0.264	0.272	0.274	0.275
A-49	Vp	A	3.664	4.122	4.452	4.704	4.903	5.454	5.799	5.938	6.032	6.105
	Vp	B	4.720	5.104	5.372	5.571	5.724	6.121	6.357	6.459	6.529	6.584
	Vp	C	4.843	5.223	5.485	5.679	5.825	6.198	6.416	6.513	6.580	6.633
	Vp	mean	4.409	4.816	5.103	5.318	5.484	5.924	6.191	6.303	6.380	6.440
	Vs	A	2.535	2.748	2.897	3.008	3.093	3.316	3.450	3.508	3.549	3.580
	Vs	B	3.020	3.261	3.422	3.534	3.614	3.794	3.889	3.934	3.966	3.991
	Vs	C	3.072	3.266	3.401	3.499	3.573	3.753	3.847	3.885	3.912	3.932
	Vs	mean	2.875	3.092	3.240	3.347	3.427	3.621	3.729	3.776	3.809	3.834
	Vp/Vs	mean	1.533	1.558	1.575	1.589	1.600	1.636	1.660	1.669	1.675	1.680
	σ	mean	0.130	0.150	0.162	0.172	0.180	0.202	0.215	0.220	0.223	0.225

Table 2. (Continued)

#		core	Pressure (MPa)									
			20	40	60	80	100	200	400	600	800	1000
A-50	Vp	A	4.107	4.587	4.917	5.153	5.325	5.715	5.897	5.972	6.025	6.067
	Vp	B	4.091	4.627	5.002	5.274	5.474	5.937	6.146	6.224	6.278	6.321
	Vp	C	4.393	4.854	5.176	5.411	5.584	5.994	6.187	6.262	6.314	6.354
	Vp	mean	4.197	4.689	5.032	5.279	5.461	5.882	6.076	6.153	6.206	6.247
	Vs	A	2.637	2.902	3.077	3.197	3.280	3.454	3.530	3.564	3.589	3.608
	Vs	B	2.656	2.983	3.197	3.343	3.444	3.647	3.731	3.768	3.794	3.815
	Vs	C	2.503	2.801	2.999	3.135	3.232	3.443	3.545	3.594	3.628	3.655
	Vs	mean	2.599	2.895	3.091	3.225	3.319	3.515	3.602	3.642	3.670	3.693
	Vp/Vs	mean	1.615	1.620	1.628	1.637	1.646	1.674	1.687	1.689	1.691	1.692
	σ	mean	0.189	0.192	0.197	0.202	0.207	0.222	0.229	0.230	0.231	0.232
A-51	Vp	A	3.563	4.129	4.542	4.857	5.101	5.734	6.064	6.176	6.279	6.307
	Vp	B	5.388	5.735	5.973	6.146	6.276	6.598	6.780	6.862	6.919	6.964
	Vp	C	4.464	4.909	5.225	5.462	5.643	6.103	6.349	6.442	6.506	6.556
	Vp	mean	4.472	4.924	5.247	5.488	5.673	6.145	6.398	6.493	6.568	6.609
	Vs	A	2.612	2.794	2.920	3.013	3.083	2.606	3.362	3.406	3.437	3.460
	Vs	B	3.117	3.341	3.490	3.594	3.667	3.827	3.903	3.938	3.963	3.982
	Vs	C	3.134	3.332	3.459	3.543	3.600	3.714	3.763	3.786	3.803	3.815
	Vs	mean	2.954	3.156	3.290	3.383	3.450	3.382	3.676	3.710	3.734	3.753
	Vp/Vs	mean	1.514	1.561	1.595	1.622	1.644	1.817	1.740	1.750	1.759	1.761
	σ	mean	0.113	0.152	0.176	0.194	0.207	0.283	0.254	0.258	0.261	0.262
A-52	Vp	A	4.253	4.787	5.150	5.412	5.604	6.067	6.325	6.447	6.533	6.601
	Vp	B	5.483	5.804	6.019	6.174	6.290	6.575	6.746	6.828	6.886	6.931
	Vp	C	5.817	6.083	6.263	6.394	6.493	6.746	6.903	6.977	7.028	7.068
	Vp	mean	5.185	5.558	5.811	5.993	6.129	6.463	6.658	6.750	6.816	6.867
	Vs	A	2.873	3.017	3.117	3.192	3.250	3.414	3.525	3.575	3.610	3.636
	Vs	B	3.447	3.550	3.362	3.677	3.721	3.850	3.941	3.978	4.002	4.020
	Vs	C	2.947	3.097	3.196	3.264	3.313	3.423	3.483	3.513	3.535	3.551
	Vs	mean	3.089	3.222	3.225	3.378	3.428	3.562	3.650	3.689	3.715	3.736
	Vp/Vs	mean	1.678	1.725	1.802	1.774	1.788	1.814	1.824	1.830	1.834	1.838
	σ	mean	0.225	0.247	0.277	0.267	0.272	0.282	0.285	0.287	0.289	0.290
A-53	Vp	A	2.955	3.855	4.490	4.951	5.290	6.070	6.419	6.557	6.654	6.732
	Vp	B	4.462	5.086	5.539	5.882	6.147	6.831	7.817	7.311	7.392	7.456
	Vp	C	4.764	5.311	5.709	6.013	6.250	6.881	7.229	7.352	7.432	7.494
	Vp	mean	4.060	4.751	5.246	5.616	5.896	6.594	7.155	7.073	7.159	7.227
	Vs	A	2.269	2.683	2.962	3.156	3.292	3.582	3.707	3.756	3.793	3.822
	Vs	B	2.953	3.285	3.514	3.679	3.799	4.073	4.197	4.245	4.279	4.306
	Vs	C	2.323	2.699	2.960	3.148	3.285	3.599	3.740	3.797	3.836	3.867
	Vs	mean	2.515	2.889	3.145	3.328	3.459	3.751	3.881	3.933	3.969	3.998
	Vp/Vs	mean	1.615	1.645	1.668	1.688	1.704	1.758	1.843	1.798	1.804	1.808
	σ	mean	0.189	0.207	0.219	0.229	0.238	0.261	0.291	0.276	0.278	0.279

Table 2. (Continued)

#		core	Pressure (MPa)									
			20	40	60	80	100	200	400	600	800	1000
A-54	Vp	A	3.700	4.564	5.114	5.470	5.702	6.112	6.235	6.286	6.322	6.351
	Vp	B	4.744	5.399	5.831	6.124	6.324	6.721	6.870	6.932	6.976	7.011
	Vp	C	4.454	5.205	5.683	5.993	6.196	6.557	6.672	6.721	6.756	6.783
	Vp	mean	4.299	5.056	5.543	5.862	6.074	6.463	6.592	6.646	6.685	6.715
	Vs	A	2.667	2.900	3.057	3.166	3.242	3.399	3.454	3.474	3.488	3.499
	Vs	B	2.999	3.266	3.442	3.561	3.642	3.803	3.862	3.887	3.904	3.917
	Vs	C	2.644	2.886	3.048	3.160	3.238	3.400	3.462	3.486	3.503	3.516
	Vs	mean	2.770	3.017	3.182	3.295	3.374	3.534	3.593	3.616	3.632	3.644
	Vp/Vs	mean	1.552	1.676	1.742	1.779	1.800	1.829	1.835	1.838	1.841	1.843
	σ	mean	0.145	0.224	0.254	0.269	0.277	0.287	0.289	0.290	0.291	0.291
A-55	Vp	A	3.025	3.661	4.104	4.427	4.669	5.262	5.597	5.753	5.866	5.955
	Vp	B	4.606	5.054	5.364	5.590	5.759	6.174	6.399	6.497	6.566	6.619
	Vp	C	4.724	5.107	5.368	5.560	5.704	6.068	6.285	6.385	6.456	6.511
	Vp	mean	4.118	4.607	4.945	5.192	5.377	5.835	6.094	6.212	6.296	6.362
	Vs	A	2.115	2.470	2.713	2.887	3.013	3.306	3.457	3.526	3.575	3.614
	Vs	B	2.822	3.092	3.275	3.404	3.498	3.706	3.803	3.845	3.874	3.894
	Vs	C	2.122	2.426	2.643	2.802	2.921	3.204	3.332	3.377	3.407	3.420
	Vs	mean	2.353	2.663	2.877	3.031	3.144	3.405	3.531	3.582	3.619	3.642
	Vp/Vs	mean	1.750	1.730	1.719	1.713	1.710	1.713	1.726	1.734	1.740	1.747
	σ	mean	0.258	0.249	0.244	0.242	0.240	0.242	0.247	0.251	0.253	0.256
A-79	Vp	A	3.448	4.131	4.616	4.971	5.235	5.857	6.136	6.238	6.308	6.363
	Vp	B	4.473	5.070	5.485	5.785	6.004	6.506	6.729	6.815	6.875	6.922
	Vp	C	4.756	5.225	5.566	5.824	6.023	6.531	6.781	6.862	6.914	6.954
	Vp	mean	4.226	4.809	5.222	5.527	5.754	6.298	6.549	6.638	6.699	6.746
	Vs	A	2.526	2.816	3.015	3.156	3.258	3.491	3.601	3.648	3.680	3.061
	Vs	B	2.831	3.125	3.327	3.469	3.572	3.802	3.904	3.946	3.975	3.998
	Vs	C	2.150	2.367	2.519	2.629	2.712	2.908	3.004	3.041	3.067	3.088
	Vs	mean	2.502	2.770	2.953	3.085	3.181	3.400	3.503	3.545	3.574	3.382
	Vp/Vs	mean	1.689	1.736	1.768	1.792	1.809	1.852	1.870	1.873	1.874	1.994
	σ	mean	0.230	0.252	0.265	0.274	0.280	0.294	0.300	0.301	0.301	0.332
A-90	Vp	A	3.164	3.434	3.612	3.747	6.859	4.230	4.611	4.821	4.964	5.074
	Vp	B	4.816	4.983	5.093	5.176	5.244	5.459	5.653	5.748	5.810	5.857
	Vp	C	5.017	5.223	5.342	5.423	5.484	5.668	5.853	5.964	6.044	6.107
	Vp	mean	4.332	4.546	4.682	4.782	5.862	5.119	5.372	5.511	5.606	5.679
	Vs	A	2.067	2.207	2.297	2.364	2.417	2.577	2.722	2.803	2.862	2.908
	Vs	B	2.975	3.103	3.185	3.247	3.288	3.412	3.514	3.571	3.613	3.645
	Vs	C	3.271	3.344	3.394	3.433	3.464	3.555	3.620	3.648	3.666	3.679
	Vs	mean	2.771	2.885	2.959	3.015	3.056	3.181	3.285	3.341	3.380	3.411
	Vp/Vs	mean	1.564	1.576	1.582	1.586	1.918	1.609	1.635	1.650	1.659	1.665
	σ	mean	0.154	0.163	0.168	0.170	0.313	0.185	0.201	0.209	0.214	0.218

Table 2. (Continued)

#		core	Pressure (MPa)									
			20	40	60	80	100	200	400	600	800	1000
A-301	Vp	A	5.344	5.436	5.490	5.530	5.563	5.672	5.790	5.860	5.910	5.949
	Vp	B	5.198	5.301	5.364	5.411	5.449	5.575	5.712	5.793	5.847	5.888
	Vp	C	5.179	5.293	5.361	5.411	5.451	5.581	5.725	5.818	5.888	5.945
	Vp	mean	5.240	5.343	5.405	5.451	5.488	5.609	5.742	5.824	5.882	5.927
	Vs	A	3.192	3.227	3.247	3.261	3.273	3.309	3.349	3.372	3.389	3.402
	Vs	B	3.176	3.204	3.221	3.235	3.246	3.287	3.338	3.373	3.400	3.420
	Vs	C	3.171	3.204	3.225	3.241	3.255	3.302	3.352	3.381	3.402	3.419
	Vs	mean	3.180	3.212	3.231	3.246	3.258	3.299	3.346	3.375	3.397	3.414
	Vp/Vs	mean	1.648	1.664	1.673	1.679	1.684	1.700	1.716	1.725	1.731	1.736
	σ	mean	0.209	0.217	0.222	0.225	0.228	0.236	0.243	0.247	0.250	0.252
A-302	Vp	A	3.562	3.876	4.097	4.269	4.407	4.824	5.139	5.277	5.368	5.438
	Vp	B	3.912	4.212	4.398	4.454	4.654	4.999	5.271	5.398	5.483	5.550
	Vp	C	3.803	4.089	4.286	4.437	4.557	4.912	5.180	5.304	5.389	5.455
	Vp	mean	3.769	4.059	4.260	4.387	4.539	4.912	5.197	5.326	5.413	5.481
	Vs	A	2.394	2.525	2.613	2.678	2.729	2.874	2.981	3.034	3.072	3.101
	Vs	B	2.446	2.567	2.651	2.714	2.764	2.900	2.990	3.029	3.056	3.077
	Vs	C	2.509	2.611	2.679	2.731	2.772	2.895	2.990	3.036	3.066	3.090
	Vs	mean	2.450	2.568	2.648	2.708	2.755	2.890	2.987	3.033	3.065	3.089
	Vp/Vs	mean	1.539	1.581	1.609	1.620	2.011	1.700	1.740	1.756	1.766	1.774
	σ	mean	0.134	0.166	0.185	0.192	0.336	0.235	0.253	0.260	0.264	0.267
A-303	Vp	A	5.328	5.391	5.433	5.466	5.494	5.587	5.679	5.724	5.750	5.768
	Vp	B	5.414	5.519	5.580	5.621	5.650	5.731	5.804	5.848	5.879	5.903
	Vp	C	5.343	5.408	5.451	5.483	5.551	5.602	5.691	5.734	5.761	5.780
	Vp	mean	5.362	5.439	5.488	5.523	5.565	5.640	5.725	5.769	5.797	5.817
	Vs	A	3.179	3.200	3.211	3.219	3.225	3.245	3.270	3.287	3.300	3.309
	Vs	B	3.301	3.312	3.321	3.329	3.335	3.353	3.359	3.359	3.358	3.356
	Vs	C	3.178	3.197	3.211	3.223	3.232	3.258	3.272	3.276	3.278	3.280
	Vs	mean	3.219	3.236	3.248	3.257	3.264	3.285	3.300	3.307	3.312	3.315
	Vp/Vs	mean	1.665	1.681	1.690	1.696	1.705	1.717	1.735	1.744	1.750	1.755
	σ	mean	0.218	0.226	0.231	0.233	0.238	0.243	0.251	0.255	0.258	0.260

Note: seismic velocities in km/sec, Vp/Vs and Poisson (σ) ratios are dimensionlesss. For densities see Table 1.

Typical velocity-pressure curves are shown in Figure 1 for a high-grade Alpine schist (A-1), two low-grade greywackes (A-302 and A-23) and Darran gabbro (A-46). These characteristic curves of velocity versus pressure show rapid decreases in slope at pressures near 100 MPa and approximate linearity at pressures between 400 and 1000 MPa. The initial steep slopes for the gabbro demonstrate that the rock from which the cores were extracted contained abundant microcracks. Many of these cracks may have originated from hammering on the massive gabbro outcrop during the sample collection process.

COMPARISONS WITH PUBLISHED LABORATORY AND FIELD MEASUREMENTS

Classification schemes for most rocks allow for fairly wide ranges in mineralogy for a specific rock type, resulting in significant ranges in physical properties. This is especially true for metamorphic rocks classified according to facies, which can have a wide variability in parent rock chemistries. Many of the rocks included in Tables 1–3 are metamorphic, ranging in grade from zeolite to amphibolite facies. The following discussion concentrates on two of the major South Island lithologies, the relatively high grade garnet-bearing Haast schist and the Torlesse greywacke.

In the Eastern subprovince some of the highest grade quartzofeldspathic mica schists are of amphibolite facies grade and often contain garnet and oligoclase. At 600 MPa, a mid crustal pressure, compressional wave velocities for the eight samples of schist (samples A-1 through A-5, A-10 through A-12) from the garnet zone range from 6.09 to 6.25 km/s and average 6.19 km/s (Table 2). The average density of these rocks is 2710 kg/m^3. Garnet is not abundant in these rocks and averages only about 1% by volume. At a similar pressure *Birch* [1960] reported a mean compressional wave velocity of 6.59 km/s and a density of 2800 kg/m^3 for a garnet schist from Woodsville, Vermont, and *Christensen* [1965] found a mean velocity of 6.50 km/s and a density of 2760 kg/m^3 for a garnet schist from Thomaston, Connecticut. Both of these are higher in velocity and density than the New Zealand schists. The differences appear to be due to higher garnet content (approximately 5%) within the Vermont and Connecticut samples.

Based on laboratory measurements, *Hughes et al.* [1993] reported a compressional wave velocity of 6.21 km/s at 600 MPa and a density of 2686 kg/m^3 for a garnet schist from Lincoln Gap, Vermont, in excellent agreement with the average New Zealand samples. *Iida et al.* [1967] measured compressional wave velocities ranging from 6.10 to 6.45 km/s for 12 garnet-bearing quartz mica schists from Japan. The

mean velocity of this suite (6.30 km/s) and density (2730 kg/m^3) are in reasonable agreement with the New Zealand garnet schists.

Comparisons of the garnet-bearing Haast schist velocities with seismically measured velocities in South Island New Zealand are possible due to several recent field studies. The garnet schists are likely to be in equilibrium at mid crustal depths. *Davey et al.* [1998] found a velocity of 6.2 km/s for the midcrust (20 km depth) of the Western subprovince of New Zealand. This velocity is similar to the 6.2 to 6.4 km/s velocities at 20 km reported by *Eberhart-Phillips* [1995] and *Leitner et al.* [2001] based on passive seismic studies and by *Scherwath et al.* [2003] and *van Avendonk et al.* [2004] based on active-source seismic profiling. These values are all in reasonable agreement with the laboratory measurements of garnet-bearing quartz mica schists at 600 MPa summarized above.

Torlesse rocks are major constituents of New Zealand, occupying approximately a quarter of the surface of South Island (Plate 1). In eastern South Island the Torlesse consists of marine clastic sediments with less common interlayered basaltic lava, tuffaceous sediments and limestone. The thickness of the Torlesse is unknown, but probably significant because of its widespread distribution. Compressional wave velocities measured for seven Torlesse greywackes (samples A-21 through A-25, A-31B, A-32B; Table 2) average 6.08 km/s at 100 MPa . The average density is 2704 kg/m^3. *Birch* [1960] reported a slightly lower velocity and density of 5.87 km/s and 2679 kg/m^3 for a New Zealand greywacke. The specific locality of this sample was not given. Velocities at 100 MPa and densities for greywacke from the Valdez Group in Southern Alaska of 6.17 km/s and 6.12 km/s and 2767 kg/m^3 and 2716 kg/m^3 [*Fuis et al.*, 1991; *Brocher et al.*, 1989] are similar to the Torlesse samples. An average compressional wave velocity of 5.89 km/s at 100 MPa, and density of 2738 kg/m^3 for a suite of ten California greywackes [*Stewart and Peselnick*, 1977] are also in reasonable agreement with the New Zealand samples, considering the variable mineralogy common to greywackes.

Recent field based measurements of seismic velocities below exposures of Torlesse are in good agreement with the laboratory measurements of greywacke in Table 2. Near surface velocities are quite low, presumably due to fracturing. However at depths of a few kilometers crustal velocities appear to be comparable to that of fracture free Torlesse greywacke. Modeling of both explosion and onshore-offshore data by *Davey et al.* [1998] gives a velocity of 5.9 km/s at 5 to 10 km depths At similar depths *Leitner et al.* [2001] found velocities of 6.1 km/s. *Eberhart-Phillips et al.* [2001] measured velocities in Torlesse of 6.1 km/s at depths below 1 km.

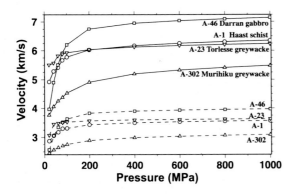

Figure 1. Measured acoustic velocities for representative samples. Solid lines are mean compressional velocity; dashed lines are mean shear velocities. Mean velocities determined by averaging of measurements in principal directions. See Table 2 for velocity numerical values.

Table 3. Whole rock chemical analyses in weight percent. Digital version of this table is on the CDROM which accompanies this volume.

	SiO$_2$	TiO$_2$	Al$_2$O$_3$	Fe$_2$O$_3$*	MgO	CaO	MnO	Na$_2$O	K$_2$O	P$_2$O$_5$	H$_2$O[†]	Total
A-1	68.71	0.69	15.37	5.36	1.72	1.55	0.08	2.86	2.95	0.15	1.20	100.65
A-2	64.16	0.70	16.73	6.88	1.99	1.83	0.07	2.59	3.77	0.19	1.80	100.73
A-3	70.03	0.65	14.96	4.76	1.46	1.51	0.07	2.71	3.03	0.11	1.50	100.80
A-4	73.55	0.51	12.89	3.86	1.06	2.41	0.07	3.68	1.06	0.10	0.70	99.90
A-5	62.90	0.79	17.72	6.31	1.79	1.31	0.07	2.12	3.79	0.14	3.70	100.65
A-6	73.61	0.57	10.65	6.76	2.01	1.11	0.06	1.40	2.84	0.14	0.50	99.67
A-7	74.81	0.57	11.94	4.23	1.53	0.18	0.05	1.33	2.78	0.13	2.10	99.66
A-8	62.41	0.73	18.45	6.16	2.57	0.28	0.05	0.16	5.09	0.11	4.50	100.52
A-9	69.18	0.57	12.52	5.68	1.96	3.14	0.11	2.16	2.42	0.14	2.80	100.69
A-10	70.25	0.57	14.62	4.97	1.63	1.75	0.05	2.92	2.84	0.13	0.80	100.54
A-11	70.03	0.58	15.19	4.21	1.38	1.54	0.06	2.81	3.44	0.14	1.20	100.59
A-12	62.57	0.83	17.82	6.39	2.27	2.22	0.04	3.65	3.32	0.18	1.40	100.70
A-13	72.74	0.47	13.51	3.96	1.30	1.07	0.04	3.25	2.47	0.11	1.50	100.43
A-14	74.49	0.41	13.00	3.09	0.91	1.38	0.04	3.11	2.66	0.06	1.40	100.55
A-15	69.92	0.57	13.99	5.11	1.36	1.82	0.09	3.89	1.74	0.23	1.90	100.62
A-16	65.36	0.78	16.17	6.48	1.84	1.22	0.06	2.70	2.66	0.18	2.80	100.26
A-17	68.44	0.63	14.64	4.84	1.60	2.32	0.08	3.81	1.87	0.16	2.10	100.50
A-18	70.03	0.56	13.94	4.26	1.50	2.61	0.07	3.75	1.93	0.14	1.40	100.20
A-19	69.21	0.57	14.55	4.92	1.48	2.21	0.07	3.81	1.89	0.10	1.70	100.52
A-20	74.59	0.46	12.20	3.59	0.95	1.15	0.05	2.38	2.65	0.07	1.60	99.70
A-21	67.45	0.57	14.02	5.93	1.58	2.37	0.06	3.41	2.63	0.13	1.40	99.56
A-22	69.82	0.44	14.11	5.15	1.05	1.74	0.06	3.71	2.74	0.09	0.60	99.52
A-23	71.55	0.50	13.19	5.73	1.31	1.12	0.06	3.43	2.28	0.09	1.10	100.38
A-24	73.10	0.34	13.90	3.05	0.78	1.56	0.04	4.12	2.53	0.09	1.00	100.53
A-25	68.24	0.47	13.97	7.20	1.53	1.76	0.07	3.14	3.05	0.12	1.00	100.57
A-26	53.42	0.06	2.78	6.53	22.28	10.54	0.19	0.20	0.21	0.09	3.20	99.74
A-28	72.25	0.27	14.16	2.20	0.56	1.14	0.03	3.94	4.12	0.11	1.00	99.80
A-29	71.11	0.38	14.09	3.07	0.85	2.10	0.04	2.66	4.77	0.10	0.40	99.57
A-30	70.08	0.57	13.63	4.64	2.13	1.76	0.06	3.39	2.14	0.12	1.90	100.44
A-31B	69.27	0.44	14.66	4.56	1.27	1.70	0.06	3.48	3.09	0.13	1.20	99.87
A-32B	66.19	0.87	14.49	7.29	2.06	2.89	0.10	3.35	1.36	0.18	1.70	100.49
A-33	70.65	0.49	13.88	4.35	1.13	1.45	0.05	4.51	1.73	0.13	1.60	99.98
A-34	71.50	0.41	13.23	3.83	1.13	2.26	0.06	3.80	2.20	0.08	1.30	99.81
A-35	69.20	0.46	14.13	3.90	1.02	2.65	0.06	3.33	2.77	0.12	1.40	99.04
A-36	61.43	0.78	15.54	6.65	2.70	4.47	0.09	4.02	1.62	0.10	2.40	99.81
A-37	63.07	0.75	14.98	6.74	2.43	3.70	0.10	3.62	1.83	0.11	2.20	99.54
A-38	64.48	0.62	15.97	6.69	1.57	1.66	0.09	2.14	4.38	0.12	1.90	99.63
A-39	60.09	0.79	17.11	6.83	2.54	2.85	0.07	3.12	2.79	0.15	2.60	98.94
A-40	67.09	0.55	15.41	4.59	1.70	1.85	0.06	3.93	2.63	0.16	2.10	100.07
A-41	48.01	2.83	14.52	12.65	5.86	9.75	0.18	3.54	0.15	0.34	2.20	100.05
A-42	68.50	0.53	14.28	5.14	1.59	1.51	0.07	3.28	2.92	0.10	2.10	100.03
A-43	63.91	0.69	14.30	6.31	2.21	3.09	0.10	5.17	1.33	0.16	1.60	98.88
A-44	64.26	0.79	16.07	6.27	1.99	1.73	0.20	2.83	3.12	0.16	2.40	99.82

Table 3. (Continued)

	SiO$_2$	TiO$_2$	Al$_2$O$_3$	Fe$_2$O$_3$*	MgO	CaO	MnO	Na$_2$O	K$_2$O	P$_2$O$_5$	H$_2$O†	Total
A-45	63.94	0.78	16.02	6.40	1.95	1.51	0.09	2.84	3.04	0.17	2.20	98.95
A-46	51.30	0.73	19.36	8.08	5.55	10.17	0.12	3.41	0.21	0.08	0.50	99.52
A-47	44.51	1.17	16.38	13.87	8.25	11.86	0.13	1.68	0.33	0.10	1.50	99.79
A-48	53.65	0.93	17.20	8.83	5.25	8.17	0.13	3.46	1.14	0.14	0.20	99.11
A-49	65.63	0.63	14.67	5.41	2.07	3.30	0.07	3.37	1.84	0.13	2.00	99.12
A-50	70.01	0.40	13.58	4.91	1.73	2.61	0.08	3.83	1.72	0.12	0.60	99.60
A-51	59.89	0.82	16.66	6.85	3.07	2.91	0.10	4.51	1.38	0.22	2.50	98.92
A-52	44.84	2.96	14.10	16.55	6.52	8.36	0.17	2.42	0.13	0.34	3.60	100.00
A-53	46.95	2.81	15.54	12.35	4.50	12.08	0.13	2.98	0.23	0.35	2.00	99.93
A-54	43.45	3.47	13.92	14.45	5.30	7.46	0.15	3.10	0.42	0.37	7.90	100.00
A-55	64.08	0.83	16.75	5.86	1.89	1.43	0.07	3.03	3.13	0.14	2.70	99.92
A-79	46.74	2.11	19.16	14.63	3.04	7.74	0.36	3.44	1.11	0.22	1.50	100.09
A-90	63.30	0.67	17.10	5.69	1.49	0.62	0.06	2.75	3.49	0.12	4.30	99.60
A-301	64.06	0.69	15.66	7.82	2.01	1.48	0.08	3.14	2.02	0.17	2.60	99.74
A-302	59.10	0.86	16.68	7.14	2.74	2.73	0.09	3.93	1.67	0.12	4.70	99.77
A-303	67.32	0.50	14.14	7.60	0.96	1.41	0.07	2.74	3.41	0.09	2.15	100.40

*Total iron as Fe$_2$O$_3$.

†Loss on ignition for ICP analyses.

VELOCITY-DENSITY AND VELOCITY-CHEMISTRY SYSTEMATICS

Relationships between seismic velocity and density have important implications for multidisciplinary geophysical studies involving seismic and gravity investigations [e.g., *Birch*, 1960; *Brocher*, 2005]. Velocity-density correlations can be useful in the determination of velocity from gravity data or vice versa. These relationships can also be used to estimate lithology from either gravity or seismic data. Velocity-density relationships also provide information about the acoustic impedances of different rock types, which are important in the interpretation of reflection seismic data.

The following discussion is limited to rocks from the Torlesse Supergroup and the Haast Schist Group, which make up much of the Southern Alps of New Zealand and have been the subject of several recent seismic investigations [see *Davey et al.*, this volume]. These rocks have parent lithologies ranging from sandstone, siltstone and shale to mafic volcanics and volcanoclastic sediments. Metamorphic grade ranges from prehnite-pumpyllite through pumpyllite-actinolite facies and greenschist facies to amphibolite facies. The lowest grade rocks occur in the Torlesse. The Haast Schist Group has been divided into three mineralogic zones [e.g., *Landis and*

Coombs, 1967]; a Chlorite Zone, Biotite Zone and Garnet-Oligoclase Zone (Plate 1). The Chlorite Zone has in turn been subdivided into subzones II, III and IV. Torlesse rocks grade into subzone II rocks and subzone IV rocks grade into the Biotite Zone.

Velocities at 600 MPa are shown versus density in Figure 2. Compressional wave velocities show a steady increase with increasing density with a relatively low density (2600 to 2800 kg/m^3) cluster of metasediments. The higher density points are mafic volcanic rocks from the Caples Terrane. These mafic rocks have significantly higher acoustic impedances and thus if of sufficient thickness should produce strong reflections when interlayered with metasediments. The slope of the shear wave velocity-density least squares fit is much lower. This is due to the high shear wave velocity of quartz, which is abundant in the metasediments. In Figure 3 we have plotted velocity versus density at 600 MPa for the metasedimentary rocks (exclusive of metavolcanics) of the Torlesse and Haast. The metamorphic grade of each data point is labeled. An important finding here is that there is little correlation between metamorphic grade and velocity or density for these rocks (see Figure 1; *Godfrey et al.* [2000]). Part of this is due to their chemical similarities (Table 3). Thus seismic studies will be unable to distinguish between the various subdivisions of the Haast Schist.

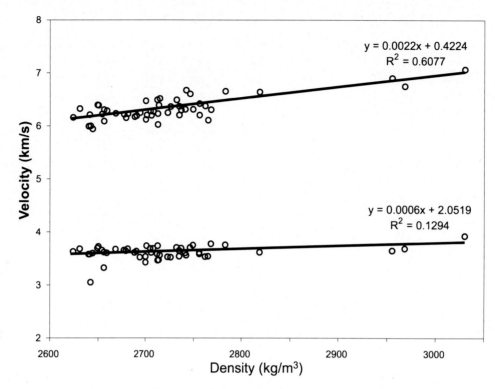

Figure 2. Compressional and shear wave velocities at 600 MPa plotted against density for rocks from the Torlesse Supergroup and the Haast Schist Group. The higher velocity rocks are metamorphosed mafic volcanics. Linear regression lines and parameters are shown in the figure.

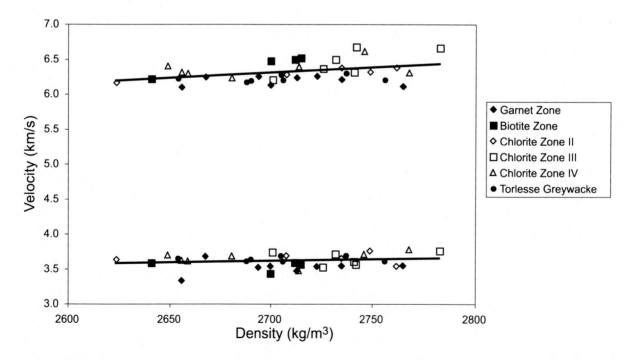

Figure 3. Compressional and shear wave velocities at 600 MPa plotted against density for metasedimentary rocks from the Torlesse Supergroup and the Haast Schist Group. The Haast Schist rocks are separated according to metamorphic grade. There is little correlation between metamorphic grade and velocity or density for these rocks.

Whole rock geochemical analyses for the rocks in Table 3 were made by Acme Analytical Laboratories, Vancouver, British Columbia, by inductively coupled plasma (ICP) analyses.

Beginning with the pioneering work of *Birch* [1961], several studies have attempted to correlate velocity with various geochemical parameters. Igneous complexes which have undergone differentiation such as the Kohistan Arc [*Miller and Christensen*, 1994] seem to show the best correlations. Some metasedimentary sequences such as phyllonites from the Brevard fault zone in North Carolina show a simple relationship between velocity and SiO_2 content [*Christensen and Szymanski*, 1988]. This correlation is evident in a plot of velocity versus weight percent SiO_2 of samples from the Torlesse and Haast Schist (Figure 4). As with the velocity-density relation, the mafic volcanics are separated from the metasediments and the slope of the shear wave velocity-%SiO_2 relation is practically flat. Note also that SiO_2 contents vary widely within the various metamorphic zones (Figure 5).

Another geochemical index that shows significant correlation with velocities of the Torlesse and Haast Schist is Mg number (Mg#), which is calculated from the mole proportion Mg divided by the sum of the mole proportion Mg + mole proportion Fe. In Figure 6 it can be seen that increasing Fe content (decreasing Mg#) correlates with decreasing compressional wave velocity. Shear wave velocity shows no correlation with Mg#. Also the Torlesse samples seem to have lower Mg#'s than the Haast Schists and samples of Haast Schist of similar metamorphic grade often show wide ranges in Mg#.

Poisson's Ratio and Vp/Vs

Often various rock elastic parameters such as the bulk modulus, compressibility and Poisson's ratio are of interest in geophysics. With the formulas of isotropic elasticity [*Birch*, 1960] any two elastic parameters can be used to calculate all others. The elastic parameters also exhibit a functional relationship to compressional and shear wave velocities and rock density.

Poisson's ratio, calculated from Vp/Vs, has been shown to be an important parameter for determining crustal composition [e.g., *Christensen*, 1996]. Values of Vp/Vs and Poisson's ratio are given in Table 1. The mean Vp for three directions has been combined with the mean Vs of the rock samples to obtain the calculated values. The Poisson's ratios in Table 1 are more reliable for the rock samples with relatively low anisotropies. For the highly anisotropic rocks the data are

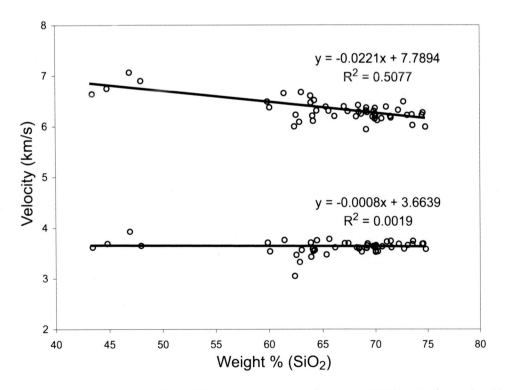

Figure 4. Compressional and shear wave velocities at 600 MPa plotted versus whole rock wt% SiO_2. The four rocks with SiO_2 contents of 40 to 50% are metamorphosed mafic volcanics.

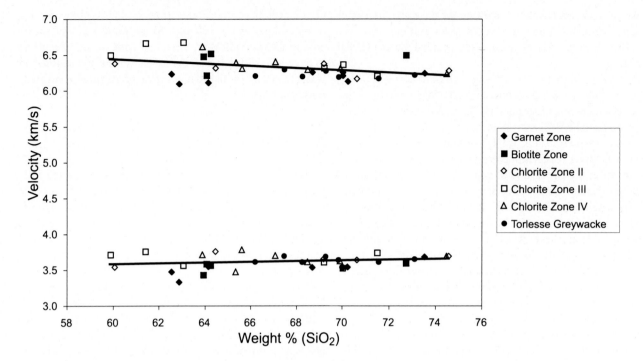

Figure 5. Compressional and shear wave velocities at 600 MPa plotted against whole rock wt% SiO_2 for metasedimentary rocks from the Torlesse Supergroup and the Haast Schist Group. The Haast Schist rocks are separated according to metamorphic grade.

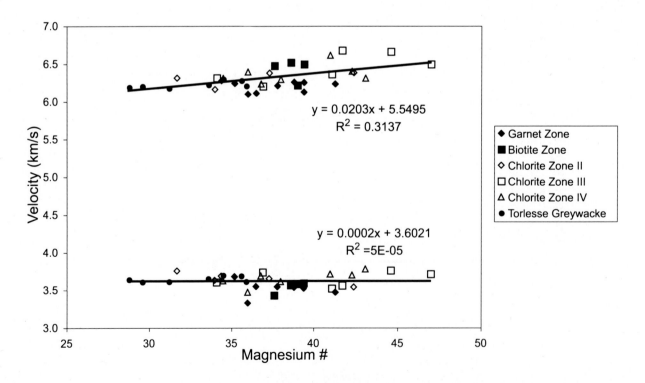

Figure 6. Laboratory measured compressional and shear wave velocities at 600 MPa plotted against whole rock magnesium number (mole proportion Mg/ mole proportion Mg + mole proportion Fe) for metasedimentary rocks from the Torlesse Supergroup and the Haast Schist Group.

presumably representative of rocks with similar mineralogy but a random mineral orientation.

The lithologies of Table 1 with the lowest anisotropies include granite, gabbro and greywacke. At high pressures, where microcracks are closed, the Tuhua granite samples (A-28 and A-29) have Poisson's ratios between 0.24 and 0.25. This compares favorably with an average of 0.24 at similar pressures for 38 granites reported by *Christensen* [1996]. The three gabbros in Table 1 (A-46, A-47 and A-48) have Poisson's ratios at high pressures between 0.28 and 0.29, in good agreement with *Christensen*'s 0.296 average for 58 gabbroic rocks. The seven samples of Torlesse greywacke have Poisson's ratios at high pressures of 0.24 to 0.25 compared to a slightly higher 0.26 average for 36 samples in *Christensen*'s compilation.

ELASTIC ANISOTROPY AND ROCK FABRIC

Seismic anisotropy is an important property of the rocks from the Haast Schist Terrane. This has been covered in detail in earlier papers [*Okaya et al.*, 1995; *Godfrey et al.*, 2000, 2002; *Okaya and Christensen*, 2002] and will only be briefly addressed here.

The three propagation directions listed in Table 2 for each sample provide information on maximum compressional wave anisotropy. Minimum velocities were measured normal to foliations (A core). Maximum compressional wave velocities were recorded for propagation in the foliations (B and C cores), usually parallel to lineations, if present. The shear wave velocities for the B and C cores give an estimate of maximum shear wave splitting for most rocks. As was discussed earlier, the B core shear wave velocities were measured with propagation and vibration directions parallel to foliations, whereas C core shear wave velocities were measured with propagations in the foliations and vibration directions normal to foliations. The results of these measurements, summarized by *Godfrey et al.* [2002], are shown in Figure 7.

Petrographic examinations of thin sections prepared from the sample core ends show that mineral preferred orientations (principally layer lattice silicates) are primarily responsible for the observed anisotropy. Since this anisotropy persists at high pressures, it is not related to crack orientation. The magnitudes of the anisotropies in Figure 7 are consistent with laboratory anisotropy measurements of similar rocks from other localities [e.g., *Christensen*, 1965, 1966; *Ji and Salisbury*, 1993; *Kern and Wenk*, 1990].

The overall symmetries of the New Zealand rocks are of prime significance in understanding their seismic anisotropies. Symmetries of rock fabrics are quite similar to those of single crystals and both can be defined in terms of point-group symmetries. Many of the metamorphic rocks included

in this study have well developed foliations and cleavages related to preferred mineral orientations. Symmetry of these rocks often approximate axial, with symmetry axes normal to the foliation or cleavage. Their elastic properties are similar to hexagonal crystals with five independent elastic constants with similar velocities for all directions of propagation within the foliation or cleavage.

Often foliations show associated linear elements. Several lineations have been observed in the New Zealand rocks including lineations formed by intersecting foliations, stretching of grains, lineations defined by the preferred orientation of elongate amphibole and crenulations of the foliation planes. For the micaceous rocks included in this study crenulation lineations are the most common, whereas lineations in the amphibolites originate from preferred orientation of prismatic hornblende. The elasticity of the lineated rocks often approximates that of orthorhombic single crystals with nine independent elastic constants. This is evident for samples in Table 2 where the velocities are different in orthogonal directions within the foliation or cleavage planes. Fast compressional wave velocities parallel the lineations.

The velocity data of Table 2 demonstrates that several rock types are, to a first approximation, isotropic. These include Tuhua granite, Torlesse greywacke, Murihuku greywacke, and Mackay gabbro. As expected, the most highly anisotropic samples are from the Alpine and Otago schist belts. Symmetries of the schists and greenschists vary from axial (samples A-1 through A-4, A-36, A-37, A-44, A-49) to orthorhombic (A-5, A-8, A-12 through A-15, A-19, A-35, A-38, A-40, A-42, A-43, A-51). The amphibolites also show axial symmetry (A-52, A-53, A-79) and orthorhombic symmetry (A-41, A-54). A complete transition from pure axial symmetry (only a foliation) to robust orthorhombic symmetry (a well-developed lineation) exists in the data suite, as shown by the magnitudes of the anisotropies (Table 2).

The velocity measurements presented in Table 2 provide details of wave propagation along major symmetry directions and usually give maximum P wave anisotropy and S wave splitting. Velocity measurements in non-symmetry directions are required to calculate complete three-dimensional velocity surfaces [*Auld*, 1990; *Johnston and Christensen*, 1995; *Godfrey et al.*, 2000]. For rocks with axial symmetry, quasi-compressional wave velocity measurements are required in one direction at 45 degrees to the symmetry axis, whereas measurements of three quasi-compressional wave velocities within the symmetry planes at 45 degrees to the symmetry axes provide velocity surfaces for rocks with orthorhombic symmetry.

Velocities have been measured at 45 degrees to the symmetry axis for an Alpine schist sample (A-1) with axial symmetry [*Godfrey et al.*, 2000]. To describe three-dimensional

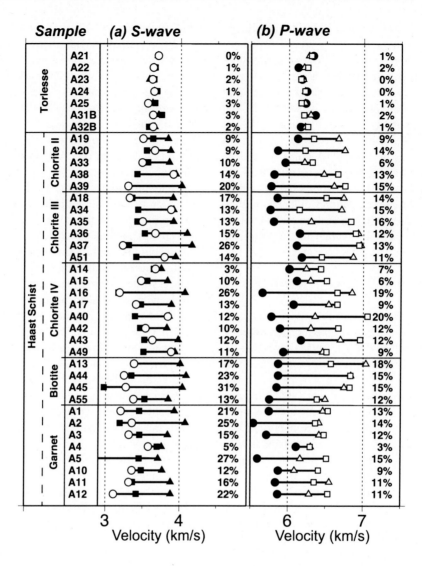

Figure 7. Velocities for Torlesse and Haast Schist metasedimentary rocks illustrating seismic anisotropy. Shear wave velocities (a) and compressional wave velocities (b) are at 600 MPa. For S-waves (a), open circle denotes propagation normal to foliation, solid square is vibration direction normal to foliation, and solid triangle is propagation and vibration parallel to foliation. For P-waves (b), solid circle denotes propagation normal to foliation, open triangle is propagation parallel to foliation but normal to lineation, and open square is propagation parallel to foliation plus lineation. Percent anisotropy is calculated from maximum velocity minus minimum velocity divided by the average of the three orthogonal velocities. Descriptions of the rock samples are provided in Tables 1–3.

wave propagation in this rock, phase velocity surfaces were calculated using the Kelvin–Christoffel equations [e.g., *Auld*, 1990] and elastic constants of the schist determined from the velocity and density measurements [*Okaya and Christensen*, 2002]. These surfaces (Figure 8) describe variations in phase velocity as a function of angle to foliation. Three velocity surfaces are calculated, one for the quasi-compressional wave (Vp), one for the shear wave vibrating parallel to the foliation (Vsh) and one for the quasi-shear wave vibrating in a plane perpendicular to the foliation (Vsv). For propagation

parallel and perpendicular to the foliation all wave modes are pure.

The velocity surfaces in Figure 8 show several interesting features about elastic wave propagation in the Alpine schist. First, compressional wave velocity does not increase significantly until propagation directions greater than about 40 degrees from foliation normal are reached. At an angle greater than 50 degrees compressional wave velocity increases rapidly and reaches a maximum for propagation parallel to the foliation. Shear wave singularities occur for propagation

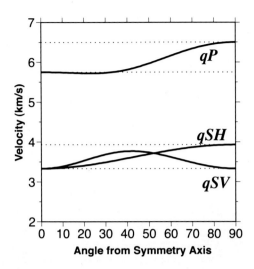

Figure 8. Phase velocities of the Haast schist as a function of angle from the symmetry axis normal to the foliation plane. The quasi-shear waves have equal velocity at approximately 50 degrees. Detectable shear wave splitting requires propagation at approximately 70–90 degrees from the symmetry axis (within 20 degrees of the foliation plane).

parallel to and at approximately 50 degrees to the symmetry axis (normal to the foliation). Although shear wave splitting occurs for all other propagation directions it does not exceed 0.2 km/s until propagation directions are greater than 65–70 degrees from the symmetry axis. Seismic sampling through this optimal but narrow range of propagation angles is made more difficult where geological structures exhibit 3D heterogeneity in the orientations of its internal fabrics [*Okaya and Christensen*, 2002]. Outside of this narrow range of propagation angles, the minimal production of shear wave splitting may misleadingly appear as isotropic seismic signals even thought the schist material is strongly anisotropic. Thus field observations of crustal anisotropy by shear wave splitting will be difficult to observe unless the propagation direction is favorably oriented with respect to foliation (i.e., approximately parallel to foliation).

Significant measured material anisotropy in schist samples suggest that anisotropic signatures should be present in seismic data collected above the Haast schist terrane. While upper mantle anisotropy has been identified using *Pn* phases within SIGHT data [e.g., *Scherwath et al.*, 2002; *Baldock and Stern*, 2005; *Bourguignon et al.*, 2007, this volume], observations of crustal seismic anisotropy are difficult to identify [see *Savage et al.*, this volume]. *Pulford et al.* [2003] found limited evidence of shear wave splitting in SIGHT seismic onshore-offshore data collected within the Southern Alps. *Stern et al.* [2001] examined the role of schist anisotropy to produce

P-wave traveltime delays within the Alpine fault zone before attributing the delays as due to elevated fluid pressures.

Because the production of a seismic anisotropic signal is directly dependent on the relative angles between the seismic wavepath and the material symmetry axes encountered along the full path, the three dimensional complexity of the Haast terrane internal structure is a contributing factor to whether the schist produces detectable seismic anisotropy. This complexity exists on three scales. At the broadest scale, in map view the Haast terrane is oroclinally bent by dextral 90 degrees as it approaches the Alpine fault (Plate 1). On a vertical crustal scale, geological models across central South Island suggest a subhorizontal Haast terrane which becomes upturned at the Alpine fault due to convergent exhumation and erosion – the "Wellman model" [*Adams*, 1979; *Wellman*, 1979; see *Cox and Sutherland*, this volume]. However, the orientation of the internal material fabric within this subhorizontal-to-upturning terrane is not well constrained and may be (a) internally subhorizontal or subvertical and (b) reoriented during exhumation by the mechanism of exhumation (e.g., "Wellman" upturning versus a backshear escalator [*Little et al.*, 2002, this volume]). These superimposed scales of geometrical complexity define the local and bulk regional orientation patterns of schist fabrics and hence influence the amount of seismic anisotropy which may be accumulated within seismic data.

The SIGHT experiments were not originally designed to collect seismic anisotropy. Onshore cross lines above the schist terrane were not collected orthogonal to the two main transects; analyses cannot be made to identify differences in schist-penetrating *Pg* velocities in perpendicular directions within the Southern Alps, eastern foothills, or (Canterbury) coastal plains. The deployment of three-component instruments during SIGHT were concentrated in the Southern Alps. Offshore seismic airgun signals and widely spaced explosion sources recorded by these instruments have raypaths with a more limited range of angles relative to local (and not well-constrained) schist fabric orientation. The SAPSE experiment did not employ dense transect or areal arrays across central South Island. Regional velocity structures obtained using isotropic seismic tomography methods [e.g., *Eberhart-Phillips and Bannister*, 2002] may have inherent larger uncertainties, necessitating the need for anisotropic tomography approaches [e.g., *Eberhart-Phillips and Henderson*, 2004]. In addition, high spatial resolution short period receiver function studies which are sensitive to crustal anisotropy have yet to be performed in central South Island within the Haast terrane.

The one set of seismic observations which was designed specifically as an attempt to observe schist seismic anisotropy was a SIGHT piggyback marine experiment in the

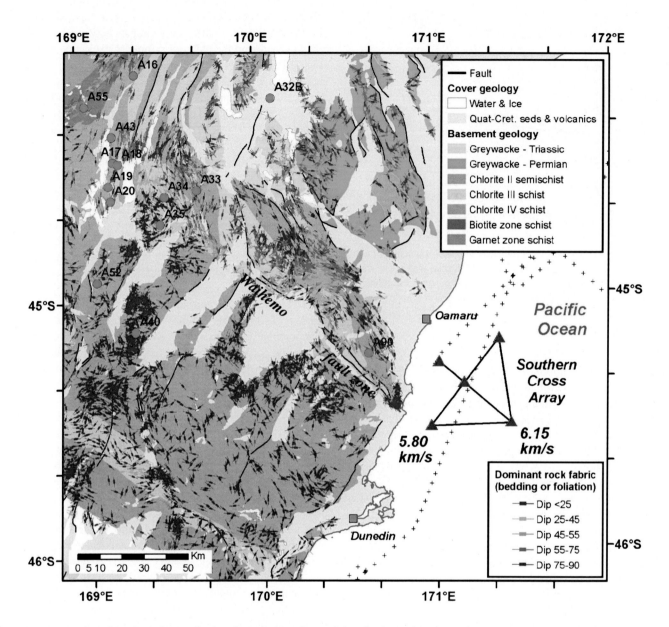

Plate 2. Location of Southern Cross seismic anisotropy experiment. Triangles denote locations of ocean bottom seismometers; solid lines connecting seismometer locations are seismic airgun ship tracks. Dominant rock fabric (schistosity) is sub-horizontal to gently dipping (warm color dip bars) for the Otago portion of the Haast schist southwest of the Waihemo fault zone and near-vertical (cool color dip bars) for the lower grade Torlesse greywacke and semischist further to the north. Labeled velocities represent upper crustal schist Pg velocities parallel to regional bulk foliation (6.15 km/s) and perpendicular (5.80 km/sec). Green circles are locations of rock samples as described in Tables 1–3. Lines of crosses denote SIGHT marine ship tracks. Geology map compiled from *Bishop and Turnbull* [1996], *Turnbull* [2000], *Forsyth* [2001], and *Cox and Barrell* [in press] and was extracted from QMAP 1:250,000 geological map of New Zealand, courtesy of Simon Cox.

Pacific Ocean south of Oamaru. Nicknamed the "Southern Cross" due to its seismic array configuration, this experiment was positioned at the offshore extension of the Otago portion of the Haast schist (Plate 2). Land exposures near the coastline indicate this schist's bulk foliation projects under the marine experiment with regional oriented NW to NNW strike and steep to vertical dips [*Mutch*, 1975; *Cox and Sutherland*, this volume]. Geologic field observations of this regional trend are denoted in Plate 2 (cool colored symbols). In contrast, southwest of the nearby Waihemo fault zone the schist defines a region of markedly differing schist foliation orientation. This latter portion of schist forms a structural northwest-trending anticlinorium with subhorizontal to moderate foliation dips (red symbols) [*Cox and Sutherland*, this volume]. The Waihemo fault zone clearly separates these domains of different rock fabric. The Southern Cross array used five ocean bottom seismometers to collect orthogonal reversed seismic refraction profiles using R/V *Ewing* airgun sources as deployed for SIGHT marine profiling. One refraction profile was approximately parallel to the regional schist foliation strike (NW-SE); its orthogonal profile is thus perpendicular to the regional foliation (Plate 2). *Smith* [1999] performed raytrace forward modeling of refracted and reflected P-waves and determined that upper crustal *Pg* phases penetrated the Haast schist basement at depths of approx. 5 km. Velocity analysis of these data defined P-wave velocities of 6.15 and 5.80 km/s along the profiles parallel and perpendicular to the bulk foliation strike, respectively. This represents P-wave seismic anisotropy of 5.9%. *Smith* [1999] conducted a reliability test to conclude that the difference in P-wave velocities was due to schist anisotropy and not due to unrelated factors such as phase mis-identification, travel time picking errors, or inversion parameter searches. The ability of a targeted seismic experiment to observe seismic anisotropy caused by schists suggests that a future combined passive- and active-source experiment design using azimuthal arrays may succeed in providing information about the subsurface schist's structure and internal fabric characteristics. This information may provide constraints to tectonic or geodynamic models which describe Pacific plate upper and middle crustal deformation paths during transpressional exhumation at the Alpine fault plate boundary [e.g., *Gerbault et al.*, 2002; *Upton and Koons*, this volume]

CONCLUSIONS

Laboratory measurements of the velocities of elastic waves to confining pressures of 1000 MPa for a variety of South Island, New Zealand rocks provide a comprehensive database for future interpretations of seismic velocities in terms of petrology. The most useful velocities for this purpose are those measured at sufficiently high pressures to close grain boundary microcracks. Velocities in the pressure range of 100 to 300 MPa for Torlesse greywacke are consistent with shallow crustal seismic measurements over exposures of Torlesse. To the northeast of the Alpine fault zone crustal velocities of 6.2 to 6.4 km/s correlate with laboratory measurements of the Haast Schist Group. It is shown that the various subfacies of the Haast have similar velocities, so it is impossible to discern metamorphic grade of these rocks from seismic field experiments. Compressional wave velocities do however show a significant lowering with increasing iron content expressed in terms of Mg#. Propagation of seismic waves in the Haast Schist Group is further complicated by strong anisotropy expressed as compressional wave azimuthal variations and shear wave splitting. Future field experiments designed specifically to measure this anisotropy will be valuable for the determination of rock composition and structure of the New Zealand crust. This is a frontier research area that is certain to provide new and exciting insights into crustal structural geology.

Recent seismic velocity studies across central South Island [e.g., *Eberhart-Phillips*, 1995; *Smith et al.*, 1995; *Leitner et al.*, 2001; *Scherwath et al.*, 2003; *van Avendonk et al.*, 2004] have found the existence of a relatively high velocity basal layer with a thickness of 5 to 11 km and a compressional wave velocity of 6.7 to 7.3 km/s. These velocities are higher than most metasedimentary rocks of the Haast Schist Group, even if anisotropy is taken into account (Table 2, Figure 7). It has been suggested that this high velocity layer represents former oceanic crust [e.g., *Holbrook et al.*, 1998], which is consistent with the layer thickness. If this layer is indeed oceanic crust, it is unlikely that original crustal lithologies (e.g., pillow basalts and sheeted diabase dikes) are now preserved. Rheological models of the Southern Alps predict that high shear strains propagate into the ductile lower crust during mountain building [e.g., *Gerbault et al.*, 2002]. These shear strains will promote recrystallization resulting in strong foliations. Amphibolite facies mafic rocks, such as A-41 (average Vp = 7.01 km/s at 800 MPa) and A-53 (average Vp = 7.16 km/s at 800 MPa), are likely to be stable at depths of 25 to 35 km [e.g., *Grapes*, 1995]. Note that both samples show significant compressional and shear wave anisotropy, which is common in amphibolites [*Christensen and Mooney*, 1995], suggesting that in addition to the overlying crust this lower crustal seismic layer may also be anisotropic.

Acknowledgements. This study was supported by NSF grants EAR-9219496 and EAR-9418530. We thank Simon Cox who generated the geological map of eastern South Island using the digital QMAP 1:250,000 Geological map of New Zealand. We thank

Ryan Smith and W. Steven Holbrook for fruitful discussions regarding offshore seismic anisotropy. Tom Brocher and an anonymous reviewer provided excellent comments. We also thank Sabrina Bradshaw and Anna Coutier for their assistance in assembling the petrophysical laboratory tables that are also provided in digital form in the CDROM which accompanies this volume.

REFERENCES

Adams, C. J. D. (1979), Age and origin of the Southern Alps, in *The Origin of the Southern Alps, R. Soc. N. Z. Bull.*, vol. 18, edited by R. I. Walcott and M. M. Cresswell, pp. 73–78.

Auld, B. A. (1990), *Acoustic Fields and Waves in Solids*, vol. 1, Robert E. Krieger, Malabar, FL, 435 pp.

Baldock, G., and T. Stern (2005), Width of mantle deformation across a continental transform: evidence from upper mantle (Pn) seismic anisotropy measurements, *Geology, 33*, 741–744.

Birch, F., (1960), The velocity of compressional waves in rocks to 10 kilobars, 1, *J. Geophys. Res., 65*, 1083–1102.

Birch, F. (1961), The velocity of compressional waves in rocks to 10 kilobars, 2, *J. Geophys. Res., 66*, 2199–2224.

Bishop, D. G., J. D. Bradshaw, and C. A. Landis (1985), Provisional terrane map of South Island, New Zealand, in *Tectonostratigraphic Terranes*, edited by D. G. Howell, Circum-Pacific Council for Energy and Mineral Resources, Houston, TX, pp. 515–521.

Bishop, D. G. and I. M. Turnbull (1996), *Geology of the Dunedin area, Institute of Geological & Nuclear Sciences 1:250,000 geological map 21 Scale 1:250,000*, Institute of Geological and Nuclear Sciences, Lower Hutt, New Zealand, 1 sheet + 52 pp.

Bourguignon, S., T. Stern, and M. K. Savage (2007), Crust and mantle thickening beneath the southern portion of the Southern Alps, New Zealand, *Geophys. J. Int., 168*, 681–690.

Bourguignon, S., M. K. Savage, and T. Stern (this volume), Crustal thickness and Pn anisotropy beneath the Southern Alps oblique collision, New Zealand.

Brocher, T. M. (2005), Empirical relations between elastic wavespeeds and density in the Earth's crust, *Bull. Seismol. Soc. Am., 95*, 2081–2092.

Brocher, T. M., M. A. Fisher, E. L. Geist, and N. I. Christensen (1989), A high resolution seismic reflection/refraction study of the Chugach-Peninsular terrane boundary, southern Alaska, *J. Geophys. Res., 94*, 4441–4555.

Christensen, N. I. (1965), Compressional wave velocities in metamorphic rocks at pressures to 10 kbar, *J. Geophys. Res., 70*, 6147–6164.

Christensen, N. I. (1966), Shear wave velocities in metamorphic rocks at pressures to 10 kilobars, *J. Geophys. Res., 71*, 3549–3556.

Christensen, N. I. (1996), Poisson's ratio and crustal seismology, *J. Geophys. Res., 101*, 3139–3156.

Christensen, N. I., and W. D. Mooney (1995), Seismic velocity structure and composition of the continental crust: A global view, *J. Geophys. Res., 100*, 9761–9788.

Christensen, N. I., and Szymanski (1988), Origin of reflections from the Brevard fault zone, *J. Geophys. Res., 93*, 1087–1102.

Cox, S. C., and D. J. A. Barrell (in press), *Geology of the Aoraki area, Institute of Geological and Nuclear Sciences 1:250,000 geological map 15*, Institute of Geological and Nuclear Sciences, Lower Hutt, New Zealand, 1 sheet + 71 pp.

Cox, S. C., and R. Sutherland (this volume), Regional geological framework of South Island, New Zealand, and its significance for understanding the active plate boundary.

Davey, F. J., T. Henyey, W. S. Holbrook, D. Okaya, T. Stern, A. Melhuish, S. Henrys, D. Eberhart-Phillips, T. McEvilly, R. Urhammer, H. Anderson, F. Wu, G. Jiracek, P. Wannamaker, G. Caldwell, and N. Christensen (1998), Preliminary results from a geophysical study across a modern continent-continent collisional plate boundary-the Southern Alps, New Zealand, *Tectonophysics, 288*, 221–235.

Davey, F. J., D. Eberhart-Phillips, M. D. Kohler, S. Bannister, G. Caldwell, S. Henrys, M. Scherwath, T. Stern, and H. van Avendonk (this volume), Geophysical structure of the Southern Alps orogen, South Island, New Zealand.

Eberhart-Phillips, D. (1995), Examination of seismicity in the central Alpine fault region, South Island, New Zealand, *N. Z. J. Geol. Geophys., 38*, 571–578.

Eberhart-Phillips, D., and S. Bannister, (2002), Three-dimensional crustal structure in the Southern Alps region of New Zealand from inversion of local earthquakes and active source data, *J. Geophys. Res.*, doi:10.1029/2001JB000567.

Eberhart-Phillips, D., and M. Henderson, (2004), Including anisotropy in 3-D velocity inversion and application to Marlborough, New Zealand, *Geophys. J. Int., 156*, 237–254.

Forsyth, P. J. (2001), *Geology of the Waitaki area, Institute of Geological and Nuclear Sciences 1:250,000 geological map 19*, Institute of Geological and Nuclear Sciences, Lower Hutt, New Zealand, 1 sheet + 64 pp.

Frost, C. D., and D. S. Coombs (1989), Nd isotope character of New Zealand sediments: Implications for terrane concepts and crustal evolution, *Am. J. Sci., 289*, 744–770.

Fuis, G. S., et al. (1991), Crustal structure of accreted terranes in southern Alaska, Chugach Mountains and Copper River Basin from seismic refraction results, *J. Geophys. Res., 96*, 4187–4227.

Gerbault, M., F. Davey, and S. Henrys (2002), Three-dimensional lateral crustal thickening in continental oblique collision: an example from the Southern Alps of New Zealand, *Geophys. J. Int., 150*, 770–779.

Godfrey, N. J., N. I. Christensen, and D. A. Okaya (2000), Anisotropy of schists: contribution of crustal anisotropy to active source seismic experiments and shear wave splitting observations, *J. Geophys., Res., 105*, 27,991–28,007.

Godfrey, N. J., N. I. Christensen, and D. A. Okaya (2002), The effect of crustal anisotropy on reflector depth and velocity determination from wide-angle seismic data: a synthetic example based on South Island, New Zealand, *Tectonophysics, 355*, 145–161.

Grapes, R. H. (1995), Uplift and exhumation of Alpine schist, Southern Alps, New Zealand: thermobarometric constraints, *N. Z. J. Geol. Geophys., 38*, 525–533.

Holbrook, W. S., D. Okaya, T. Stern, F. Davey, S. Henrys, and H. van Avendonk (1998), Deep seismic profiles across the Pacific-Australian plate boundary, South Island, New Zealand, *Eos Trans. AGU, 79,* F901.

Hughes, S., J. H. Luetgert, and N. I. Christensen (1993), Reconciling deep seismic refraction and reflection data from the Grenvillian-Appalachian boundary in western New England, *Tectonophysics, 225,* 255–269.

Iida, K., T. Sugio, H. Furuhashi, and M. Kumazawa (1967), Elastic wave velocity in crystalline Schists from Sanbagawa metamorphic terrane, Shikoku, Japan, *J. Earth Sci. Nagoya Univ.,* 15, 112–123.

Ji, S., and M. H. Salisbury (1993), Shear wave velocities, anisotropy and splitting in high-grade mylonites, *Tectonophysics, 221,* 453–473.

Johnston, J. E., and N. I. Christensen (1995), Seismic anisotropy of shales, *J. Geophys, Res., 100,* 5991–6003.

Kern, H., and H.-R. Wenk (1990), Fabric related velocity anisotropy and shear wave splitting in rocks from the Santa Rosa mylonite zone, California, *J. Geophys. Res., 95,* 11,212–11,223.

Landis, C. A., and D. S. Coombs (1967), Metamorphic belts and orogenesis in southern New Zealand, *Tectonophysics, 4,* 501–518.

Leitner B., D. Eberhart-Phillips, H. Anderson, and J. L. Nabelek (2001), A focused look at the Alpine fault, New Zealand: seismicity, focal mechanisms, and stress observations, *J. Geophys. Res., 106,* 2193–2220.

Little, T. A., R. J. Holcombe, and B. R. Ilg (2002), Ductile fabrics in the zone of active oblique convergence near the Alpine Fault, New Zealand: identifying the neotectonic overprint, *J. Struct. Geol., 24,* 193–217.

Little, T., R. Wightman, R. J. Holcombe, and M. Hill (this volume), Transpression models and ductile deformation of the lower crust of the Pacific Plate in the central Southern Alps, a perspective from structural geology.

Miller, D. J., and N. I. Christensen (1994), Seismic signature and geochemistry of an island arc: a multidisciplinary study of the Kohistan accreted terrane, northern Pakistan, *J. Geophys. Res., 99,* 11,623–11,642.

Mutch, A. R. (1975), *Geological map of New Zealand 1:250,000, sheet 23: Oamaru,* Dept. Sci. Ind. Res., Wellington, New Zealand.

O'Connell, R. J., and B. Budiansky (1974), Seismic velocities in dry and saturated cracked solids, *J. Geophys. Res., 79,* 5412–5426.

Okaya, D., N. I. Christensen, D. Stanley, and T. S. Stern (1995), Crustal anisotropy in the vicinity of the Alpine Fault Zone, *N. Z. J. Geol. Geophys., 38,* 579–583.

Okaya, D., and N. I. Christensen (2002), Anisotropic effects of non-axial seismic wave propagation in foliated crustal rocks, *Geophys. Res. Lett.,* doi:10.1029/2001GL014285.

Pulford, A., M. Savage, and T. Stern (2003), Absent anisotropy: the paradox of the Southern Alps orogen, *Geophys. Res. Lett., 30,* 20, doi:10.1029/2003GL017758.

Savage, M., M. Duclos, and K. Marson-Pidgeon (this volume), Seismic anisotropy in South Island, New Zealand.

Scherwath, M., A. Melhuish, T. Stern, and P. Molnar (2002), Pn anisotropy and distributed upper mantle deformation associated with a continental transform fault., *Geophys. Res. Lett.,* doi:10.1029/2001GL014179.

Scherwath, M., T. Stern, F. Davey, D. Okaya, W. S. Holbrook, R. Davies, and S. Kleffmann, (2003), Lithospheric structure across oblique continental collision in New Zealand from wide-angle P-wave modeling, *J. Geophys. Res., 108,* 2566, doi:10.1029/2002JB002286.

Smith, E. G. C., T. Stern, and B. O'Brien (1995), A seismic velocity profile across the central South Island, New Zealand, from explosion data, *N. Z. J. Geol. Geophys., 38,* 565–570.

Smith, R. A. (1999), Macroscopic P-wave anisotropy in the Haast Schist, New Zealand: implications for errors in wide-angle seismic studies of metamorphic terranes, Masters thesis, University of Wyoming, Laramie, WY, 147 pp.

Stern, T., S. Kleffmann, D. Okaya, M. Scherwath, and S. Bannister (2001), Low seismic-wave speeds and enhanced fluid pressure beneath the Southern Alps of New Zealand, *Geology, 29,* 679–682.

Turnbull, I. M. (2000), Geology of the Wakatipu area, Institute of Geological and Nuclear Sciences 1:250,000 geological map 18. Lower Hutt, Institute of Geological and Nuclear Sciences Ltd. 1 sheet + 72 p.

Upton, P., and P. O. Koons (this volume), Three-dimensional geodynamic framework for the Central Southern Alps, New Zealand: integrating geology, geophysics and mechanical observations.

Stewart, R. and L. Peselnick (1977), Velocity of compressional waves in dry Franciscan rocks to 8 kilobar and 300°C, *J. Geophys. Res., 82,* 2027–2039.

van Avendonk, H. J. A., W. S. Holbrook, D. Okaya, J. K. Austin, F. Davey, and T. Stern (2004), Continental crust under compression: a seismic refraction study of South Island Geophysical Transect I, South Island, New Zealand, *J. Geophys. Res., 109,* B06302, doi: 10.1029/2003JB002790.

Wellman, H. W. (1979), An uplift map for the South Island of New Zealand, and model for the uplift of the Southern Alps, in *The Origin of the Southern Alps, R. Soc. of N. Z. Bull.,* vol. 18, edited by R. I. Walcott and M. M. Cresswell, pp. 12–20.

N. I. Christensen, Department of Geology and Geophysics, University of Wisconsin-Madison, Madison, WI, USA.

D. A. Okaya, Department of Earth Sciences, University of Southern California, Los Angeles, CA, USA.

The Alpine Fault, New Zealand:
Surface Geology and Field Relationships

Richard J. Norris and Alan F. Cooper

Department of Geology, University of Otago, Dunedin, New Zealand

The Alpine Fault forms part of the on-land Pacific-Australian plate boundary. Generally the fault has a simple straight trace, striking ca. 055° and dipping moderately eastwards, but in central areas it is serially partitioned. Late Quaternary strike-slip rates are 23–25 mm/yr, but dip-slip rates range from ca. 10 mm/yr in the central part, reducing north and south and reaching zero southwest of Jackson Bay. Due to oblique-slip, mylonites are exhumed from ca. 25 km in a ca. 1 km wide fault zone east of the present trace, undergoing intense cataclasis close to the fault at shallow levels. Pseudotachylytes are generated by preferential melting of phyllosilicate-rich assemblages. Ductile shear strain is high, reaching 200-300 in ultramylonites. Displacement of metamorphic zones and a pegmatite swarm, with respect to their location in the hanging wall is consistent with ca. 100 km of dextral ductile shear within the mylonites since ca. 5 Ma.

INTRODUCTION

The Alpine Fault is the principal structural feature of the Australia/Pacific plate boundary in South Island of New Zealand, and currently accommodates around two thirds of the inter-plate slip (Figure 1). The fault therefore represents major localization of inter-plate displacement in at least the upper crust. The SIGHT program was concerned with geophysical imaging of the entire deforming zone along the South Island plate boundary. Clearly the Alpine Fault is a critical component of the orogen and in this paper we review aspects of the surface geology and field relationships of the fault zone.

In 1942 *Wellman and Willett* adopted the name Alpine Fault from *Henderson* [1937] and used it for the fault extending 500 km along strike from Lake Rotoroa, Nelson to Milford Sound. The concept of a 480 km dextral strike slip displacement, using as a marker a unit now termed the Dun Mountain Ophiolite Belt (Figure 1), was presented by Wellman in 1949 at the Pacific Science Congress [*Benson*, 1952].

A Continental Plate Boundary: Tectonics at South Island, New Zealand
Geophysical Monograph Series 175
Copyright 2007 by the American Geophysical Union.
10.1029/175GM09

The timing of this offset generated a major controversy for the next two or three decades before a post-Oligocene age for the fault was finally accepted [*Carter and Norris*, 1976; *Kamp*, 1986; *Cooper et al.*, 1987].

In 1953 *Wellman* assembled geological evidence for Recent and Pleistocene displacement on the fault, compiled from fieldwork and aerial photograph interpretation. These data and results of individual field studies, chiefly by members of the New Zealand Geological Survey [e.g., *Bowen*, 1954], were summarized by *Suggate* [1963] and *Reed* [1964]. Systematic mapping of the Alpine Fault and structural-petrological investigation of its fault rocks was initiated by the University of Otago in the late 1970s resulting initially in a publication by *Wallace* [1976] on pseudotachylyte from the Moeraki River area. The program has continued to the present day with the mapping and characterization of Alpine Fault relationships between the Taramakau River [*Rattenbury*, 1987, 1991] and Milford Sound [*Cooper and Norris*, 1990]. Fault rocks have been most intensively studied in the Gaunt Creek, Waitangi-taona River [*Cooper and Norris*, 1994] and Hare Mare Creek, Waikukupa River [*Norris and Cooper*, 1997] areas, where cross sections of the fault zone have been outlined and segmentation of the surface trace of the fault [*Norris and Cooper*, 1995] has been described.

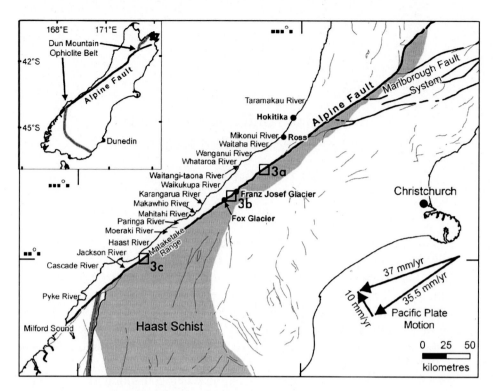

Figure 1. Regional South Island map, showing Alpine Fault and localities referred to in text. Boxes show locations of detailed maps in Figure 4. Inset shows location map with offset of Dun Mountain ophiolite belt indicated. Arrows give Nuvel 1A [*Demets et al.*, 1994] interplate velocity vector in the central South Island and its components parallel and perpendicular to the Alpine Fault.

This paper is intended to compliment other studies in this volume that deal with both the regional (southwest Pacific) context and more localized (South Island, New Zealand) plate tectonic setting of the Alpine Fault, its geophysical character and paleoseismicity. Herein we detail the structural field relationships and outcrop characteristics of the fault and its associated fault rocks. We summarize the evidence for present day slip rates and provide estimates of the total shear strain that has accumulated.

STRUCTURAL FEATURES OF THE ALPINE FAULT

The Alpine Fault from Hokitika to Milford Sound appears as a remarkably straight trace on a satellite photograph, with an average strike of 055° (Figure 1). In outcrop, the fault in general dips to the east at a moderate angle, although south of Haast, the fault tends more to vertical. Over most of its length, the fault has a single active trace and only rarely are there multiple fault strands and a wide zone of deformation [*Berryman et al.*, 1992]. Most of the slip occurs within a zone of gouge and cataclasite no more than 50m wide, and commonly less than 10m wide. Secondary faulting with

zones of gouge a few centimeters wide is found either side of the main trace, although mostly within the hangingwall rocks, for a distance of about 100 m. Occasional faults occur beyond this distance, but the zone of intensive fracturing, or damage zone, is usually less than 100 m in width [e.g., *Cooper and Norris*, 1994; *Norris and Cooper*, 1997; *Wright*, 1998]. Figure 2(a) is a plot of basal gouge zones measured between Hokitika and Haast. The average orientation is 054°/30°SE. This is subparallel to the overall fault strike; the dip, which is highly variable, reflects the development of thrust sections of the fault extending to the west over Quaternary deposits.

Slip on the fault, as determined from offset features such as gravel deposits, river terraces, stream channel deposits, etc., is dominantly oblique along most of its length, with components of dextral strike-slip and reverse dip-slip [*Wellman*, 1953; *Sibson et al.*, 1979; *Norris and Cooper*, 2001]. Shear sense indicators on shears throughout the fault zone, although variable, are generally consistent with a dextral-reverse, east side up, sense of displacement [e.g., *Cooper and Norris*, 1994; *Norris and Cooper*, 1997]. Due to the uplift of the eastern side, deep-seated fault rock have been

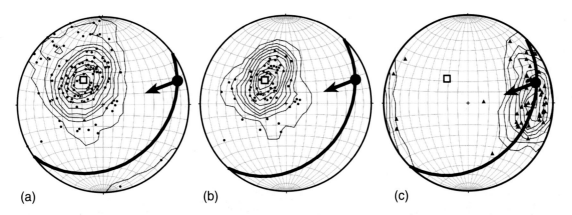

Figure 2. Lower hemisphere, equal area stereographic projections of structural data from the Alpine Fault gouge zone. Contours are by Kamb method at a spacing of 2σ. Open square is pole to mean surface represented by heavy great circle. Arrow on this great circle represents the projection of the Nuvel 1A interplate velocity vector. (a) Poles to basal gouge shear surfaces, Hokitika to Haast. (b) Poles to basal gouge shear surfaces, central section (Whataroa to Karangarua rivers). (c) Wear striations on basal gouge shear surfaces, central section (Whataroa to Karangarua rivers). The mean surface shown is the same as in (b). (Plots constructed using Stereonet 6.3.2X by R. Allmendinger).

exhumed adjacent to the present surface trace. The fault zone passes eastwards from gouge and cataclasite in contact with Late Quaternary sediments or Western Province bedrock, through ultramylonite, mylonite and protomylonite over a distance of up to a kilometer, before passing into Alpine Schist [*Sibson et al.*, 1979; *Norris and Cooper*, 2003] (Figure 3).

Measurements of the basal fault gouge within the central part of the fault (between the Whataroa and Karangarua rivers, Figure 1) have a mean orientation of 043°/30°SE

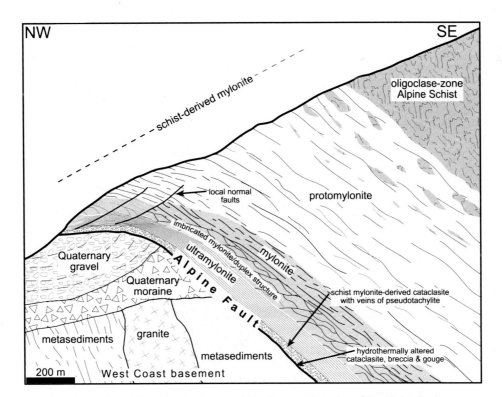

Figure 3. Schematic cross section of an oblique thrust segment within the central section of the Alpine Fault.

(Figure 2b). This is a more northerly strike than the fault average, and the mean dip is again lower than that of the cataclasite at depth. The low dip is partly due to the best basal gouge zones being exposed where the fault has formed an overthrust above late Quaternary gravels, and has flattened out as it has done so [e.g., *Cooper and Norris*, 1994]. Where the basal cataclasites are observed in deeply eroded sections, the dip tends to be a little steeper at around 40°. Figure 2(c) is a plot of wear striations measured on gouge surfaces within the basal gouge/cataclasite. The striations show a spread between 050° and 110°, with a mean at 084°. On the most recent gouge surfaces at the contact of the fault rocks with underlying deposits, the orientation tends to be between 070° and 080°. The data are compatible with slip on the fault being approximately parallel to the Nuvel 1A interplate velocity vector [*DeMets et al.*, 1994], although clearly the scatter is sufficient to allow for a divergence between the two. The mean orientation might indicate a slightly greater dip-slip component on the fault than expected from the plate motion vector, although the data are sufficiently variable (as might be expected with a fault gouge zone at the surface) to be equivocal.

Mapping of the fault in detail has revealed that the geometry of the fault trace has three principal styles. In many places, particularly between the Whataroa and Hokitika Rivers, it forms essentially a straight trace on a strike of around 055°, with an oblique dextral-reverse slip (Figure 4a). Locally overthrusts extend out to the west for short distances (Figure 3a). In the central section, between the Whataroa and Haast rivers, the fault tends to break up into a zig-zag pattern of more northerly-striking oblique thrust zones linked by near vertical, more easterly striking, dextral strike-slip faults [*Norris et al.*, 1990; *Norris and Cooper*, 1995; Figure 4b]. Most of the measurements of basal gouge in Figure 2(b) were from outcrops of the thrust sections, which have a more northerly strike and lower angle of dip. The strike-slip faults are only rarely exposed. Where they are, they are narrow (< one meter) zones of clay gouge and fractured rock. They do not extend far beyond the linked thrust zones and the movement on them is essentially parallel to that on the oblique thrust sections [*Norris and Cooper*, 1995]. In the southern section south of Haast (Figure 4c), the fault is essentially straight with recent short en-echelon traces and step-overs typical of a strike-slip fault [*Berryman et al.*, 1992; *Sutherland and Norris*, 1995; *Wright*, 2000]. An important exception is the area around the Cascade River [*Berryman et al.*, 1992; *Campbell*, 2005] where the fault zone widens to include several older reactivated structures (see below).

The structural geometry in the central part of the fault, where the obliquely convergent fault is partitioned into serially linked sections of thrust and strike-slip character, was termed "serial partitioning" (Figure 5a) by *Norris and Cooper* [1997]. This behavior appears linked to rapid erosion by rivers crossing the fault. The effect of river erosion across the westward-propagating thrust wedges is to reduce their effective surface slope rendering them subcritical [*Dahlen*, 1984]. The thrust wedge will respond by internal thickening by imbrication, but continuing erosion by the river will eventually result in formation of an out-of-sequence thrust on the northeast side of the river with an associated eastward jump in the active trace [*Simpson et al.*, 1994; *Norris and Cooper*, 1997]. Linking of these offset thrust segments by east-west striking strike-slip faults leads to the zig-zag pattern observed. The spacing of offsets in the serially partitioned section is generally less than 3 km, so that it is unlikely that this pattern extends to the base of the seismogenic zone, or forms an obstacle to seismic rupturing [*Norris and Cooper*, 1995].

A more common pattern globally of partitioning of strike-slip and convergence onto different structures at the surface is when the two sets of structures form parallel arrays [e.g., *Wentworth and Zoback*, 1989]. *Norris and Cooper* [1997] termed this style "parallel partitioning" (Figure 5b), and it is the style of the southern part of the fault, where the main trace is mainly strike-slip in character (Figure 4c) and the convergent component is accommodated on other parallel structures in the mountains to the east and off the coast [*Sutherland et al.*, 2006]. South of Milford Sound, the Alpine Fault zone passes offshore. Here it is characterized by a prominent zone of dominantly strike-slip displacement near to the coast with a series of subparallel thrust faults farther offshore [*Barnes et al.*, 2005]. This is a classic example of parallel partitioning, similar to that described from many subduction zones around the world [e.g., *McCaffrey*, 1992].

The area around the Cascade River mentioned above is perhaps the most complex part of the Alpine Fault zone. Here, Mesozoic fault zones along terrane boundaries in the basement rocks have been reactivated and rotated towards parallelism with the Alpine Fault [*Campbell*, 2002, 2005]. The whole system forms a large anti-dilational jog or positive flower structure along the fault, with motion partitioned among several major strands. To the south, other faults carry the motion back onto the main trace via a dilational zone.

Within the Alpine Schists of the hangingwall within the central portion, east-west dextral strike-slip faults form a prominent set of structures visible on aerial photographs [*Hanson et al.*, 1990]. Together with less common conjugate faults, they represent Coulomb or Riedel shears within the zone of transpression above the main fault. They are suitably orientated to be activated as the linking strike-slip sections during serial partitioning of the main fault trace. Close to the active fault trace, west-dipping normal faults also occur, due to the collapse of the thrust front as it is translated westwards. Commonly, the dextral strike-slip faults, or reverse faults

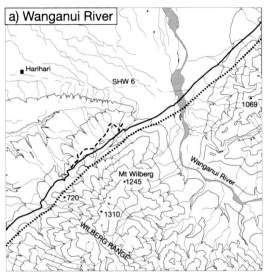

a) Wanganui River

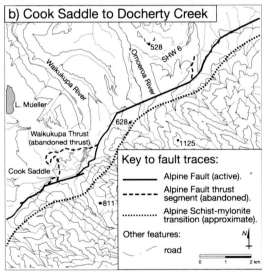

b) Cook Saddle to Docherty Creek

Key to fault traces:

———— Alpine Fault (active).

– – – Alpine Fault thrust segment (abandoned).

·········· Alpine Schist-mylonite transition (approximate).

Other features:

～～ road

0   1   2 km

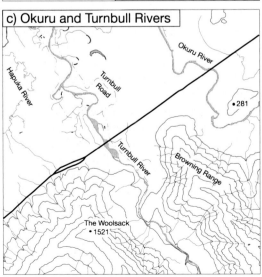

c) Okuru and Turnbull Rivers

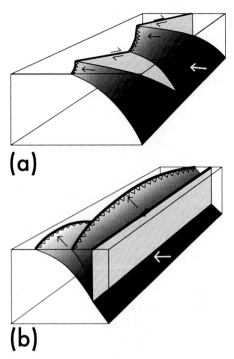

Figure 5. Diagrams showing idealized forms of slip partitioning on a dextral-reverse oblique-slip fault zone at the surface [after *Norris and Cooper*, 1997] (a) Serial partitioning; (b) Parallel partitioning.

conjugate to the Alpine thrust, may be reactivated as normal faults [*Hanson et al.*, 1990; *Cooper and Norris*, 1994].

LATE QUATERNARY SLIP RATES

Wellman [1953] compiled data on the offset of prominent topographic features along the Alpine Fault as identified on aerial photographs. Using these data, he made estimates of rates of displacement. Lack of information on the age of these features meant that ages had to be estimated from

Figure 4. Maps showing examples of surface trace patterns: (a) northern section of Alpine Fault near the Wanganui River [*Wright, C.A.*, 1998], characterized by an oblique slip fault with local development of thrusts out to the west; (b) central section between Fox and Franz Josef glaciers [*Norris and Cooper*, 1997]; the fault here shows serial partitioning into oblique thrust sections connected by more east-west strike-slip sections; (c) southern section south of Haast [*Wright, T.D.*, 2000]; the fault here is essentially a straight strike-slip fault with minor step-overs. Parallel partitioning has led to the convergence being accommodated on structures in the mountains to the east and offshore to the west (not shown on this map). Locations of maps shown on Figure 1. See text for discussion. The figure is based on maps of the whole Alpine Fault trace available at http://www.otago.ac.nz/geology/af/alpinefault.htm.

possible correlations to late Quaternary glaciations. Despite these large uncertainties, *Wellman* [1953] clearly identified significant rates of both strike-slip and vertical offset on the Alpine Fault during the Late Quaternary. *Suggate* [1968] used fossils within an uplifted silt unit at Paringa to determine uplift rates relative to sea level over the last 13 kyr. Later work in this area by *Simpson et al.* [1994] indicated that Suggate's rate of 10.7 mm/yr, while correct, only applied to a rapidly growing anticline along the fault. The regional rate of uplift was 7–8 mm/yr.

In 1979, *Wellman* published an uplift map of the Southern Alps based on data of variable reliability. Maximum uplift rates of ca. 10 mm/yr were shown to be attained in central Westland, reducing to the northeast and southwest. *Bull and Cooper* [1986] used benches on the front of the Alps, interpreted as cut by wave action, to estimate uplift rates along the length of the fault over the last 130 kyr. Again, these reached a maximum around the glaciers of ca. 8 mm/yr, reducing to 5 mm/yr to the northeast near Hokitika and to the southwest at Haast. Using fission track measurements, *Tippett and Kamp* [1993] also calculated an exhumation rate over the last ca. 1 Myr of ca. 9 mm/yr in the vicinity of the glaciers, reducing to the northeast and southwest.

Since 1990, a number of studies have attempted to estimate Late Quaternary slip rates on the Alpine Fault [e.g., *Berryman et al.*, 1992; *Cooper and Norris*, 1994, 1995; *Sutherland and Norris*, 1995; *Yetton and Nobes*, 1998]. *Norris and Cooper* [2001] compiled all published, and some unpublished, data on offset of Late Quaternary (1–65 ka) geomorphic features such as terraces, channels, etc. across the fault and direct observation of displacement in fault outcrops. They recalculated ages of features where necessary and attempted to assign uncertainties on the estimates. Their best estimates of strike-slip and dip-slip rates are shown in Figure 6, and Figure 7 shows the data plotted against distance along the fault. Strike-slip rate on the fault diminishes markedly north of the confluence of the Hope Fault, which has a high documented slip rate [*Cowan and McGlone*, 1991; *Langridge et al.*, 2003]. Whereas they were able to fit a mean strike-slip rate of 27±5 mm/yr (1σ) along the whole length of the fault south of the Hope Fault, the individual dip-slip rate estimates varied between localities well beyond their likely error bars, with a maximum dip-slip rate of ca. 10 mm/yr in the central portion, diminishing to ca. 6 mm/yr the northeast and to zero to the southwest. This pattern is in accordance with the earlier estimates discussed above. The strike-slip rate is also consistent with a long term rate of ca. 26 mm/yr reported by *Sutherland* [1994] for offset since the Pliocene.

More recently, *Sutherland et al.* [2006] have recalculated the southernmost two estimates of the strike-slip rate in

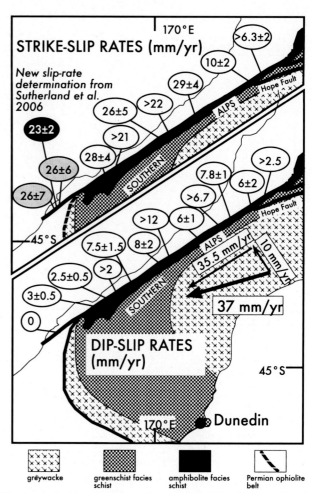

Figure 6. Estimates of Late Quaternary strike-slip and dip-slip rates along the Alpine Fault [data from *Norris and Cooper*, 2001, and *Sutherland et al.*, 2006]

Figure 6 [from *Sutherland and Norris*, 1995], using extensive offset data with new age measurements of late Quaternary moraines to provide a tightly constrained estimate of the strike-slip rate of 23 ± 2 mm/yr (2σ). There are a number of ways these new data may be incorporated. We can still fit a constant strike-slip rate of 24 mm/yr to the whole fault that lies within the error bars of all the estimates including the new data (Figure 7). Alternatively, we could adopt the new rate of 23 ± 2 mm/yr for the whole fault, given the much smaller error limits on this result. Given the large uncertainties of the earlier data, this approach could be justified. Plate tectonic calculations based on the Nuvel 1A model [*DeMets et al.*, 1994] indicate little change in fault-parallel plate motion through South Island (although there is a reduction in convergence southwards; Figure 7). Therefore a constant strike-slip rate on the fault is a reasonable scenario. *Norris*

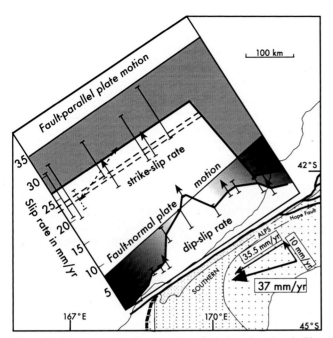

Figure 7. Slip rate versus distance along fault, based on data in Figure 6. Solid I-bars are estimates with uncertainties from *Norris and Cooper* [2001]; dotted I-bar is revised estimate from *Sutherland et al.,* (2006). Solid thick lines represent best-fit average slip rates adopted by *Norris and Cooper* [2001]; heavy dashed line is revised uniform strike slip rate of 24 mm/yr; thin dashed line is strike-slip rate of *Sutherland et al.* (2006); sloping dashed arrow represents possible gradient in strike-slip rate along fault. The lines indicating fault parallel and fault normal plate motion are calculated from the Nuvel-1 global model [*DeMets et al., 1994*].

and Cooper [2001] had no grounds, given the error bars on the data, to suggest anything other than a constant rate of strike-slip. With the increased accuracy of the *Sutherland et al.* [2006] rate, however, there is the possibility that the strike-slip rate is lower to the southwest, possibly due to 2–3 mm/yr being accommodated by faults around the eastern margin of Fiordland [cf. *Campbell*, 2005]. The reduction in dip-slip rate to the southwest may be accompanied by a small reduction in strike-slip rate, with increased bending and shortening in northwest Otago [cf. *Sutherland et al.*, 2006]. Better data are still required from the central part of the fault to demonstrate a variable strike-slip rate.

Implications of the data are that between 9 and 12 mm/yr of strike-slip and between 0 and 10 mm/yr of dip-slip (or convergence) must be accommodated away from the Alpine Fault. It is possible that a small proportion of this is distributed among minor structures immediately east of the fault, but most must be taken up on structures within the Southern Alps or on their eastern foothills. The component of strike-slip distributed east of the fault must result in rotation and bending of western South Island [cf. *Norris, 1979; Little et al.*, 2002b]. There is little evidence of large amounts of active deformation west of the Alpine Fault through much of Westland, and most seismic activity is currently to the east [*Leitner et al.*, 2001]. South of Jackson Bay, earthquakes become common offshore, and north of Hokitika, earthquakes and active faulting occur to the west of the Alpine Fault.

The central section of the Alpine Fault, where the dip-slip component reaches a maximum, is the area of greatest relief and the narrowest part of the orogen. It coincides with the youngest mineral ages, the highest rates of exhumation, and the highest heat-flow [*Little et al.*, 2005; *Sutherland et al.*, this volume]. It also coincides with the characteristic serial partitioning pattern of the Alpine Fault trace discussed previously. To the north, coinciding with a reduction in dip-slip rates, convergence is progressively distributed eastwards into Canterbury [e.g., *Litchfield et al.*, 2003; *Howard et al.*, 2005] as the plate motion is transferred towards the Hikurangi subduction zone. To the south, the convergence is also partly redistributed to the east, with contraction extending as far as the Otago coast. Long term shortening rates and preliminary GPS data, however, suggest that only around 3 mm/yr of contraction is accommodated in central and east Otago [*Norris and Nicolls*, 2004; P. Denys, personal communication, 2005], so the remaining 5 mm/yr must be taken up within the Southern Alps, or in part transferred offshore onto the thrust systems mapped by *Barnes et al.* [2005] west of Fiordland.

The highly localized convergence on the central part of the Alpine Fault is in the centre of the zone of oblique continental collision, with convergence becoming more distributed towards each end of the fault as it passes into areas of continent-ocean interaction. Interaction of intense erosion, uplift and exhumation in this central region are likely to lead to thermal weakening and increasing localization of deformation [*Ellis et al.*, 2001; *Koons et al.*, 2003].

FAULT ROCKS

Introduction

Fault rocks associated with the Alpine Fault were first recognized by *Bell and Fraser* [1906] and *Morgan* [1908] from the area between Hokitika and Ross in Westland. *Morgan* [1908, p.71] wrote: "On the western side of Mt Bonar a fault zone is shown by the disturbed and crushed nature of the gneissic schists" and (pp. 72) "at both Cowhide Creek and at Hende's the crushed rock has been pushed over recent gravels".

A major treatise on Alpine Fault related rocks by *Reed* [1964] recognized three groups of fault rocks occurring principally in what is now referred to as the Fraser Complex [*Rattenbury*, 1991], immediately west of the Alpine Fault trace. Reed described fault gouge and breccia which he inferred were related to Quaternary movements, the cataclasite series conspicuous in granite and attributed to late Tertiary movement, and the mylonite series, again said to be 'largely within granite' and formed contemporaneously with Rangitata (Jurassic-Cretaceous) metamorphism. East of the Alpine Fault trace, mylonitisation of the Alpine Schist was also described. *Rattenbury* [1991] inferred that most of the "mylonite" west of the Alpine Fault, the Fraser Complex, was in fact part of the gneissic and granitic basement of the West Coast. The foliation, lineation and sense of shear of a local mylonitic overprint, however, are all consistent with dextral strike slip movement on the Alpine Fault [*Rattenbury*, 1987].

The mylonites described by *Reed* [1964] on the west side of the Alpine Fault, extending south-west to the Wanganui River, were correlated with mylonites on the southeast side of the fault, extending north-east as far as the Pyke River [*Grindley*, 1963]. However, the correlation was made on the assumption of a 'virtual absence of true mylonites along the Alpine Fault' in the intervening area, an assumption that we now know is incorrect.

Sibson et al. [1979] reported the results of a field investigation of fault rocks distributed along 450 km of the Alpine Fault. They recognized an essentially similar range of rock types to *Reed* [1964], but were able to relate them spatially to a schematic cross section of a one km wide fault zone. In contrast to Reed's inferred sequential development of fault rocks from the Rangitata orogeny through to the present, *Sibson et al.* [1979] interpreted the three fault rock series as having formed concurrently at different depth zones on the Alpine Fault during the late Cenozoic to Recent phase of dextral-reverse slip. *Sibson et al.*'s [1979] schematic section is interpreted in terms of derivation of cataclasites and augen mylonites from the Tuhua Granites in the west, through green mylonites and curly schists (mylonites) formed from an Alpine Schist protolith in the east. No major tectonic break is shown, although the section is cut by a number of gouge zones, slightly discordant to the mylonitic foliation.

Subsequently, a number of detailed studies have been made, and a regional program mapping the Alpine Fault trace has been completed (http://www.otago.ac.nz/geology/af/alpinefault. htm). The results of this work is summarized below.

Cataclasites and Mylonites

As a consequence of its reverse oblique slip character, the Alpine Fault is remarkable in that it exhibits ductilely de-formed mylonite series rocks alongside those formed, and still forming, by brittle deformation at the present day. The mylonite zone in the hangingwall of the fault zone is 1 to 1.5 km wide, with the mylonitic foliation (056°/33°SE, Figure 8) essentially parallel to the moderately dipping basal Alpine Fault plane (Figures 2, 3) [*Norris and Cooper*, 1995]. The most spectacular development of fault rock sequences occurs in the thrust segments of the fault, where there is a progressive westward increase in the degree of deformation and mylonitisation of the Alpine Schist protolith towards a zone of ultramylonite that, in turn, is overprinted by intense cataclasis above the fault plane (Figure 3). In this discussion of fault rocks, the nomenclature adopted is predominantly that of *Sibson* [1977]. One modification is that we use the term cataclasite for all rocks deformed predominantly by cataclasis irrespective of the degree of induration, which may vary even on a hand-specimen scale.

Cataclasite

West Coast basement is only rarely exposed in the central Westland fault sections and a green cataclasite, typically 20-30 m thick, representing the base of the hangingwall (Pacific plate) sequence, is thrust westwards across fluvio-glacial sediments on the footwall (Australian plate). At the base of this cataclasite, the Alpine Fault plane is knife sharp with ~30 cm of indurated gouge passing upwards into massive to weakly foliated cataclastic gouge (a classification not recognized in the classification of *Sibson* [1977]) of the overriding Pacific plate. The fault orientation is generally moderately east-dipping, with a mean orientation measured in outcrop of 054/30° SE, although there is considerable variation along the fault (see previous discussion). Slickenside lineations in the basal gouge within the central part of the fault range plunge at 20° on a trend ranging from 050° to 110°, with a mean of 084° (Figure 2c). Orientations of the fault and the thickness of the basal cataclasite are variable, particularly where the overthrust sheet advances across and is controlled by topography in the underlying land surface, and the cataclasite imbricates internally in order to achieve the necessary critical thickness and strength for continued westward propagation. Some of the overthrusts extend westwards across the footwall for distances of up to 1.5 km from the line of the fault. The basal horizons of the cataclasite are commonly sites where wood fragments can be collected, scraped from the vegetated (?) footwall surface and preserved in the relatively impermeable clay-rich cataclasite. Platy schist clasts in gravels beneath the basal thrust are rotated to a steep orientation, as are occasional fine sand interbeds in fluvial sequences [e.g., *Norris and Cooper*, 1997, Figure 6]. Cataclasites contain angular clasts of mylonite in a finely comminuted clay-rich matrix containing combinations of muscovite, chlorite, smectite, illite and epidote. Based on the composition of included fragments of garnet, *Cooper and Norris* [1994] determined that the cataclasite has been derived in part from the hangingwall Alpine Schists, although a contribution from the footwall granitoids at depth cannot be discounted. Pods of broken quartz represent hydrothermal veins disrupted by subsequent cataclasis.

Mineralogical determinations by *Warr and Cox* [2001] were interpreted as indicating three stages of development in Alpine Fault cataclasites (1) anhydrous cataclasis and frictional melting, (2) hydrous chloritization under sub-greenschist facies conditions (<320°C), and (3) growth of swelling clays in the matrix (smectite formation typically occurs under hydrothermal conditions of <120°C).

Pseudotachylyte

At Gaunt Creek, the upper part of the basal cataclasite is overlain by a dark-colored, strongly cataclastic ultramylonite. The dark color is in part a reflection of the content of multi-generation pseudotachylyte veins. Pseudotachylytes range from concordant, subparallel to the mylonitic foliation, to slightly discordant, shear-hosted veins. Pseudotachylytes, as recognized by *Sibson et al.* [1979], occur throughout the Alpine Fault zone except within gouge. However, it is our observation that pseudotachylytes are most common with 200 m of the fault plane.

First analysed by *Wallace* [1976], pseudotachylytes have also been studied by *Bossière* [1991], *Warr et al.* [2003] and *Warr and van der Pluijm* [2005]. Most studies have been of fallen blocks whose field relationships are unknown, or they are derived from Harold Creek [*Bossière* 1991], near the Wanganui River (Figure 1), where the fault rock sequence has been mapped subsequently in great detail and has been shown to contain slices of both Western Province and Alpine Schist derived mylonites (V. Toy, pers. comm., 2005). Uncertainty exists, therefore, in many studies about the protolith that has partially melted to produce the pseudotachylyte.

Bossière claims that his microprobe analyses 'were always the result of mixed analyses', a consequence of the crowding of pseudotachylyte glass by abundant microcrystalline porphyroclastic debris. Most of the Alpine schist-derived pseudotachylytes examined by us are devitrified and pseudomorphed by lepidoblastic aggregates of muscovite. One pseudotachylyte at Gaunt Creek is, however, fresh and an analysis is presented in Table 1. It is clearly not an equilibrium minimum partial melt, being characterized by a relatively Fe-Mg-poor, highly potassic composition. The origin is interpreted to be by flash melting of muscovite and subordinate amounts of other minerals, from the host Alpine Schists. *Bossière* [1991], *Warr et al.* [2003] and *Warr*

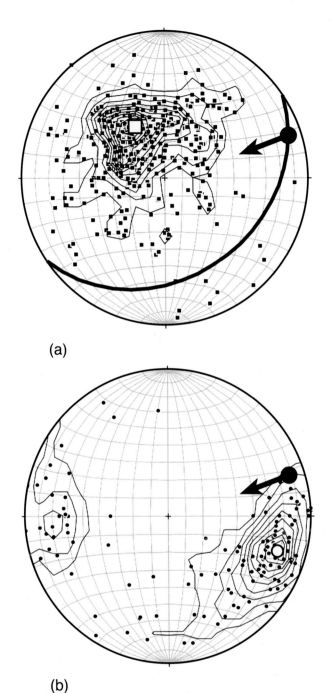

(a)

(b)

Figure 8. Lower hemisphere, equal area stereographic projections of structural data from the mylonite zone along the Alpine Fault between Hokitika and Haast. Contours are by the Kamb method at a spacing of 2σ. (a) Poles to mylonitic foliation. Open square is pole to mean surface represented by heavy great circle, mean orientation 056°/33°SE. Arrow on this great circle represents the projection of the Nuvel 1A inter-plate velocity vector. (b) Mylonitic lineations; open circle is mean orientation 20°/098°. The arrow is the Nuvel 1A vector from (a). (Plots constructed using Stereonet 6.3.2X by R. Allmendinger).

Table 1. Analyses of pseudotachylytes from the Alpine Fault zone. QF-p is quartzofeldspathic pseudotachylyte and QF-m quartzofeldspathic mylonite from Harold Creek near the Wanganui River [*Warr et al.*, 2003; *Warr and van der Pluijm*, 2005]; GC-5 is a pseudotachylyte glass from Gaunt Creek.

Reference	*Warr et al.* [2003]; *Warr and van der Pluijm* [2005]				This work	*Wallace* [1976]	
Alpin Fault loc.	Wanganui River area, Harold Creek				Gaunt Ck	Moeraki River area	
No of analyses	4	15	2	10	8	15	2
Sample	QF-p, margin	QF-p, center	QF-m	QF-m	GC 5	Hyalomylonite	microlitic u'myl
SiO_2	64.22	72.17	70.90	78.80	58.38	62.20	56.25
TiO_2	7.00	0.46	0.00	0.00	0.99	0.98	1.42
Al_2O_3	16.42	17.45	19.80	8.80	18.28	15.80	14.44
FeO	1.72	0.67	1.20	3.80	4.22	7.42	8.83
MnO	0.33	0.04	0.00	0.10	0.07	0.19	0.14
MgO	2.63	2.95	1.50	0.60	2.15	2.05	4.19
CaO	2.18	2.32	3.10	5.00	0.72	4.37	5.43
Na_2O	2.63	3.73	3.50	3.00	1.92	2.17	0.79
K_2O	2.64	0.20	0.00	0.20	9.59	2.42	7.17
P_2O_5							
Cr_2O_3	1.18	0.00	0.00	0.00			
Total	100.95	99.99	100.00	100.30	96.32	97.60	98.66

and van der Pluijm [2005] also concluded that the hydrous minerals in the protolith dominate the compositions of the melt phase. The Gaunt Creek glass is compared with other pseudotachylytes occurring in mylonites in the Alpine Fault zone in Table 1. Pseudotachylytes differ widely in composition, influenced probably in part by the local mineralogical environment where melting took place (a reflection of derivation from both footwall Western Province rocks, and eastern hangingwall Alpine Schists), and the degree of melting achieved. The highly ephemeral nature of earthquake-generated, small volume pseudotachylyte melts make crystal fractionation [*Warr and van der Pluijm*, 2005] a less attractive mechanism of explaining compositional variation than the factors outlined above.

Ultramylonite

Above the green cataclasite horizon the degree of cataclasis decreases, although the overlying ultramylonites, mylonites and protomylonites are cut by cataclastic shears and gouge-filled fault zones throughout the 1 km wide fault zone. Several of the shears and fault planes are marked by narrow (cm wide) bleached zones in the surrounding mylonite that have been chloritized during greenschist facies retrogressive fluid flow. Such fractures are filled with cal-

cite and, in metabasites, with pyrite. Fault zones exhibit combinations of dextral strike slip, reverse, and normal character with single slickenside lineations, in gouge, continuous for up to 1m in length. Faults of normal dip-slip character (Figure 3) probably relate to bending and local extensional collapse at the eastern end of extensive overthrust sheets.

In the dominant quartzofeldspathic lithologies, ultramylonites are dark colored, very fine grained, finely laminated, homogeneous rocks where only sporadic feldspar or garnet porphyroclasts and rare muscovite 'mica-fish' [*Lister and Snoke*, 1984] are recognizable in hand specimen. The dominant component of the rock is a microcrystalline matrix composed of recrystallized quartz, biotite and minor phases. In thin section, plagioclase porphyroclasts show incipient boudinage with extensional fractures infilled with quartz or calcite. Other rock types respond differently to crystal-plastic deformation and recrystallisation. Many amphibolites in this zone are still relatively coarse-grained aggregates of plagioclase and hornblende and have clearly acted more competently to shearing and recrystallisation than the quartzofeldspathic rock types. In the Waikukupa thrust of *Norris and Cooper* [1997] occur the remnants of highly disrupted, intensely strained and sheared granite pegmatites that originate in the Alpine Schists, approximately 75 km to the southwest. Ultramylonite pegmatites pinch and swell around porphyroclasts of feldspar and muscovite whose dimensions are commonly greater than the overall thickness of the pegmatite band along strike [*Norris and Cooper*, 2003]. Matrix to the ovoid feldspar and lenticular muscovite mica-fish comprises a swirling fine-grained aggregate of ribbon quartz.

Mylonite

Further to the east, the ultramylonites grade into, are interbanded with, and are ultimately replaced by mylonites that contain a higher proportion of porphyroclasts of relict schist minerals to recrystallized mylonite grains [*Sibson*, 1977]. The mylonites are characterized by an S-C fabric that consistently indicates a dextral plus top-to-the-northwest sense of shear. The dominant foliation has a mean orientation of 056/33° SE (Figure 8a), effectively identical to the average orientation of the Alpine Fault. The mean lineation direction on the foliation plunges 20° towards 098° (Figure 8b). Shear surfaces are marked by the realignment and partial recrystallisation of mica grains, and the formation of elongate aggregates of ribbon quartz with strongly sutured grain boundaries. Biotite is much less resistant to recrystallisation than muscovite, and is generally totally recrystallized at the western (high-strain) margin of the mylonite zone. In contrast, some of the larger muscovite grains, although marginally recrystallized, survive into the ultramylonite zone.

Protomylonite

The protomylonites grade eastwards into the Alpine Schists. Schist, retaining early (Rangitata?) fold structures, persists as elongate pods up to several meters in length, enveloped by and either truncated or marginally transposed by the mylonitic foliation of the protomylonite. Pods are aligned parallel or subparallel to the mylonitic foliation. Petrographically, the protomylonites are characterized by incipient development of microshears that in terms of geometry resemble those that control the external form of the schist pods. Shears are discontinuous and step-like, eventually developing into a penetrative S-C fabric. Analysis of the orientations of 63 pairs of S and C planes in distal (proto-) mylonites near Franz Josef [*Little et al.*, 2002a] indicates an average mean slip direction of 078° ± 6° (± two standard errors) that is statistically indistinguishable from the Nuvel-1 and G.P.S. plate motion vectors. Strongly undulose extinction in quartz and micas is associated with a subgrain and shape-fabric development that is oblique to the external foliation. Rare intra-folial folds develop. Garnet in places forms winged porphyroclasts or is rotated or sheared

into tablets, which are then stacked into aggregates resembling 'fish'.

Wellman [1955, Figure 7] and *Reed* [1964] refer to some mylonites as 'curly schists', in reference to their uneven, undulating foliation. This texture is common throughout the protomylonite zone, and probably reflects the combination of intersecting foliations in a mesoscopic S-C fabric and the strongly porphyroclastic nature of garnet (and in places feldspar) crystals about which the mylonitic foliation is flattened and draped.

Fault Zone Protolith and Structure of the Fault Zone.

In the south, the Alpine Fault truncates the Dun Mountain Ophiolite Belt, Maitai, and Murihiku terranes and suites of mylonites derived from these terranes occur immediately east of the fault trace [*Campbell*, 2005]. North of the Livingstone Fault, the mylonite sequence is derived from the Alpine Schists whose metamorphic grade ranges from greenschist facies (biotite zone) and greenschist-amphibolite facies transition (garnet zone) at Jackson River [*Ransley*, 1983], in the south, to the dominant amphibolite facies (oligoclase zone) further north (Figure 9). In the Mataketake Range, immediately south of the Moeraki River, the oligoclase zone Alpine Schists have undergone partial melting prior to Alpine Fault inception [*Chamberlain et al.*, 1995, *Batt et al.*, 1999], producing granitic pegmatite dykes and sills. This zone of pegmatites can be traced northwards into the Alpine Fault zone, where it is progressively sheared, mylonitized and translated ~70 km north-eastwards by strike slip displacement, occurring in the ultramylonite zone at the Waikukupa River [*Norris and Cooper*, 2003]. An isolated pegmatite vein occurs a further 22 km further northeast at Gaunt Creek, Waitangi-taona River (Figure 9).

In the Alpine Schists, regionally metamorphosed greenschist facies rocks are characterized by the presence of plagioclase whose composition is close to end-member albite. In the upper biotite zone [*Grapes and Otsuki*, 1983; *Grapes and Watanabe*, 1992], or at the garnet isograd [*Cooper*, 1972], albite is joined by a more calcic plagioclase (typically An_{20-25}, oligoclase), and the two felspars coexist as peristerites throughout the garnet zone. The closure of the peristerite solvus, and the elimination of albite marks the onset of amphibolite facies, oligoclase zone conditions. Towards higher metamorphic grade approaching the Alpine Fault, maximum An-contents remain fairly constant, An_{30-32} [*Grapes and Otsuki*, 1983], or increase, An_{35-40} [*Grapes and Watanabe*, 1994]. *Grapes and Watanabe* [1994] and *Grapes* [1995] record retrograde blebs or patches of albite within oligoclase zone schists, but these apparently disappear within K-feldspar zone schists closer to the Alpine Fault. A pale

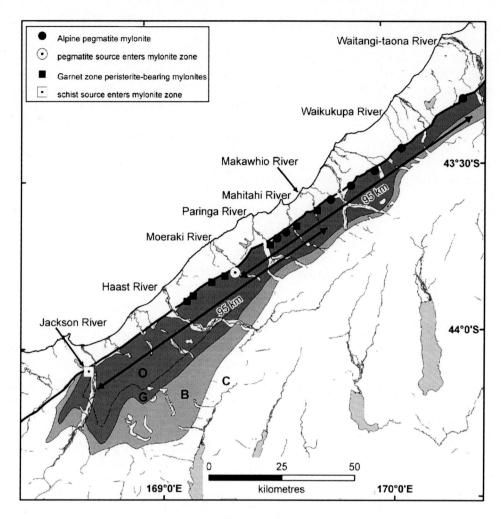

Figure 9. Map of the south Westland section of the Alpine Fault showing localities within the mylonite zone of Alpine pegmatites and anomalously low grade, peristerite-bearing, garnet zone Alpine Schists. Mylonitic strain within a unit increases to the northeast from point where non-mylonitic parent rock strikes into the mylonite zone. O-oligoclase zone, G-garnet zone, B-biotite zone, C-chlorite zone (all within Alpine Schists)

colored, retrogressed quartzofeldspathic mylonite from near the Waitangi-taona River, in which biotite is replaced by chlorite, has oligoclase (An_{19}) apparently replaced by plagioclase ranging in composition from $An_{0.96}$ to $An_{13.3}$ [*Read*, 1994]. In the Fox Glacier area, overlapping with and extending a further 10 km south-west of the area studied by Grapes and co-workers, *Upton* [1995] studied samples across the schist-mylonite transition, and through the mylonite zone to within 350 m of the Alpine Fault trace. She found a rapid transition from greenschist facies albite-bearing schists to oligoclase-bearing schists and ultimately to mylonites. No albite occurred in the mylonites.

In contrast, ultramylonites and mylonites cropping out close to Alpine Fault cataclasites in the area immediately north of Haast River are characterized by the presence of,

and in places the preponderance of, low-Ca albitic plagioclase as porphyroclasts enveloped by the mylonitic matrix. Sporadic oligoclase feldspar occurs in some specimens as a partial rim around albite, and as a corona around epidotegroup inclusions in albite. Both textural features are characteristic of relationships in lower garnet zone Alpine Schists achieved during progressive regional metamorphism, as described from the Haast River section [*Cooper*, 1972]. Albite-bearing mylonites have also been described from the Paringa [*Simpson*, 1992] and Mahitahi Rivers [*Higham*, 1996]. A re-examination of a quartzofeldspathic ultramylonite (OU 69775) collected from the Makawhio River [*McClintock and Cooper*, 2003] was found to contain zoned porphyroclasts of plagioclase enclosed within an anastomosing fine grained mylonitic fabric defined by new dynamically recrystallized biotite. The plagioclase zoning analysed by electron microprobe shows cores of albite ($An_{0.23}Ab_{99.43}Or_{0.33}$ to $An_{0.47}Ab_{99.03}Or_{0.50}$) overgrown by oligoclase ($An_{25.40}Ab_{74.15}Or_{0.50}$ to $An_{19.20}Ab_{80.30}Or_{0.50}$). This peristerite-bearing metamorphic assemblage is characteristic of the prograde metamorphic textures of the Alpine Schist garnet zone, east of the fault-related mylonite zone.

These albite-bearing mylonites occur geographically west of higher-grade oligoclase zone schists, a reversal of the regional trend of increasing metamorphic grade and greatest exhumation adjacent to the Alpine Fault. The garnet zone rocks in the fault zone may represent (1) the western limbs of pre-mylonitic, late-metamorphic folds structurally repeating lower grade rocks that occur further east or (2) low-grade rocks from a protolith that abuts the Alpine Fault zone in the vicinity of Jackson River (Figure 9) physically translated along strike due to the high dextral-reverse shear strain gradient within the mylonite zone. Late-metamorphic structures are known to duplicate metamorphic zonal sequences at Haast River, east of the fault [*Cooper*, 1974], but there is no structural evidence to support post-metamorphic folding in the Paringa-Makawhio area.

The Makawhio River mylonite sequence differs radically from the typically quartzofeldspathic Alpine Schists in that it is dominated by amphibolite, and also contains significant calc-schist and a pod of kyanite-bearing ultramafic schist [*McClintock and Cooper*, 2003]. Metabasite-rich Alpine Schists with garnet zone assemblages, potential sources of the Makawhio sequence, occur in the middle reaches of the Haast River [*Cooper*, 1972], and in the Jackson River area further to the southwest [*Ransley*, 1983]. If the Makawhio garnet zone rocks were derived from Haast River metabasites, one would expect them be have been replaced by oligoclase zone assemblages as they were traced along strike into the higher grade rocks adjacent to the Alpine Fault at Makawhio River. Given that the Alpine pegmatites have an established mylonite shear zone length of 90-100 km, the most likely interpretation for the peristerite-bearing, garnet zone, metabasite-rich sequence is that it has been sheared approximately 95 km within the Alpine Fault mylonite zone from a source in the Jackson River area (Figure 9).

Other distinctive rock types in the mylonite zone include serpentinite at Waitaha River [*Wright*, 1998] (which may be far-travelled from the Dun Mountain Ophiolite Belt, or of more local derivation from the Pounamu Ultramafic Belt), and a distinctive marble unit cropping out between the Mikonui and Waitaha rivers [*Rattenbury*, 1987; *Wright*, 1998], for which there is no unequivocal Alpine Schist equivalent. Amphibolites are well represented in many mylonite sequences, but their mineralogy and geochemical characteristics are not sufficiently distinctive for them to be used as marker horizons.

Strain Within the Mylonite Zone

Within the mylonites, strain intensity as suggested by intensity of foliation and grain-size reduction increases towards the surface trace of the Alpine Fault. Parallelism of structural features with the foliation and destruction of the original schist fabric indicate high strains within the mylonites and ultramylonites. Quantifying shear strain within mylonites is difficult because most potential strain markers have been destroyed by the high strain and recrystallization. *Norris and Cooper* [2003] measured thicknesses of the pegmatite veins referred to above and compared the thickness distributions of pegmatites within the ultramylonites, mylonites and protomylonites with their unmylonitized counterparts within the hangingwall schists. Depending on the amount of pure shear component assumed, shear strains ranged from 12–22 in the protomylonites, 120–200 in the mylonites, and 180–300 in the ultramylonites. These represent very high strains and some of the highest measured shear strains reported in the literature.

Shear sense indicators on both mesoscopic and microscopic scales consistently record dextral-reverse shear in accordance with current displacement on the fault. There is no indication of a reversal of shear sense within the mylonites as would be expected if they are being actively extruded from a low-viscosity channel, as suggested by *Walcott* [1998] and as proposed for the higher Himalaya [e.g., *Grujic et al., 2002; Searle and Szulc, 2005*]. Preliminary quartz fabric data [M. Walrond, V. Toy, pers. comm., 2005] is consistent with high simple shear strains within the mylonite zone. Strain indicators of late Cenozoic age within the hangingwall Alpine Schists, on the other hand, represent more of a horizontal contraction or transpression [*Little et al.*, 2002a; *Holm et al.*, 1989]. Thus the mylonite zone appears to represent localization of intense non-coaxial shear below the Southern Alps [cf. Little et al., this volume].

Exhumation of the mylonites must have taken place within the last 5–6 Myr according to most plate tectonic models [*Sutherland*, 1995]. During this time, motion on the plate boundary became more convergent as reflected by the oblique stretching fabrics within the mylonites. An exponential function may be fitted to the shear strain estimates across the mylonite zone [*Norris and Cooper*, 2003] to produce an integrated displacement of 55–60 km (Figure 10). At an average slip rate of 24 mm/yr (see earlier discussion), a total of 120 km of displacement would have occurred over the last 5 Myr. Since the brittle–ductile transition is at a depth of around 10 km [*Leitner et al.*, 2001], only two-thirds of this movement would be recorded as ductile shear within the mylonites. In addition, the ultramylonites are exposed adjacent to the present fault trace. It is likely that at least a portion of the mylonite zone, and possibly an equal thickness to that exposed, is still buried

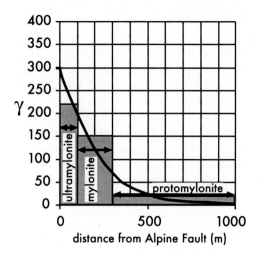

Figure 10. Strain estimates within mylonites from *Norris and Cooper* [2003] plotted against distance from the present fault trace. An exponential curve is fitted to the data and integrated to calculate the total displacement across the mylonite zone (see text for discussion).

on the footwall side of the trace. Thus the amount of ductile shear displacement expected across the currently exposed mylonite zone if all the fault offset were accommodated at depth by ductile shear within a 1–2 km wide mylonite zone would be somewhere between 40 and 80 km, consistent with that calculated from the strain estimates in Figure 10. This is also consistent with the lateral distribution of horizons within the mylonites relative to their protoliths in the hangingwall Alpine Schists (see earlier discussion).

A displacement of 60 km across a 1 km wide shear zone corresponds to an average shear strain of 60. Over ca. 5 Myr, this translates to an average shear strain rate of ca. 4.10^{-13}/s. For the ultramylonites, an average strain of 250 over the same period results in a strain rate of ca. 2.10^{-12}/s. The average strain rate is an order of magnitude higher than that commonly assumed for geological deformation [e.g., *Pfiffner and Ramsay*, 1982] while the strain rate in the ultramylonites is two orders of magnitude higher.

Timing of Metamorphism and Mylonitisation

In the Otago Schist, recent compilations of Ar-Ar ages [*Gray and Foster*, 2004] suggest the main metamorphism and deformation on the low grade flanks of the schist belt occurred in the Late Jurassic, between 160 and 140 Ma, with exhumation and cooling of the high grade core between 109 and 100 Ma. This contrasts with arguments by *Adams et al.* [1985] for Early Jurassic and *Little et al.* [1999] for Middle Jurassic ages of metamorphism.

In the Alpine Schists, most dating has been determined by K-Ar or Ar-Ar techniques [*Adams*, 1979; *Adams and Gabites*, 1985; *Chamberlain et al.*, 1995] but the ages are strongly discordant due to recent (< 7 Ma) [*Tippett and Kamp*, 1993] (< 5Ma) [*Sutherland*, 1995, 1996; *Batt et al.*, 1999] and rapid (*Wellman*, 1979; *Bull and Cooper*, 1986; *Norris and Cooper*, 2001] uplift through argon-retention temperatures following prolonged burial at various depths. As a consequence of release and redistribution of argon, there is a spectrum of very young and very old ages, none of which necessarily relates to the timing of metamorphism.

Early Rb-Sr dating by Grapes and Watanabe, summarized in *Vry et al.* [2004], suggests a late Cretaceous phase of metamorphism, with ages of 86 Ma, and an unexplained date of 150 ± 8.5 Ma. Anatectic pegmatites in the Alpine Schists of the Mataketake Range, dated by U-Th-Pb techniques at 68–80 Ma (*Chamberlain et al.*, 1995; *Batt et al.*, 1999], intrude the Alpine Schists as both concordant sills and as dykes discordant to the foliation [*Norris and Cooper*, 2003]. Schists that host the pegmatites contain sporadic monazite, which was dated by *Mortimer and Cooper* [2004] at 71 Ma, an inferred metamorphic recrystallisation within the range of intrusion ages of the pegmatites.

Recent application of Nd-Sm, and Lu-Hf dating of garnet in schists has given a range of ages that cluster between 68 and 86 Ma for metamorphic garnet growth [*Vry et al.*, 2004]. Very young ages (with very large errors) were determined for syn- to post-mylonitic recrystallisation (12 ± 37 Ma) [*Vry et al.*, 2004] with the latest zones in garnet porphyroclast growth that overprints the mylonitic foliation, inferred to have occurred 6 ± 2 Ma.

Pseudotachylytes have been dated in a number of studies. *Adams* [1981] determined the K-Ar age of pseudotachylyte from Harold Creek at 9.8 Ma, and *Seward and Sibson* [1985] determined a fission track age of 0.43 ± 0.17 Ma from schist-derived pseudotachylyte from northeast of Haast River. *Warr et al.* [2003] determined the ^{40}Ar/^{39}Ar total gas ages along a transect through a pseudotachylyte vein, supplied from a fallen block in the Wanganui River. Ages of 11.92 ± 0.030 and 13.260 ± 0.040 Ma were determined from the host schist, 6 cm from the vein wall. The schist wall of the vein gave ages of 7.186 ± 0.020 to 11.473 ± 0.027 Ma, and the centre of the pseudotachylyte vein returned 1.11 ±0.04 Ma for the earthquake that is inferred to have triggered the melting event. *Warr et al.* [2003] argue that, based on current exhumation rates, the melting would have occurred at a crustal depth of 11 km. This depth would have been close to the brittle-ductile transition [*Leitner et al.*, 2001] if the age determination and uplift rate assumptions are correct.

Radiometric dating reinforces the concept of polyphase metamorphic, mylonitic and cataclastic recrystallisation inferred from petrographic and structural observations.

P-T Conditions of Metamorphism and Mylonitisation

Kyanite-talc-zoisite mylonites, associated with the Makawhio River metabasites, equilibrated at 10 kb, 650-700°C [*Cooper*, 1980]. Allowing for 11-15 km of post-metamorphic, but pre-Alpine Fault uplift, based on the presence of prehnite-pumpellyite facies greywacke detritus in West Coast Tertiary basins, the balance of the exhumation, approximately 25 km, was ascribed to the oblique convergent phase of Alpine Fault movement [*Cooper*, 1980].

On the basis of fission track ages, *Tippett and Kamp* [1993] suggested a continental crustal section at least 19 km thick has been uplifted on the Alpine Fault. The amount of Mesozoic uplift ranged from minimal amounts north of Arthur's Pass, to approximately 3 km near Mount Cook, and 10 km in the south.

Using geothermobarometric analysis of the schist assemblages, *Grapes and Watanabe* [1994] determined the maximum pressure of oligoclase zone metamorphism as 9.5 kb at 615°C, and deduced a very similar exhumation history to that of *Cooper* [1980], with 11 to 13 km of Mesozoic and 19 to 22 km of Cenozoic uplift in the central section of the Alpine Fault.

Vry et al. [2004] analysed the outermost, Ca-enriched rim of zoned garnet in mylonite from the Taramakau River, 35 km east of Hokitika, and on the basis of geothermobarometry of the associated assemblage determined a range of P-T conditions from 580°C/10.1 kb to 620°C/11.4 kb. Based on petrographic relationships, with garnet overprinting the mylonitic foliation, this would appear to be the most robust estimate for conditions of mylonitisation in the northern section of the fault, and indicates approximately 41 km of exhumation occurred in the past 5 Myr. However, present day crustal thicknesses determined by the SIGHT seismic survey for both the northern transect 1 [Figure 13 of *van Avendonk et al.*, 2004] and the southern transect 2 [Figure 4 of *Scherwath et al.*, 2003] across the Alpine fault zone show only 32-35 km of schist overlying oceanic crust beneath the Southern Alps. Thus either the interpretation of the garnet texture as post-mylonitic, or the results of the geobarometric measurements, appear at odds with the possible depth of exhumation of the mylonites.

The rapid recent uplift of rocks east of the Alpine Fault has led to advection of hot rocks with a resulting increased thermal gradient in the top several kilometers of crust, as evidenced by hot springs [*Allis et al.*, 1979], bore-hole measurements [*Shi et al.*, 1996], fluid inclusion data [*Holm et al.*, 1989; *Craw*, 1997] and thermal modeling [*Koons*, 1987; *Shi et al.*, 1996]. The fluid inclusion data allow determination of a thermal exhumation path of rocks now at the surface, rather than an instantaneous thermal gradient. They indicate thermal gradients of 70-95°C/km in the top 2-3 km of the crust. These estimates are comparable with or higher than

those from a borehole near Franz Josef glacier of 90°C/km uncorrected, 60 ± 15°C/km after correction for topography [*Shi et al.*, 1996], but are consistent with the modeling of *Koons* [1987]. *Shi et al.* [1996], on the other hand, model lower gradients due to cooling by downward advection of crust into the root. All models agree that the thermal gradient decreases with depth. The fluid inclusion data of *Holm et al.* [1989] and *Craw* [1997] predict average temperature gradients of 70-95°C/km in the top 3 km, reducing to 35-45°C/km over the top 8 km, and to around 30-40°C/km over the top 10 km. *Holm et al.* [1989] suggest an average gradient of ca. 20°C/km down to >20 km depth. From a starting temperature of around 500°C at 25 km, the rocks cool slowly over the initial 15 km of uplift and exhumation at a rate of around 10°C/km and then begin to cool more rapidly as they approach the surface. In the *Shi et al.* [1996] model, the rate of cooling at depth during exhumation is more rapid and the final near-surface cooling somewhat less than indicated by fluid inclusion data, but the general pattern of thermal history is similar. The thermal exhumation models all show that the mylonites are likely to have remained at fairly high temperatures during the early stages of uplift and so were likely to continue to deform under amphibolite facies temperatures and falling pressures as exhumation took place.

SUMMARY

- The Alpine Fault forms a simple straight trace on a South Island map, striking ca. 055° and dipping moderately to steeply east. In detail, however, the trace is more complex, exhibiting a serially partitioned pattern in the central section on a scale of a few kilometers, and a relatively straight, steeply dipping, parallel partitioned section south of Haast.
- Average Late Quaternary strike-slip rates are 23–25 mm/yr, with the possibility of a small reduction in rate from north to south. Dip-slip rates reach a maximum of ca. 10 mm/yr in the central part, reducing north and south and reaching zero south of Jackson Bay. These changes in dip-slip are mirrored by the differences in surface trace geometry and probably reflect the transition to ocean-continent interaction from the central continent-continent oblique collision.
- Due to the oblique-slip component, mylonites are exhumed from ca. 25-30 km depth in a ca. 1 km wide zone immediately east of the present trace. Cataclastic deformation is intense in the first 50 m or so, with thick cataclasite and gouge zones. Pseudotachylytes occur in many locations and represent friction melting during seismic slip, mainly on small faults adjacent to the main trace. Most appear to be generated by preferential melting of phyllosilicates.

- Ductile shear strain in the mylonites is high, reaching 200-300 in the ultramylonites, and is consistent with the bulk of the surface offset being accommodated by ductile shear in a 1-2 km wide mylonite zone at depth.
- Displacement of metamorphic zones and a pegmatite swarm, together with other lithological units, with respect to their location in the hanging wall is also consistent with ca. 100 km of dextral ductile shear within the mylonites since ca. 5 Ma.

Acknowledgments. We thank the many students who have worked on the Alpine Fault over the last 20 years, in particular Heather Campbell, Carl Hanson, Caroline Higham, Daniel Holm, Murray McClintock, Stephen Read, Guy Simpson, Rupert Sutherland, Virginia Toy, Mark Walrond, Craig Wright and Trevor Wright. We also thank colleagues for their collaboration in Alpine Fault research and their ongoing discussion, especially Kelvin Berryman, Bill Bull, Dave Craw, Peter Koons, Tim Little, Dave Prior, Rick Sibson, Rupert Sutherland and Phaedra Upton. Permits for work on the fault were provided by the NZ Department of Conservation. Research was funded by the New Zealand Public Good Science Fund, the New Zealand Earthquake Commission Research Committee, and the Otago University Research Committee.

REFERENCES

Adams, C.J.D. (1979), Age and origin of the Southern Alps, In: *The Origin of the Southern Alps: Bull. Roy. Soc. N. Z.,* 18, edited by R.I. Walcott and M.M. Cresswell, pp. 73–77, Roy. Soc. N.Z. Wellington.

Adams, C.J. (1981), Uplift rates and thermal structure in the Alpine Fault zone and Alpine Schists, Southern Alps, New Zealand, *Geol. Soc. London, Special Publication 9,* 211–222.

Adams, C.J., and J.E. Gabites (1985), Age of metamorphism and uplift in the Haast Schist Group at Haaast Pass, Lake Wanaka and Lake Hawea, South Island, New Zealand, *N. Z. J. Geol. Geophys., 28,* 85–96.

Adams, C.J., D.G. Bishop, and J.E. Gabites (1985), Potassium-argon age studies of low-grade, progressively metamorphosed greywacke sequence, Dansey Pass, South Island, New Zealand, *Geol. Soc. London J., 142,* 339–349.

Allis, R.G., R.W. Henley and A.F. Carman (1979), The thermal regime beneath the Southern Alps, In: *The Origin of the Southern Alps. Bull. Roy. Soc. N. Z.,* 18, edited by R.I. Walcott and M.M. Cresswell, pp. 55–65, Roy. Soc. N.Z. Wellington.

Barnes, P.M., R. Sutherland, and J. Delteil (2005), Strike-slip structure and sedimentary basins of the southern Alpine Fault, Fiordland, New Zealand, *Geol. Soc. Am. Bull., 117,* 411–435.

Batt, G.E., B.P., Kohn, J. Braun, I. McDougall, and T.R. Ireland (1999), New insight into the dynamic development of the Southern Alps, New Zealand, from detailed thermochronological investigation of the Mataketake Range, In: *Exhumation Processes:*

normal faulting, ductile flow and erosion, Geol. Soc. London, Spec. Pub., 154, Edited by U. Ring, M. Brandon, G. Lister, S. Willett, pp. 261–282, Geol. Soc. London, London.

Bell, J.M., and C. Fraser (1906), The geology of the Hokitika Sheet, north Westland quadrangle, *N. Z. Geol. Surv. Bull.,* 1101 pp., N. Z. Geol. Surv. Wellington

Benson, W.N. (1952), Meeting of the Geological Division of the Pacific Science Congress in New Zealand, February 1949, *Interim Proc. Geol. Soc. Am., 1950,* 11–13.

Berryman, K.R., S. Beanland, A.F. Cooper, H.N. Cutten, R.J. Norris, and P.R. Wood (1992), The Alpine Fault, New Zealand: variation in Quaternary tectonic style and geomorphic expression, *Ann. Tect., VI,* 126–163.

Bossière, G. (1991), Petrology of pseudotachylytes from the Alpine Fault of New Zealand, *Tectonophysics, 196,* 173–193.

Bowen, F.E. (1954), Late Pleistocene and Recent vertical movement at the Alpine Fault, *N. Z. J. Sci. Tech., B35,* 390–397.

Bull, W.B., and A.F. Cooper (1986), Uplifted marine terraces along the Alpine fault, New Zealand, *Science, 234,* 1225–1228.

Campbell, H. (2002), Strain partitioning on the Alpine Fault, South Westland, *Geol. Soc. N. Z. Misc. Pub. 110A,* p18.

Campbell, H. (2005), Partitioning of plate boundary deformation in South Westland, New Zealand: controls from reactivated structures, PhD thesis, Univ. of Otago, Dunedin, N. Z.

Carter, R.M., and R.J. Norris (1976), Cainozoic history of southern New Zealand: an accord between geological observations and plate tectonic predictions, *Earth Planet. Sci. Lett., 31,* 85–94.

Chamberlain, C.P., P.K. Zeitler, and A.F. Cooper (1995), Geochronologic constraints of the uplift and metamorphism along the Alpine Fault, *N. Z. J. Geol. Geophys., 38,* 515–523.

Cooper, A.F. (1972), Progressive metamorphism of metabasic rocks from the Haast Schist Group of southern New Zealand, *J. Petrol., 13,* 457–492.

Cooper A.F. (1974), Multiphase deformation and its relationship to metamorphic crystallisation at Haast River, South Westland, New Zealand, *N. Z. J. Geol. Geophys., 14,* 855–880.

Cooper, A.F. (1980), Retrograde alteration of chromian kyanite in metachert and amphibolite whiteschist from the Southern Alps, New Zealand, with implications for uplift on the Alpine Fault, *Contrib. Mineral. Petrol., 75,* 153–164.

Cooper, A.F. and R.J. Norris (1990), Estimates for the timing of the last coseismic displacement on the Alpine fault, northern Fiordland, *New Zealand, N. Z. J. Geol. Geophys., 33,* 303–308.

Cooper, A.F., and R.J. Norris (1994), Anatomy, structural evolution and slip rate of a plate-boundary thrust: The Alpine fault at Gaunt Creek, Westland, New Zealand, *Geol. Soc. Am. Bull., 106,* 627–633.

Cooper, A.F., and R.J. Norris (1995), Displacement on the Alpine Fault at Haast River, South Westland, New Zealand, *N. Z. J. Geol. Geophys., 38,* 509–514.

Cooper, A.F., B.A. Barreiro, D.L. Kimbrough, and J.M. Mattinson (1987), Lamprophyre dyke intrusion and the age of the Alpine Fault, New Zealand, *Geology, 15,* 941–944.

Cowan, H.A., and M.S. McGlone (1991), Late Holocene displacements and characteristic earthquakes on the Hope River segment of the Hope Fault, New Zealand, *J. Roy. Soc. N. Z., 21,* 373–384.

Craw, D. (1997), Fluid inclusion evidence for geothermal structure beneath the Southern Alps, *N. Z. J. Geol. Geophys., 40,* 43–52.

Dahlen, F. A. (1984), Non-cohesive critical Coulomb wedges: an exact solution, *J. Geophys. Res., 89,* 10125–10133.

DeMets, C., R.G. Gordon, D.F. Argus, and S. Stein (1994), Effect of recent revisions to the geomagnetic reversal time scale on estimates of current plate motions, *Geophys. Res. Lett., 21,* 2191–2194.

Ellis, S., S. Wissing, and A. Pfiffner (2001), Strain localisation as a key to reconciling experimentally derived flow-law data with dynamic models of continental collision, *Int. J. Earth Sci., 90,* 168–180.

Grapes, R.H. (1995), Uplift and exhumation of Alpine Schist, Southern Alps, New Zealand: thermobarometric constraints, *N. Z. J. Geol. Geophys., 38,* 525–533.

Grapes, R.H., and M. Otsuki (1983), Peristerite compositions in quartzofeldspathic schists, Franz Josef-Fox Glacier area, New Zealand, *J. Metamorph. Geol., 1,* 47–62.

Grapes, R.H., and T. Watanabe, (1992), Metamorphism and uplift of Alpine Schist in the Franz Josef-Fox Glacier area of the Southern Alps, New Zealand. *J. Metamorph. Geol., 10,* 171–180.

Grapes, R.H., and T. Watanabe (1994), Mineral composition variation in Alpine Schist, Southern Alps, New Zealand: Implications for recrystallisation and exhumation, *Island Arc, 3,* 163–181.

Gray, D.R., and D.A. Foster (2004), $^{40}Ar/^{39}Ar$ thermochronologic constraints on deformation, metamorphism, and cooling/exhumation of a Mesozoic accretionary wedge, Otago Schist, New Zealand, *Tectonophysics, 385,* 181–210.

Grindley, G.W. (1963), Structure of the Alpine schists of south Westland, New Zealand, *N. Z. J. Geol. Geophys., 6,* 872–930.

Hanson, C.R., R.J. Norris, and A.F. Cooper (1990), Regional fracture patterns east of the Alpine Fault between the Fox and Franz Josef Glaciers, Westland, New Zealand, *N. Z. J. Geol. Geophys., 33,* 617–622.

Henderson, J. (1937), The West Nelson Earthquakes of 1929 (with notes on the geological structure of west Nelson), *N. Z. J. Sci. Tech., 19,* 65–144.

Higham, C.H.R. (1996), A structural and mineralogical study of the Alpine Fault zone, and an investigation into the post-glacial environment, BSc Hons. dissertation, Univ. of Otago, Dunedin, N. Z.

Holm, D.K., R.J. Norris, and D. Craw (1989), Brittle/ductile deformation in a zone of rapid uplift: central Southern Alps, New Zealand, *Tectonics, 8,* 153–168.

Howard, M., A. Nicol, J. Campbell, and J.R. Pettinga (2005), Holocene earthquakes on the strike-slip Porters Pass Fault, Canterbury, New Zealand, *N. Z. J. Geol. Geophys., 48*(1), 59–74.

Kamp, P.J.J. (1986), The mid-Cenozoic Challenger rift system of western New Zealand and its implication for the age of Alpine Fault inception, *Geol. Soc. Am. Bull., 97,* 255–281.

Koons, P.O. (1987), Some thermal and mechanical consequences of rapid uplift: an example from the Southern Alps, New Zealand. *Earth Planet. Sci. Lett., 86,* 307–319.

Koons, P.O., R.J. Norris, D. Craw, and A.F. Cooper (2003), Influence of exhumation on the structural evolution of transpressional plate boundaries: An example from the Southern Alps, New Zealand, *Geology, 31,* 3–6.

Langridge, R., J. Campbell, N. Hill, V. Pere, J. Pope, J.R. Pettinga, B. Estrada, and K. Berryman (2003), Paleoseismology and slip rate of the Conway segment of the Hope Fault at Greenburn Stream, South Island, New Zealand, *Ann. Geophys., 46*(5),1119–1139.

Leitner, B., D. Eberhart-Philips, H. Anderson, and J.N. Nabelek (2001), A focused look at the Alpine Fault, New Zealand: seismicity, focal mechanisms and stress inversions, *J. Geophys. Res., 106,* 2193–2220.

Lister, G.S., and A.W. Snoke (1984), S-C mylonites, *J. Struct. Geol., 6,* 617–638.

Litchfield, N.J., J. Campbell, and A. Nicol (2003), Recognition of active reverse faults and folds in North Canterbury, New Zealand, using structural mapping and geomorphic analysis, *N. Z. J. Geol. Geophys., 46*(4), 563–579.

Little, T.A., N. Mortimer, and M. McWilliams (1999), An episodic Cretaceous cooling model for the Otago-Marlborough Schist, New Zealand, based on $^{40}Ar/^{39}Ar$ white mica ages, *N. Z. J. Geol. Geophys., 42,* 305–325.

Little, T.A., R.J. Holcombe, and B.R. Ilg (2002a), Kinematics of oblique collision and ramping inferred from microstructures and strain in middle crustal rocks, central Southern Alps, New Zealand, *J. Struct. Geol., 24,* 219–239.

Little, T.A., M.K. Savage, and B. Tikoff (2002b), Relationship between crustal finite strain and seismic anisotropy in the mantle, Pacific-Australia plate boundary zone, South Island, New Zealand, *Geophys. J. Int., 151,* 106–116.

Little, T.A., S. Cox, J.K. Vry, and G.E. Batt (2005), Variations in exhumation level and uplift rate related to oblique-slip ramp geometry, Alpine Fault, central Southern Alps, New Zealand, *Geol. Soc. Am. Bull., 117*(4) ,707–723.

McCaffrey, R. (1992), Oblique plate convergence, slip vectors, and forearc deformation, *J. Geophys. Res., 97,* 8905–8915.

McClintock, M.K. and A.F. Cooper (2003), Geochemistry, mineralogy, and metamorphic history of kyanite-orthoamphibole-bearing Alpine Fault mylonite, South Westland, New Zealand, *N. Z. J. Geol. Geophys., 46,* 47–62.

Morgan, P.G. (1908), The geology of the Mikonui subdivision, North Westland, *N. Z. Geol. Surv. Bull., 6,* 175 pp., N. Z. Geol. Surv., Wellington.

Mortimer, N. and A. F. Cooper (2004), U-Pb and Sm-Nd ages from the Alpine Schist, *N. Z. J. Geol. Geophys.,* 47, 21–28, 2004.

Norris, R.J. (1979). A geometrical study of finite strain and bending in the South Island, in: *The origin of the Southern Alps. Wellington, New Zealand: Roy. Soc. N. Z. Bull.,* vol. 18, edited by R.I. Walcott, and M.M. Cresswell, pp. 21–28, Roy. Soc. N. Z., Wellington.

Norris, R.J., and A.F. Cooper (1995), Origin of small-scale segmentation and transpressional thrusting along the Alpine fault, New Zealand, *Geol. Soc Am. Bull., 107,* 231–240.

Norris, R.J., and A.F. Cooper (1997), Erosional control on the structural evolution of a transpressional thrust complex on the Alpine Fault, New Zealand, *J. Struct. Geol., 19,* 1323–1342.

Norris, R.J. and A.F. Cooper (2001), Late Quaternary slip rates and slip partitioning on the Alpine Fault, New Zealand, *J. Struct. Geol., 23,* 507–520.

Norris, R.J. and A.F. Cooper (2003) Very high strains recorded in mylonites along the Alpine Fault, New Zealand: implications for the deep structure of plate boundary faults, *J. Struct. Geol., 25,* 2141–2157.

Norris, R.J., and R. Nicolls (2004), Strain accumulation and episodicity of fault movements in Otago, 1 ed. Wellington, New Zealand, *EQC Res. Rep. 01/445,* N. Z. Earthquake Commission, Wellington, N. Z.

Norris, R.J., P.O. Koons, and A.F. Cooper (1990), The obliquely-convergent plate boundary in the South Island of New Zealand: implications for ancient collision zones, *J. Struct. Geol., 12,* 715–725.

Pfiffner, O.A., and J.G. Ramsay (1982), Constraints on geological strain rates: arguments from finite strain states of deformed rocks. *J. Geophys. Res., 87,* 311–321.

Ransley, J.F. (1983), The geology of the Jackson River catchment, BSc Hons. dissertation, Univ. of Otago, Dunedin, N. Z.

Rattenbury, M.S. (1987), Timing of mylonitisation west of the Alpine Fault, central Westland, New Zealand, *N. Z. J. Geol. Geophys., 30,* 287–297.

Rattenbury, M.S. (1991), The Fraser Complex: high-grade metamorphic, igneous and mylonitic rocks in central Westland, New Zealand, *N. Z. J. Geol. Geophys., 34,* 23–33.

Read, S.E. (1994), Alpine Fault segmentation and range front structure between Gaunt Creek and Little Man River, near Whataroa, central Westland, New Zealand, BSc Hons. dissertation, Univ. of Otago at Dunedin, N. Z.

Reed, J.J. (1964), Mylonites, cataclasites, and associated rocks along the Alpine Fault, South Island, New Zealand, *N. Z. J. Geol. Geophys., 7,* 645–684.

Scherwath, M., T. Stern, F. Davey, D. Okaya, W.S. Holbrook, R. Davies, and S. Kleffmann (2003), Lithospheric structure across oblique collision in New Zealand from wide-angle P wave modelling, *J. Geophys. Res., 108*(B12), 2566, doi:10.1029/2002JB002286.

Searle, M. P. and A. G. Szulc (2005), Channel flow and ductile extrusion of the high Himalayan slab – the Kangchenjunga – Darjeeling profile, Sikkim Himalaya. *J. Asian Earth Sci., 25,* 173–185.

Seward, D., and R.H. Sibson (1985), Fission-track age for a pseudotachylyte from the Alpine Fault zone, *N. Z. J. Geol. Geophys., 28,* 553–557.

Shi, Y.L., R.G. Allis and F. Davey (1996), Thermal modeling of the Southern Alps, New Zealand, *PAGEOPH, 146,* 469–501.

Sibson, R.H. (1977), Fault rocks and fault mechanisms, *Geol. Soc. London J., 133,* 191–213.

Sibson, R.H., S.H. White, and B.K. Atkinson (1979), Fault rock distribution and structure within the Alpine Fault Zone: a preliminary account, In: *The Origin of the Southern Alps. Bull. Roy. Soc. N. Z.,* 18, edited by R.I. Walcott and M.M. Cresswell, pp. 55–65, Roy. Soc. N.Z. Wellington.

Simpson, G.D.H. (1992), Quaternary evolution of the Alpine Fault zone and a mineralogical/microstructural study of the schist mylonite transition, BSc Hons. Dissertation, Univ. of Otago, Dunedin, N. Z.

Simpson, G.D.H., A.F. Cooper, and R.J. Norris (1994), Late Quaternary evolution of the Alpine fault zone at Paringa, South Westland, *N. Z. J. Geol. Geophys., 37,* 49–58.

Suggate, R.P. (1963), The Alpine Fault, *Trans. Roy, Soc, N. Z., 2,* 105–129.

Suggate, R. P. (1968), The Paringa Formation, Westland, New Zealand, *N. Z. J. Geol. Geophys., 11,* 345–355.

Sutherland, R. (1994), Displacement since the Pliocene along the southern section of the Alpine fault, New Zealand, *Geology, 22,* 327–331.

Sutherland, R. (1995), The Australia-Pacific boundary and Cenozoic plate motions in the SW Pacific: some constraints from the Geosat data, *Tectonics, 14,* 819–831.

Sutherland, R. (1996), Transpressional development of the Australia-Pacific boundary through southern New Zealand; constraints from Miocene-Pliocene sediments, Waiho-1 borehole, South Westland, *N. Z. J. Geol. Geophys., 39,* 251–264.

Sutherland, R., and R.J. Norris (1995), Late Quaternary displacement rate, paleoseismicity, and geomorphic evolution of the Alpine Fault: evidence from Hokuri Creek, South Westland, New Zealand, *N. Z. J. Geol. Geophys., 38,* 419–430.

Sutherland, R., K. R. Berryman and R. J. Norris (2006), Quaternary slip rate and geomorphology of the Alpine fault: implications for kinematics and seismic hazard in southwest New Zealand, *Geol. Soc. Am. Bull. 118,* 464–474.

Tippett, J.M., and P.J.J. Kamp, (1993), Fission track analysis of the Late Cenozoic vertical kinematics of continental Pacific crust, South Island, New Zealand, *J. Geophys. Res., 98,* 16119–16148.

Upton, P. (1995), Mechanics, reaction and fluid flow associated with continental collision along the Alpine Fault, Southern Alps, New Zealand, PhD thesis, Univ. of Otago, Dunedin, N. Z.

van Avendonk, H.J.A., W.S. Holbrook, D. Okaya, J.K. Austin, F. Davey, and T. Stern (2004), Continental crust under compression: A seismic refraction study of South Island Geophysical Transect, South Island, New Zealand, *J. Geophys. Res., 109,* B06302, doi:10.1029/2003JB002790.

Vry, J.K., J. Baker, R. Maas, T.A. Little, R. Grapes, and M. Dixon (2004), Zoned (Cretaceous and Cenozoic) garnet and the timing of high grade metamorphism, Southern Alps, New Zealand, *J. Metamorph. Geol., 22,* 137–157.

Walcott, R. I. (1998), Modes of oblique compression: late Cenozoic tectonics of the South Island, New Zealand, *Rev. Geophys., 36,* 1–26.

Wallace, R.C. (1976), Partial fusion along the Alpine Fault zone, New Zealand, *Geol. Soc. Am. Bull., 87,* 1225–1228.

Warr, L.N., and S. Cox (2001), Clay mineral transformations and weakening mechanisms along the Alpine Fault, New Zealand, In: *The nature and tectonic significance of fault zone weakening, Geol. Soc. London Spec. Pub.,* 186, edited by R.E. Holdsworth, R.A. Strachan, J.F. Magloughlin, and R.J. Knipe, pp. 85–101, Geol. Soc. London, London.

Warr, L.N., and B.A. van der Pluijm (2005), Crystal fractionation in the friction melts of seismic faults (Alpine Fault, New Zealand), *Tectonophysics, 402,* 111–124.

Warr, L.N., B.A. van der Pluijm, D.R. Peacor, and C.M. Hall (2003), Frictional melt pulses during a ~1.1 Ma earthquake along the Alpine Fault, New Zealand, *Earth Planet. Sci. Lett., 209*, 39–52.

Wellman, H.W. (1953), Data for the study of Recent and late Pleistocene faulting in the South Island of New Zealand, *N. Z. J. Sci. Tech., 34B*, 270–288.

Wellman, H.W. (1955), The geology between Bruce Bay and Haast River, South Westland, *N. Z. Geol. Surv. Bull., 48*, 46 pp., N. Z. Geol. Surv. Wellington.

Wellman, H.W. (1979), An uplift map for the South Island of New Zealand, and a model for uplift of the Southern Alps, In: *The Origin of the Southern Alps, Bull. Roy. Soc. N. Z.*, 18, edited by R.I. Walcott and M.M. Cresswell, pp. 13–20, Roy. Soc. N. Z., Wellington.

Wellman, H.W. (1979), An uplift map for the South Island of New Zealand, and a model for uplift of the Southern Alps, *Roy. Soc. N. Z. Bull., 18*, 13–20.

Wellman, H.W., and R.W. Willett (1942), The geology of the West Coast from Abut Head to Milford Sound – Part I, *Trans. Roy. Soc. N. Z., 71*, 282–306.

Wentworth, C.M., and M.D. Zoback (1989), The style of late Cenozoic deformation at the eastern front of the California Coast Ranges, *Tectonics, 8*(2), 237–246.

Wright, C.A. (1998), Geology and paleoseismology of the Central Alpine Fault, New Zealand, MSc thesis, Univ. of Otago at Dunedin, N. Z.

Wright, T.D. (2000), Paleoseismicity and structure of the Alpine Fault between Haast River and Arawhata River, MSc thesis, Univ. of Otago, Dunedin, N.Z.

Yetton, M. D., and D.C. Nobes (1998), Recent vertical offset and near-surface structure of the Alpine Fault in Westland, New Zealand, from ground penetrating radar profiling, *N. Z. J. Geol. Geophys, 41*, 485–492.

Richard J Norris, and Alan F. Cooper, Department of Geology, University of Otago, Dunedin, New Zealand (richard.norris@stonebow.otago.ac.nz)

Deformation of the Pacific Plate Above the Alpine Fault Ramp and its Relationship to Expulsion of Metamorphic Fluids: an Array of Backshears

Ruth H. Wightman[1,2] and Timothy A. Little[1]

A ~2 km-wide array of near-vertical backshears in the central Southern Alps, New Zealand, is interpreted to have slipped in an escalator-like way to up-ramp the Pacific Plate onto the Alpine Fault ramp, and to play an important role in channelling metamorphic fluids upward through this active orogen. The oblique-slip backshears formed in the lower crust, are evenly spaced (~30 cm), and have an average offset of 14 cm that is brittle to ductile and extend over 500 m in vertical length. Cumulative vertical displacements suggest that the causative ramp-step in the Alpine Fault at depth had an angle of $22 \pm 8°$. Microscale shearing between the backshears probably accomplished additional crustal tilting to ~45°. We infer this shearing was focused above the basal ramp-step, was transient, and aseismic. Focal mechanisms of earthquakes in the Southern Alps suggest that similar backshearing may be accumulating at depth today, where it is linked to seismic-slip on upper crustal faults. Fluid was integral to the formation and accumulation of shear along the backshears. Near-lithostatic fluid pressures triggered deep, brittle shear failure (>20 km). The steep, dilative backshears allowed these fluids to escape upwards through low permeability (1×10^{-18} m^2) schist. Fluid expulsion may thus have accomplished a devolatilisation and rheological strengthening along the Alpine mylonite source region at depth, while also causing a hydrolytic weakening of the fluid-invaded rocks (especially quartz veins) in the Pacific Plate. These coupled strength changes may have enhanced the local partitioning of deformation onto steep planes in the Alpine Fault hangingwall.

INTRODUCTION

Deformation during continental collision occurs at a range of scales from crustal through to microscopic, and can be accommodated through a variety of small- and large-scale structures. Commonly occurring structures include crustal-scale thrust faults, consisting of a decollement or 'flat' at depth and a more steeply dipping ramp, up which the upper crustal rocks are displaced. As material is displaced over the bend between the fault-flat and the fault ramp, it experiences a strain related to the change in trajectory. Two particulars of this bending can be addressed: (1) the amount of strain that is accumulated and (2) the way in which the strain is accommodated. The first of these, the amount of strain that is

[1]School of Earth Sciences, Victoria University of Wellington, Wellington, New Zealand

[2]Department of Earth Sciences, University of Durham, Durham, United Kingdom

A Continental Plate Boundary: Tectonics at South Island, New Zealand
Geophysical Monograph Series 175
Copyright 2007 by the American Geophysical Union.
10.1029/175GM10

generated in the hangingwall rocks by displacement across a bend, depends chiefly on two factors: the geometry or dip, α, of the fault ramp, and the radius of curvature or width, ω, across which this bending occurs [*Knipe*, 1985]. Numerical modelling of footwall ramp deformation demonstrates that a zone of high differential stress and strain-rate is localised at the bend in the footwall ramp, extending upward into the overlying hangingwall rocks [*Braun and Beaumont*, 1995; *Beaumont et al.*, 1996]. This zone remains fixed to the bend in the footwall ramp as the hangingwall rocks move through it experiencing the elevated stresses and strain-rate only transiently before being passively displaced up the ramp.

In order to accommodate this strain the hangingwall must undergo some degree of deformation; how this deformation is accomplished depends not only on the rate and direction of convergence between the hangingwall and foot-wall and the geometry of the ramp, but also on the lithologies involved and the conditions under which the deformation must occur [*Knipe*, 1985; *Axen et al.*, 2001]. If the ramp angle, α, is small and/or the rate of hangingwall displacement is slow, or if the hangingwall has weak anisotropic layering that lies parallel to the fault, the rocks may accommodate this ramp-related bending by a flexure and/or by layer-parallel internal shearing within the rock mass [*Knipe*, 1985]. On the other

hand, if the rate of displacement is high and/or α is large, the strain may instead be accommodated by slip across a series of inclined backshears that lift the hangingwall onto the ramp. Two-dimensional analogue and numerical modelling of backshear development during thrust faulting under compressional regimes generally predict backshears to develop in the hangingwall with a moderate dip inclined in the opposite direction to the underlying foot-wall ramp [*Berger and Johnson*, 1980; *Chester et al.*, 1991; *Erickson and Jamison*, 1995; *Strayer and Hudleston*, 1997; *Bonini et al.*, 2000; *Erickson et al.*, 2001].

This paper examines in detail a natural array of exhumed backshears exposed in the transpressional setting of the central Southern Alps, New Zealand. These natural backshears are interpreted to have initiated sequentially, in a broadly escalator-like fashion, to accommodate the neotectonic tilting and uplift of the Pacific Plate up onto the Alpine Fault ramp at depth (Figure 1). Possibly for the first time, the geometry, characteristics, and kinematics of naturally occurring backshears formed under geological conditions during oblique collision have been documented in detail. We firstly document variations in their fundamental characteristics as a function of depth towards the base of the footwall ramp, the amount of strain accommodated by these structures and

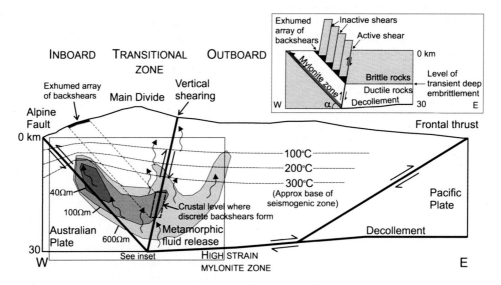

Figure 1. Schematic cross-section of Southern Alps orogen based on fluid flow model of *Koons et al.* [1998] showing the main structural features including the Alpine Fault, basal decollement, frontal thrust fault, and the area of backshearing in the transition zone of the orogen that acts as vertical conduit for fluid escape. Metamorphic fluid flow is represented by wiggly closed arrows. Isotherms are approximate, after thermal modelling by *Batt and Braun* [1999]; 300°C isotherm approximately coincident with base of seismogenic zone of *Leitner et al.* [2001]. Contours of apparent resistivity are superimposed in grey shades after *Wannamaker et al.* [2002]. Inset: enlarged schematic cross-section showing escalator model for formation of brittle-ductile fault array above the footwall step (proshear) in the Alpine Fault ramp [after *Little et al.*, 2002a, 2002b]. Active brittle-ductile fault coincides spatially with the zone of vertical backshearing in the larger diagram.

will attempt to make some inferences about how the back-tilting deformation was accommodated at structurally higher levels in the crust. These backshears allow insight not only into how transpressional deformation is accommodated at different structural levels in the crust, but also enable us to quantify that deformation in terms of strain, strain-rates, and magnitude and directions of shearing in the mid-crust of the Pacific Plate hangingwall.

Quartz-calcite veins that infill the near-vertical backshears are interpreted to have been precipitated by lower crustal metamorphic fluids escaping upwards through these structures. This is in agreement with mechanical models [e.g., *Koons et al.*, 1998] of the Southern Alps orogen that predict fluid in the actively deforming mylonite zone at the base of the hangingwall can only effectively migrate or escape due to hydraulic fracturing and shear failure [*Sibson*, 1996]. In the absence of these processes, background permeability is thought to be inherently low in this region. In the dextral strike-slip dominated setting of the Southern Alps, numerically modelled failure planes are commonly shown as steep to vertical in dip. Such fractures might provide efficient vertical conduits for fluid flow [*Koons and Craw*, 1991a; *Koons*, 1994; *Upton et al.*, 1995; *Koons et al.*, 1998]. In the second part of this paper we use our observations of the veins that infill the backshears to quantify the volume of fluid released during deformation-driven metamorphism at the base of the Alpine Fault ramp, which has escaped into the Alpine Fault hangingwall along the backshears. We then discuss the implications of this deep metamorphic fluid flow on the crustal rheology of the Southern Alps orogen, and the feedback between fluid flow at depth and the occurrence of brittle backshearing deformation in the mid-crust of the Pacific Plate.

GEOLOGICAL SETTING

The Alpine Fault is a dominant structure in South Island, New Zealand, and represents the western edge of the obliquely convergent Australian-Pacific plate boundary (Figure 2). The Southern Alps are the topographic expression of this on-going oblique convergence. The relative plate motion across the central section of the Southern Alps is 37 ± 2 mm/yr at an azimuth of 071 ± 2°, calculated from the Nuvel-1A global plate model [*De Mets et al.*, 1990, 1994]. Estimates of late-Quaternary slip-rates indicate that two-thirds to three-quarters of the plate motion is accommodated by slip along a single structure, the dextral-reverse Alpine Fault [*Norris and Cooper*, 2001], an observation that is consistent with geodetic strain measurements by GPS surveys [*Beavan et al.*, 1999]. The remainder of the plate motion must be accommodated by deformation on structures to the east of the Alpine Fault. Dip-slip-rates along the Alpine Fault are highest through the central section of the orogen, at >8–12 mm/yr [*Norris and Cooper*, 2001], and drop off to ~<1 mm/yr to the south near Haast and Milford Sound. Late-Quaternary strike-slip-rates, however, are constant along the entire length of the fault, at 27 ± 5 mm/yr [*Norris and Cooper*, 2001]. Rock uplift-rates along the fault appear to be in the range of 5–10 mm/yr [*Wellman*, 1979; *Bull and Cooper*, 1986; *Tippett and Kamp*, 1993; *Beavan et al.*, 1999], approximately equal to rates of erosion in the Southern Alps, as is consistent with an approximately steady-state topography in the range [*Adams*, 1981].

Oblique convergence across the Alpine Fault ramp has led to the tilting and uplifting of the hangingwall rocks on the southeast side of the Alpine Fault from depths of over 25 km where rocks are detached along a mid-crustal decollement [*Norris et al.*, 1990; *Davey et al.*, 1998]. The tilted hangingwall consists of Mesozoic, predominantly quartzofeldspathic schists, termed the Alpine Schist, that range from amphibolite facies rocks adjacent to the Alpine Fault, to prehnite-pumpellyite facies rocks at the Main Divide ~15–20 km to the east of that structure [*Grapes and Watanabe*, 1992]. Intermixed, and structurally beneath the quartzofeldspathic psammitic schists (Torlesse Terrane) are layers of carbonaceous pelite, metabasite, chert, and marble, currently referred to as the Aspiring Terrane [e.g., *Hutton*, 1940; *White*, 1996; *Little et al.*, 2002a; *Cox et al.*, unpub. QMap data]. Within 1–2 km of the Alpine Fault, a mylonite zone has developed, with the degree of mylonitisation increasing inhomogeneously towards the west from the amphibolite facies schist protolith to the east through to a cataclastically overprinted ultramylonite adjacent to the fault trace [*Sibson et al.*, 1981; *Norris and Cooper*, 2003]. The mylonitisation occurred at amphibolite-facies grade at 25–30 km depth during the late Cenozoic [*Grapes and Watanabe*, 1992; *Vry et al.*, 2004].

Inferences made from the attitude of the mylonitic foliation suggest that the fault in the central section around Franz Josef and Fox glaciers is dipping at ~40–60° SE [*Sibson et al.*, 1981; *Cooper and Norris*, 1994; *Little et al.*, 2002b]. Very young (late Cenozoic) thermochronological ages for hornblende, muscovite, biotite and zircon in the hangingwall of the Alpine Fault have led some to believe that a greater degree of rock uplift and crustal exhumation has taken place in the central part of the Southern Alps, in part because the Alpine Fault has a slightly steeper dip (by 10–20°) in this central section than to the north or to the south [*Little et al.*, 2005]. These authors have argued that one result of this locally steeper Alpine Fault dip in the central section of the Southern Alps, is that the hangingwall rocks are there flexed across a more abrupt ramp step at depth, and that this boundary condition, including the locally high differential stresses related to it, has resulted in the failure in this region of a series of backshears in the Pacific Plate above the basal ramp-step

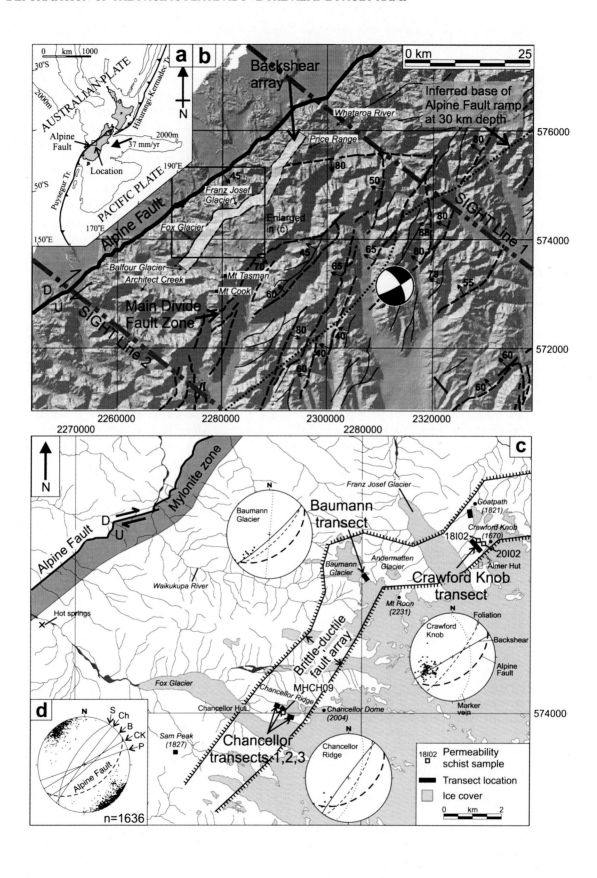

in the Alpine Fault. Slip on these, rather than a more distributed bending or layer-parallel flexing, has accommodated the ramp-related strain.

Geophysical seismic and electrical data collected in two 160 km-long transects across the central Southern Alps as part of the South Island Geophysical Transect (SIGHT) project are interpreted to record the presence of high fluid pressures at depth, inferred to have been released during prograde metamorphism in the basal high strain zone [*Stern et al.*, 1997, 2001; *Wannamaker et al.*, 2002, 2004] (Figure 1). The presence of fluids in the mid- to lower crust under the Southern Alps may suggest that this part of the Australian-Pacific plate boundary is inherently weak, and has been unable to allow differential stresses to accumulate in the mid- to lower crust [e.g., *Stern et al.*, 2001; *Wannamaker et al.*, 2004]. The suite of veins that infill the vertical backshears help to constrain both the original location and quantity of this inferred fluid flow and shed light on the crustal dynamics of this active orogen.

A WELL-EXPOSED ARRAY OF RAMP-RELATED BACKSHEARS

A systematically spaced set of faults or shear zones, most of which are brittle, but some parts of which are ductile, are exposed in biotite zone rocks in the hangingwall of the Alpine Fault. Only found in the central part of the Southern Alps (Figure 2), these structures have been observed in glaciated outcrops in five areas where they were mapped and surveyed in detail (Figure 2). The areas include, from north to south: the Price Range; glacially cut benches around Crawford Knob and Baumann Glacier, to the east and west of Franz Josef Glacier; glacially cut benches around Chancellor Ridge, Fox Glacier; and Sam Peak above the Balfour Glacier. Beyond the extent of the shear array shown in Figure 2, the structures are not observed in outcrop. Because of the plate margin-parallel strike of the shears, their dextral-oblique kinematics, and their late-stage, overprinting relationship with respect to all other ductile fabrics in the Alpine Schist, these structures have been interpreted as a late Cenozoic suite of backshears that were apparently activated in a sequential escalator-like fashion to accommodate tilting at mid-crustal depths of the Pacific Plate hangingwall onto the Alpine Fault ramp (Figure 1 inset) [*Little et al.*, 2002a, 2002b, 2005; *Wightman et al.*, 2006]. As such, these structures allow us to quantify deformation gradients and strain in the Alpine fault's hangingwall during Cenozoic uplift.

Observed Distribution and Geometry of the Shear Array

The shears occupy a broadly tabular, southeast dipping zone that is 1–2 km wide, and which lies ~6–10 km horizontally southeast of the Alpine Fault (Figure 2). Glaciated outcrop exposure of the shears is excellent throughout the central Southern Alps, except where ice and/or snow cover is present. We have extrapolated between the accessible outcrops using aerial photography to define the lateral extent of this shear zone array. The backshears can be traced on 1:25,000 aerial photographs, as they exert a strong tabular aspect to the bedrock geomorphology due to the pervasiveness, planarity, and evenly spaced nature of the associated fracturing (Figure 3a).

The five main sites that have been mapped and surveyed in detail are shown in Figure 2. The extremities of the backsheared zone can be placed due to their noted absence in both the Whataroa River valley to the north, and Architect Creek to the south. Within these limits, the backshears define a tabular zone, 30 km long, that strikes subparallel to the Alpine Fault trace (Figure 2), and which lies 5–7 km structurally above the Alpine Fault ramp (assuming a 50° SE-

Figure 2. (Opposite) Geodynamic context of brittle-ductile fault array. (a) tectonic setting of New Zealand, showing major tectonic features, including the Alpine Fault and the plate motion vector in central South Island [from *De Mets et al.*, 1990, 1994; *Walcott*, 1998]. Location of map (b) is shown as box. (b) Shaded digital elevation model image of the central Southern Alps showing location of exhumed backshear array (grey) with reference to active (heavy lines), inferred active (heavy dashed liSnes) and inactive faults (fine lines), and approximate positions of SIGHT transects [*Stern et al.*, 1997, 2001; *Wannamaker et al.*, 2004]. The surface projection of the Alpine Fault ramp base at 30 km depth is shown by bold line; small arrows on faults represent dip and dip direction where data available. DEM image and active faults are from New Zealand Active Faults Database [*GNS Science*: http://data.gns.cri.nz/af]; inferred active and inactive faults from *Simon Cox* [unpublished data]; focal mechanism of 24/06/84 earthquake from body waveform modelling by *Anderson et al.* [1993], plotted in lower hemisphere projection where black quadrants are compressional. (c) enlarged map of study area showing Alpine Fault, site locations mentioned in text, and permeability sample locations. Lower hemisphere stereonet projections show mean brittle-ductile fault (solid line), dominant foliation (small dashed great circle), mean marker vein orientation (dotted great circle), Alpine Fault (large dashed great circle), and slip lineations on the brittle-ductile faults (filled squares). Graticules are in New Zealand Map Grid (metres). (d) lower hemisphere stereonet projection of data from all localities showing poles to brittle-ductile faults (small dots), average brittle-ductile fault orientation at different sites (thin lines) and Alpine Fault (dashed line). S=Sam Peak, Ch=Chancellor Ridge, B=Baumann Glacier, CK=Crawford Knob, and P=Price Range. See (c) for locations.

dipping Alpine Fault). The zone deflects across the topography in a way that suggests that the zone dips moderately to the southeast, subparallel to the Alpine Fault. The southeast and northwest boundaries of the zone will therefore be referred to as the structural 'top' and the 'bottom' of the zone, respectively. These boundaries of the shear array can be traced approximately using aerial photography based on the absence of shears and shear-related topographic fabric outside of the zone (Figure 2).

Observed Structural 'Bottom' and 'Top' of the Array

The top and bottom contacts of the array are difficult to locate precisely in the field and are defined by the maximum observed extent of discrete, systematically spaced planar shears displacing features in the Alpine Schist. The contacts displayed on the map (Figure 2) were constrained with the help of distant views across the valleys and aerial photography due to the flaggy characteristic of shear-bounded outcrops within the backshear array. The structurally lowest sites where backshears have been observed include Chancellor Ridge and Baumann Glacier (Figure 2). There, psammite-rich quartzofeldspathic schist of the Torlesse Terrane near the bottom of the array contacts with a structurally deeper package (Aspiring Terrane) chiefly of dark pelitic schists and including some metabasites or chert that dominate the highest grade, basal parts of the Alpine Schist. This terrane boundary is marked by a distinctive zone of a broken formation with psammite phacoids (rods) that lie in a pelitic matrix and lies at or near the base of the array in most observed areas. We suggest that the rheological change that occurs across the terrane boundary controls the location of the base of the backshear array.

Towards the base (the bottom ~500 m) of the observed array (e.g., towards the Goatpath at Crawford Knob; Figure 2c), the backshears begin to anastomose with one another and with the dominant foliation in pelitic interbeds within the schist. There, the otherwise brittlely faulted quartzofeldspathic schist begins to show clear evidence of crystal-plastic deformation, including thin (1–3 cm) strongly foliated mylonite zones that flank the individual shears (Figure 3f). Especially in pelitic rocks, the strike of the backshears also deflects anticlockwise in this region away from subparallelism with the Alpine Fault to become subparallel with the dominant foliation in the Alpine Schist (~033°) rather than cross-cutting that fabric at a $27 \pm 9°$ strike angle as typically is the case.

The structural 'top' of the array has not been pin-pointed in outcrop chiefly due to permanent ice and snow cover. Areas have been identified that are structurally higher than the shear array and which lack the conspicuous shear structures (e.g., to the southwest of Almer Glacier at Crawford Knob, at Mt Roon above the Baumann Glacier, and the headwall of the Andermatten Glacier; Figure 2c) but the detailed characteristics of this transition remain unclear. In quartz veins in the schist that have been deformed by the backshears, the ratio of ductile-slip to total slip decreases structurally upwards in the array (see Figure 7b in next section) implying that brittle behaviour becomes increasingly dominant upwards towards the top of the array, not just in the quartzofeldspathic schist, but also in the generally much more ductilely deformed quartz veins that are embedded in that schist.

Outcrop Scale Observations and Organisation of Backshears

The backshears are observed to be dominantly very planar, smooth faults that are laterally and vertically continuous for 10s to 100s of metres. These structures brittlely offset pre-existing veins and the foliation in their biotite-grade psammitic (metagreywacke) host. On average, the faults strike subparallel to the Alpine Fault (~055°) and, although there is systematic variability from site to site, each area surveyed within the zone is characterised by backshears that are remarkably uniform in their local attitude (Figure 2c). A gradual change

Figure 3. (Opposite) Photographs of late Cenozoic backshears exposed in the central Southern Alps, illustrating characteristics of the shears: the thickness of the infilling vein (*Th*), width of the deformation zone (ω_{def}), spacing of the backshears (δ_i), and the brittle (*b*) and ductile (*d*) components of offset. Note the systematic spacing of the backshears, and their obliquity to the dominant foliation. (a) aerial photograph (1:25000) scale of Chancellor area showing large scale geomorphic control the backshears have on edges of cliffs and ridges. Oriented north to the top. (b) View from a distance of backshears on a vertical cliff face below Baumann Glacier, facing southwest. (c) Backshears exposed on a vertical cliff face at Sam Peak, facing southwest. (d) Close-up of outcrop face at Baumann Glacier. (e) Outcrop at Crawford Knob, facing northeast. Note the brittle to ductile offset of quartz vein and entirely brittle offset of psammitic schist. (f) Structurally deep exposure of backshears west of Crawford Knob, showing typically anastomosing nature of backshears towards the bottom of the array and ductile deformation of Alpine Schist wall rock in proximity to the shear zones. Photograph is oriented top to the northwest. (g) Close-up of 8 mm thick vein infilling a backshear at Chancellor Ridge. Photograph facing southeast, looking down on horizontal outcrop. (h) Close-up of quartz fibre lineation on fault surface vein at Crawford Knob, facing southeast. Lineation pitches moderately to the southwest.

Table 1. Summary of shear zone properties

Site	n	strike	dip	spacing δ (cm)	slip (cm)[b] χ_{net}	obliquity (°)[c]	strike-normal rate (mm/yr)[d]
Price Range	5	071	90	—	—	0	0
Crawford Knob	868	060	89 NW	60	15	11	7
Baumann	305	049	83 SE	14	18	22	14
Chancellor[a]	399	037	87 NW	15	11	34	21
Sam Peak	59	033	68 NW	20	10	38	23

[a]Data combined from three transects: 2 from *Hill* [2005], 1 this study.
[b]Calculated using a net-slip vector pitching 36°SW.
[c]Obliquity of backshear strike to trend of plate motion vector (anticlockwise angle).
[d]Strike-normal component of the plate motion vector across backshears using pole of *De Mets et al.* [1994].

in mean strike of the shears is apparent from the north end of the zone at the Price Range where shears strike 071°, through to Sam Peak at the southern extent of the zone where shears strike 033° (Figure 2d, Table 1). The faults are commonly near-vertical structures (dips of 85–90°), dipping on average very steeply to the southeast (e.g., at Baumann Glacier) or to the northwest (all other locations; Figures 2d, 3b, Table 1). Only at the extreme southwest end of the zone (i.e., Sam Peak) are the structures not near-vertical in dip. Here, and only here, they dip moderately (average 68°) to the northwest (Figures 2d, 3c).

The brittle faults generally cut and offset the steeply southeast-dipping and slightly more north-striking dominant foliation in the Alpine Schist (S_3 of *Little et al.*, 2002a) at a low angle (Figures 2c, 3b, c). Towards the structural bottom of the array of backshears (i.e., the northwest boundary), many backshears locally refract into parallelism with this foliation and reactivate it in slip (Figure 3f), especially in

incompetent pelitic or graphite-rich layers, rather than cutting it at an acute angle.

Remarkably, the backshears show no appearance of frictional wear products such as cataclasite or gouge (Figure 3d), but are commonly infilled by syntectonic quartz-calcite veins that have a thickness, *Th*, of 1–20 mm (Figures 3e, g, h, 4), although ~10% of the faults contain no infilling vein. These veins must, at least in part, post-date the initiation of shear failure. The infilling veins are predominantly composed of quartz but contain up to 50% calcite and other phases (less than 5% total) such as biotite, chlorite, and muscovite, and trace amounts (less than 1%) of phases such as TiO_2, ilmenite, siderite, epidote, apatite, pyrite or pyrrhotite. Chlorite and biotite are often present as a selvage along the vein margins, but are rarely seen incorporated into the central part of the veins, indicating the majority of veins have not seen multiple fluid flow events. This is supported by the uniform distribution of ^{18}O isotopes within vein quartz and calcite, as well has homogeneous densities of fluid inclusions trapped in individual veins [*Wightman*, 2005; *Wightman et al.*, in review]. Apart from the biotite and chlorite selvage along the margins of the veins, the host rocks surrounding these fault-infilling veins show no evidence for retrogression during fluid infiltration.

Where the fault intersects pre-existing quartz veins in the schist, the veins are typically displaced in a coherent and ductile manner rather than brittlely offset (Figures 3d, e). These marker veins range from 1 mm to several centimetres thick, and are predominantly oriented at a high angle to the backshears (Figure 2c), with a majority of the veins striking 340–030° and dipping 55–90° E. The deformed (sheared) veins can be subdivided into two groups: 1) early quartz and quartz-siderite veins that pre-date the late Mesozoic metamorphism and related ductile strain of the Alpine Schist (these are folded or crenulated about the dominant foliation

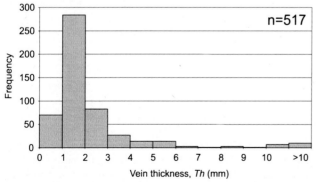

Figure 4. Frequency histogram showing distribution of infilling-vein thicknesses in mm. Data collected from four sites along strike of the backshear array (n = 517). The average thickness of infilling vein is 1.4 ± 4 mm.

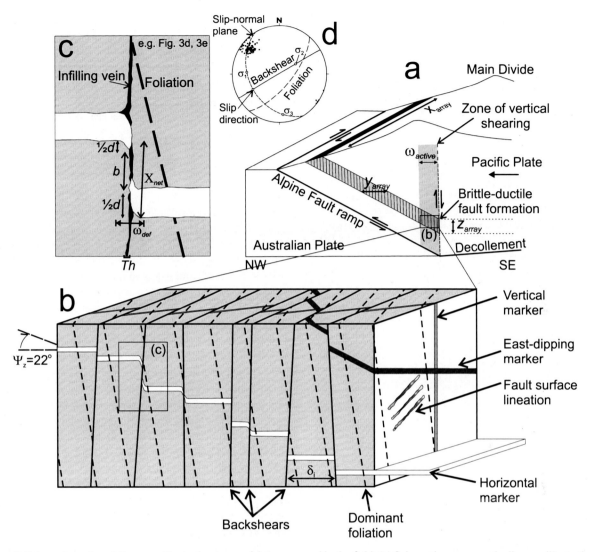

Figure 5. Schematic series of diagrams illustrating types of data measured in the field. (a) Schematic orogen-scale diagram illustrating the strike-length, x_{array}, the horizontal width, y_{array}, and the vertical extent, z_{array} of the array, as well as the active zone of backshearing deformation, ω_{active}. (b) Schematic block diagram of a short hypothetical transect across part of the backshear array. These include: orientation of backshears and sheared marker veins, spacing between backshears, observed separation of marker and the attitude of the outcrop surface, and pitch of any fibre lineations on the fault surface. The shear angle of the accumulated bulk vertical shear strain, ψ_z, is illustrated for a hypothetical horizontal marker. (c) enlargement of box shown in (b) with details of the observed separation of a quartz vein marker across a backshear, including the total (χ_{net}), brittle (b), and ductile (d) components of separation, the width of the ductile deformation zone (ω_{def}), and the thickness (Th) of the vein infilling the backshear. (d) Lower hemisphere stereographic plot showing orientation of average backshear and foliation planes at Crawford Knob and relationship to stress orientations. Principal stress directions were calculated assuming Coulomb failure theory and a coefficient of internal friction of 0.6. Filled squares represent poles to dominant foliation.

(S_3) in the schist) and 2) later, probably late Cenozoic, quartz and quartz-carbonate veins that post-date the main "Alpine" foliation (these are smooth, planar and uncrenulated). Both of these vein types have been obliquely sheared across the brittle to ductile backshears in the late Cenozoic. Intersections between the marker veins and the backshears vary in pitch (Figure 2c) from moderately NE-pitching (e.g., Craw-

ford Knob), near vertical around Baumann Glacier, to subhorizontal (e.g., Chancellor Ridge). Where the quartz veins are truncated against the backshear plane they commonly are decorated by a moderately SW-pitching fault surface lineation along the surface of the offset vein. This lineation is defined by elongate quartz fibres that are finely dusted with chlorite (Figure 3h).

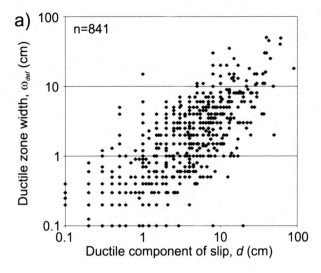

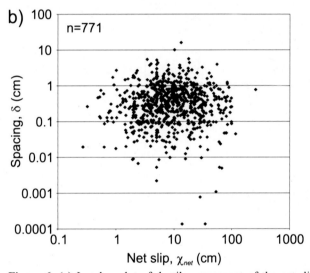

Figure 6. (a) Log-log plot of ductile component of the net-slip against the width of the ductile deformation compiled from all five transects. (b) Log-log plot of net-slip against intershear spacing from Crawford Knob transect.

Ductile deformation of the sheared quartz veins does not affect the host quartzofeldspathic schist into which they are embedded but is confined to a fault-parallel zone inside the vein that is up to 10 cm wide, with the average width of this zone, ω_{def}, being 1–2 cm (Figures 5, 6). The width of this ductile deformation zone, and the ratio of ductile/total separation experienced by a sheared quartz vein scale in an approximately linear way with the thickness of the deformed vein; veins greater than ~2 cm thick are commonly completely ductilely deformed, whereas veins less than ~2 cm thick may be ductilely deformed, brittlely deformed, or both (Figures 3d, e). Faults in the brittlely deformed host psam-

mitic schist commonly refract around the ductilely sheared marker veins, suggesting that brittle fault tips in quartzofeldspathic schist propagated along the margins of the ductilely sheared vein, apparently exploiting the rheologically weak boundary between the quartz vein and its stronger, quartzofeldspathic schist host (Figure 3d).

The remarkably systematic spacing, δ, of the backshears and repeated sense and magnitude of their offsets, χ_{net}, strongly suggests that all of the backshears are characterised by a similar direction and magnitude of finite slip, an assumption that we will make in our subsequent analysis. On approximately horizontal glacially cut benches, steeply dipping deformed veins everywhere show a dextral sense of strike-separation (Figure 5). Likewise, on vertical outcrop faces (Figure 5), sub-horizontal marker veins everywhere show a NW-up sense of dip-separation. Most deformed veins strike north and dip moderately east, are oriented at a high angle to the shears, and show both dextral and northwest-up senses of strike- and dip-separations, respectively (Figure 5). These, and other basic field observations, indicate that the backshear slip vectors must pitch between 0 and 90° SW (Figure 5), although a steeply NE-pitching slip vector cannot be ruled out to explain offsets of some gently dipping marker veins.

A fibrous chlorite ± quartz fault surface lineation is commonly observed along the brittle fault surfaces in metagreywacke schist along the margins of the infilling quartz-calcite veins and also on the exposed cutoffs where ductilely deformed quartz veins intersect the backshears. These fibre lineations typically pitch 36 ± 5° to the southwest (Figures 3h, 5). An outcrop at Baumann Glacier (Figure 2c), containing a 3-dimensional exposure of the shear plane and several differently oriented deformed marker veins, allowed the finite-slip vector to be precisely calculated by utilising the multiple planar markers and the observed intersections of their cutoffs on both sides of the shear zone. This slip vector was found to pitch at 51 ± 10° SW, somewhat steeper than the mean pitch of fault surface vein fibres at that site, although within the 22–65° SW range of lineation pitches observed at that site. The fault-surface lineations have an orientation (pitch) that is consistent with the repetitive pattern of marker vein offsets on outcrop faces of all orientations. The interpretation that these lineations are parallel to the finite slip vector on the shears is further supported by statistical and geometrical modelling of the full data set of offset markers on the several fault-slip transects analysed for this study (n = 1230 offsets, see next section). These data were used to attempt to resolve the mean direction of net-slip, assuming that there is one (repeated) mean SW-pitching slip direction that is representative of each transect as a whole [for details see *Wightman*, 2005]. This kinematic modelling, together

Table 2. Summary of transect data and calculated shear strains

Location	Transect length (m)	n	Mean spacing δ (cm)	Mean spacing 2σ	Δt (years)[a]	Bulk shear strain γ_x	Bulk shear strain γ_z	Bulk shear strain ψ_z (°)
Crawford Knob	443	795	53	4.7	84.4	0.17	0.18	10
Baumann	15	111	14	2.8	13.4	0.88	0.64	33
Chancellor 1[e]	4.5	100	4.5	0.9	2.4	2.18	2.08	64
Chancellor 2[e]	21	155	15	2.4	7.3	0.60	0.59	31
Chancellor 3	25	69	30	6.0	27.4	0.19	0.23	13

Location	Shear strain γ (per shear)[b]	Shear strain 2σ	Ductile shear strain γ_d (per shear)[c]	Ductile shear strain 2σ	Ductile shear strain-rate (s^{-1}) per shear (e_{max})	Ductile shear strain-rate ω_{active} = 100 m	Ductile shear strain-rate ω_{active} = 500 m	no. data with slip vector pitch ≠36°[d]
Crawford Knob	14.3	2.8	3.1	0.5	1.2×10^{-9}	7.0×10^{-12}	1.4×10^{-12}	66
Baumann	24.0	7.5	6.8	1.8	2.1×10^{-8}	3.0×10^{-11}	6.0×10^{-12}	0
Chancellor 1[e]	27.0	9.2	13.3	7.8	4.6×10^{-7}	8.8×10^{-11}	1.8×10^{-11}	20
Chancellor 2[e]	18.1	5.6	6.4	2.6	7.9×10^{-8}	4.2×10^{-11}	8.4×10^{-12}	22
Chancellor 3	5.0	1.5	2.6	1.0	2.1×10^{-8}	1.7×10^{-11}	3.5×10^{-12}	9

Note: net slip and shear strains are calculated using an input slip vector of 36° SW in all transects except where fault surface lineations were locally observed.

[a]Average time taken to translate distance between shears at 11 mm/yr.

[b]Average total shear strain recorded in each shear zone.

[c]Average ductile shear strain recorded in each shear zone.

[d]Input slip vector ≠36°SW in order to preserve a dextral-oblique sense of shear on calculated offset (excluding those data with observed fault surface lineations).

[e]Raw data from *Hill* [2005].

with the lineation data, strongly suggests that the net-slip vector pitches between 35° and 50° SW, within error of the observed mineral lineation of 36 ± 5° SW. It is therefore likely, and we will hereafter assume, that the chlorite and quartz fibre lineations, pitching moderately to the southwest, are approximately parallel to the mean direction of finite dextral-oblique slip on the backshears.

In our finite-slip calculations, below, we will assume that the slip vectors on any individual backshear that has an observed offset that is either 1) parallel to the mineral fibres observed on that shear, or 2) parallel to the mean pitch of all the mineral fibres observed on that transect. Errors that we will report on net-slip magnitudes derived from these offset observations are those relating to the 95% uncertainty in pitch of the slip-vector on any given backshear, which we took to include the range 35° to 50° SW. On some individual slip observations, the local slip direction had to be increased to a pitch of up to ~70° SW in order to conform with the locally observed sense of marker separations on the outcrop while maintaining a dextral-oblique sense of slip on the backshears (see Table 2 for statistics).

Field Methodology and Analyses of Transect Data

Quantitative transects of fault-slip data were undertaken at four different sites (Figure 2c) in order to quantify the 3-dimensional geometry and kinematics of the zone. Each linear survey was measured approximately perpendicular to the strike of the Alpine Fault, and the data were projected onto a plane perpendicular to that structure. The longest transect (443 m; 795 backshears) was undertaken at Crawford Knob (Figure 2) in the centre of the array, and is a composite of two shorter transects (Table 2). The transects from the other sites are shorter (<25 m). Measurements undertaken along these transects (Figure 5) include: the planar attitude of each backshear and of each marker offset by it (usually a pre-existing quartz vein); the attitude of the outcrop surface; the separation of the marker across the trace of the shear on that outcrop plane ($b + d$), the fraction of brittle (b) versus ductile (d) offset accomplishing the total separation; the deformed and undeformed width of the quartz veins, and the thickness of the ductile deformation zone in each sheared vein (ω_{def}); the thickness of the syntectonic fault-infilling quartz vein (*Th*); host rock lithology; the pitch of any observed fault surface fibre lineations; and the position of the backshear structure along the transect line.

This distance between one backshear to the next was used to calculate the spacing, δ_i between the shears, by correcting for the plunge of the transect line and the attitude of the shears so that spacing measurements quoted here refer to the true perpendicular distance between adjacent backshears in three dimensions (Table 1). Using the above data as inputs, vector algebra on an Excel spreadsheet (details given in the work of *Little* [1995]) was used to solve for the three-dimensional net-slip vector, χ_{net}, for each backshear. Backshears measured along the transect for which there was no data concerning marker offset were simply excluded from the transect and subsequent bulk strain calculation, with the transects being correspondingly shortened in the cumulative sample length.

The three-dimensional slip vector, χ_{net} for each backshear was divided up into X, Y, Z components relative to an Alpine Fault-based coordinate system: (χ_X) the dextral-slip parallel to the strike of the Alpine Fault, where a positive direction is dextral; (χ_Y) the heave perpendicular to the strike of the Alpine Fault where a positive direction is specified motion of the western block towards the Alpine Fault; and (χ_Z) the throw, the portion of the slip vector that is vertical, where positive values record vertical upthrow of the northwest block. Summing these slip components in an accumulative fashion along the array allow us to evaluate gradients in bulk displacement accommodated by shearing along the transect and to calculate the 3-dimensional finite strain associated with them (see *Little* [1995]). The inverse tangent of the bulk vertical shear-strain (γ_z) is a measure of the degree of net tilting (ψ_z) of originally horizontal surfaces as accommodated by dip-slip on backshears over the length of the transect (Figure 5).

Bulk Geometric and Kinematic Properties of the Shear Array

Intershear spacing, δ_i. The backshears are very systematically spaced (average δ =34 ± 3 cm (2σ), all transect sites), with little variation in this spacing as a function of structural depth in the array (Figures 3e, 7d). The longest survey of transect data at Crawford Knob yielded δ = 53 ± 4.7 cm (Table 1). This site is characterised by an alternation of psammitic and pelitic schist at a scale of 2–5 m. Baumann Glacier and Chancellor Ridge transects to the south of Crawford Knob yielded average backshear spacings in psammitic schist of 14 ± 2.8 cm and 15 ± 2.5 cm, respectively (Table 2). One of the three Chancellor transects recorded a finer average spacing of 4.5 ± 0.9 cm. Field measurements (not

transect data) of spacing at Sam Peak at the southern end of the zone (δ = 20 ± 4 cm; Table 1), seem to confirm that the finer spacing (10–20 cm) is more representative of the true spacing in uniformly psammitic schists than the coarser spacing seen in heterogeneous lithologies in the transects at Crawford Knob. In addition, there was a sampling bias at Crawford Knob where backshears with very small offsets (<5 mm) were ignored at this site in the interest of sampling the longest possible transect length. Backshear spacing at Crawford Knob also demonstrates an abrupt increase (average spacing increases from 48 cm to 99 cm) at ~300 m along the transect (Figure 7d). This increase coincides with a change from a psammitic to pelitic lithology and is also associated with a change in strike of the backshears along the transect to subparallel with the dominant foliation in the pelitic rocks (~033°).

As such, we infer that lithology exerts an important control on the spacing and attitude of the backshears. The observed spacing of the backshears is a strongly non-fractal, non-scale invariant relationship. The average observed lateral and vertical extent of the backshears (>100s m) seems consistent with a non-fractal δ_i. The capability of the rocks to fracture into thin but laterally and vertically extensive and continuous blocks suggests there is a minimum limit to δ_i that is dependent on rock strength. Below this minimum δ_i limit, backshears must become less laterally continuous. δ_i above ~2–3 m are also very rare in the Southern Alps except through large pelitic horizons where distributed deformation makes the concept of spacing questionable. Backshears formed in pelite and bands of graphite-rich schist have much coarser mean spacing than those formed in psammite – a relationship that is accompanied by what appears to be a larger fraction of foliation-parallel slip and other types of distributed (i.e., not spaced at mesoscale) deformation to accommodate the imposed shearing. This is also displayed in the difference between field transects taken in different areas, e.g., the Crawford Knob transect through metre-scale alternating psammitic and pelitic schist has a much coarser average spacing than the backshears developed through predominantly psammite at Chancellor Ridge to the south (Table 1). We interpret this difference in mean spacing to be due to the relative weakness of pelitic horizons, a rheology that favours a more finely spaced and "pervasive" deformation. In these areas (i.e., Crawford Knob) it is easier for the backshearing process to reactivate the existing foliation in the pelitic bands than to initiate a new backshear that cross-cuts that pre-existing anisotropy.

Ductile deformation zone width and brittle/ductile slip ratio. The width of the ductile deformation zone, ω_{def}, in sheared quartz veins averages 3.01 ± 0.33 cm and does

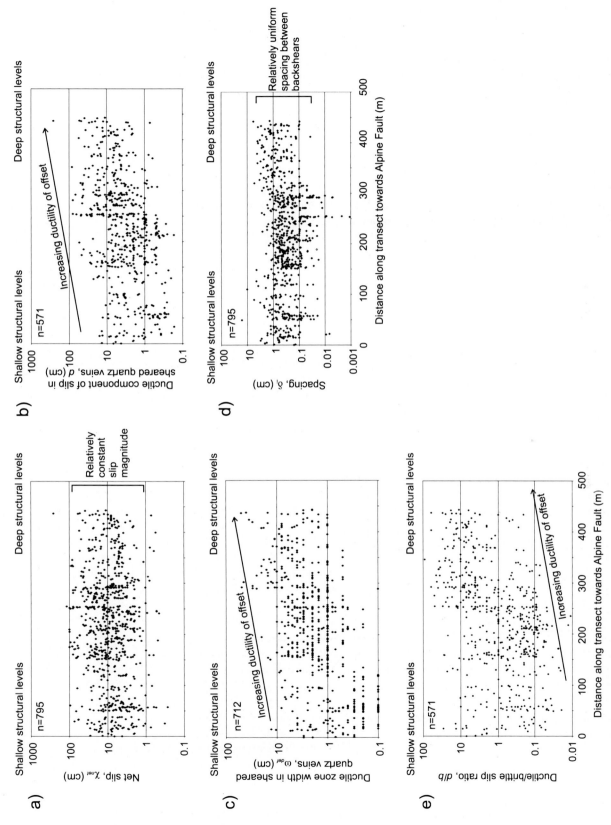

Figure 7. Log-normal plots of geometrical characteristics of shear zones in Crawford Knob transect, plotted against distance along the transect perpendicular to the Alpine Fault trace (0 m = furthest from Alpine Fault and structurally highest). (a) Net-slip along each backshear; (b) Ductile component of the net-slip along each backshear; (c) Width of the ductile zone of deformation of each backshear; (d) Backshear-perpendicular spacing between adjacent backshears; (e) Ratio of ductile to brittle component of slip along each backshear.

not seem to scale with δ_i. There is a strong linear correlation, however, between the ratio of ductile/total separation of sheared quartz veins, $d/(b + d)$ (i.e., the ductility of the offset) and ω_{def} (Figure 6a). Larger ω_{def} are associated with more ductile offsets and both parameters scale approximately linearly with the undeformed thickness of the veins. In other words, originally thicker quartz veins are both sheared across a wider zone and are deformed more ductilely; thinner veins are more narrowly deformed with a higher fraction of brittle slip. This correlation between the width of quartz veins and the width of ductile shear zones cutting through them implies that the pre-existing quartz veins are a weaker and more ductile lithology than the surrounding, generally brittlely deformed, quartzofeldspathic schist [Ellis et al., work in progress; Little et al., 2004]. As expected, both $d/(b + d)$ and ω_{def} increase with increasing distance along the transect towards the Alpine Fault, i.e., with increasing structural depth and inferred deformation temperatures (Figures 7b, c, e).

Average slip magnitude, χ_{net}. The magnitude of net-slip on an individual backshear, χ_{net}, can be calculated by assuming a pitch for the net-slip vector. The average slip magnitude calculated using the entire data set is 14.1 ± 1.2 cm for a net-slip vector pitching 36° SW. This average slip magnitude is relatively insensitive to variations in the assigned mean slip vector pitch across the full range in its uncertainty. As is expectable given the systematic nature of the repeated vein offsets, χ_{net} magnitudes are remarkably constant both structurally across the zone (Figure 7a), and laterally along strike from site to site (Table 1), with only slight variability between different areas. No reliable correlation can be made between χ_{net} and δ_i of individual backshears due to the remarkably small amount of variation in any of these quantities at the various study sites in the array (Figure 6b; cf. Figures 7a, d).

Finite shear strain of veins deformed by individual backshears, γ. Finite shear-strains, γ, for quartz veins ductilely offset across individual backshears can be estimated by $\gamma = \chi_{net}/\omega_{def}$, where $\chi_{net} = b + d$, representing the brittle and ductile components of slip, respectively. Moreover, a ductile shear-strain, γd can be specified by $\gamma_d = d/\omega_{def}$. These total and fractional ductile shear-strain estimates vary slightly between transects (Table 2), but average 16.6 ± 2.5 and 4.8 ± 3.8, respectively, combining data for all the sites. Total shear-strain decreases with increasing structural depth, in part because of the widening of the ductile shear zones with increasing depth (i.e., increasing ω_{def}) as χ_{net} remains approximately constant (Figures 7a, c, e). γd, however, displays little variation with increasing structural depth due to the congruent increase in both d and ωdef (Figures 7b, c). This approximately constant

magnitude of ductile shear-strain (i.e., ~5.0) is expressed on the graph of ωdef against d by a positive correlation of these two quantities to one another (Figure 6a).

These data of both increasing ω_{def} and increasing $d/(b+d)$ with increasing structural depth (Figures 7c, b) suggest that this scaling relationship is due to an overall increase in quartz vein ductilely with increasing depths. The width of the ductile shear zone in a sheared quartz vein is therefore partially controlled by the temperature of deformation. This is supported by the localised onset of crystal-plastic deformation in the quartzofeldspathic wall rocks of the backshears in the deeper parts of the array where mylonitic fabrics are seen in the metagreywacke host (Figure 3f).

Conditions of Backshear Formation

The field observations described above imply that the backshears were formed under brittle and ductile conditions. Previous work by Wightman [2005] and Wightman et al. [2006] established the temperature, depth, fluid pressure, and stress conditions under which this brittle-ductile deformation occurred, and is briefly summarised below. We will then use the field transect data to quantify the shear strain and shear strain-rates of the backshear deformation.

Fluid inclusion analyses and stable isotope geothermometry of veins that infill the backshears indicate that fracture initiation and vein deposition occurred at temperatures of $450 \pm 50°C$ and post-failure fluid pressures were ~310 MPa [*Wightman et al.*, in review]. The shear zones formed in a systematic array as near-vertical planar fractures through the previously intact quartzofeldspathic schist, and have a dextral-oblique sense of shear. Given their vertical attitude and context in a transpressive plate boundary zone, striking sub-parallel to the margin, Coulomb failure theory predicts that the intermediate principal stress direction, σ_2, should be close to vertical (Figure 5d), an inference that is also consistent with the numerical mechanical modelling of *Koons et al.* [2003] and focal mechanisms for earthquakes in the central Southern Alps [*Leitner et al.*, 2001]. Therefore we can estimate that σ_2 was approximately equal to the vertical overburden. By assuming a dip-slip rate along this central part of the Alpine Fault of 10 mm/yr [e.g., *Norris and Cooper*, 2001] and interpreting 3–4 Ma $^{40}Ar-^{39}Ar$ cooling ages in white mica as the time since these rocks passed through the ~400°C closure temperature for that mineral, a corresponding depth in the crust of >20–25 km is inferred for the (higher temperature) backshear initiation (i.e., $450 \pm 50°C$) [*Wightman et al.*, 2006]. This is equivalent to an overburden pressure, σ_v, at the time of backshear development of 560 MPa. Therefore $\sigma_v = \sigma_2 = 560$ MPa. For an ideally oriented fault, forming in previously unfractured rock, the

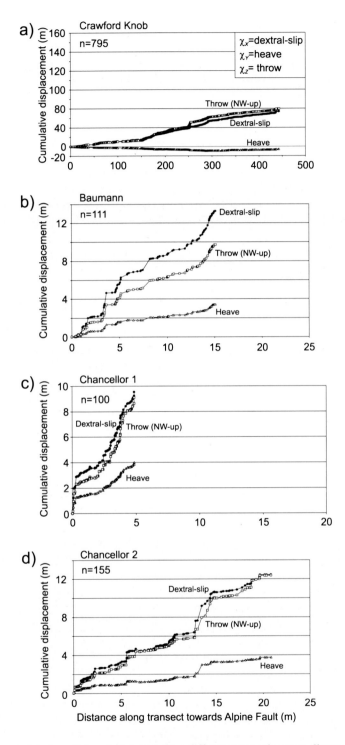

Figure 8. Cumulative displacement plots from four main transects from different areas using a net slip vector pitching 36°SW. See Figure 2c for location of each transect site. Displacements are oriented relative to Alpine Fault trace and separated out into dextral strike-slip component (filled circles), heave component (unfilled triangles), and throw component (unfilled squares). Ratio of horizontal to vertical scale is constant between sites so that bulk shear-strains (slopes of cumulative slip data) are directly comparable. (a) Crawford Knob; (b) Baumann Glacier; (c) Chancellor transect 1; (d) Chancellor transect 2.

maximum compressive stress, σ_1, is predicted to lie within the plane that contains the shear direction and the pole to the fault (Figure 5d). For a coefficient of friction of ~0.6, σ_1 will lie ~30° away from the shear plane (Figure 5d). An interesting corollary to the near-vertical orientation of these backshears, combined with their oblique-slip kinematics, is that the greatest and least principal compressive stresses, σ_1 and σ_3, respectively, are non-horizontal with respect to the Earth's surface (Figure 5d).

Differential stresses resulting in shear-extensional (mixed mode I-II, I-III) or Griffith failure are inferred to have been >100 MPa, and possibly as high as 300 MPa during back-shear failure based on Mohr-Coulomb analysis [*Wightman et al.*, in review]. This would have required lithostatic or even supra-lithostatic fluid pressures at the time of failure, i.e., a fluid pressure ratio, λ, ≥ 1. Fluid pressures during vein deposition recorded by the fluid inclusions in the infilling veins (e.g., 310 MPa) imply a post-failure fluid pressure of $\lambda = 0.55$–0.6 [*Wightman et al.*, in review]. These authors interpret that the backshears formed in a zone of transiently elevated differential stresses above the corner of the Alpine Fault ramp. A combination of high differential stresses and high fluid pressures led to a short-lived pulse of brittle failure at depths of 20–25 km, well below the current seismically determined brittle-ductile transition zone at 8–10 km [*Leitner et al.*, 2001]. Fluid pressure dropped significantly post-failure, suggesting a cycling of fluid pressure during the active "life" of a backshear.

QUANTIFYING DEFORMATION AND STRAIN IN THE HANGINGWALL OF THE ALPINE FAULT

Shear-Related Bulk Deformation and Finite Strain

Cumulative displacement plots of the three components of slip along the four main field transects are displayed in Figure 8 using an assigned uniform net-slip vector for all the transects, pitching 36° SW. The difference between positive and negative heave values reflects the change in dip of the backshear from a southeast to a northwest dip, respectively, with a corresponding change from nominal mild extension to mild compression in the horizontal direction perpendicular to strike. For example, the change in incremental heave polarity (sign of the slope on the cumulative plot) seen in Figure 8a at ~300 m reflects a change from predominantly northwest dipping (reverse-slip) shear zones to predominantly southeast dipping (normal-slip) shears at that site. Apart from this variation in heave-polarity, the cumulative slip curves are generally quasi-linear with only minor steps in the curve where single, relatively large offsets have locally been calculated.

These three displacement profiles at each site, χ_x, χ_y, χ_z (Figure 8) can be used to estimate graphically the bulk shear-strains (slopes) in the three coordinate directions, e.g., dextral, heave, and throw; that is, to quantify the rate of slip accumulated along each transect. The shear strains inferred from these slopes (Table 2) describe the bulk-scale strain accommodated by shearing in the hangingwall of the Alpine Fault and ignores any deformation in the wall rocks to these shears (Table 2) and so are minimum values. The bulk shear-strains determined for the five transect data sets presented here show no significant variation along strike of the array (Table 2). The bulk horizontal shear-strain parallel to X (Alpine Fault strike direction), γ_x, reflects the amount of dextral-slip accommodated by the shears. This component has a mean value of 0.46 ± 0.12, which corresponds to a horizontal angular shear strain, ψ_x, of $25 \pm 6°$. The vertical shear-strain, γ_z, is inferred to be a measure of the dip-slip shearing accomplished across the toe of the Alpine Fault ramp to effect an east-tilting of the Pacific Plate crustal section across that bend. This component has a mean value of 0.41 ± 0.14 (Figure 5, Table 2). The small variations in displacement gradients between transects preclude any inference from site to site as to any changes in dip of the underlying fault ramp over which those rocks have been displaced.

Ramp-Related Bulk 3D Strain and Strain Partitioning

The degree of tilting that the backshears would have accommodated on any original horizontal marker surfaces over the length of the transect can be described by $\psi_z = \tan^{-1}(\gamma_z)$, where ψ_z is the tilt, or shear angle (Figure 5). ψ_z is relatively consistent between transects ($\psi_z = 22 \pm 8°$), except for the first Chancellor transect where ψ_z is locally as high as 64°, and can be viewed as a minimum estimate of the ramp-step angle, α, at the toe of the Alpine Fault ramp. Apart from this one anomalously high ψ_z value at the Chancellor 1 transect set, the apparent angles of bulk tilting of an originally horizontal surface that the backshears accommodate along the transects suggest a minimum angle of $22 \pm 8°$ for the change in dip between the decollement in the Pacific Plate beneath the Southern Alps and the dipping Alpine Fault ramp along which they are exhumed. The anomalously high vertical (z) shear-strain and tilt angle recorded in the Chancellor 1 transect (Table 2), could be due to sampling problems. For instance, at 4.5 m, this transect is by far the shortest of the transects, and although 100 backshears were measured along it, it may have traversed a zone of anomalously closely-spaced backshears, or a particularly high bulk strain area for the Southern Alps backshear array as a whole. Given that the other four transects yield consistently lower bulk angular shear-strains of $\psi_z = 10–33°$, we will consider this transect to constitute an outlier.

Given that previous geological studies of the Alpine Fault mylonites and geophysical imaging of the crustal structure beneath the Southern Alps have led workers to infer a variable mean southeast dip on the Alpine Fault of $\alpha = 45–60°$ [e.g., Sibson et al., 1981; Norris et al., 1990; Davey et al., 1995], the backshears accommodate an eastward tilting of the Pacific Plate crust onto the Alpine Fault of only $22 \pm 8°$. If up-ramping has been accomplished by an escalator-like process, this leads us to ask where or how the potential 20–40° (e.g., $\alpha - \psi_z$) of additional vertical (west-side-up) shearing at depth has been accommodated during the translation of the Pacific Plate across the Alpine Fault. One possibility that could explain this apparently absent vertical shear-strain recorded by the backshear array, is that the Alpine Fault ramp at depth may not be as steep as has previously been inferred [e.g., Davey et al., 1995; Kleffman et al., 1998]. Another possibility might be that the actual net-slip vector along the backshears was steeper pitching than 36° SW, in which case more vertical shearing would have been accommodated on the backshears than that which we calculated using a slip vector pitch of $36 \pm 5°$. Even if the net-slip vector of the backshears was as high as 45° SW (the maximum net-slip estimated from the kinematic modelling of the transect data) [Wightman, 2005], the calculated ψ_z would only increase to $28 \pm 23°$.

A third possibility that could explain the "missing" vertical strain is that the Alpine Fault, although dipping at 45–60° SE near the surface, may gradually steepen to this attitude from the approximately flat-lying decollement in a series of two or more ramp-steps. The vertical deformation recorded by the backshears may only be reflecting deformation occurring over one of these bends. If this were the case, and the Alpine Fault gradually increased in dip at a shallower level in the crust, then the remaining vertical shearing needed to tilt the rocks onto a 45–60° dipping ramp may have been accommodated by flexure of the Pacific Plate during continued uplift rather than the discrete backshear deformation at the base of the ramp.

Our preferred interpretation is that the Alpine Fault does dip 45–60° SE and accomplishes this change in dip over a single bend located at the base of the ramp where the Alpine Fault merges with the decollement. The extra ramping-related shear-strain is partitioned as distributed ductile deformation into the schist between the discrete and spaced shear zones whose slip was documented on the fault-offset transects, although it is difficult to distinguish this interpretation from the possibility of a two-step bend in the Alpine Fault ramp where deformation is accumulated further up the ramp by flexural bending. Several

lines of evidence support the interpretation of distributed deformation partitioned between the discrete backshears: (1) the total strain, γ, recorded by slip on the discrete backshears decreases with structural depth (c.f. Figures 7a–c). Field relationships indicate that this decrease in γ is accompanied by an increasing width of the ductile deformation zones surrounding the shears (Figure 7c), including into the host quartzofeldspathic schist, while χ_{net} along the backshears does not significantly change (Figure 7a); (2) at deeper levels of the array the backshears are observed to anastomose and merge with one another and locally with the dominant foliation in the wall rock schist (Figure 3f) and become less laterally continuous than higher in the structural rock package; (3) previous work on shear sense indicators in distributed, widely sheared garnet-zone schist at structural levels below the strongly localised shearing in the backshear array invariably shows a northwest-up sense of shear, similar to the shear accommodated more discretely in the backshear array, and opposite to that recorded in the mylonite zone of the Alpine fault [Little et al., 2002a; Little, 2004; Little et al., this volume]. Holcombe and Little [2001] estimated a bulk vertical shear-strain related to this shearing in the garnet-zone schists of $\gamma_z = {\sim}0.6$, equivalent to a tilting angle of $\psi_z = 31°$. This result suggests that at structural depths beneath the backshear array most of the ramp-related tilting required for an Alpine Fault dipping 45–60° SE could be accounted for by distributed ductile deformation within the rheologically weaker, higher grade schist as underestimation is always likely due to the possibility of highest strain zones not being sampled [Holcombe and Little, 2001]. It is worth noting, however, that an angular shear strain of 31° recorded in the garnet-zone rocks is not significantly greater than $\psi_z = 22°$ recorded in the structurally higher backshear array, and may simply reflect a similar amount of angular shear being accommodated by different mechanisms at different structural levels.

These lines of evidence lead us to infer that the slip accommodated by backshears diffuses downward into the host schist where it is accommodated by an increasing proportion of distributed ductile shear involving microscale slip along foliation planes in the wall rocks [see Little et al., this volume]. In other words, the 'missing' vertical shear-strain suggested by the transect data (i.e., $\alpha - \psi_z$) is probably accommodated by a distributed component of oblique shear (up-to-the-northwest) that is not measurable by the discrete offsets on the spaced backshears, and is additional to the more obvious deformation. Such shear distributed throughout the bulk rock (between the spaced backshears) could account for the remaining $(\alpha - \psi_z) = 20–40°$ of ramp-related tilting of Pacific Plate onto the dipping Alpine Fault ramp as is supported by the shear sense indicators observed by *Little et al.* [2002a] and *Little* [2004], and the shear strain estimate of Holcombe and *Little* [2001]. It is not possible from these data to determine where on the Alpine Fault ramp this extra vertical shearing is accommodated. The deformation may occur over a single sharp bend, α, at the base of the ramp, contemporaneous with the backshearing process (our preferred interpretation), or it may occur during tilting of the Pacific Plate over one or more additional bends in the Alpine Fault (α_1, α_2, α_3, etc.) located at shallower crustal levels.

Ramp-Related Strain-Rate Estimates

The spacing, δ_i, between backshears, together with known plate convergence velocities, can be used to calculate the lifespan of the backshears assuming an escalator-like model of sequential activation such as that reproduced by many analogue and numerical modelling studies [e.g., *Chester et al.*, 1991; *Beaumont et al.*, 1996; *Erickson et al.*, 2001]. Using a plate motion velocity of 37 mm/yr towards 071° (*De Mets et al.*, 1994), the boundary-normal convergence-rate, v_{normal}, across the backshears can be calculated. This convergence-rate would be slower, and the resulting lifespan of an individual backshear subsequently longer, if the plate motion velocity of *Cande and Stock* [2004] were used. Whatever the absolute velocity, this boundary-normal convergence-rate will be variable along strike of the array, being a function of the obliquity of the ramp-strike to the plate vector as here inferred to be reflected in the corresponding strike of the hangingwall ramp backshears. By using the strike of the backshears as a proxy for Alpine Fault ramp strike at depth, we can calculate site-specific plate convergence-rates by application of Nuvel-1A PAC-AUS pole [*De Mets et al.*, 1990, 1994]. The average active "life" of a backshear, Δt, is represented by $\Delta t = \delta_{local} / v_{normal}$ where δ_{local} is the local inter-shear spacing. Average Δt over the entire array is ~30 years (Table 2). An estimate of the maximum strain-rate, $\dot{e}_{max}$, of the deformed quartz veins in the array can therefore be calculated by

$$\dot{e}_{max} = \frac{\gamma}{\Delta t} = \frac{\chi_{net} / \omega_{def}}{\delta / v_{normal}} \qquad (1)$$

The average strain-rate between all transects calculated in this way is $1.0 \times 10^{-8}\,\text{s}^{-1} \pm 1.7 \times 10^{-7}\,\text{s}^{-1}$ with some variation between strain-rates calculated from different transects. This is a maximum estimate because this escalator model assumes that only one backshear was active at any time and that all deformation accrued on this backshear must have been accomplished prior to the adjacent shear becoming activated.

If, on the other hand, multiple backshears were simul-

taneously active, then the active lifespan of an individual backshear may be longer, resulting in slower calculated strain-rates. A minimum estimate of shear strain-rate, $\dot{e}_{min}$, through the deformed quartz veins can be made by estimating a maximum width of rock across which all backshears might have been active simultaneously, ω_{active}. Thus,

$$\dot{e}_{max} = \frac{\gamma}{\omega_{active} / v_{normal}} \qquad (2)$$

As a minimum estimate of strain-rate, we take ω_{active} to be equal to the entire width of the exposed brittle-ductile backshear array (e.g., $\omega_{active} = y_{array} = 2$ km). By widening the putative deformation zone, ω_{active}, more time is available for the same amount of finite shear-strain to accumulate across a ramp-step compared to the model of instantaneously activated (and deactivated) backshears. Thus this latter technique yields a slower shear-strain-rate estimate of $1.8 \times 10^{-12}\,\text{s}^{-1}$ for an assumed deformation width of 2 km of simultaneously active backshears. This deformation width is arbitrary and smaller blocks of Pacific Plate up to this maximum length of y_{array} may be sheared simultaneously. The numerical models of *Erickson et al.* [2001] were created using a finite-element grid with a spacing of 50 m, but the models predict backshears forming with a ω_{active} of ~400 m and spaced approximately 250 m apart. Therefore, deforming zones on the order of 100s of metres, as presented here, are not an unreasonable maximum estimate of deformation width. This maximum ω_{active} gives an indication, however, as to the magnitude of the minimum strain-rates permissible in the zone ($\sim 1.8 \times 10^{-12}\,\text{s}^{-1}$), which are still relatively fast by geological standards ($1 \times 10^{-14}\,\text{s}^{-1}$) [*Pfiffner and Ramsey*, 1982].

Elevated strain-rates above the toe of the Alpine Fault ramp have been modelled by various workers [e.g., *Braun and Beaumont*, 1995; *Beaumont et al.*, 1996] who anticipate elevated differential stresses and strain-rate nucleating from such a fault-ramp bend. Rocks only transiently experience elevated strain-rates (and differential stresses) as they are moved into the Southern Alps orogen, across the fault ramp bend where backshearing occurs, and then are passively translated up the Alpine Fault to the surface. The extreme strain-rates that occur during the backshearing process are inferred to be a short-lived phenomenon, and the rocks undergo little further deformation during their subsequent uplift history.

Some of the deformed quartz marker veins, although now microstructurally annealed in appearance, preserve crystallographic preferred orientation fabrics that are very weak to random despite having macroscopic ductile shear-strains of 5–10, indicating that the ductile deformation was partially accommodated by mechanisms other than dislocation creep

[*Wightman et al.*, 2006]. Diffusion creep-accommodated grain boundary sliding (GBS) has been invoked to explain the weak fabrics in these veins [*Wightman et al.*, 2006]. Quartz is unable to deform ductilely at temperatures much below 300°C [e.g., *Hirth and Tullis*, 1994; *Stockhert et al.*, 1999], which is the temperature that approximately coincides with the seismogenic brittle-ductile transition zone in the crust. This "cut-off" temperature for ductile quartz deformation allows the estimation of a maximum period of time, over which backshear deformation could take place if the shearing initiated, as we have inferred, at 21 ± 5 km depth [*Wightman et al.*, in review] and was arrested at the base of seismogenic zone under the central Southern Alps, at a depth of ~8–10 km [*Leitner et al.*, 2001]. Assuming a late-Quaternary dip-slip-rate on the Alpine Fault of ~10 mm/yr [*Norris and Cooper*, 2001], this simple approximation yields a maximum available time window of 1.5 Myrs to accumulate the ductile shear-strain in these rapidly exhumed veins. If time to anneal these quartz veins is subtracted from the exhumation period (i.e., ~1 Myr) [*Wightman*, 2005], as based on an analysis of published experimental grain-growth rates in quartz, then the time available to accumulate the macroscopic ductile shear-strain reduces to a maximum of 0.5 Myrs. This limit implies minimum strain-rates of 3×10^{-13} s^{-1}, an estimate that is totally independent of the escalator model-based strain-rate estimates.

Provided that differential stresses did not exceed the threshold for brittle failure in the quartz vein, a condition that is partially dependent on the fluid pressure in the immediately surrounding rock, the quartz could deform by ductile processes, including combinations of dislocation and diffusion creep. At fluid pressure ratios of 0.95, i.e., near-lithostatic, as required to achieve the original shear-extensional failure in the schist at 21 km depth, the maximum differential stress possible in the quartz without causing brittle failure is ~100 MPa, assuming a coefficient of internal friction of 0.6 [*Wightman et al.*, in review]. Using published flow laws for quartz, this differential stress would be sufficient to create shear strains of ~5 in the available time window if all ductile shear was to accumulate by dislocation creep mechanisms (a strain-rate of ~3×10^{-13} s^{-1}) over the entire 0.5 Myrs available. The differential stress would have to have been higher than this, however, if any significant percentage of shear-strain in the veins was to be accommodated by a late increment of diffusion creep-accommodated GBS, as this is a lower differential stress / slower strain-rate deformation mechanism [*Wightman et al.*, 2006].

One of the more intriguing results that has been revealed by this analysis of the deformation conditions under which the quartz marker veins underwent the backshearing event is that quartz in nature appears to be significantly weaker than

that deformed experimentally in laboratories under elevated temperatures and non-geologically realistic strain-rates. Two potential reasons are proposed to explain this apparent disparity in quartz strength, both attributable to the presence of a fluid phase: 1) a reduction in fluid pressure that occurred either in the quartz veins themselves, or at the boundary between the veins and the tip of the brittle fault, permitted differential stresses to increase significantly above the background differential stresses in the bulk rock without initiating brittle failure of the quartz; or 2) the water trapped as intragranular fluid in the quartz vein had a high fugacity enabling dislocation creep deformation to occur at geologically fast strain-rates (~1×10^{-11} s^{-1}) but at differential stresses below that required to initiate brittle failure. Both of these scenarios could account for the apparent weakness of the quartz marker veins in the Southern Alps relative to the strength predicted by experimentally determined quartz flow laws and Byerlee failure theory [*Wightman et al.*, in review]. Put another way, the experimental flow laws predict that quartz undergoing deformation at the strain-rates apparently required to achieve the observed shear-strain in the time-frame available, should have been deforming at differential stresses large enough to induce brittle failure, yet such fractures typically do not exist. We infer, therefore, that the presence of water during this deformation of the quartz veins was integral in the accumulation of shear-strain by ductile deformation mechanisms without allowing the transition to brittle deformation to occur. The backshears are infilled by syntectonic quartz veins, allowing us to attempt to quantify the volume of fluid that flowed through the backshear system, the presence of which was fundamental in allowing this ductile deformation of the marker veins in the schist to occur.

ESTIMATION OF FLUID VOLUMES REQUIRED TO PRECIPITATE INFILLING-VEINS AND ITS TECTONIC IMPLICATIONS

Due to the systematic spacing of the brittle-ductile faults and the relatively constant thickness of the veins that infill them, an estimate of the total volume of quartz in these veins throughout the fault array can be made. From this information one can estimate the total volume of fluid that must have flushed through the system to result in that quartz precipitation. The total volume of quartz, V_q, present in a 1-mm thick infilling vein in the brittle-ductile fault array can be estimated by:

$$V_q = Th \times x_{array} \times z_{array} \qquad (3)$$

where *Th* is the average width of a vein (~1 mm), x_{array} is the strike-length of the array (~30 km), and z_{array} is 1 km,

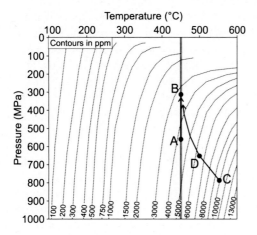

Figure 9. Contours of silica solubility (in ppm) in pure water as a function of pressure and temperature after *Bons* [2000]. Contours are based on the equation for silica solubility as a function of temperature and specific volume of pure water [*Fournier and Potter*, 1982]. Point A represents the backshears formation temperature and fluid pressure (i.e., lithostatic); point B represents conditions after backshear failure with sublithostatic fluid pressures ($\lambda = 0.6$); point C represents the maximum original depth and temperature of the fluid that precipitated the shear-infilling veins (i.e., fluid originating in the mylonite zone at the base of the hangingwall); point D represents a more reasonable estimate for the original source of the fluids, at depth and temperatures structurally just below the observed backshear array. See text for details.

the elevation difference between the top and bottom of the array where veins were observed in field exposure (Figure 5a). This simple approximation yields V_q of a 1 mm-thick vein of 3×10^4 m^3. Throughout the entire array width, y_{array}, assuming a mean spacing of brittle-ductile faults of 0.5 m, the total volume of quartz present is 1.2×10^8 m^3. The volume of fluid, V_f, required to precipitate this volume of quartz, depends on the solubility of silica in the fluid. Precipitation of quartz from solution would have been due to a change in fluid chemistry. This was most likely caused by changes in fluid temperature and/or pressure as the solubility of quartz is strongly temperature dependent with only a slight pressure dependency, particularly at high temperatures and low pressures (Figure 9). Fluid inclusions trapped in the infilling veins record a H_2O-NaCl fluid, with salinites up to 5.2 wt% (equivalent) NaCl [*Wightman et al.*, in review]. Such a low salinity value does not significantly alter the solubility of quartz in an aqueous fluid at the temperatures and pressures under consideration here [*Newton and Manning*, 2000; *Shmulovich et al.*, 2001]; therefore, for simplicity's sake, we will assume a pure aqueous fluid.

The vein-precipitating fluids have been shown to be metamorphic in origin [*Wightman*, 2005], and have presumably risen through the crust from the base of the Pacific Plate hangingwall where fluids were released due to deformation-driven dehydration metamorphism within, and immediately above, the mylonite zone [e.g., *Koons et al.*, 1998]. Upward migration of these fluids in the absence of fractures is a function of the background intrinsic permeability of the schist. Preliminary permeability measurements on the host biotite-grade schist were undertaken using a gas permeameter at Woods Hole Oceanographic Institution and are presented in Table 3 (see Figure 2c for sample locations). These permeabilities represent the background permeability in the host rock at room temperature and pressure, and would have been lower under the overburden pressures present during formation of the brittle-ductile faults. Although indicative only, the wall-rock schist appears to be more permeable parallel to the foliation than in a direction perpendicular to it, but it is quite low overall, averaging around 1×10^{-17} to 1×10^{-18} m^2 (Table 3). This permeability is sufficiently low to suggest that the wall rock schist could support a fluid pressure close to lithostatic without significant fluid being expelled out of the rock, and upward migration of fluid through the rocks, in the absence of any fractures, would be inherently slow. Geophysical seismic and electrical data beneath the central Southern Alps record the presence of interconnected fluids at depth that appear to escape once they reach the seismically defined brittle-ductile transition zone in the crust (Figure 1) [*Stern et al.*, 1997, 2001; *Wannamaker et al.*, 2002, 2004]. Once fluids are in the upper, brittle region of the crust they are inferred to approach the surface through induced hydrofractures [*Sibson and Scott*, 1998].

We infer that these metamorphic fluids were driven upwards through the Alpine Fault hangingwall from depth to the brittle upper crust due to hydraulic pressure gradients established by failure of the brittle-ductile faults, by a mechanism similar to seismic pumping through faults at shallower crustal levels [e.g., *Sibson*, 1996]. This drop in fluid pressure within the fracture immediately following failure would have created a pressure gradient between the backshear and its wall rock. Any fluid pressure gradient in the crust will try to equilibrate, accomplished by flow of fluid through the wall rocks towards the fracture, with fluid moving both laterally through the rocks (e.g., point A to B in Figure 9), as well as flowing along the fracture from deeper structural levels (e.g., points C to D to B in Figure 9). The rate at which this equilibration process can have occurred is a function of the pressure differential, the permeability of the host rock, and the rate at which the fracture was resealed by infilling mineralisation.

It is possible that the fractures and infilling veins may have formed in a cyclic crack-seal process where fractures were incrementally created and subsequently healed by pre-

Table 3. Permeability of host schist parallel and perpendicular to dominwant foliation.

Sample ID	Location	Easting	Northing	Elevation (m)	Permeability (m^2) Parallel	Perpendicular
18I02	NE of Franz Josef Glacier	2284179	5745690	1600	3.9×10^{-15}	7.3×10^{-18}
20I02	NE of Franz Josef Glacier	2283924	5745733	1540	3.1×10^{-17}	2.0×10^{-18}
MHCH09	NE of Fox Glacier	2277249	5740071	1685	1.1×10^{-17}	3.9×10^{-17}

Permeability measurements were made at Woods Hole Oceanographic Institution by Wenlu Zhu, measured parallel and perpendicular to the dominant foliation in the samples using a gas permeameter and are indicative only.

cipitation from a stationary fluid in the fracture due to multiple small drops in fluid pressure [e.g., *Fisher et al.*, 1995]. Geological field and microstructural evidence supporting such crack-seal processes during deposition of the infilling veins is rarely preserved. Very few infilling veins show any record of multiple fluid events during deposition, such as margin-parallel biotite or chlorite selvages incorporated into the veins, suggesting that crack-seal processes were not frequently active during fluid flow through the brittle-ductile faults. Uniform ^{18}O isotopes across the veins and the homogeneous densities of fluid inclusion preserved in vein quartz and calcite, support this interpretation [*Wightman*, 2005]. Although fluid pressure drops would have occurred within a fluid present at the time of backshear failure, leading to some precipitation of silica, we infer that flow of fluid from the surrounding rocks through the fractures must have occurred in order to precipitate the volume of quartz evident in the infilling veins in the absence of any multiple crack-seal process.

A model for fluid flow during the lifetime of a single backshear implies two (albeit related) sources for the fluid that precipitated the infilling veins, summarised in Figure 9: (1) fluid drains from the wall rocks immediately adjacent to the backshear fracture (flow path from A to B); and (2) fluid is tapped by the fracture from deeper structural levels in the crust (flow from point D to B). Precipitation of the infilling veins occurs at $450 \pm 50°C$ at fluid pressures of 310 MPa ($\lambda = 0.6$; point B in Figure 9). The first scenario has fluid originating at (A), at the same structural level (and therefore the same temperature) as the precipitating infilling vein (B) but at lithostatic fluid pressures. Precipitation must be caused by variation in quartz solubility related to reduction in fluid pressure, not fluid temperature (Figure 9). In this scenario, fluid does not necessarily flow through the fracture; a stationary fluid experiencing a pressure drop will precipitate silica into the fracture. However, due to the lack of evidence for crack-seal processes, as outlined above, there must be some flow of fluid from the surrounding wall rocks.

Reduction in fluid pressure from lithostatic to $\lambda = 0.6$ leads to a 1000 ppm (~25%) reduction in silica solubility at 450°C. The precipitation of V_q would require $\sim 3 \times 10^7$ m^3 of fluid to flow through the backshear.

The second scenario involves fluid flowing from deeper structural levels (e.g., points (C) and (D), Figure 9), and precipitating at point (B). Point (C) represents the maximum depth and temperature from which the fluid could flow, as the mylonite zone at the base of the Pacific Plate hangingwall is estimated to be undergoing metamorphism at 550–600°C at ~30 km depth [*Vry et al.*, 2004]. Point (D) is a depth between the base of the hangingwall and the location of the brittle-ductile shears. Its precise location may be variable but represents a depth where extension of the brittle fracture downward through the crust may tap deeper crustal fluids. Fluid travelling from anywhere along the flow of (C) to (B) will experience changes in silica solubility due to variation in both fluid temperature and pressure. Fluids originating at deeper (and therefore hotter) levels in the crust have dramatically higher initial solubility, resulting in a larger reduction in solubility and greater precipitation of silica during flow (Figure 9).

These two scenarios are end members for the source of the fluid, one locally derived, the other external to the backshear system. The fluid that deposited the infilling veins is likely to have been sourced from a combination of the two scenarios, with quartz being precipitated from fluid migrating laterally into the backshear fracture from the wall rocks due to a pressure differential, as well as a component of fluid being channelised upwards from deeper (and hotter) levels in the hangingwall.

GEODYNAMIC IMPLICATIONS AND RHEOLOGICAL FEEDBACK OF FLUID FLOW

As the Pacific Plate steps up onto the base of the Alpine Fault ramp, the hangingwall rocks experience vertical shearing and elevated differential stress levels (Figure 1)

[*Koons and Craw*, 1991b; *Braun and Beaumont*, 1995; *Koons et al.*, 1998]. Modelling by *Koons et al.* [1998] and *Koons and Craw* [1991b] of this transitional region suggest that shear strains will be high, and that the deformation will cause failure on steep to vertical faults. Any metamorphic fluids released by metamorphism at the base of the Alpine Fault ramp and along the mid-crustal decollement (Figure 1) due to thickening of the crust during convergence should migrate upwards initially along low-permeability foliation planes and then focussed along these high permeability fracture planes, driven by transient near-hydraulic fluid pressure gradients.

The near-vertical backshears in the Southern Alps are interpreted to be failure planes that indeed formed as the hangingwall crust was displaced through a zone of high differential stress immediately above the Alpine Fault ramp. In their vertical orientation, the backshears acted as conduits to allow the escape of significant quantities of metamorphic fluids that were released during prograde metamorphism of the lower crust that were otherwise unable to escape through ductilely deforming lower crustal rocks. The presence of these fluids under lithostatic pressures, located in the lower and mid- crust above the decollement and Alpine Fault plane, can be considered in terms of the geodynamic implications for the Southern Alps orogen. The rheology of the crust is weakened by the presence of fluids, both at the intergranular scale by decreasing the differential stresses required for ductile deformation (e.g., hydrolytic weakening) [*Tullis and Yund*, 1989; *Tullis et al.*, 1996] and on a macroscopic scale where fluids reduce the effective normal stresses acting across fractures [e.g., *Cox*, 2002]. The fluid flow recorded by veins infilling the exhumed brittle-ductile backshears indicates metamorphic fluids were present in significant quantity during backshear deformation at depth under the paleo-Southern Alps, whereas seismic and electrical data [*Stern et al.*, 2001; *Wannamaker et al.*, 2002, 2004] record the presence of connected fluids in the modern orogen. These two data sets suggest that metamorphic fluids under the Southern Alps are released during prograde metamorphism of the lower crust driven by continued crustal thickening at the base of the Alpine Fault ramp [*Koons et al.*, 1998; *Stern et al.*, 2001], leading to embrittlement of the mid- to lower crust.

The brittle-ductile faults are inferred to have formed above the Alpine Fault ramp in the transition zone between the outboard and inboard regions of the orogen at 21 ± 5 km depth (Figure 1), and to have acted as conduits to tap deep to mid-crustal metamorphic fluids. This proposed vertical flux of fluid is approximately coincident with one of the three vertical "conduits" of high conductivity observed in the magnetotelluric data (Figure 1), which has previously been interpreted to represent fluids becoming interconnected

and escaping upwards through the brittle-ductile transition zone by hydrofracturing [*Wannamaker et al.*, 2004]. Assuming that a similar backshearing process is currently active beneath the modern Southern Alps orogen, the vertical backshears may be acting to channelise fluid flow from depth, causing the observed conductivity anomaly, until the fluid reaches sufficiently shallow depths to allow hydrofracturing to occur, i.e., the steady-state brittle-ductile transition zone at ~8–10 km. The escape to shallow depths of these lower crustal metamorphic fluids is reliant on the formation of the vertical backshear structures to channelise flow as the background permeability of the biotite-grade schist is sufficiently low to inhibit fluid migration, an interpretation supported by the presence of the high conductivity anomaly beneath the Southern Alps [*Wannamaker et al.*, 2004].

The presence of this fluid at lithostatic pressures also plays an integral role in the development of the vertical backshears, allowing transient embrittlement of the mid- to lower crust under the central Southern Alps, which, in turn, allow fluid channelisation and escape. In strike-slip or compressional tectonic regimes, lithostatic or near-lithostatic fluid pressures are required in order to achieve brittle shear failure at 20–25 km depth, and lithostatic or supra-lithostatic fluid pressures are required to achieve extensional failure [e.g., *Sibson and Scott*, 1998; *Sibson*, 2000]. At lower fluid pressures, the effective normal stresses are too low to induce brittle failure at that depth, and the rocks will instead undergo ductile deformation. The presence of high fluid pressures is therefore a requirement for backshear formation under the central Southern Alps, particularly as the build-up of differential stresses in the mid- to lower crust is inhibited due to the weakening effect of fluid.

EVIDENCE FOR RAMP-RELATED DEFORMATION ACCOMMODATED AT STRUCTURALLY HIGHER CRUSTAL LEVELS IN THE PACIFIC PLATE

Much of the discussion above has focussed on how the shear strain associated with tilting over the Alpine fault ramp was accommodated at crustal levels within and beneath the observed backshear array. The possibility exists that the strain we have documented along the backshears was accomplished seismically, particularly at higher levels in the crust. If so, earthquakes with focal mechanisms consistent with backshearing deformation may be observable in the modern day Southern Alps orogen. Focal mechanisms for these predicted earthquakes can be calculated assuming a net-slip direction along the backshears of 36° SW. Figure 10a displays the four different focal mechanisms that we would expect from earthquakes along faults with exactly the same slip kinematics as the backshears as these localities, as constrained by the mean pitch of the fault-surface fibre vein lineations at Crawford Knob, Baumann Glacier, Chancellor Ridge, and Sam Peak field sites. Focal mechanisms calculated from real data tend to be restricted to earthquake events greater than M_W 6, as earthquakes smaller than this tend to be too small for their first motions to be unambiguously recorded due to background noise [*Anderson et al.*, 1993]. Despite this, a search of the focal mechanism literature reveals an earthquake event, on 24/06/84, that shows remarkably similar focal mechanism and fault plane solution to that

predicted for a backshear event (Figure 2b) [*Anderson et al.*, 1993]. This focal mechanism is anomalous in this area, as most earthquakes around the central part of the Southern Alps have first motions more consistent with Alpine Fault transpression [*Anderson et al.*, 1993]. This focal mechanism tensor, derived from body waveform modelling, calculates that the fault plane of the earthquake was striking 062° and dipping 83° SE, and that the slip vector had a pitch of 26° SW within this plane (the preferred solution of *Anderson et al.* [1993]). The location of the earthquake hypocentre, at 13 ± 3 km depth, plots within close proximity to the surface pro-

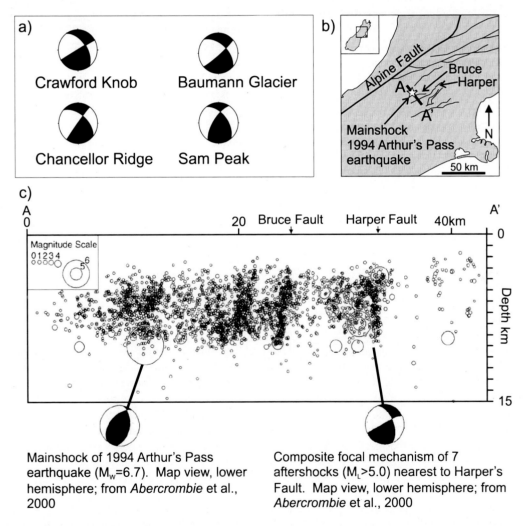

Figure 10. (a) Ideal focal mechanisms plotted in lower hemisphere projection for an earthquake along the average orientation backshear at Crawford Knob, Baumann Glacier, Chancellor Ridge, and Sam Peak, with a net-slip direction pitching 36° SW. Compressional quadrants are shaded. (b) Location map of the 1994 Arthur's Pass M_W 6.7 earthquake, showing location of cross section and the Bruce and Harper faults. (c) Cross section line shown in (b), taken from *Abercrombie et al.* [2000], showing location of mainshock and aftershocks of 1994 Arthurs Pass earthquake near the Harper Fault. Focal mechanism of the mainshock earthquake and composite focal mechanisms derived from the first motions of seven $M_L \geq 5.0$ aftershocks near the Harper Fault are also shown.

jection of the base of the Alpine Fault ramp at 30 km depth assuming a 45°SE dipping fault ramp (Figure 2b).

The earthquake of 24/06/84 was of magnitude M_W 6.1. The regression by *Dowrick and Rhoades* [2004] can be used to estimate the equivalent mean displacement along a fault surface of an earthquake of this magnitude:

$$\log(D) = a + bM_W \qquad (4)$$

where D is the mean displacement in m, M_W is the moment magnitude, and a and b are constants, with a equal to 6.09 ± 0.06 and b constrained to 2.0. The 24/06/84 earthquake is equivalent to ~1 m mean displacement using this regression. This calculated displacement of ~1 m is significantly larger than the mean displacement on the backshears of 15 cm. However, the focal depth of the earthquake was 13 km, ~10 km shallower than our depth estimate for backshear deformation. This is not unexpected, however, as this earthquake reflects seismogenic deformation in the upper, brittle crust, whereas the exhumed backshears formed in the lower crust and show no evidence for gouge, pseudotachylite or other likely products of seismic fault slip. The larger displacement recorded by the M_W 6.1 earthquake may reflect the accumulation, on the time-scale of the seismic cycle, of the slip contributed by aseismic creep on approximately six or seven backshears (i.e., 6 × 15 cm = 90 cm displacement), with the displacement zone narrowing upward onto a single brittle rupture plane in the mid-upper crust.

The focal mechanism of the hypocentre of the 24/06/84 earthquake is also approximately coincident with the subsurface extrapolation of the surface trace of an inferred active fault [*Cox*, unpublished data] that strikes subparallel to the backshear array, and dips at 80° NW (Figure 2b). This dominantly dextral fault also shows some evidence for reverse-slip motion [*Cox*, unpublished data] and its surface trace lies only 5 km to the southeast of the projected base of the Alpine Fault ramp at 30 km depth (assuming an Alpine Fault dip of ~45°SE; Figure 2b), as is consistent with a mean backshear dip of 80° NW.

A M_W 6.7 earthquake at Arthur's Pass, ~ 100 km to the northeast of Franz Josef Glacier, occurred on 18/06/94 at a depth of 5 km [*Abercrombie et al.*, 2000]. It was a predominantly reverse-slip event which is typical for earthquakes in that region, and was located ~25 km to the southeast of the Alpine Fault trace in that area (Figure 10b). The aftershocks of this earthquake, however, show a complex spatial distribution, and have been relocated by *Abercrombie et al.* [2000] using both 1D and 3D velocity inversions (Figure 10c). Although on a much different scale to the exhumed backshears described in this paper in the central Southern Alps, with the Arthur's Pass aftershocks spread out over a ~40 km

wide area predominantly to the southeast of the mainshock, the aftershock activity was concentrated on a spaced array of subparallel NE-SE striking faults that dip steeply to the NNW (Figure 10c), some of which can be correlated with active faults exposed on the surface, including the Harper and Bruce faults. A composite focal mechanism of seven aftershocks greater than M_L 5.0 reveals an oblique-reverse sense of motion with a focal mechanism that is remarkably similar to those predicted for the backshears in the Franz Josef-Fox glaciers region (Figure 10) [*Abercrombie et al.*, 2000].

Despite the lack of geophysical data on fault focal mechanisms in the central Southern Alps, an array of "inferred active" and inactive faults are present between the projected surface trace of the base of the Alpine Fault ramp and the exhumed backshear array (Figure 2b), including the Main Divide Fault Zone, a series of predominantly inactive reverse faults that outcrop to the east of the Main Divide [*Cox and Findlay*, 1995; *Cox et al.*, 1997]. It is possible that these faults represent existing and/or previously active late Quaternary faults that accomplished the ramp-related brittle deformation in the upper crust as the Pacific Plate was upramped and tilted onto the underlying Alpine Fault ramp. If true, this would imply a systematic activation of the faults as the rocks are tilted onto the ramp at depth, in a similar manner to the escalator model for backshear formation (Figure 1). Although "out of sequence" faults may have occurred, a temporal sequence of fault activation and abandonment would be expected, from oldest nearest the exhumed array in the northwest, to the active fault to the southeast where deformation is currently being accommodated.

DEFORMATION PARTITIONING IN THE SOUTHERN ALPS OROGEN

The kinematic shear direction of the backshears indicates that they accommodate a significant component of strike-slip, despite being formed primarily as a result of vertical tilting over a bend in the underlying fault ramp. This slip-partitioned mode of hangingwall deformation (i.e., resolution of dip-slip shear on vertical planes) would be favoured by any rheological process that either weakens the crust of the Pacific Plate in the hangingwall of the Alpine Fault, or that strengthens the basal shear resistance on that slab. In this paper we have shown that a net mechanical weakening of the hangingwall probably resulted from high fluid pressures possibly driven by the upward flux of metamorphic fluid into vertical backshears. The backshears acted as hydraulic conduits that carried a fluid into the mid-crust, and that effected a rheological change on the surrounding rocks. In particular, they apparently hydrolytically weakened the quartzose veins that were deformed at high strain-rates by the ductile backshearing process without undergoing any brittle fracturing. Conversely, such an upward fluid flow might also cause the original source area for these fluids, inferred to be the actively deforming and dewatering Alpine mylonite zone at the base of the Pacific Plate, to become drier and thus potentially stronger. Evidence for a residually dry Alpine Fault mylonite zone is observed in stable isotopes (both oxygen and deuterium), the typical paucity of hydrous phases or other evidence of late-stage retrogression in the rocks, and a general lack of late-stage quartz throughout most of the mylonite zone [*Vry et al.*, 2001].

The net effect of the fluid expulsion and flow, therefore, might have been to cause a net rheological strengthening of the Alpine Fault and associated mylonite zone at the same time that it was weakening the upwardly ramping hangingwall of that fault. The sense of these changes would dynamically favour resolution of plate motion being partitioned, at least temporarily (i.e., during the ramping), from the underlying Alpine Fault and mylonite zone into the near-vertical backshears in the hangingwall. This partitioning of plate motion between the underlying fault ramp and its hangingwall may explain why the vertical backshears accommodated a significant component of strike-slip deformation despite forming to accommodate vertical shearing over the fault ramp. The oblique slip direction of the backshears may also be assisted by the dilatant nature of many of the backshears, as this mode of failure implies total or near-total loss of effective normal stress during failure. This would allow the backshears to slip in whatever direction the maximum resolved stress required.

The evolution of the backshear array suggests the following feedback: deep embrittlement associated with formation of the brittle-ductile failure planes (backshears) opened up efficient vertical conduits through which deeper and suprahydrostatically pressured metamorphic fluid could escape upward from potentially as deep as the basal mylonites and into the Alpine Fault's hangingwall. The rheological effect of this fluid was to weaken the hangingwall further and possibly also to increase the basal traction (differential stress) in the underlying mylonite zone. The nature of these coupled strength changes would have dynamically favoured resolution of dextral-strike slip away from the mylonite zone onto the vertical backshear planes [e.g., *McCaffrey*, 1992], if only temporarily.

GLOBAL SIGNIFICANCE OF BACKSHEARING DEFORMATION

A similar evolution of deformation has been documented by *Axen et al.* [2001] for two major extensional fault systems in the European Alps, the Simplon and Brenner shear zones.

There, brittle-ductile shear zones (the 'early post-mylonitic structures' or EPS, of *Axen et al.* [2001]) develop in the footwall of the fault zone, with brittle failure attributed by *Axen* and coworkers to increasing strain-rates related to the bend in the footwall ramp, under conditions of high fluid pressure [*Axen et al.*, 2001, Figure 4]. These fluids are interpreted to have flowed preferentially along the EPS structures, into the overlying shear zone of the fault itself at the top of the footwall. This fluid flux drains overpressurised fluid from the footwall while weakening the overlying shear zone and allowing ductile deformation to focus into a narrow mylonite zone [*Axen et al.*, 2001].

This rheological evolution of an extensional fault zone and its footwall during exhumation is remarkably similar to that we infer here for the evolution of the backshear array and the transpressional Alpine Fault. This suggests variations in rheology of fault zones and their surrounding hangingwalls and/or footwalls due to fluctuations in stress and strain-rate around major fault bends may not be limited to only one tectonic setting. Such uniform, systematic failure of the crust in a backshear-like fashion, and the associated fluid flux related to that fracturing, may be possible wherever there is a significant bend in a major fault zone, be it a mid-crustal extensional shear zone, such as the Brenner and Simplon faults, or a plate boundary-scale orogen, such as the transpressional Alpine Fault system.

CONCLUSIONS

Backshears in the obliquely convergent Southern Alps orogen formed due to an escalator-like upramping and tilting of the Pacific Plate onto the Alpine Fault ramp at depth. These backshears are near-vertical structures, contrary to predictions made from purely convergent laboratory models, and strike subparallel to the Alpine Fault. The shears are systematically spaced, on average at 34 ± 4 cm (2σ). This spacing between the backshears does not have a fractal or scale invariant distribution and does not vary with increasing depth through the backshear array. At their inferred depth of formation in the lower crust of >20 km, the backshears were expressed by variably brittle to ductile deformation styles in different rock types. They brittlely offset the quartzofeldspathic schist host as narrow (<1–2 mm wide) faults, but ductilely offset pre-existing quartz veins in the schist across 1–2 cm wide ductile shear zones, into which these faults were apparently blunted. The shears become increasingly ductile with paleodepth in the crust, a result of the temperature dependence of the operative flow laws.

Net-slip on the backshears averages 14.1 ± 1.2 cm and does not show any strong correlation with either mean spacing between the backshears, or structurally with depth in the crust, but is relatively uniform. The backshears are covered by syntectonic quartz-calcite veins that display a fault surface lineation along their surface that pitches 36 ± 5° SW, interpreted to represent the mean net-slip direction of the backshears. This non-horizontal pitch reveals that the backshears accommodate significant strike-slip, in addition to up-ramping the Pacific Plate onto the Alpine Fault plane (by dip-slip antithetic to that structure).

The constancy in spacing and net-slip on the shears is inferred to be a reflection of the ramp angle that governed the transient activation and slip of these backshears in overlying rocks of the Pacific Plate that were displaced across this foot-wall ramp. Cumulative displacement gradients recorded through five measured field transects indicate the backshears accommodated a finite horizontal dextral shear-strain of $\gamma_x = 0.46 \pm 0.12$, corresponding to a horizontal angular shear-strain of $\psi_x = 25 \pm 6°$, and a bulk vertical shear-strain of $\gamma_z = 0.41 \pm 0.14$. The minimum vertical shear angle accommodated by slip on the backshears suggests that the ramp step at the base of the Alpine Fault effects a dip change of $22 \pm 8°$. Given other estimates for the dip on the Alpine Fault of 45–60°, an additional 20–40° of net vertical shearing is inferred to be accommodated by a distributed (micro-scale) shearing along the schistosity planes between the discrete backshears or by crustal flexure. The proportion of distributed (inter-backshear) deformation increases with depth until all the vertical shearing is accomplished in a pervasive way below the structural base of the backshear array. At structurally higher levels in the brittle-frictional upper crust of the Pacific Plate, focal mechanisms of historical earthquakes indicate that something akin to backshearing may be still be taking place on active faults, at least locally, perhaps in conjunction with aseismic creep on more numerous, deeper shear structures.

High strain-rates are recorded in the field transects made through the backshear array, with a maximum of $1.0 \times 10^{-8}\,\mathrm{s^{-1}}$ estimated for individually active backshears. Elevated strain-rates are likely to be a transient, short-lived event focused around the bend in the Alpine Fault ramp. The maximum possible duration of ductile deformation that could affect a piece of up-ramped crust prior to its exhumational cooling is 0.5 Myrs. This maximum duration indicates that in order to accomplish the observed ductile shear strains in the backshears of 5–10, the deformed quartz marker veins need to be hydrolytically weakened, deforming at strain-rates faster than those predicted by existing, experimentally-derived quartz flow laws without initiating brittle failure.

We infer that the backshears acted as vertical conduits for the escape of metamorphic fluids from depth during convergence and uplift along the Alpine Fault ramp. The formation of the brittle-ductile faults and the corresponding fluid flow that occurred is coincident with a region imaged by seismic and electrical data that indicate the presence of fluid under high pressure under the modern Southern Alps orogen. Metamorphic fluids flushed through the brittle-ductile faults from depth until the brittle-ductile transition zone was reached at ~10 km depth where fluids are inferred to have escaped by hydrofracturing processes. Lithostatic fluid pressures are required in order for failure of the brittle-ductile faults to occur at 20–25 km depth. This upward draining of metamorphic fluid from the base of the hangingwall would have left the underlying source region near its base residually drier. Fluid expulsion may thus have accomplished a net rheological strengthening along the Alpine mylonite zone at the same time that hydrolytic weakening affected structurally higher rocks of the Pacific Plate crust above the ramp. These fluid-related strength changes in the lower- to mid-crust play an important role in allowing deformation to localise in discrete shear zones in the Alpine Fault hangingwall away from the hot, 'weak' mylonite zone. It is likely that similar variations in crustal rheology occur in other orogens worldwide, in both transpressional and extensional settings, due to fluxing of fluid through the crust. Contemporaneous failure of the crust under both brittle and ductile conditions in a manner comparable to Southern Alps backshearing may be more common than is currently documented, particularly associated with high differential stresses surrounding major fault bends.

Acknowledgements. The authors would like to thank the New Zealand Foundation for Research, Science and Technology for funding this research, and Gary Axen and Tim Byrne for useful comments during revision of this manuscript.

REFERENCES

Abercrombie, R. E., T. Webb, R. Robinson, P. J. McGinty, J. J. Mori, and R. J. Beavan (2000), The enigma of the Arthur's Pass, New Zealand, earthquakes. Reconciling a variety of data for an unusual earthquake sequence, *J. Geophys. Res., 115*(B7), 16,199–16,137.

Adams, C. J. (1981), Uplift rates and thermal structure in the Alpine Fault Zone and Alpine Schists, Southern Alps, New Zealand. *Geological Society of London Special Publication, 9,* 211–222.

Anderson, H., T. Webb, and J. Jackson (1993), Focal mechanisms of large earthquakes in the South Island of New Zealand: implications for the accommodation of Pacific-Australia plate motion, *Geophys. J. Int., 115,* 1032–1054.

Axen, G. J., J. Selverstone, and T. Wawrzyniec (2001), High-temperature embrittlement of extensional Alpine mylonite zones in the midcrustal ductile-brittle transition, *J. Geophys. Res., 106*(B3), 4337–4348.

Batt, G. E., and J. Braun (1999), The tectonic evolution of the Southern Alps, New Zealand: insights from fully thermally coupled dynamical modeling, *Geophys. J. Int., 136*(2), 403–420.

Beaumont, C., P. J. J. Kamp, J. Hamilton, and P. Fullsack (1996), The continental collision zone, South Island, New Zealand: Comparison of geodynamical models and observations, *J. Geophys. Res., 101*, 3333–3359.

Beavan, J., M., Moore, C. Pearson, M. Henderson, B. Parsons, S. Bourne, P. England, R. I. Walcott, G. Blick, D. Darby, and K. Hodgkinson (1999), Crustal deformation during 1994–1998 due to oblique continental collision in the central Southern Alps, New Zealand, and implications for seismic potential of the Alpine Fault, *J. Geophys. Res., 104*, 25,232–25,255.

Berger, P., and A. M. Johnson (1980), First-order analysis of deformation of a thrust sheet moving over ramp, *Tectonophysics, 70*, T9–T24.

Bonini, M., D. Sokoutis, G. Mulugeta, and E. Katrivanos (2000), Modelling hanging wall accommodation above rigid thrust ramps, *J. Struct. Geol., 22*, 1165–1179.

Bons, P.D. (2000), The formation of veins and their microstructures, in *Stress, Strain and Structure, A volume in honour of W. D. Means*, edited by Jessell, M.W., and J. L. Urai, *Journal of the Virtual Explorer, 2*.

Braun, J. and C. Beaumont (1995), Three-dimensional numerical experiments of strain partitioning at oblique plate boundaries: Implications for contrasting tectonic styles in the southern Coast Ranges, California, and central South Island, New Zealand, *J. Geophys. Res., 100*(B9), 18,059–18,074.

Breeding, C. M. and J. J. Ague (2002), Slab-derived fluids and quartz-vein formation in an accretionary prism, Otago Schist, New Zealand, *Geology, 30*(6), 499–502.

Bull, W. B. and A. F. Cooper (1986), Uplifted marine terraces along the Alpine Fault, New Zealand, *Science, 234*, 1225–1228.

Cande, S. C., and J. M. Stock (2004). Pacific-Antarctic-Australia motion and the formation of the Macquarie Plate, *Geophys. J. Int., 157*(1), 399–414.

Chester, J. S., J. M. Logan, and J. H. Spang (1991), Influence of layering and boundary conditions on fault-bend and fault-propagation folding, *Geol. Soc. Am. Bull., 103*, 1059–1072.

Cooper, A. F., and R. J. Norris (1994), Anatomy, structural evolution, and slip rate of a plate-boundary thrust: The Alpine Fault at Gaunt Creek, Westland, New Zealand, *Geol. Soc. Am. Bull., 106*, 627–633.

Cox, S. F., V. J. Wall, M. A. Etheridge, and T. F. Potter (1991), Deformational and metamorphic processes in the formation of mesothermal vein-hosted gold deposits - examples from the Lachlan Fold Belt in central Victoria, Australia, *Ore Geology Reviews, 6*, 391–423.

Cox, S. C., and R. H. Findlay (1995), The Main Divide Fault Zone and its role in the formation of the Southern Alps, New Zealand, *New Zealand J. Geol. and Geophys., 38*, 489–499.

Cox, S. C., D. Craw, and C. P. Chamberlain (1997), Structure and fluid migration in a late Cenozoic duplex system forming the Main Divide in the central Southern Alps, New Zealand, *New Zealand J. Geol. and Geophys., 40*, 359–373.

Cox, S. F., (2002), Fluid flow in mid- to deep crustal shear systems: Experimental constraints, observations on exhumed high fluid flux shear systems, and implications for the seismogenic process, *Earth Planets Space, 54*, 1121–1126.

Davey, F., T. Henyey, S. Kleffman, A. Melhuish, D. Okaya, T. A. Stern, and D. J. Woodward (1995), Crustal reflections from the Alpine Fault zone, South Island, New Zealand, *New Zealand J. Geol. and Geophys., 38*, 601–604.

Davey, F., T. Henyey, W. S. Holbrook, D. Okaya, T. A. Stern, A. Melhuish, S. Henrys, H. Anderson, D. Eberhart-Phillips, T. V. McEvilly, R. Uhrhammer, F. T. Wu, G. R. Jiracek, P. E. Wannamaker, T. G. Caldwell, and N. Christensen (1998), Preliminary results from a geophysical study across a modern continent-continent collisional plate boundary — the Southern Alps, New Zealand, *Tectonophysics, 288*, 221–235.

De Mets, C., R. G. Gordon, D. F. Argus, and S. Stein (1990), Current plate motions, *Geophys. J. Int., 101*, 425–478.

De Mets, C., R. G. Gordon, D. F. Argus, and S. Stein (1994), Effect of recent revisions to the geomagnetic reversal time scale on estimates of current plate motions, *Geophys. Res. Lett., 21*, 2191–2194.

Dowrick, D. J., and D. A. Rhoades (2004), Relations between earthquake magnitude and fault rupture dimensions: how regionally variable are they? *Bull. Seismol. Soc. Am., 94*(3), 776–788.

Erickson, S. G., and W. R. Jamison (1995), Viscous-plastic finite-element models of fault-bend folds, *J. Struct. Geol., 17*(4), 561–573.

Erickson, S. G., L. M. Strayer, and J. Suppe (2001), Initiation and reactivation of faults during movement over a thrust-fault ramp: numerical mechanical models, *J. Struct. Geol., 23*, 11–23.

Ferry, J. M., and G. M. Dipple (1991), Fluid flow, mineral reactions, and metasomatism, *Geology, 19*, 211–214.

Fisher, D.M., S. L. Brantley, M. Everett, and J. Dzvonik (1995), Cyclic fluid flow through a regionally extensive fracture network within the Kodiak accretionary prism, *J. Geophys. Res., 100*, 12,881–12,894.

Fournier, R.O., and R. W. Potter (1982), An equation correlating the solubility of quartz in water from 25° to 900°C at pressures up to 10,000 bars, *Geochem. Cosmochem. Acta, 46*, 1969–1973.

Graham, C. M., K. M. Greig, S. M. F. Sheppard, and B. Turi (1983), Genesis and mobility of the H_2O-CO_2 fluid phase during regional greenschist and epidote amphibolite facies metamorphism: a petrological and stable isotope study in the Scottish Dalradian, *J. Geol. Soc. Lond., 140*, 577–599.

Grapes, R. H., and T. Watanabe (1992), Metamorphism and uplift of Alpine schist in the Franz Josef - Fox Glacier area of the Southern Alps, New Zealand, *J. Met. Geol., 10*, 171–180.

Hedenquist, J. W., and R. W. Henley (1985), Effect of CO2 on freezing point depression measurements of fluid inclusions — evidence from active systems and implications for epithermal ore deposition, Econ. Geol., 80, 1379–1406.

Hirth, G., and J. Tullis (1994), The brittle-plastic transition in experimentally deformed quartz aggregates, *J. Geophys. Res., 99*(B6), 11,731–11,747.

Holcombe, R. J., and T. A. Little (2001), A sensitive vorticity gauge using rotated porphyroblasts, and its application to rocks adjacent to the Alpine fault, New Zealand, *J. Struct. Geol., 23*(6–7), 979–989.

Hutton, C. O. (1940), Metamorphism in the Lake Wakatipu region, western Otago, New Zealand, in *Geological Memoirs, 5*, Department of Scientific and Industrial Research.

Kleffman, S., F. Davey, A. Melhuish, D. Okaya, and T. A. Stern (1998), Crustal structure in the central South Island, New Zealand, from the Lake Pukaki seismic experiment, *New Zealand J. Geol. and Geophys., 41*, 39–49.

Knipe, R. J. (1985), Footwall geometry and the rheology of thrust sheets, *J. Struct. Geol., 7*(1), 1–10.

Koons, P. O. (1994), Three-dimensional critical wedges; tectonics and topography in oblique collisional orogens, *J. Geophys. Res., 99*, 12,301–12,315.

Koons, P. O., and D. Craw (1991a), Gold mineralization as a consequence of continental collision: an example from the Southern Alps, New Zealand, *Earth Planet. Sci. Lett., 103*, 1–9.

Koons, P. O., and D. Craw (1991b), Evolution of fluid driving forces and composition within collisional orogens. *Geophys. Res. Lett., 18*(5), 935–938.

Koons, P. O., D. Craw, S. C. Cox, P. Upton, A. S. Templeton, and C. P. Chamberlain (1998), Fluid flow during active oblique convergence: A Southern Alps model from mechanical and geochemical observations, *Geology, 26*(2), 159–162.

Leitner, B., D. Eberhart-Phillips, H. Anderson, and J. L. Nabelek (2001), A focused look at the Alpine fault, New Zealand: seismicity, focal mechanisms, and stress observations, *J. Geophys. Res., 106*(B2), 2193–2220.

Little, T. A. (1995), Brittle deformation adjacent to the Awatere strike-slip fault in New Zealand: Faulting patterns, scaling relationships, and displacement partitioning, *GSA Bull., 107*(11), 1255–1271.

Little, T. A. (2004), Transpressive ductile flow and oblique ramping of lower crust in a two-sided orogen: insight from quartz grain-shape fabrics near the Alpine Fault, New Zealand, *Tectonics, 23*(TC2013), doi:10.1029/2002TC0011456.

Little, T. A., R. J. Holcombe, and B. R. Ilg (2002a), Ductile fabrics in the zone of active oblique convergence near the Alpine Fault, New Zealand; identifying the neotectonic overprint, *J. Struct. Geol., 24*(1), 193–217.

Little, T. A., R. J. Holcombe, and B. R. Ilg (2002b), Kinematics of oblique continental collision inferred from ductile microstructures and strain in mid-crustal Alpine Schist, central South Island, New Zealand, *J. Struct. Geol., 24*(1), 219–239.

Little, T. A., R. H. Wightman, M. Hill, and S. Ellis (2004), Transient deformation processes in the lower crust of the Southern Alps, New Zealand, in *Symposium on "Tectonics on Human Time Scales"*, Ruhr-University, Bochum, Germany.

Little T., R. Wightman, R. J. Holcombe, & M. Hill (this volume), Transpression models and ductile deformation of the lower crust of the Pacific Plate in the central Southern Alps, a perspective from structural geology.

Manning, C. E. (1994), The solubility of quartz in H_2O in the lower crust and upper mantle, *Geochem. Cosmochim. Acta, 58*, 4831–4839.

McCaffrey, R. (1992), Oblique plate convergence, slip vectors, and forearc deformation, *J. Geophys. Res., 97*(B6), 8905–8915.

Newton, R. C., and C. E. Manning (2000), Quartz solubility in H_2O-NaCl and H_2O-CO_2 solutions at deep crust-upper mantle pressures and temperatures: 2–15 kbar and 500–900°C, *Geochem. Cosmochim. Acta, 64*(17), 2993–3005.

Norris, R. J., and A. F. Cooper (2001), Late Quaternary slip rates and slip partitioning on the Alpine Fault, New Zealand, *J. Struct. Geol., 23*, 507–520.

Norris, R. J., and A. F. Cooper (2003), Very high strains recorded in mylonites along the Alpine Fault, New Zealand: implications for the deep structure of plate boundary faults, *J. Struct. Geol., 25*, 2141–2157.

Norris, R. J., P. O. Koons, and A. F. Cooper (1990), The obliquely-convergent plate boundary in the South Island of New Zealand: implications for ancient collision zones, *J. Struct. Geol., 12*(5/6), 715–725.

Pfiffner, O. A., and J. G. Ramsay (1982), Constraints of geological strain rates: arguments from finite strain states of naturally deformed rocks, *J. Geophys. Res., 87*(B1), 311–321.

Shmulovich, K., C. M. Graham, and B. Yardley (2001), Quartz, albite and diopside solubilities in H_2O-NaCl and H_2O-CO_2 fluids at 0.5-0.9 GPa, *Contrib. Min. Pet., 141*(1), 95–108.

Sibson, R. (1996), Structural permeability of fluid-driven fault-fracture meshes, *J. Struct. Geol., 18*(8), 1031–1042.

Sibson, R. (2000), A brittle failure mode plot defining conditions for high-flux flow, *Econ. Geol., 95*, 41–48.

Sibson, R., and J. Scott (1998), Stress/fault controls on the containment and release of overpressured fluids: Examples from gold-quartz vein systems in Juneau, Alaska; Victoria, Australia and Otago, New Zealand, *Ore Geol. Reviews, 13*, 293–306.

Sibson, R., S. H. White, and B. K. Atkinson (1981), Structure and distribution of fault rocks in the Alpine Fault Zone, New Zealand, *Special Publication of the Geological Society of London 9*, 197–210.

Stern, T. A., P. E. Wannamaker, D. Eberhart-Phillips, D. Okaya, and F. Davey (1997), Mountain building and active deformation studied in New Zealand, *Trans. Am. Geophys. Union, 78*(32), 329, 335–336.

Stern, T. A., S. Kleffman, D. Okaya, M. Scherwath, and S. Bannister (2001), Low seismic wave speeds and enhanced fluid pressure beneath the Southern Alps of New Zealand, *Geology, 29*(8), 679–682.

Stockhert, B., M. R. Brix, R. Kleinschredt, A. J. Hurford, and R. Wirth (1999), Thermochronometry and microstructures of quartz - a comparison with experimental flow laws and predictions on the temperature of the brittle-plastic transition, *J. Struct. Geol., 21*, 351–369.

Strayer, L. M., and P. J. Hudleston (1997), Numerical modeling of fold initiation at thrust ramps, *J. Struct. Geol., 19*(3–4), 551–566.

Thompson, A. B. (1997), Flow and Focusing of Metamorphic Fluids, in, *Fluid Flow and Transport in rocks*, edited by Jamtveit, B. and B. Yardley, Chapman and Hall, London, 297–314.

Tippett, J. M., and P. J. J. Kamp (1993), Fission track analysis of the late Cenozoic vertical kinematics of continental Pacific crust, South Island, New Zealand, *J. Geophys. Res., 98*(B9), 16, 119–16,148.

Tullis, J., and R. A. Yund (1989), Hydrolytic weakening of quartz aggregates: the effects of water and pressure on recovery, *Geophys. Res. Lett., 16*(11), 1343–1346.

Tullis, J., R. A. Yund, and J. R. Farver (1996), Deformation-enhanced fluid distribution in feldspar aggregates and implications for ductile shear zones, *Geology, 24*, 63–66.

Upton, P., P. O. Koons, and C. P. Chamberlain (1995), Penetration of deformation-driven meteoric water into ductile rocks: isotopic and model observations from the Southern Alps, New Zealand, *New Zealand J. Geol. and Geophys., 38*, 535–543.

Vry, J. K., A. C. Storkey, and C. Harris (2001), Role of fluids in the metamorphism of the Alpine Fault Zone, New Zealand, *J. Met. Geol., 19*, 1–11.

Vry, J., J. Baker, R. Maas, T. A. Little, R. H. Grapes, and M. Dixon (2004), Zoned (Cretaceous and Cenozoic) garnet and the timing of high grade metamorphism, Southern Alps, New Zealand, *J. Met. Geol., 22*(3), 137–157.

Walcott, R. I. (1998), Modes of oblique compression: late Cenozoic tectonics of the South Island of New Zealand, *Reviews of Geophys., 36*(1), 1–26.

Wannamaker, P. E., G. R. Jiracek, J. A. Stodt, T. G. Caldwell, V. M. Gonzalez, J. D. McKnight, and A. D. Porter (2002), Fluid generation and pathways beneath an active compressional orogen, the New Zealand Southern Alps, inferred from magnetotelluric data, *Geophys. J. Int., 107*(6), 1–22.

Wannamaker, P. E., T. G. Caldwell, W. M. Doerner, G. R. Jiracek (2004), Fault zone fluids and seismicity in compressional and extensional environments inferred from electrical conductivity: the New Zealand Southern Alps and U.S. Great Basin, *Earth Planets Space, 56*, 1171–1176.

Wellman, H. W. (1979), An uplift map for the South Island of New Zealand and a model for the uplift of the Southern Alps, *Royal Society of New Zealand Bulletin, 18*, 13–20.

White, S. H. (1996), Composition and zoning of garnet and plagioclase in Haast Schist, northwest Otago, New Zealand: implications for progressive regional metamorphism, *New Zealand J. Geol. and Geophys., 39*, 515–531.

Wightman, R. H. (2005), Deformation and orogenesis: the geodynamic significance of a brittle-ductile fault array in the central Southern Alps, New Zealand, Unpublished PhD thesis, Victoria University of Wellington, Wellington, New Zealand.

Wightman, R. H., D. J. Prior, and T. A. Little (2006), Quartz veins deformed by diffusion creep-accommodated grain boundary sliding during a transient, high strain-rate event in the Southern Alps, New Zealand, *J. Struct. Geol., 28*, 902–918

Wightman, R. H., T. A. Little, S. L. Baldwin, and J. W. Valley (in review), Stress, fluid pressure cycling, and transient deep embrittlement of the lower crust recorded in a paleobrittle-ductile transition zone, Southern Alps, New Zealand, *J. Geophys. Res.*

T.A. Little and R.H. Wightman, School of Earth Sciences, Victoria University of Wellington, Wellington, New Zealand.

Geophysical Exploration and Dynamics of the Alpine Fault Zone

Tim Stern[1], David Okaya[2], Stefan Kleffmann[1,3], Martin Scherwath[1,4],
Stuart Henrys[5], and Fred Davey[5]

The Alpine Fault of central South Island New Zealand, can be tracked with seismic reflection methods to depths of ~35 km as a listric-shaped surface with strong reflectivity. Maximum dips of the surface are ~60 degrees at 15 km depth and the dip then lessens with depth until the reflectivity is sub-horizontal at ~35 km. Wide-angle seismic methods are used to show that the P-wave velocities of the rocks are up to 10% less than normal in the zone above the fault surface. In cross-section this low-velocity Alpine Fault Zone is elongate, sits above the fault surface, and has dimensions roughly 45 by 20 km. A magnetotelluric study shows a low-resistivity anomaly that is roughly coincident with the zone of low seismic velocity. A straightforward interpretation is that both the electrical and seismic anomalies are caused by interconnected fluids at lithostatic pressure. The inference of fluids in the lower crust is supported by an attribute analysis of seismic reflections on specific shot gathers where the Alpine Fault reflections can unequivocally be identified. We reference both the amplitude and phase of the fault-zone reflections to the distinctive side-swipe reflections generated at the far shore of Lake Pukaki. High reflection coefficients of ~0.25 are estimated for the Alpine Fault reflections, which may require both anisotropy and fluid to explain. We interpret the source of water to be metamorphic dewatering of the schist-greywacke rocks that thicken into the orogen. A detachment surface along which the greywacke-schist rocks are obducted is recognised as a zone of strong reflectivity on an 80-km-long, unmigrated seismic reflection section. This zone of strong reflectivity, which apparently merges into the Alpine Fault reflections, does not correlate with depth to the Moho but rather with the boundary between the base of the schist-greywacke rocks (Vp ~6–6.2 km/s) and the lower crust (Vp ~7–7.2 km/s). We interpret the strong reflectivity on this boundary as being due to a shear fabric. Both geological and geophysical observations imply deformation in the lower and mid-crust and mantle that appears to be caused by a combination of ductile and brittle behaviour, with no evidence of lithospheric flexure. We interpret the Alpine Fault Zone as a profoundly hot, wet, and weak region of continental crust.

[1]School of Earth Sciences, Victoria University of Wellington, Wellington, New Zealand.

[2]Dept of Earth Sciences, University of Southern California, Los Angeles, California.

[3]Now at NZ Oil and Gas, Wellington, New Zealand.

[4]Now at GEOMAR, Kiel, Germany.

[5]Institute of Geological and Nuclear Sciences, Lower Hutt, New Zealand.

A Continental Plate Boundary: Tectonics at South Island, New Zealand
Geophysical Monograph Series 175
Copyright 2007 by the American Geophysical Union.
10.1029/175GM11

1. INTRODUCTION

Our knowledge of the internal structure of continental fault zones is largely based on indirect means including studies of exhumed paleo-faults, seismicity and laboratory studies [*Sibson*, 1986]. More recently, multinational teams are directly sampling fault zones to depths of a few kilometres with advanced drilling methods [*Hickman et al.*, 2004]. Directly imaging faults zones to depths > 1–2 km with seismic methods has, however, proved to be difficult, especially for faults that are near vertical [*Hole*, 1996; *Hole et al.*, 2001]. In this study a successful imaging of New Zealand's Alpine Fault to depths of 35 km is presented. This work formed part of South Island Geophysical Transect (SIGHT) project that was a joint US-New Zealand initiative to image a major continental transform fault to lower crust-upper mantle depths [*Davey*, 1997]. A major focus of the SIGHT experimental design was to take advantage of the narrowness of central South Island, and the dip of the fault, to develop seismic methods that could illuminate the structure of the fault from surface outcrop to lower crustal depths.

The term "Alpine Fault Zone" is here defined as the subsurface downward projection of the Alpine Fault and the triangular area of crustal rocks that constitute the hanging wall of the Alpine Fault. Linked to the main goal of imaging the fault geometry at depth was an additional aim of learning about the mechanical properties of rocks within the broader Alpine Fault Zone. In particular, we set out to use modern geological and geophysical methods so we could detect the presence or otherwise of fluids within the zone. Since the SIGHT project, further questions concerning continental transforms and crustal fluids have been highlighted that may also be addressed by exploration of the Alpine Fault. For example, the structure and strength of the San Andreas Fault of California has been the focus of recent controversy [*Scholz*, 2000; *Zoback*, 2000]. Low heat-flow and principal stress directions being perpendicular to the fault have been argued as evidence that the San Andreas Fault is "weak" [*Townend and Zoback*, 2001]. What the Alpine Fault offers, compared to other continental transform faults, is a fault that penetrates to greater crustal depths as an inclined to listric-shaped zone of shearing and deformation. Such inclined structures are easier to image with seismic methods than vertically orientated faults [*Hole et al.*, 2001].

Mid- to lower-crustal rocks immediately east of the surface trace of the Alpine Fault are being rapidly exhumed via an Alpine Fault "ramp" system [*Wellman*, 1979; *Little*, 2004]. This geometry facilitates the application of seismic methods to extrapolate structures from deep in the crust to the surface of the earth. Furthermore, because of the component of convergence at the Alpine Fault, the fault zone is ef-

fectively being loaded by mountain building at the surface of the earth, and also loaded by thickened, and therefore cold, mantle lithosphere pulling from below. This allows us to make another more direct, and general, measure of strength via flexural rigidity of the lithosphere.

In this paper we present a summary of geophysical data that pertains to the structure of the Alpine Fault Zone at depth. We quantitatively estimate its strength from flexural rigidity due to loading. We also make an assessment of fluid pressures, from seismic and electrical observations, within the Alpine Fault Zone. We also discuss our results in light of recent research on strain release by low–frequency earthquakes in other regions that are dominated by high fluid pressures. Finally, we comment on the potential of the Alpine Fault Zone to be a source of exploitable energy.

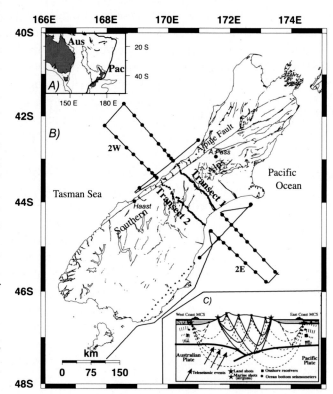

Figure 1. a. Plate setting for the South Pacific and Australian (Aus) and Pacific (Pac) plates. b. Location map for Transect 1 and 2 (referred to as T1 and T2 in text). Off-shore lines are tracks for the multichannel seismic lines and solid dots represent locations of ocean-bottom seismographs. On both transects 1 and 2 about 400 portable seismographs were deployed. Twenty-one land shots with 300–1000 kg of explosives per shot were detonated in 50 m deep bore holes. Including cross lines, about 1000 seismograph deployments were made for the whole project. c. A cartoon depicting the geometry of ray paths for onshore–offshore shooting, onshore shots, and teleseismic arrivals.

2. CRUSTAL STRUCTURE FROM THE SIGHT PROGRAMME

Programs of passive seismology, magnetotelluric and electrical studies, petrophysics as well as crustal-scale reflection profiling, and refraction-wide-angle reflection profiling were associated with the SIGHT study [*Davey*, 1997; *Stern et al.*, 2000; *Okaya et al.*, 2002]. Buried shots of ~1000 kg of explosives were used on land for refraction shooting, whereas the reflection-wide angle work was anchored by marine multichannel seismic (MCS) profiling; both onshore seismographs and ocean-bottom seismometers (OBS) recorded the MCS airgun sources. These data were collected in 1996 along two parallel transects across central South Island (Figure 1).

What made the SIGHT programme both unique and effective was its two-sided onshore–offshore component that allowed efficient coast-to-coast imaging. Onshore–offshore seismic methods across "continental islands", of dimensions similar to that of central South Island, allows high resolution of crust and upper mantle structure [*Okaya et al.*, 2002]. With island widths of 100-250 km, powerful ship-mounted air-gun arrays, deployed each side of the island, provide optimal coast to coast coverage of structure beneath the island (Figures 1 and 2). Combining the onshore-offshore experiment on each side of the island with

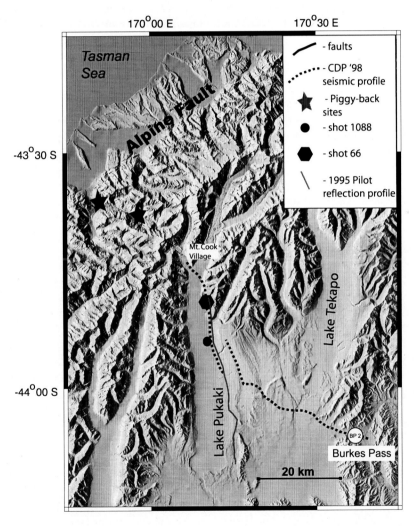

Figure 2. Location of other seismic reflection profiles associated with transect T2. Thin black line shows position of the Pukaki'95 experiment described in *Kleffmann et al.* [1998]. Black dashed-line shows position of the CDP'98 profile described in the text, which is in two portions separated by a large topographic scarp. The position of "piggy-back" sites on the west coast used to record wide-angle reflections shots from CPD'98 are shown [*Stern et al.*, 2001]. Positions of shot 1088 (Pukaki'95) and shot 66 (CDP'98) are shown. White circle at Burkes Pass is position of shot BP2 discussed in text.

the land refraction profiles permitted shot "super gathers" of approximately 600 km in length to be created [*Okaya et al.*, 2002].

Teleseismic events from the western Pacific were serendipitously recorded across the onshore seismograph arrays during the 7-day window of seismic profiling. These data provide important constraints on structure of the mantle beneath South Island [*Stern et al.*, 2000]. Because the teleseismic waves from the western Pacific come up through the crust at a steep angle, they also illuminated structure within the Alpine Fault Zone (Figure 1c).

Most of the data collected by SIGHT possess high signal to noise ratio with strong Pn phases recorded to offsets of ~300 km [*Scherwath et al.*, 2003]. Strong PmP, SmS, and PmS phases from both air guns and land shots were recorded from the airgun shots [*Pulford et al.*, 2003].

Crustal Structure Images

A combination of forward modelling and tomographic inversion was used to create crustal-upper mantle structure images of the plate boundary [*Scherwath et al.*, 2003; *van Avendonk et al.*, 2004]. Principal features and interpretations to note are (Plate 1a and b):

1. A crust that increases in thickness from 27 km at the coasts of South Island to a maximum of 44 km just east of the main divide of the Southern Alps on transect 2. This increase (~17 km) in crustal thickness is nearly twice that required to isostatically support the average topography of South Island.

2. Seismic P-wave speeds of 6–6.3 km/s throughout most of the crust except for two regions of lower wave speed directly beneath the Southern Alps on both transects. These zones of low Vp are interpreted as being due to enhanced fluid pressure associated with the release of metamorphic fluids into the lower crust [*Stern et al.*, 2001; *Wannamaker et al.*, 2002].

3. Seismic wave speeds of ~8.1 km/s in the mantle, except for a zone that extends 100 km on each side of the Alpine Fault in which strong anisotropy in Pn is detected [*Scherwath et al.*, 2003; *van Avendonk et al.*, 2004]; see later section.

These observations underpin the following compilation and analysis of data related to the Alpine Fault that includes: analyses of the geometry and phase and amplitude of fault zone reflections; delay times of wide angle reflections that pass through the Alpine Fault Zone; and an analysis of both

unmigrated and migrated seismic reflection section from the eastern side of the Alps.

3. FAULT ZONE REFLECTIONS

In March 1995, a pilot seismic reflection and piggyback wide-angle seismic reflection experiment was carried out along the eastern shore of Lake Pukaki (Figure 2). Advantage was taken of a straight and isolated road along the NE shore of the lake. Here a 6 km long, 120 channel, spread was laid out along the road and shots of 25 kg were detonated at a water depth of 20 m in the adjacent lake. Shots were offset by 4 km to the south of the 120-channel spread [*Kleffmann et al.*, 1998]. At the northern end of the lake, the seismic line is about 45 km from the Alpine Fault and at an azimuth of about 60° to the strike of the fault (Figure 2).

A shot gather from the northern end of the line shows good signal to noise ratio where the seismic array was north of the edge of the lake and extended into an area floored by river gravels rather than glacial till (Figure 3a). A series of deeper crustal reflections can be seen in the last 60 channels of the shot gathers where the geophones were in near-surface river gravel. At 9 s two-way-travel-time (twtt) there is a prominent reflection with "reverse moveout"; i.e., the reflection occurs at progressively earlier times on geophones with increasing offset from the shot. Such reverse-moveout is typical of reflections that have come from a dipping reflector where the dip is towards the geophone spread (Figure 4a and b). After filtering in the F-K domain the reverse moveout reflections are clearer (Figure 3b). In particular, a further stronger event is now evident at about 13 s. Together with the 9 s reflection these define a zone that is ~4 s twtt thick. If a mean P-wave speed of ~5.5–6 km/s is assumed for the zone, its thickness is on the order of 11–12 km.

Position and Dip of Alpine Fault Reflectors

The arrival time for a reflection from a planar reflector with dip ϕ is described by a hyperbolic equation given by:

$$T^2 = 4\, Z^2/V^2_{rms} + X^2/V^2_{rms} + 4\, (Z/V^2_{rms})\, X \sin \phi \qquad (1)$$

where T is the travel time, Z is the depth to the reflector and V_{rms} is the "root mean square" velocity between the reflector and the ground surface [*Sheriff and Geldart*, 1995] and X is geophone offset either positive or negative for shooting down-dip and up-dip, respectively.

If Equation 1 is plotted for values of X between 4 and 10 km and V_{rms} = 5900 m/s, it can be seen that the magnitude of negative moveout across a spread is a sensitive measure of dip of the reflector (Figure 4a). We measure the nega-

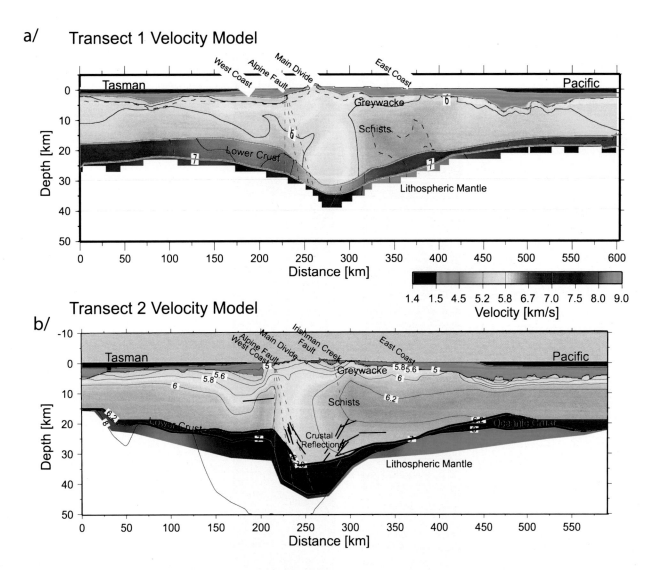

Plate 1. a. Crustal structure image of transect 1 [after *Okaya et al.*, 2002; *van Avendonk et al.,* 2004]. 5× vertical exaggeration. Note low-velocity zone in crust beneath Southern Alps. Dashed lines emanating from surface trace of Alpine fault represent theoretical dips of 0, 60 and 45 degrees. b. Crustal structure cross-section for transect T2 at 5× vertical exaggeration [*Scherwath et al.*, 2003]. Black bars show positions of strong seismic reflectors discussed later. Note region of low seismic velocity in crust beneath Southern Alps and low upper mantle seismic velocities beneath and west of Alpine Fault.

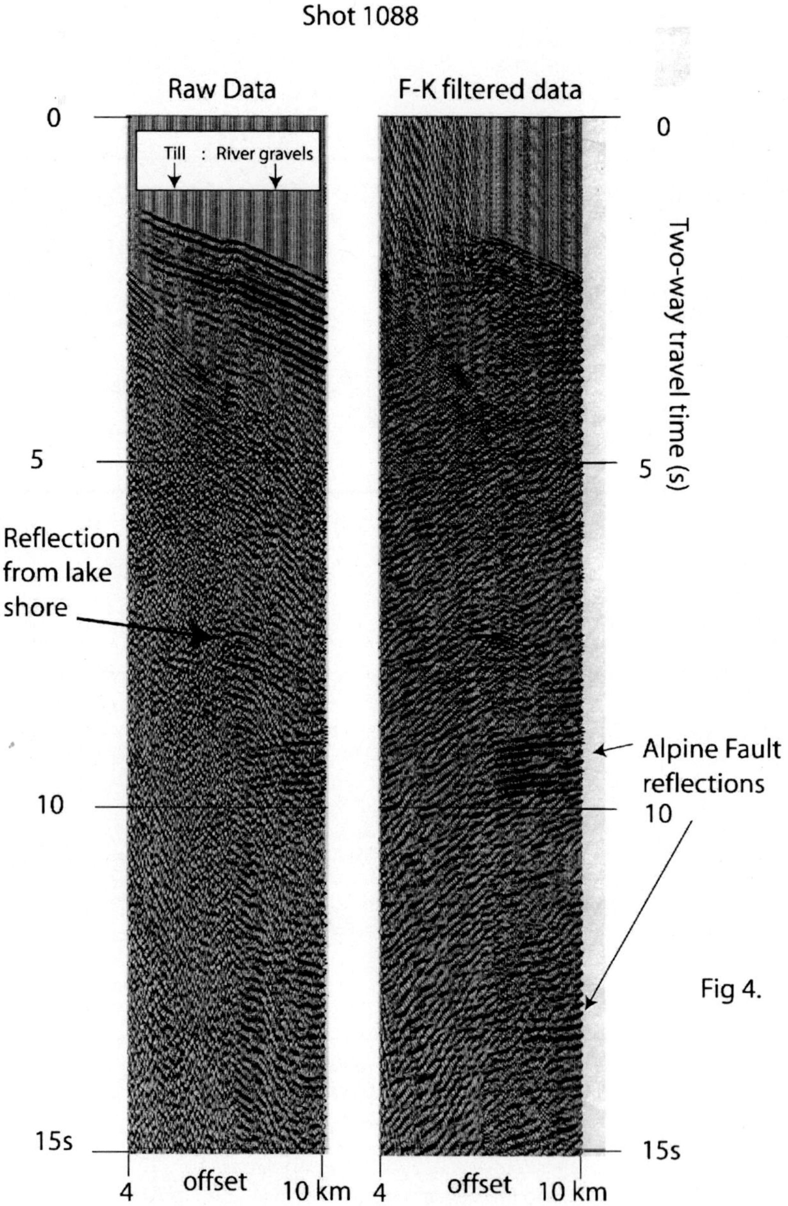

Figure 3. Shot gather shot 1088. Location shown in Figure 2. Spacing between phones is 50 m, which gives a 5950 m spread length for 120 channels. Offset to shot is 4 km and in the lake to south of the spread. Panel on left shows data passed through 20 s AGC and broad band-pass filter. Note how data improve for far offsets as the geophone spread spanned a transition from glacial tills (first 60 channels) to sorted river gravels (last 60 channels). Curved event ~7 s twtt is interpreted to be a side-swipe reflection from the far side of lake. Deeper reverse moveout reflection events at 9–13 s twtt are interpreted to be from the dipping Alpine Fault zone. Panel on left shows same data filtered with an F-K filter that suppresses all events that dip to the right. This filter enhances the 9 s event and brings out an event at 13 s that would not otherwise have been seen.

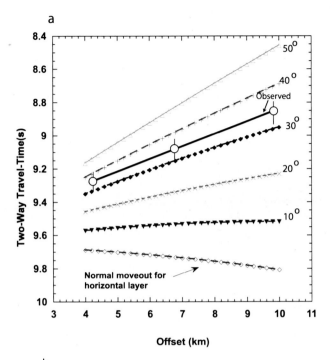

a

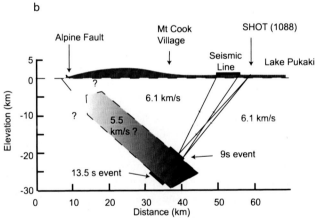

b

Figure 4. a. A plot showing theoretical moveout for a reflector that is 30 km deep in a constant velocity 6 km/s medium. Different curves are shown for dips between 10 and 50 degrees in the offset range of 4 to 10 km, appropriate for shot 1088. Also shown is the observed moveout for the 9 s event on shot 1088 including estimated uncertainties. Interpreted apparent dip of ~34 ± 5°. It is an apparent dip because the shooting line is at an angle to the dip-plane of the fault plane (see text). b. Schematic interpretation of reflectors from shot gather 1088 based on just ~4 km of shot gather data. Note that the projection of the reflectors to the surface approximately corresponds with surface trace of Alpine Fault. Spacing between geophones = 50 m.

tive moveout, across the 6 km long spread, on a number of adjacent shot gathers to be 380 ± 20 ms and plot it against the calculated negative moveouts for dips ranging between 10 and 50 degrees (Figure 4a). The best fit for the reflector dip from measuring individual segments on shot gathers range between 33–40° [*Davey et al.*, 1995; *Kleffmann et al.*, 1998]. For shot 1088 we estimate the dip as 35 ± 5° at a depth of ~25 km beneath Mt. Cook Village (Figure 4b). If the reflection is from the downward projection of the Alpine Fault then the dip estimate is only an apparent dip given that the azimuth of the seismic array was aligned at 60° to the Alpine Fault (Figure 1). For a shooting line that is not in the direction of dip, the angle δ is the apparent dip, which is related to the true dip φ by *Dobrin* [1976]:

$$\sin \delta = \cos \theta \sin \phi \qquad (2)$$

where θ is the angle between the shooting line and the direction in which the reflector dips. Thus for δ = 35° and θ = 30° the true dip of the reflector is φ ≈ 40°. A simple interpretation is that the reflector is a perpendicular distance of about 30 km from the shot, and thus at a depth of about 25 km directly below Mt. Cook (Figure 4b). When projected upward the dipping reflector intersects the surface close to the surface trace of the Alpine Fault.

It is not clear if the reflection is due to an acoustic impedance contrast between two different rock types each side of the fault, or if it is due to physical property changes within the fault zone itself. Field observations from the Alpine Fault outcrop [*Grapes and Watanabe*, 1992; *Norris and Cooper*, 2003; *Little*, 2004] shows a 1–2 km wide zone of strongly foliated mylonite dipping at ~45 degrees, which is overlain by ~15 km thick section of high grade schist. The mylonites are one possible source for the strong reflections given their strong foliation. They are, however, only observed to be 1–2 km thick in surface outcrop [*Norris and Cooper*, 2003] and the zone of reflectivity we see is more than 10 km thick. In order to resolve these differences we need to try to quantify the acoustic impedance contrast of the reflector.

4. ATTRIBUTE ANALYSIS OF ALPINE FAULT REFLECTION

An 80-km-long crustal scale seismic reflection profile (CDP'98) was collected in 1998 as part of SIGHT in an attempt to track the Alpine Fault to depth (Figure 2). CDP'98 was carried out in the Southern Alps foothills within the Pacific plate using 50 kg explosive sources and a commercial 850 channel seismic recording system. At a nominal 40 m

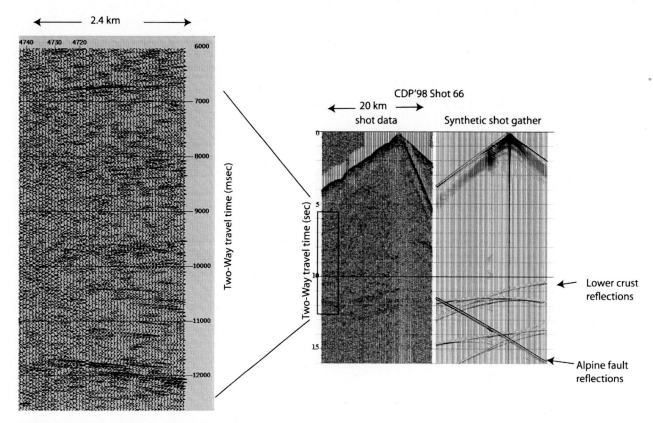

Figure 5. Data from shot 66 of the CDP'98 project. Spacing between geophones = 40 m. Location shown on Figure 2. Blow up on left shows the 2.4 km long part of the shot gather where both reflections from lake shore (6500 msec twt) and Alpine Fault (1200 msec twt) occur on common traces. Synthetic on right shows the predicted position for a reflection from a 60° southeast-dipping reflector that outcrops at the surface trace of the Alpine Fault [*Okaya et al., 2007*].

spacing for geophones, the length of the seismic array here was ~34 km, compared to 6 km for the Pukaki'95 experiment (Figure 3). Shot-gathers from this experiment also showed the reverse-moveout Alpine Fault reflections but over a larger offset range (Figure 5).

From both the 1995 and 1998 experiments a strongly curved event between 6 and 7 s twtt is observed (Figures 3 and 5). This event had such large moveout (high curvature) that it initially proved puzzling. Its normal moveout (NMO) velocity is about 1.7 ± 0.3 km/s, which is close to the speed of seismic P-waves in water [*Kleffmann et al., 1998*]. Furthermore, the timing of 7 s twtt, combined with the NMO velocity, makes it impossible for it to be a reflection from a horizon within the solid earth. Our interpretation of this large moveout event is that it represents a horizontally travelling ray through Lake Pukaki that has reflected off the far western side of the lake and come back to the geophone spread. The distance across the lake is about 5 km and the western and eastern lake-shores are remarkably parallel (Figure 2). The travel time from shot to reflection point and back to geophone is approx. $2 \times$ distance /velocity $= (2 \times 5 \text{ km})/1.5$ km/s

≈ 6.7 s, which is in the range of the observed reflection. This "side-swipe" event is particularly useful as we can readily calculate the acoustic impedance contrast between lake water and rock, thereby providing a standard of comparison for the amplitude and phase of deeper events (Plate 2).

Where both lake-shore and Alpine Fault events occur on the same trace we perform attribute analyses on their relative phase and amplitude (Plate 2). The phase of an event in a seismogram represents the polarity of the particle motion of the seismic wave, which can be positive or negative. A reflection will have a positive or a negative phase depending on whether the acoustic impedance at the reflector is positive or negative for a down-going wave. For near vertical incidence waves, the acoustic impedance contrast at a horizontal interface between an upper layer 1 and a lower layer 2 is defined by:

$$\Delta Z = (Z_2 - Z_1)/(Z_1 + Z_2) \qquad (3)$$

where $Z_1 = V_{P1}\rho_1$ is the acoustic impedance of layer 1, where V_p and ρ are the P-wave velocity and density. ΔZ is a quan-

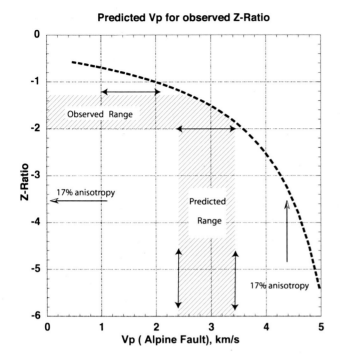

Predicted Vp for observed Z-Ratio

Figure 6. Iterative solution to Equation 4 showing the relationship between the range of observed ΔZ ratio and the required value range of P-wave velocity in the Alpine Fault Zone (Equation 4). Adopted parameters for velocity and density shown in Table 1. Also shown is the predicted, maximum, drop in P-wave velocity for 17% anisotropy (i.e., 5.5 to 4.5 km/s) and the consequent ΔZ ratio. Note that this ratio is still roughly 2 times what is observed. In order to get down to observed ΔZ ratio, V_p in the Alpine Fault Zone has to be lowered by ~50% or other processes need to be proposed (see text).

tity that expresses the ratio of the reflected and transmitted amplitudes often called R, the reflection coefficient [*Sheriff and Geldart*, 1995].

The phase of the first coherent lake-shore reflection appears to be positive on the attribute plot (pink in Plate 2a), which is consistent with the contrast between low P-wave speed in water and the high P-wave speed in rock. In comparison, the first distinctive pulse from the Alpine Fault has opposite polarity (blue in Plate 2a). We therefore interpret the Alpine Fault zone reflector as having a negative impedance contrast. This dual visual comparison was performed on several gathers and on many traces. In general, the analysis gave a reverse polarity for the Alpine Fault reflector, but not always. In about 15% of the cases the same polarity was evident. Because some events have normal polarity we conclude that the reflections arise from a fault zone that consists of alternating negative and positive impedance contrasts but the zone is laterally variable.

We make an estimate of A_{lake}/A_{Afault} (ratio of amplitude as measured in arbitrary units on a common trace) from an amplitude comparison between the lakeshore reflection and the Alpine Fault reflection (Plate 2b and 7c). Shot gathers were analysed with little processing apart from a broad band-pass filter and an Automatic Gain Control (AGC) window of 20 s. This serves to even out the gain over 20 s but will have negligible effect on the true relative amplitudes between the reflections of interest at 7 and 11 s twtt.

We make the approximation [*Sheriff and Geldart*, 1995]:

$$A_{lake}/A_{Afault} = \Delta Z_{lake}/\Delta Z_{Afault} \qquad (4)$$

The assumption here is that there is only a small change in the incident wave amplitude between 7 and 11 s twtt due to spherical divergence [*Sheriff and Geldart*, 1995].

As for the phase analysis we observe a variety of amplitude responses with the two extreme ones shown in the example of Plate 2b and c. This range of A_{lake}/A_{Afault} is 1.3 to 2.0. Thus for ΔZ_{lake} ~0.5 (Table 1) we estimate ΔZ_{AF} to be between 0.25 and 0.33. For a lower crustal reflection this is high compared to other estimates of reflectivity from the lower crust. Most maximum observed values of the reflection coefficient for reflectors in the lower crust are around 0.2 [*Warner*, 2004]. Assuming near vertical incidence, the ratio of impedance contrasts (Z ratio) is:

$$\Delta Z_{lake}/\Delta Z_{Afault} = [(Z_{tg}-Z_{w})/(Z_{tg}+Z_{W})]$$
$$* [(Z_{af} - Z_{g})/(Z_{af} + Z_{g})]^{-1} \approx 1.5-2.0 \qquad (5)$$

Here subscripts tg, w, af, and g on the acoustic impedance (Z) refer to till-gravel, water, Alpine Fault and greywacke respectively. Values of Z for all quantities except Z_{af} are known (see Table 1) and thus Z_{af} can be solved for by iteration.

Petrophysical work shows a negligible variation in density, ρ, for schist-greywacke rocks from within the Alpine Fault Zone [*Garrick and Hatherton*, 1973]. Accordingly, as $Z_{af} = V_{af}\rho_{af}$ a range of V_p between 2.5 and 3.5 km/s is predicted for the Alpine Fault Zone (Figure 6). This is a drop of between 30 and 60% of its host rock (greywacke-schist) values (Table1). Observed P-wave seismic velocities within the Alpine Fault Zone, as measured along ray paths tens of km in length, have dropped by only 10% compared to the host rock [*Stern et al.*, 2001], so we seek additional explanations for the strong reflectivity within the Alpine Fault Zone. Two possibilities are anisotropy and fluids.

Maximum measured anisotropy of a suite of high grade Haast Schists and mylonites is 17% [*Okaya et al.*, 1995]. A 17% anisotropy would produce a Z-ratio (Equation 5) of ~3.5 (Figure 6), which is greater than what is observed.

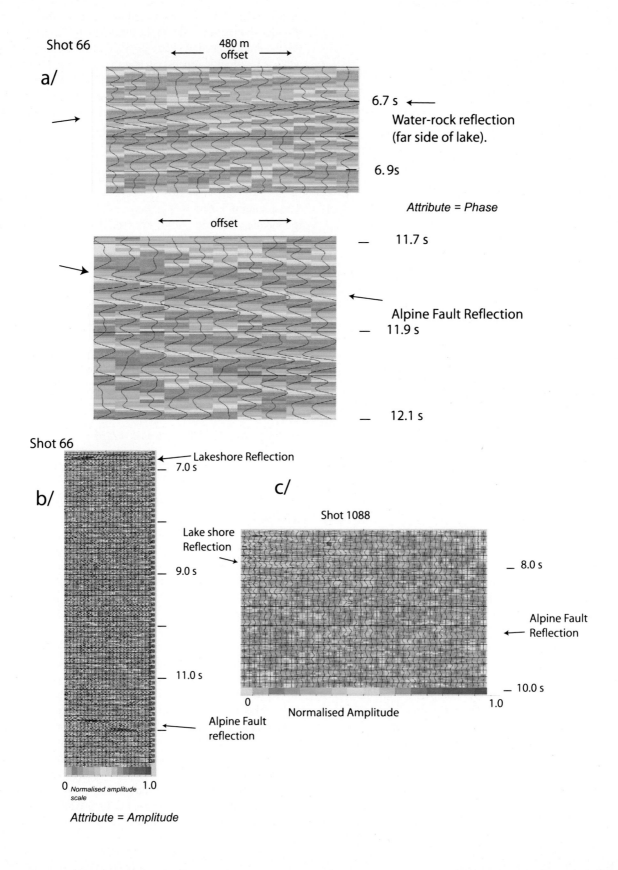

Table 1. Data used as input to impedance contrast analysis. [After *Kleffmann et al.*, 1998; *Kleffmann*, 1999].

Rock Type	P-wave Velocity km/s	Density Mg/m^3	Acoustic impedance Z = velocity*density (Mg m^{-2} s^{-1})
Greywacke-Schist	4.5–6	2.7	13.5
Shallow Glacial till	1.8–2.3	2.3	4.6
Water	1.5–1.55	1.0	1.5
Alpine Fault Zone rocks	?	2.7	?

Other means of enhancing the reflectivity could be constructive interference within layers of contrasting schist grade, or pockets of fluid saturated rock within the Alpine Fault Zone [*Warner*, 1990]. The source rock for this strong reflectivity is discussed further after other seismic indicators of structure are considered.

5. DETECTION OF LOW SEISMIC WAVE SPEEDS WITHIN THE ALPINE FAULT ZONE

Evidence for a low-speed region in the hanging wall of the Alpine Fault initially came from a 1983 reconnaissance seismic refraction survey of the central Southern Alps [*Smith et al.*, 1995]. This analysis was based on three large shots recorded by 17 analogue seismographs distributed along Transect 2 (T2) (Figure 1). The identification of the low-speed region was based on anomalous delays at just one seismograph. Later work, however, within the main SIGHT project collected seismic refraction data on 400 seismographs along T2 from shot 27 west of the Alpine Fault (Figure 7). These data show that, with respect to a well–determined velocity model for the upper-crust [*Kleffmann et al.*, 1998], delays for Pg occur for rays that turn through the top of Alpine Fault Zone [*Kleffmann*, 1999]. The delays increase with offset out to 70 km beyond which they are constant at ~0.3–0.4 s (Figure 7a).

Based on forward modelling, the zone where these delays start to occur can be isolated to the shaded box in Figure 7a [*Kleffmann*, 1999]. The length of the shaded region is ~16 km. For a time delay Δt, along a path Δx where the velocity has been reduced from V$_1$ to V$_2$, we have:

$$\Delta t = \Delta x(1/V_1 - 1/V_2) \qquad (6)$$

If we let V$_1$ be 6.2 km/s, the value of V$_2$ can be solved to be 5.5 km/s, which is a reduction of 11%.

A larger delay is seen for the lower-crust reflection (P$_i$P) phase (Figure 7b) [*Kleffmann*, 1999]. From shot 27 there are systematic delays of 0.6–0.8 s for reflections from a lower crust reflector recorded on seismographs to the east of the Southern Alps. The position of, and velocities above, the lower crustal reflector are well controlled from shots east of the Southern Alps. Hence the delay is confined to a path sector along the inclined path from the shot down to the reflector.

These delays were confirmed by a separate experiment that reversed shot 27 (Figure 7b) and "piggy-backed" off shots from the CDP'98 profile (Figures 2 and 8) [*Stern et al.*, 2001]. Two high-resolution, 48-channel, exploration seismographs were located in the valleys of the west coast and recorded the 50 kg shots from the CDP experiment on eastern South Island (Figure 8). If the regular average, eastern South Island, P-wave speeds of 6–6.2 km/s are used for the crust, then the arrivals at the west coast seismographs are 0.8 ± 0.1 s late (Figure 8b); a similar delay found from shooting in the opposite direction (Figure 7b). The length of the wide-angle ray path from the lower crustal reflector to the surface is about 50 km. Using Equation 6 with Δx = 50 km, Δt = 0.8s, and V$_1$ = 6.1 km/s gives the predicted velocity in the low-velocity zone to be ~5.6 km/s, similar to that calculated from delays in Pg (Figure 7a).

Plate 2. (opposite) a. Attribute analysis for phase. Colour is linked to whether the seismic trace deflects left or right. On top plot the deflection, or "first break", of the first coherent and energetic reflection from the far side of the lake is shown in red. This reflection is from water to rock, from low to high acoustic impedance, and is thus designated to be a positive impedance contrast. On the lower plot the first impulsive reflection from the Alpine Fault can be seen to deflect in the opposite direction and is shown as blue. This indicates a negative impedance contrast for this part of the Alpine Fault reflector. (Attribute analysis carried out with Claritas™ software). b. Attribute analysis of shot 66 for amplitude. The colour scale is normalised as shown at the bottom in an arbitrary linear scale between 0 and 1. Note that lake-shore and Alpine Fault reflectors occur on common traces. Using this record we estimate that the lake-shore reflection is 1.5 times larger than that of the Alpine Fault reflection (amplitude analysis carried out with Claritas™ software). c. As for 7b but on shot 1088. Where there are common overlapping traces the lake-shore reflection is estimated to be ~2 times greater than the Alpine Fault reflector in amplitude.

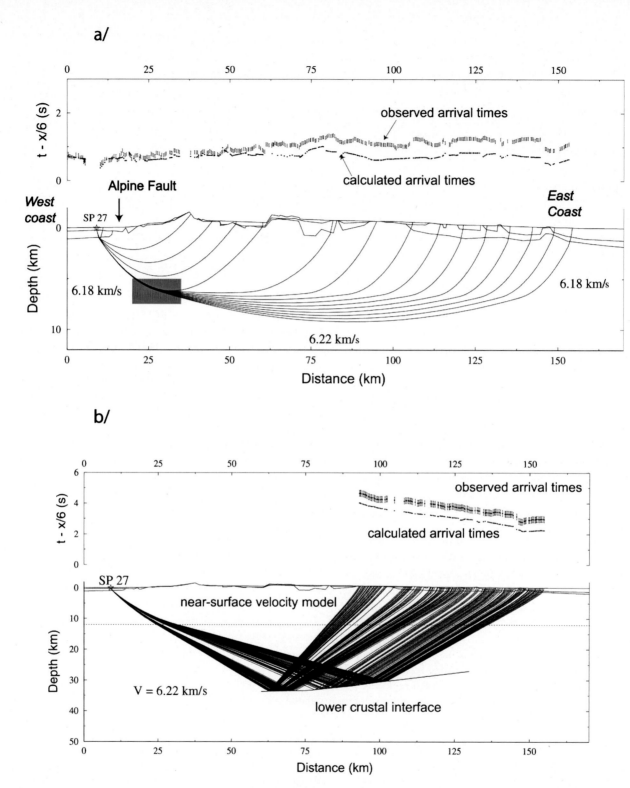

Figure 7. a. A plot of observed and calculated arrivals for shot 27. A simple velocity model [*Kleffmann et al.*, 1998] and then 6.18 km/s below 5 km is used. The shallow velocity structure is based on analysing data from the series of 7 shots along transect T2 [*Kleffmann*, 1999]. Note that a progressive misfit develops out to offset of about 75 km and then stays constant at about 0.35 s. The source zone for the misfit lies in the shaded box shown. A ~11% drop in velocity is required within shaded box (see text). b. Misfits for the arrivals for wide-angle reflections recorded on the eastern side of South Island from shot 27. The wide-angle reflections come from a strong lower crustal interface whose position has been well located by a series of shots east of the Alps. The same shallow velocity structure as in Figure 7a was used but 6.22 km/s was used below 10 km.

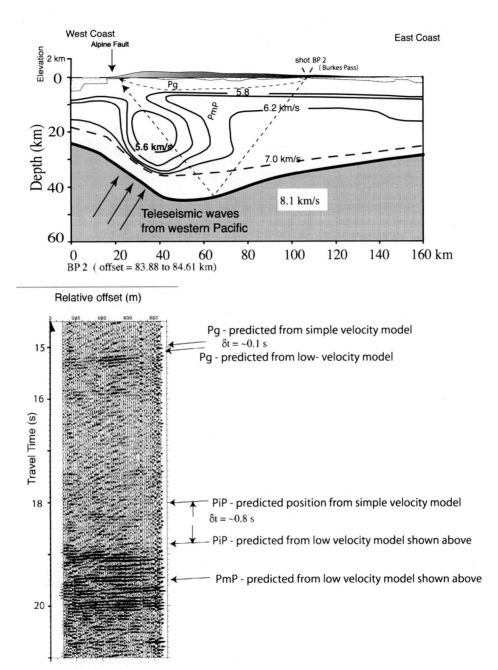

Figure 8. a. A summary model for the low-velocity zone beneath the Southern Alps based on delayed wide-angle reflections and delayed teleseismic P-wave signals. The shot gather shown is from shot BP2 of the CDP'98 project. Location of the shot is shown in Figure 2. Note the wide-angle reflection profile is reversed, i.e., shot 27 from west to east and CDP'98 shots to piggy-back array (Figure 2) from east to west. b. Shot gather from shot BP2 recorded on the piggy-back array (Figure 2) on the west coast, just east of the Alpine Fault [*Stern et al.*, 2001]. The geometry of the shot and receiver is shown in Figure 10a. Geophone spacing is 16 m, and spread length 752 m. Offset from shot is 85 km. The arrows indicate where arrivals are predicted for a simple model (no low-velocity zone) and a model with the low velocity zone shown in Figure 10a. A clear misfit of ~0.8 s is evident.

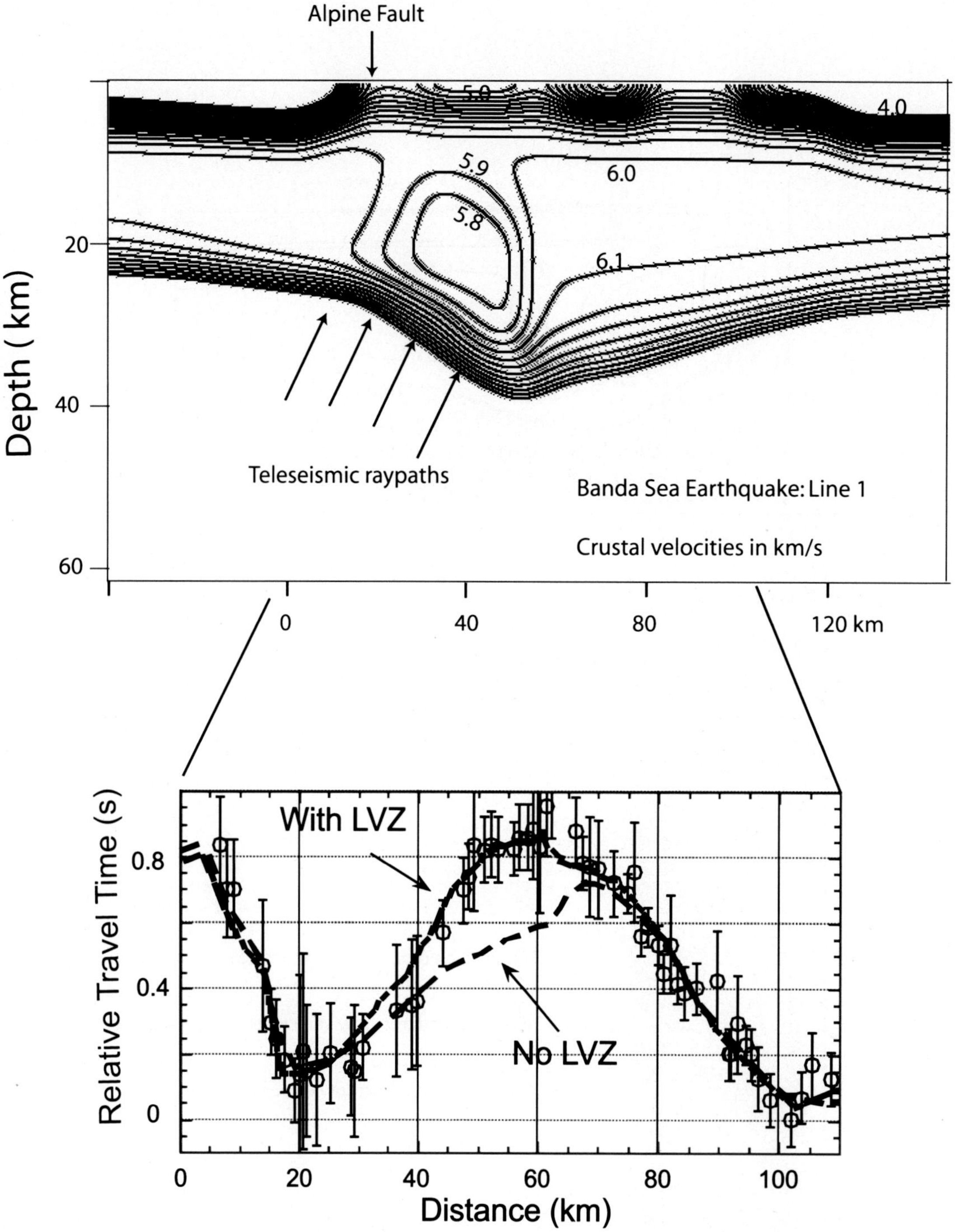

Figure 9. Plot of teleseismic P-wave delays for ray paths that pass through the Alpine Fault Zone. Delays are shown for an earthquake from the Banda Sea recorded on T1. The long-dashed line fit is for a model showing a deep body of thickened mantle as described more fully by *Stern et al.* [2000]. That study noted the shorter wavelength misfit for all models where the ray-paths travelled through the Alpine Fault Zone. Here we show a solution (short-dash lines) that include the low velocity regions shown by closed iso-velocity contours. Similar delays due to a low speed zone are also seen on T2. This set of teleseismic delays are combined with delays of PmP and Pg to produce the crustal low velocity model shown in Figure 8a.

The low-velocity zone is also detected by teleseismic arrivals that pass perpendicularly through the Alpine Fault and Alpine Fault Zone [*Stern et al.*, 2000]. Two earthquakes from the western Pacific recorded on both SIGHT transects give four independent data sets that show delays of ~0.2 s. An example is shown in Figure 9. The forward modelling solution to the low velocity zone is shown in Figure 8a, which incorporates the teleseismic delays of ~0.2 s perpendicular to the LVZ, the Pg delay of ~0.3 s (Figure 7a) and the wide-angle delay of 0.8 s parallel to the low velocity zone is shown in Figure 8b. This low-velocity structure has an apparent maximum thickness of 30 km, a depth extent of 35 km and a slant length of about 45 km.

High fluid pressure is postulated as the source of the low-velocity zone [*Stern et al.*, 2001]. Laboratory data show that for low porosity crustal rocks the compressional velocity will by reduced by10% when the pore pressure approaches lithostatic [*Jones and Nur*, 1984]. A fluid origin for the low-speed zone is further supported by a magnetotelluric study [*Wannamaker et al.*, 2002] along Transect T1 that shows a region of low (~40 ohm-m) resistivity in the crust that is approximately coincident with the low-speed zones [*Henrys et al.*, 2004; *Stern et al.*, 2001]. Interconnected fluid is required to produce the magnetotelluric anomaly, whereas fluid under a pressure approaching lithostatic is necessary to reduce seismic wave speeds by the requisite 10%.

Immediately beneath the trace of the Alpine fault fluid-flow is dominated by local convection of meteoric water with little metamorphic signature [*Vry et al.*, 2001]. In contrast, geological [*Koons et al.*, 1998] and geophysical studies [*Wannamaker et al.*, 2002] confirm that deep metamorphic fluids migrate up on near vertical paths from a high strain zone in the root zone of the Southern Alps. Thus the location of the low-speed zone, and hence high pore pressure zone, directly above the region of highest reflection intensity is no coincidence (Plate 4). Permeability is enhanced by deformation and when the fluid pressure gets high enough fluid can advance along hydraulic fractures.

One of the consequences of enhanced fluid pressures in the crust is the effect on fault strength. The condition for sliding to occur on a pre-existing plane is [*King Hubbert and Rubey*, 1959]:

$$\tau_f = F (T_n - P_f) \qquad (7)$$

where τ_f is the shear traction at failure, F is the coefficient of friction, T_n is the normal traction on the fault and P_f is the fluid pressure. This well known relationship shows an increase in fluid pressure can cause a decrease in the differential stress required for shear failure on a fault, or for an array of faults that make up a fault zone [*Rice*, 1992].

6. LOWER CRUSTAL REFLECTIVITY AND DECOLLEMENT

An unmigrated, stack of CDP'98 shows two bands of reflections that converge westwards towards a central low point just to the west of Lake Pukaki (Plate 3a). Within the eastern part of the section we interpret the lower of two sub-parallel bands of strong reflectivity between 7 and 12 s twtt (20-35 km depth approximately) as a decollement surface between the mid- and lower crust for the following reasons. First, this reflectivity zone is prominent on our near vertical image but occurs at shallower depths than the Moho as determined from refraction and wide-angle reflection data [*Henrys et al.*, 2004] (Plate 3a). It is unclear why the Moho is not seen in the stack of vertical reflection data, but it could be because the Moho is gradational rather than abrupt. A second reason for the strong reflectivity to represent a decollement surface is that the reflectivity lies at the boundary between higher-velocity (>7 km/s) rocks and the quartz-schist type rock velocities of 6–6.2 km/s (Plate 3b). Thirdly, the high reflectivity zone can be tracked through to rocks exposed at the surface east of the Alpine Fault. These rocks are in essence identical to the full greywacke-schist crustal section seen in eastern South Island, albeit compressed into a thinner section. No mantle and only minor amounts of lower crustal rocks are brought up along the Alpine Fault.

7. MIGRATED SEISMIC REFLECTION IMAGE OF THE ALPINE FAULT ZONE

The steepest east-dipping reflectors located west of Lake Pukaki (Figure 5 and Plate 3a) are interpreted from shot gathers to be coming from a steeply dipping Alpine Fault and need to be correctly positioned using seismic migration methods. Industry-style and other migration methods give varying degrees of success [*Okaya et al.*, 2007]. The best migrated image using the western half of the CDP'98 data (Figure 2) suggests that the seismic data are imaging the Alpine Fault Zone in the depth range of 15–30 km (Plate 4a). At a depth of 15 km the fault zone is an inclined surface with steep dips (~60°), which then flattens and broadens into a sub-horizontal zone of strong reflectivity at a depth of about 30 km (Plate 4a).

One possible interpretation of the image is that the Alpine Fault is listric at depth and the fault flattens out near the surface. Such a flattening may be due to topographical smearing as the Alps are being progressively thrust westward. Migration of deep data will, however, produce some smearing and "smiles" [*Warner*, 1987] that may be present in our image. On the other-hand, confidence that the listric

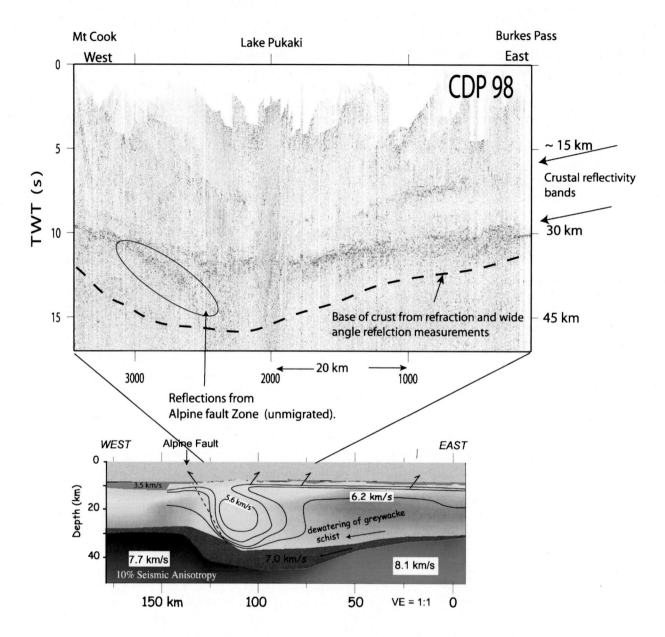

Plate 3. a. Stacked but unmigrated section of the CDP'98 survey from Burkes Pass to Mt Cook [*Okaya et al.,* 2007]. A total of 181 shots were detonated along the survey line. Sixty-five large (50 kg of explosives) shots were placed at 1 km intervals and buried 20 m deep. Smaller shots of 2.5 kg, buried 3 m deep, were located 250 m on either side of each large shot. Shot triggering was keyed on GPS time, to allow secondary piggy-back arrays to trigger without the need for direct radio communication. A total of 1000 geophone receiver group positions were used to record the shots, with a group spacing of 40 m. The number of recording channels for any one shot varied from a minimum of 400 to a maximum of 870. The maximum array aperture was nearly 35 km and the maximum shot receiver offset ~30 km. The raw seismic data for each shot were recorded to 30 s time, at a 2 ms sampling rate. Maximum fold is 12. Unmigrated Alpine Fault reflections are shown (in superposed ellipse region) cutting across the reflectivity from the lower crust. Depth scale on right is only approximate. b. Velocity model based on refraction data and wide-angle reflection inversion [*Scherwath et al.,* 2003]. Note that the strongest reflectivity in the stacked section (Plate 3a) corresponds to the lower-crust boundary that separates greywacke-schist rocks from mafic lower crust, possibly old oceanic crust. Variation in upper mantle wave speed has been interpreted as being due to ~10% mantle anisotropy

shape to the fault zone is real is gained by the strength of reflectivity within the westernmost sector.

If projected to the surface, the migrated reflections correspond to the surface trace of the Alpine Fault and the outcropping 1–2 km thick mylonite zone. It is unclear if the strong reflectivity in the depth range of 15–35 km also represents the highly foliated mylonite zone, but 2–5 times thicker than that observed at the surface. Geochemical evidence also shows the highest grade of schists thins by about 40% as they are exhumed [*Grapes and Watanabe*, 1992; *Grapes*, 1995].

A P-wave hitting the anisotropic [*Okaya et al.*, 1995] schist-mylonite rocks, perpendicular to the foliation, would produce a moderate reflection (Figure 6). Fluids being released by metamorphic reactions are, however, a more likely cause of the strong reflectivity given the close spatial relationship of the migrated reflectivity with the low velocity zone in the hanging wall of the Alpine Fault (Plate 4a), and the strength of Alpine Fault reflections (Figure 3 and Plate 2). Strong reflections can be produced by fluid being present in pockets of high concentration. Alternatively, fluids in just small concentrations (< 0.5 vol %) can weaken feldspathic rock and thus enhance and concentrate shear deformation [*Tullis and Tullis*, 1986]. A thick zone of shearing is also more likely to increase reflectivity by constructive interference [*Gough*, 1986].

Fluid enhanced shearing is our favoured interpretation of the strong reflectivity found in the lower crust beneath the Alps. The source of fluids is likely to be deep metamorphic fluids that geological [*Koons et al.*, 1998; *Vry et al.*, 2001] and geophysical [*Wannamaker et al.*, 2002] evidence suggest migrates up near vertical paths to the surface east of the Alpine Fault trace. The cause of the fluid release would be from both prograde metamorphism, linked to increased depth of burial [*Fyfe*, 1978], coupled with strain - induced metamorphism [*Koons et al.*, 1998]. The proposed pathway for the fluid is directly up through the low wave speed zone (Figure 13a, b). Thus this migrated seismic reflectivity data, combined with low-wave speed zone image, provides us with a new high-resolution image of both the source zone and pathway for fluids generated in the deep lower crust.

8. LOW SPEED, HIGH FLUID PRESSURE, STRONG REFLECTIVITY FAULTS AND SEISMICITY

As the rocks are detached from the lower crust and moved to the surface via the Alpine Fault Zone they are deformed. Mapping of exposed rocks within this zone shows it to be floored by ~1–2 km thick mylonites with shear fabrics subparallel to the dipping Alpine Fault Zone [*Wellman*, 1979; *Norris et al.*, 1990; *Norris and Cooper*, 2003; *Little*, 2004].

Above the mylonites, greywacke-schist rocks have been uplifted and sheared on near-vertical planes during the Cenozoic [*Little*, 2004] (Plate 4b). Pre-existing quartz veins within the greywacke-schist rock have been systematically offset across the shears in both brittle and ductile manners, probably during a steady (aseismic) creep process [*Little*, 2004; *Little et al.*, this volume; *Wightman and Little*, this volume].

Despite evidence for deep brittle behaviour, reconstructions of this deep region of the Alpine Fault Zone require the lithosphere here to be of low flexural rigidity since the lower crust has been deformed through a bend in transport direction with a tight radius of curvature (<10 km) (Plate 4b). This is also the region of anomalously low P-wave velocities, implying high fluid pressures and low strength, as discussed earlier. Thus an apparent contradiction exists in unusually deep reaching brittle behavior, within an otherwise weak crust.

An explanation for this apparent contradiction can be gained from an understanding of how high fluid pressure can effect the rheology of crustal rocks [*Sibson*, 1986]. Yield-stress envelopes are modified by an increase in pore pressure. For the case shown in Plate 4c, a move from hydrostatic to lithostatic fluid pressure leads to an increase in thickness of the brittle layer (embrittlement) from 8 to 22 km [*Meissner and Wever*, 1992]. At the same time there is a lowering of the maximum sustainable shear stress by nearly an order of magnitude. Thus a prediction of this fluid-related embrittlement process is that a deep region of brittle behaviour will result in more earthquakes of small magnitudes rather than fewer larger events distributed within a thinner brittle layer. This is a well known effect from injection of fluids in deep wells and impoundments of reservoirs [*Simpson*, 1986]. High strain rates may also contribute to the embrittlement process [*Little et al.*, this volume].

Recent results from active convergent plate boundaries show that fluid rich regions undergoing shear can radiate seismic energy in unusual ways. For example, non-volcanic, episodic tremor and slow-slip events have been noted in both Japan [*Obara*, 2002] and western Canada [*Rogers and Dragert*, 2003]. These appear to be linked to zones in the crust where fluids are plentiful and fluid pressure will be close to lithostatic. More specific associations are that metamorphically released fluid assists slip in shear zones at convergent margins [*Shelly et al.*, 2006] and that high pore pressure is the direct cause of silent slip events [*Kodaira et al.*, 2004].

Lateral Extent of Low Velocity Zone and Associated Weak Rheology

Seismicity for events M > 2 is restricted to the top 8 km beneath the central section of the Alpine Fault [*Leitner et al.*, 2001], roughly the depth at which the top of the low velocity

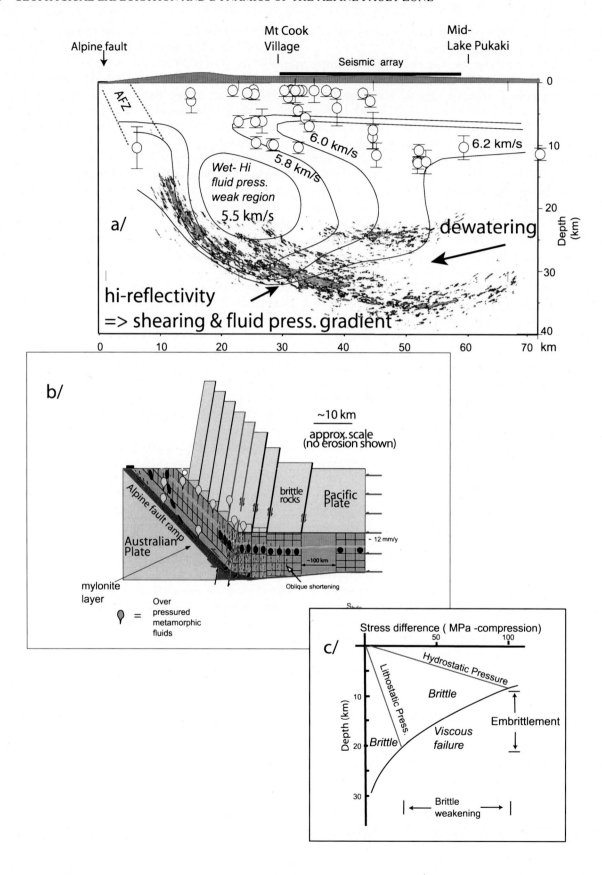

zone is observed. One interpretation of the lack of seismicity below 8 km is that strain is accumulating in this part of the fault [*Sutherland et al.*, this volume]. An alternative interpretation of the scarce seismicity is that fluids have reduced the effective normal stress within the fault zone. Strain release is either as slow earthquakes and after slip, or as earthquakes below the M > 2.0 threshold for the most recent survey.

At the northern and southern ends of the central section of the Alpine Fault topography and rock uplift diminish [*Kamp and Tippett*, 1993]. Both crustal and mantle seismicity increase to the north and south of this central section [*Anderson and Webb*, 1994]. There is, therefore, an association of low seismic wave-speeds in the crust to both a lack of seismicity and high topography. A further distinctive aspect of the central section is a localised region of high exhumation rate in the region between Transects T1 and T2 [*Little et al.*, 2005]. The tightly clustered contours of mineral cooling age in this region resemble the data describing the "tectonic aneurysm" in the Himalayan syntaxes [*Zeitler et al.*, 2001].

Little et al. [2005] interpret the localised high exhumation zone as being due to changing geometry in the ramp up which rocks are exhumed. They propose a faulted back-shear mode in the central section of the Alpine Fault (Plate 4b) compared to a broadly curved, unfaulted ramp further south. Why such a special mode of exhumation has developed in such a confined region remains unknown. We suggest the presence of excess fluids, high fluid pressure and reduced normal stress may also be factors that facilitate the development of the back shears.

9. MANTLE STRUCTURE AND DEFORMATION

During the shooting of the SIGHT program 160 seismographs were left running in continuous mode for about 7 days. Within this period three earthquakes from the western Pacific of $M_w \geq 5.0$ were recorded. As the azimuth of the path from the western Pacific to the west coast of South Island is nearly that of the seismic lines (Figure 1), arrivals from these events are treated like in–line shots with a large offset and a small incident angle (~25°) angle. These data were processed to produce plots of teleseismic P-wave delays along both transects [*Stern et al.*, 2000].

Two distinct trends can be seen in the teleseismic arrivals (Figure 10a): (a) a delay due to thickening of the crust beneath the Southern Alps (km 20–150) and (b) an advance of nearly a second between offsets of 40 to 100 km. Such a large advance in P-wave arrivals requires a region of high seismic wave-speed in the mantle. Tracing a ray back into the earth from the surface at km 100, where the maximum advance is observed, shows its path will be directly beneath the root of the Alps at a depth of 120 km (Figure 10a). This is consistent with relative high-speed mantle due to uniform thickening of a 100 km thick lithosphere that has been shortened by ~100 km [*Stern et al.*, 2000]. The shortened and relatively cold, denser and faster mantle lithosphere is displaced into the hotter, less dense and slower asthenosphere directly beneath the crustal root.

Six independent data sets were generated from the three earthquakes recorded on two transects. These data sets all suggest much the same structure - a roughly symmetric body, 2–17% faster than the surrounding mantle, which is located directly beneath the crustal root (Figure 14b). Our interpretation of the shape and position of this high-speed zone in the mantle is that it represents mantle lithosphere that has thickened and deformed in a ductile and continuous fashion. This is in contrast to intracontinental subduction that requires differential shear to be concentrated on a subduction thrust [e.g., *Koons*, 1990; *Beaumont et al.*, 1996]. Such a ductile and weak rheology may be linked to the 45 my of strike-slip shear causing both a pervasive heating [*Stern et al.*, 2000] and/or dynamic recrystallization in the mantle [*Scherwath et al.*, 2002].

A localized high-speed region in the mantle, in which P-wave speeds are 1–7% faster than in its surroundings, should also be denser than adjacent mantle and therefore contribute a gravity anomaly. Gravity analysis along T2 shows that the data are consistent with a mass anomaly of about 40 kg/m^3 in cross-section area of 120×80 km^2 [*Stern et al.*, 2000].

10. FLEXURAL RIGIDITY OF CENTRAL SOUTH ISLAND LITHOSPHERE

Much of the discussion on "strength" of faults has been based on noting that the directions of principal horizontal

---

Plate 4. (Opposite) a. Migrated version of the western portion of the stacked section shown in Plate 3 [*Okaya et al.*, 2007]. Superimposed on the stacked section are contours of the P-wave speed shown in (Figure 8a), and earthquake locations for events M >2 [*Leitner et al.*, 2001]. Interpreted regions of dewatering, shearing and high fluid pressure are shown. These data are based on the segment of profile that runs from Mt Cook Village to mid-way down Lake Pukaki (see dotted line on Figure 2). b. Cross-sectional interpretational model based on original concept of [*Wellman*, 1979] then modelled in detail by [*Little*, 2004]. Observations from the field suggest a combination of ductile and brittle behaviour to a depth of 20–25 km within the back shears. c. Yield stress envelop for compression under both hydrostatic and lithostatic fluid pressure [*Meissner and Wever*, 1992]. Note how a move from hydrostatic to lithostatic pressure creates an increase in the depth of brittle behaviour (embrittlement), and a drop in the maximum shear stress (brittle weakening).

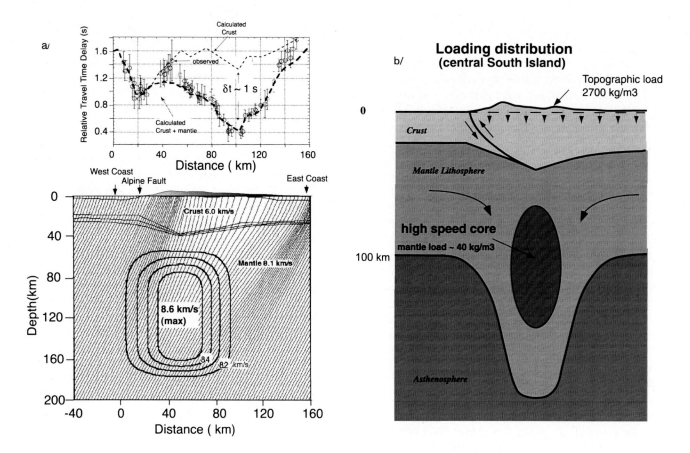

Figure 10. a. P-wave teleseismic delays from a Honshu earthquake recorded across transect 1. Delays (upper plot) are with respect to a standard earth model given by the IASP91 model [*Kennett and Engdahl*, 1991]. Note the positive delay (slowdown) associated with the crustal root then the major speed-up of (~δt = 1 s) that peaks at km 100. Delays have been corrected for elevation of stations above sea level. Distance is with respect to the west coast. Lightly dashed line represents fit for seismically determined crustal structure only. Heavy-dashed line represents fit of both crust and mantle structure. Bottom figure shows the ray-tracing of teleseismic waves from Honshu as they pass up through mantle and crust beneath central South Island. The velocity structure for the crust varies between 5.5 and 6.2 km/s, then 7.1 km/s for thin lower crustal layer. Mantle seismic wave speeds are 8.1 km/s except for symmetrical high-speed (max = 8.6 km/s) body that is required to sit directly beneath the crustal root. b. Cartoon showing tectonic interpretation of how high-speed zone in asthenosphere represents thickened, cold, and therefore fast and more dense, mantle lithosphere. Loading on the Moho is therefore in two parts: Surface topography and subsurface loading from the cold and therefore dense thickened mantle.

stress from fault plane solutions being at right angles to the fault zone [e.g., *Townend and Zoback*, 2001]. This only gives a relative sense of strength for the crust. Laboratory experiments have assisted in assessing the relative strength of various rock types, but these experiments are done at strain rates and temperatures much higher than what occurs in the earth [*Tullis and Tullis*, 1986]. We are also interested in how strong the plate boundary zone is, rather than just the Alpine Fault, and therefore seek a measure of both crust and mantle strength within and adjacent to the fault. Flexural rigidity provides such a measure of the gross strength of a

sector of lithosphere [*Watts*, 2001]. To analyze flexure we need to measure the amplitude and wavelength of deformation induced by a known load. Critical to doing this is being able to identify a horizon that was once horizontal or planar, and is now deformed. Further, we must establish what the loading on the lithosphere is from loads within both the crust and mantle.

We assume the principal, vertical load acting on South Island is the "push" of topography and the "pull" of shortened mantle lithosphere (Figure 10b). A simple 2D gravity analysis shows that these two loads, although distributed in different ways, are similar in magnitude [*Stern et al.*, 2000]. The Moho is used as a reference surface for this study of loading and flexure. Implicit in doing this is the assumption that the Moho was flat prior to collision and loading. Our justification for assuming a flat, pre-Pliocene Moho is based on the geological record of central South Island. Late Miocene shallow-marine sediments exist in all but the central alps, and shallow-water terrestrial sediments of early Pliocene age [*Gair*, 1967] are exposed over a wide region of central South Island, including the Mt. Cook region. Thus, prior to the main episode of loading at ~6 Ma muted topography and, therefore, a low relief Moho, are implied. A further assumption in using the Moho as a reference horizon is the lower and upper crust, and mantle lithosphere, all deform in unison.

Starting with a continuous elastic sheet, the effective elastic thickness (T_e), loading, and the restoring force from mantle buoyancy can be varied in an arbitrary manner via finite difference techniques (Figure 11) [*Bodine et al.*, 1981].

Principal loads on the elastic sheet are topography above sea level and the thickened mantle lithosphere. We also allow for a small load representing thrust sheet build up east of the Southern Alps into an assumed paleobathymetric depth of ~400 m. Justification for pre-Pliocene paleobathymetry east of the Southern Alps is based on geological evidence [*Stern*, 1995].

The effective elastic thickness (T_e) remains the variable to resolve from trial and error modeling [*Watts*, 2001]. The choice of a continuous, rather than a broken, or faulted, plate is more general and allows us to accommodate the concepts of a wide zone of deformation in the mantle and lower crust [*Molnar et al.*, 1999], while at the same time permitting us to approach the broken plate situation. Our best-fit variation of T_e is shown in Figure 11 and three fits to the deformed Moho are shown in Figure 12.

When both the topography and the shortened mantle are used as loads, and a constant T_e of 25 km, the resulting de-

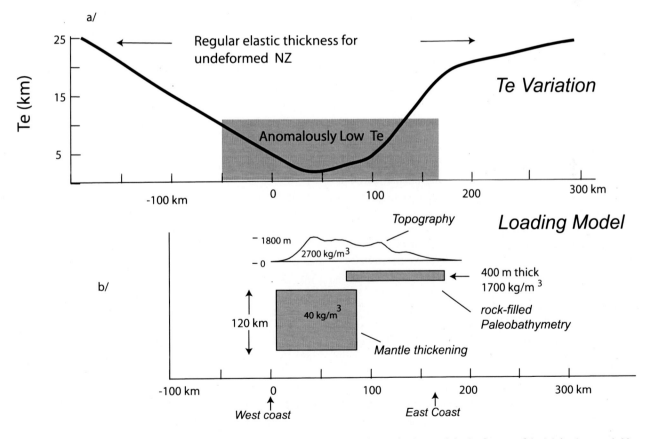

Figure 11. a. Plot showing the variation in T_e across central South Island that is required to explain the flexure of the Moho (see text). Note how T_e drops from regular value ($T_e = 25$ km) to a small value of $T_e = 1$ km beneath the central Alps. The region of anomalously small T_e (< 10 km) coincides with sub-aerial South Island. b. The two-dimensional distribution and magnitude of loads used in the flexure model. The rock filled paleobathymetry and smoothed topography is based on the gravity interpretation of *Stern* [1995] and the mantle load is based on the gravity interpretation model of *Stern et al.* [2000].

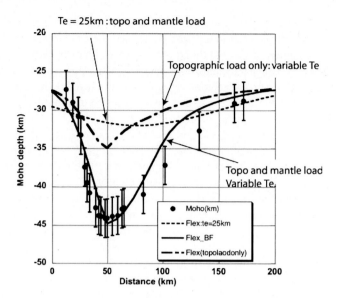

Figure 12. Flexure model for deformation of the Moho, central South Island based on the shape of the Moho (sold circles with error bars) from seismic measurements. Loading is based on model in Figure 11b. A continuous plate model is used and loading, elastic thickness (T_e) and restoring force are allowed to vary in an arbitrary fashion. Displacements (solid and dashed curves) are calculated using a finite-difference code [*Bodine et al.*, 1981] and compared to Moho picks (solid circles with ± 3 km errors bars). The Moho is assumed to have been flat and at a depth of 27 km prior to collision and loading. When both topographic and mantle loads are applied a T_e variation shown in Figure 11a is required to match (solid curve) the observed Moho displacement. Two curves are also shown that don't match the observations: dashed curve represents model where T_e is set to constant value of 25 km and loading is as shown in Figure 11b; dash-dot curve is for the topographic load using the T_e variation shown in Figure 15a.

flection does not match the Moho topography (Figure 12). Only when we allow T_e to drop to vanishingly small values in central South Island do we get a reasonable match in both wavelength and amplitude (Figure 12). Also shown is the predicted deflection of the Moho due to just topography loaded on a plate with the preferred T_e variation shown in Figure 11. Topography only provides sufficient loading to produce half the observed amplitude of the crustal root.

A value of T_e = 20–25 km is consistent with a well-constrained loading study on the western platform of New Zealand [*Holt and Stern*, 1991]. But values of T_e < 10 km that are required for most of subaerial central South Island are unusual [*Watts*, 2001], although many mountain ranges do show the same trend of a decrease of T_e towards their central regions [*Stewart and Watts*, 1997]. Such low values suggest that there is effectively no elastic strength in the mantle or crust, and strain and deformation is inferred to be accommodated by either ductile flow or upper crustal faulting. This observation is consistent with structural studies of veining and shearing discussed earlier [*Little et al.*, 2005]. What is new here is that there appears to be no contribution of flexural rigidity from the mantle. This is consistent with the mantle behaving in a ductile manner, and undergoing uniform thickening and distributed shear as advanced previously [*Molnar et al.*, 1999; *Stern et al.*, 2000; *Savage et al.*, this volume].

Ductile deformation and shortening of the mantle lithosphere is also consistent with the stark gap in subcrustal seismicity beneath central South Island [*Anderson and Webb*, 1994]. Some subcrustal earthquakes have, however, been reported [*Reyners*, 1987] beneath the central Southern Alps - the significance of which is still unclear.

11. DISCUSSION

The SIGHT program and its associated geological studies have demonstrated how central South Island is an optimal locality to image and study continental convergence. In particular, we have shown how a narrow continental island can be taken advantage of for advanced onshore-offshore seismic methods [*Okaya et al.*, 2002]. One of the central discoveries of SIGHT is the existence at depth of a broad Alpine Fault Zone where high fluid pressure and interconnected fluids in the crust are inferred. The depth and lateral extent (45 × 20 km) of the fault zone appears to be larger than for other continental transform faults [*Mooney and Ginzburg*,

1986; *Stern and McBride*, 1998]. Possible reasons for the successful image of the Alpine Fault Zone include its favorable inclined geometry that is optimal for seismic imaging, and the extreme exhumation rate east of the Alpine Fault [*Blythe*, 1998]. Strong exhumation would focus and enhance the dewatering process [*Koons et al.*, 1998] and would therefore accentuate the fluid-dependent geophysical anomalies we have described. Future experiments could be directed to exploring north and south of transects T1 and T2 to try and establish if the low–velocity, high-conductivity structures extend beyond the central Southern Alps. In addition, future experiments may benefit from more 3-component seismic instrumentation so we can target more directly the mechanical properties of the fault zone system.

Because of the discovery of a large zone of what is interpreted to be high fluid pressure in the Alpine Fault Zone, an analogy is drawn between it and the episodic slip and tremor documented in the subduction zones of Japan [*Obara*, 2002] and western Canada [*Rogers and Dragert*, 2003]. In these subduction zones the anomalous slip and tremor

events are linked to water from dehydration and associated high-fluid pressures [*Julian*, 2002]. Even though the settings of the Alpine Fault Zone and these two subduction zones are different, the processes are similar. In both cases we are dealing with the seismological detection of dewatering. Because greywacke-schist rocks liberate water at lower temperatures and pressure than oceanic crust [*Fyfe*, 1978], the seismological image of the high fluid pressure region is shallower in central South Island than under Japan and western Canada. No episodic slip or tectonic tremor have been detected in central South Island. But as both Obara and Julian point out, with the present instrumentation in most other places, including New Zealand, they would not be detected. In this regard its is noteworthy that slow slip events have been detected on large continental faults that are distant from subduction zones, but where instrumentation is more advanced [*Kanamori*, 1989; *Kanamori and Hauksson*, 1992; *Linde et al.*, 1996]. Episodic tremor has also recently been reported from the San Andreas Fault [*Nadeau and Dolenc*, 2005].

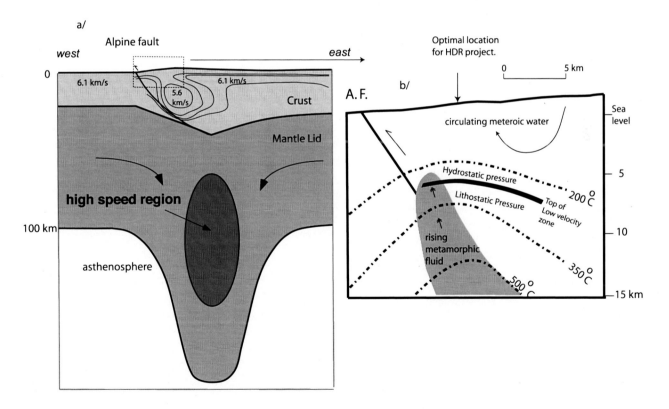

Figure 13. a. A summary of both crust-mantle structure showing crustal low velocity zone and inferred mantle thickening [*Stern et al.*, 2000]. Dashed box area shown in more detail in part b. b, A blow-up of crustal structure and fluid - phenomena in the top 10 km of crust east of the Alpine Fault Zone [*Wannamaker et al.*, 2001]. A suggested optimal position for a possible Hot-dry-rock power investigation is shown.

An important contribution of this study is an integration of data sets to give a larger picture of the strength of the plate boundary. What is seen, relative to regular parts of the continental lithosphere, is a region of vanishingly small flexural rigidity. Accordingly, most deformation in both the crust and mantle is inferred to be achieved by plastic flow and/or brittle faulting. Little or no long term elastic bending is evident. This does not, however, rule out short term (~300 yr) build up of elastic strain that could be released in large earthquakes [*Sutherland et al., this volume*]. Nevertheless, the central Southern Alps low-velocity zone is a curious feature worthy of further study. It is clearly linked to the part of the Alpine Fault where anomalously fast exhumation is taking place, where topography is at a maximum, and geophysical and geological data show evidence of water at high pressure in the crust. Such regions are unusual for the middle to upper crust. It is possible, therefore, that this zone could act as a damper, that slows throughgoing ruptures by distributing strain more evenly in both space and time, and by anelastic processes.

There are potentially new insights from SIGHT to the study of resources. For example, future research might be directed at the thermal structure of the Alpine Fault Zone with a view to assess the zone for power generation by hot-water heat exchange methods. In Australia and Europe research is presently taking place into Hot-Dry Rock (HDR) exploration as a means of generating carbon-free electrical energy [*Barbier, 2002*]. In these projects ~5 km deep bore-holes are drilled into areas of high natural heatflow. The subsurface rock is hydro-fractured, water is pumped into these deep holes, allowed to become super-heated, then returned to the surface via other bore holes.

We suggest that the Alpine Fault Zone could be a favourable site for a HDR investigation for the following reasons. Firstly, HDR systems typically need rock temperatures of 200–250 °C, which according to most thermal and geophysical models (Figure 13) will be reached at depths of ~5 km [*Koons, 1987; Shi et al., 1996; Wannamaker et al., 2002*]. Secondly, there is a ready water supply from the west coast rivers, and that at a depth of 6 km our geophysical models predict that water will be tapped at lithostatic, or higher pressure. Thus, compared to some HDR projects elsewhere in the world, one centered here would need less drilling and no importing of water from other catchments.

12. CONCLUSIONS

We conclude our study of the Alpine Fault Zone with the following key points.

1. The Alpine Fault can be traced by seismic methods to a depth of about 35 km. The fault broadens with depth and appears to develop a listric shape.

2. Strong reflectivity zones are present in the root zone of the Alpine Fault that we infer to be shear-related. A calibrated attribute analysis shows that reflection coefficients are on the order of 0.25 and largely due to negative impedance contrasts. Anisotropy of the shear fabric, water filled voids and constructive interference could all contribute to this strong reflectivity.

3. We identify on the seismic reflection section a strong mid-crustal decollement surface along which crustal rocks are obducted to the surface and exhumed.

4. A zone of inferred high fluid pressures is defined within the Alpine Fault Zone by both low seismic P-wave velocity and high electrical conductivity. We interpret this zone as being due to water released by metamorphic reactions in the greywacke-schist as the crust is being thickened. Just as for dehydration processes in subduction zones, this zone may be critical in modulating the mode of strain release in the central section of the Alpine Fault.

5. A flexural loading study indicates that no long-term elastic strength is indicated for central South Island. T_e values recovery from near-zero values beneath the central Southern Alps to more normal values of ~20 km at the east and west coasts.

6. Weakening of the crust is linked to recent convergence, crustal thickening, prograde metamorphism and excess fluid pressure. In the mantle the lack of strength is consistent with the observed seismic anisotropy and therefore a history of shearing, dissipative heating, and possible dynamic recrystallisation of olivine.

7. We suggest the thermal state of the Alpine fault Zone is worthy of closer study with a view to assessing the top few kilometers as a site for a hot-dry-rock project.

Acknowledgements. We thank Peter Malin, Tim Little and one other anonymous reviewer for thorough and helpful reviews. This work was funded by NSF project EAR-9418530 and funding from the New Zealand Science Foundation.

REFERENCES

Anderson, H., and T. H. Webb (1994), New Zealand seismicity patterns revealed by the upgraded National Seismic Network, *N. Z. J. Geol. and Geophys., 37*, 477–493.

Barbier, E. (2002), Geothermal energy technology and current status: an overview, *Renewable and Sustainable Energy Reviews, 6*, 3–65.

Beaumont, C., P. J. J. Kamp, J. Hamilton, and P. Fullsack (1996), The continental collision zone, South Island, New Zealand; comparison of geodynamical models and observations., *J. Geophys. Res., 101*, 3333–3359.

Blythe, A. E. (1998), Active tectonics and Ultrahigh-pressure rocks, in When continents collide: Geodynamics and Geochemistry of

Ultrahigh-pressure Rocks, edited by B. R. Hacker, Liou. J. G., pp. 141–160, Kluwer Academic Publishers.

Bodine, J. H., M. S. Steckler, and A. B. Watts (1981), Observations of flexure and rheology of the oceanic lithosphere, *J. Geophys. Res., 86*, 3695–3707.

Davey, F. J., T. Henyey, S. Kleffmann, A. Melhuish, D. Okaya, T. A. Stern, and D. J. Woodward (1995), Crustal reflections from the Alpine Fault Zone, South Island, New Zealand., *N. Z. J. Geol. & Geophys., 38*, 601–604.

Davey, F. D., T. Henyey, W. S. Holbrook, D. Okaya, T. A. Stern, A. Melhuish, S. Henrys, D. Eberhart-Phillips, T. McEvilly, R. Urhammer, H. Anderson, F. Wu, G. Jiracek, P. Wannamaker, G. Caldwell, and N. Christensen, (1997), Preliminary results from a geophysical study across a modern, continent-continent collisional plate boundary - the Southern Alps , New Zealand., *Tectonophysics, 288*, 221–235.

Dobrin, M. B. (1976), Introduction to Geophysical Prospecting, 630 pp., McGraw-Hill, New York.

Fyfe, W. S., N. J. Price, and A. B. Thompson (1978), Fluids in the Earth's Crust, 383 pp., Elsevier, New York.

Gair, H. S. (1967), Sheet 20 - Mt Cook. Geological map of New Zealand, 1:250,000, Department of Scientific and Industrial Research, Wellington.

Garrick, R., and T. Hatherton (1973), Seismic velocity studies in the Southern Alps, New Zealand, *N. Z. J. Geol & Geophys, 16*, 973–995.

Gough, D. I. (1986), Seismic reflectors, conductivity, water and stress in the continental crust, *Nature, 323*, 143–147.

Grapes, R. H. (1995), Uplift and exhumation of Alpine Schist, Southern Alps, New Zealand., *N. Z. J. Geol. & Geophys., 38*, 525–534.

Grapes, R. H., and T. Watanabe (1992), Metamorphism and uplift of the Alpine schist in the Franz Josef-Fox Glacier area, Southern Alps, New Zealand, *J. Metamorph. Pet., 10*, 171–180.

Henrys, S. H., D. J. Woodward, D. Okaya, and J. Yu (2004), Mapping the Moho beneath the Southern Alps continent-continent collision, New Zealand, using wide-angle reflections, *Geophys. Res. Lett., 31*, doi:10.1029/2004GLO20561.

Hickman, S., M. D. Zoback, and W. Ellsworth (2004), Introduction to special sections: Preparing for the San Andreas Fault Observatory at Depth, *Geophys. Res. Lett., 31*, L12S01, doi:10.1029/2004GL020688.

Hole, J. A. (1996), Seismic reflections from the near vertical San Andreas Fault., *Geophys. Res. Lett., 23*, 237–240.

Hole, J. A., R. D. Catchings, K. C. St Clair, M. J. Rymer, D. Okaya, and B. J. Carney (2001), Steep-dip seismic imaging of the shallow San Andreas Fault near Parkfield, *Science, 294*, 1513–1515.

Holt, W. E., and T. A. Stern (1991), Sediment loading on the Western Platform of the New Zealand continent: Implications for the strength of a continental margin, *Earth Planet. Sci. Lett., 107*, 523–538.

Jones, T. D., and A. Nur (1984), The nature of seismic reflections from deep crustal fault zones, *J. Geophys. Res., 89*, 3153–3171.

Julian, B. (2002), Seismological detection of slab metamorphism, *Science, 296*, 1625–1626.

Kamp, P. J., and J. M. Tippett (1993), Dynamics of Pacific Plate crust in the South Island (New Zealand) zone of oblique continent-continent convergence, *J. Geophys. Res., 98*, 16,105–116,118.

Kanamori, H. (1989), A slow seismic event recorded in Pasadena, *Geophys. Res. Lett., 16*, 1411–1414.

Kanamori, H., and E. Hauksson (1992), A slow earthquake in the Santa Maria Basin, California, *Bull. Seis. Soc. Am., 82*, 2087–2096.

Kennett, B. L. N., and E. R. Engdahl (1991), Travel times for global earthquake location and phase identification, *Geophys. J. Int., 105*, 429–465.

King Hubbert, M., and W. W. Rubey (1959), Role of Fluid pressure in mechanics of overthrust faulting, *Bull. Geol. Soc. Am., 70*, 115–166.

Kleffmann, S. (1999), Crustal structure studies of a transpressional plate boundary - the central South Island of New Zealand, PhD thesis, 233 pp, Victoria University of Wellington, Wellington.

Kleffmann, S., F. Davey, A. Melhuish, D. Okaya, and T. Stern (1998), Crustal structure in the central South Island from the Lake Pukaki seismic experiment., *N. Z. J. Geol. Geophys., 41*, 39–49.

Kodaira, S., T. Iidaka, A. Kato, J. O. Park, T. Iwasaki, and Y. Kaneda (2004), High Pore Fluid Pressure May Cause Silent Slip in the Nankai Trough, *Science, 304*, 1295–1298.

Koons, P. O. (1987), Some thermal and mechanical consequences of rapid uplift: an example from the Southern Alps, New Zealand, *Earth Planet. Sci. Lett., 86*, 307–319.

Koons, P. O. (1990), Two-sided orogen; collision and erosion from the sandbox to the Southern Alps, New Zealand, *Geology, 18*, 679–682.

Koons, P. O., D. Craw, S. C. Cox, P. Upton, A. S. Templeton, and C. P. Chamberlain (1998), Fluid flow during active oblique convergence: A Southern Alps model from mechanical and geochemical observations, *Geology, 26*, 159–162.

Leitner, B., D. Eberhart-Phillips, H. Anderson, and J. Nabeleck (2001), A focussed look at the Alpine fault, New Zealand: seismicity, focal mechanisms, and stress observations, *J. Geophys. Res., 106*, 2193–2220.

Linde, A. T., M. T. Gladwin, M. J. S. Johnston, R. L. Gwyther, and R. G. Bilham (1996), A slow earthquake sequence on the San Andreas fault, *Nature, 383*, 65–68.

Little, T. A. (2004), Transpressive ductile flow and oblique ramping of lower crust in a two-sided orogen: Insight from quartz grain-shaped fabrics near the Alpine Fault, New Zealand, *Tectonics*, doi:10.1029/2002TC001456.

Little, T. A., S. Cox, J. K. Vry, and G. Batt (2005), Variations in exhumation level and uplift rate along the oblique-slip Alpine fault, central Southern Alps, *GSA Bull., 117*, 707–723.

Little et al. (this volume), Transpression models and ductile deformation of the lower crust of the Pacific Plate in the central Southern Alps, a perspective from structural geology

Little, T., R. Wightman, R. J. Holcombe, and M. Hill (this volume), Transpression models and ductile deformation of the lower crust of the Pacific Plate in the central Southern Alps, a perspective from structural geology.

Meissner, R., and T. Wever (1992), The possible role of fluids for the structuring of the continental crust, *Earth Sci. Rev., 32*, 19–32.

Molnar, P., H. Anderson, E. Audoine, D. Eberhart-Philips, K. Gledhill, E. Klosko, T. McEvilly, D. Okaya, M. Savage, T. Stern, and F. Wu (1999), Continuous deformation versus faulting through the continental lithosphere of New Zealand, *Science, 286*, 516–619.

Mooney, W. D., and A. Ginzburg (1986), Seismic measurements of the internal properties of fault zones, *Pure. Appl. Geophys., 124*, 141–157.

Nadeau, R. M., and D. Dolenc (2005), Nonvolcanic tremors deep beneath the San Andreas Fault, *Science, 307*, 389.

Norris, R. J., and A. F. Cooper (2003), Very high strains recorded in mylonites along the Alpine Fault, New Zealand: implications for the deep structure of plate boundary faults, *J. of Struct. Geol., 25*.

Norris, R. J., P. O. Koons, and A. F. Cooper (1990), The obliquely convergent plate boundary in the South Island of New Zealand; implications for ancient collision zones, *J. Struct. Geology, 12*, 715–725.

Obara, K. (2002), Nonvolcanic deep tremor associated with subduction in southwestern Japan, *Science, 296*, 1679–1681.

Okaya, D., N. Christensen, D. Stanley, and T. Stern (1995), Crustal anisotropy in the vicinity of the Alpine Fault Zone , South Island New Zealand., *N. Z. J. Geol. & Geophys., 38*, 579–584.

Okaya, D., S. Henrys, and T. Stern (2002), Double-sided onshore-offshore seismic imaging of a Plate Boundary: Super-gathers across South Island of New Zealand, *Tectonophysics, 355*, 247–263.

Okaya, D., T. A. Stern, and S. H. Henrys (2007), Seismic image of continental collision at a transform plate boundary, New Zealand, *Science, submitted.*

Pulford, A., M. K. Savage, and T. A. Stern (2003), Absent anisotropy: the paradox of the Southern Alps orogen, *Geophys. Res. Lett.,* doi:10.1029/2003GL017758.

Reyners, M. (1987), Subcrustal earthquakes in the central South Island, New Zealand, and the root of the Southern Alps, *Geology, 15*, 1168–1171.

Rice, J. R. (1992), Fault stress states, pore pressure distributions, and the weakness of the San Andreas fault., in Fault mechanics and the transport properties of rocks, edited by B. Evans and T. F. Wong, pp. 475–504, Academic Press, New York.

Rogers, G., and H. Dragert (2003), Episodic tremor and slip: the chatter of slow earthquakes, *Science, 300*, 1942–1944.

Savage et al. (this volume).

Scherwath, M., T. Stern, A. Melhusih, and P. Molnar (2002), Pn anisotropy and distributed upper mantle deformation associated with a continental transform fault, *Geophys. Res. Lett.,* doi:10.1029/2001GL014179.

Scherwath, M., T. A. Stern, F. J. Davey, D. Okaya, W. S. Holbrooke, R. Davies, and S. Kleffmann (2003), Lithospheric structure across oblique continental collision in New Zealand from wide-angle P-wave modeling, *J. Geophys. Res.,* doi:10.1029/2002JB002286.

Scholz, C. (2000), Evidence for a strong San Andreas fault., *Geology, 28*, 163-166.

Shelly, D. R., G. C. Beroza, S. Ide, and S. Nakamula (2006), Low-frequency earthquakes in Shikoku, Japan, and their relationship to episodic tremor and slip, *Nature, 442*, 188–191.

Sheriff, R. E., and L. P. Geldart (1995), Exploration Seismology, 2nd ed., 592 pp., Cambridge University Press.

Shi, Y., R. Allis, and F. Davey (1996), Thermal modelling of the Southern Alps, *Pure. Appl. Geophys., 146*, 469–501.

Sibson, R. H. (1986), Earthquakes and rock deformation on a crustal fault, *Ann. Rev. Earth Planet. Sci., 14*, 149–176.

Simpson, D. W. (1986), Triggered earthquakes, *Ann Rev. Earth and Planet Sci., 14*, 21–42.

Smith, E. G., C., T. Stern, and B. O'Brien (1995), A seismic velocity profile across the central South Island, New Zealand, from explosion data, *N. Z. J. Geol. Geophys.*, 38, 565–570.

Stern, T., S. Kleffmann, D. Okaya, M. Scherwath, and S. Bannister (2001), Low seismic wave-speeds and enhanced fluid pressure beneath the Southern Alps, New Zealand, *Geology, 29*, 679–682.

Stern, T. A. (1995), Gravity anomalies and crustal loading at and adjacent to the Alpine Fault, New Zealand, *N. Z. J. Geol. Geophys., 38*, 593–600.

Stern, T. A., and J. H. McBride (1998), Seismic exploration of strike-slip zones, Special issue: Deep Seismic Profiling of the Continents: General Results and New Methods, *Tectonophysics, 286*, 63–78.

Stern, T. A., P. Molnar, D. Okaya, and D. Eberhart-Phillips (2000), Teleseismic P-wave delays and modes of shortening the mantle beneath the South Island, New Zealand, *J. Geophys. Res., 105*, 21,615–621,631.

Stewart, J., and A. B. Watts (1997), Gravity anomalies and spatial variations of flexural rigidity at mountain ranges, *J. Geophys. Res., 102*, 5327–5352.

Sutherland, R., D. Eberhart-Phillips, R. A. Harris, T. Stern, J. Beavan, S. Ellis, S. Henrys, S. Cox, R. J. Norris, K. R. Berryman, J. Townend, S. Bannister, J. Pettinga, B. Leitner, L. Wallace, T. A. Little, A. F. Cooper, M. Yetton, and M. Stirling (this volume), Do great earthquakes occur on the Alpine fault in central South Island, New Zealand?

Townend, J., and Z. Zoback (2001), Implications of earthquake focal mechanisms for the frictional strength of the San Andreas fault system, Special Publication of the geological society of London, in press.

Tullis, T. E., and J. Tullis (1986), Experimental rock deformation techniques., in Mineral and Rock Deformation: Laboratory Studies, edited by B. E. Hobbs, Heard, H. C., American Geophysical Union, Washington.

van Avendonk, H., W. S. Holbrook, D. Okaya, J. Austin, F. Davey, and T. Stern (2004), Continental crust under compression: A seismic reflection study of South Island Geophysical Transect 1, South Island, New Zealand, *J. Geophys. Res., 109*, doi:10.1029/2003JB002790.

Vry, J. K., A. C. Storkey, and C. Harris (2001), Role of fluids in the metamorphism of the Alpine Fault Zone, New Zealand, *J. Metamorphic Geol., 19*, 21–31.

Wannamaker, P. E., G. R. Jiracek, J. A. Stodt, T. G. Caldwell, V. M. Gonzalez, J. D. McKnight, and A. D. Porter (2002), Fluid generation and pathways beneath an active compressional orogen, the New ZealandSouthern Alps, inferred from magnetotelluric data, *J. Geophys. Res., 107,* 2117, doi2110.1029/2001JB000186.

Warner, M. (1987), Migration — why doesn't it work for deep continental data?, *Geophys. J. R. astron. Soc., 89*, 21–26.

Warner, M. (1990), Absolute reflection coefficients from deep seismic reflections, *Tectonophysics, 173*, 15–23.

Warner, M. (2004), Free water and seismic reflectivity in the lower continental crust, *J. of Geophysics and Engineering, 1*, 88–101.

Watts, A. B. (2001), Isostasy and flexure of the lithosphere, 458 pp., Cambridge University Press.

Wellman, H. W. (1979), An uplift map for the South Island of New Zealand, *Bull. Roy. Soc. N. Z., 18*, 13–20.

Wightman, R. H., and T. A. Little (this volume), Deformation of the Pacific Plate above the Alpine Fault ramp and its relationship to expulsion of metamorphic fluids: an array of backshears.

Zeitler, P. K., A. S. Meltzer, P. O. Koons, D. Craw, B. Hallet, C. P. Chamberlin, W. S. F. Kidd, S. K. Park, L. Seeber, M. Bishop, and J. Shroder (2001), Erosion, himalayan geodynamics, and the geomorphology of metamorphism, *GSA Today, 11*, 4–9.

Zoback, M. (2000), Strength of the San Andreas, *Nature, 405*, 31.

Fred Davey and Stuart Henrys, Institute of Geological and Nuclear Sciences, Lower Hutt, New Zealand.

Stefan Kleffmann, New Zealand Oil and Gas, Wellington, New Zealand.

David Okaya, Department of Earth Sciences, University of Southern California, Los Angeles, CA, USA.

Martin Scherwath, GEOMAR, Kiel, Germany.

Tim Stern, School of Earth Sciences, Victoria University of Wellington, Wellington, New Zealand.

Do Great Earthquakes Occur on the Alpine Fault in Central South Island, New Zealand?

R. Sutherland[1], D. Eberhart-Phillips[2], R.A. Harris[3], T. Stern[4], J. Beavan[1], S. Ellis[1], S. Henrys[1], S. Cox[2], R.J. Norris[5], K.R. Berryman[1], J. Townend[4], S. Bannister[1], J. Pettinga[6], B. Leitner[1], L. Wallace[1], T.A. Little[4], A.F. Cooper[5], M. Yetton[7], M. Stirling[1].

Geological observations require that episodic slip on the Alpine fault averages to a long-term displacement rate of 2–3 cm/yr. Patterns of seismicity and geodetic strain suggest the fault is locked above a depth of 6–12 km and will probably fail during an earthquake. High pore-fluid pressures in the deeper fault zone are inferred from low seismic P-wave velocity and high electrical conductivity in central South Island, and may limit the seismogenic zone east of the Alpine fault to depths as shallow as 6 km. A simplified dynamic rupture model suggests an episode of aseismic slip at depth may not inhibit later propagation of a fully developed earthquake rupture. Although it is difficult to resolve surface displacement during an ancient earthquake from displacements that occurred in the months and years that immediately surround the event, sufficient data exist to evaluate the extent of the last three Alpine fault ruptures: the 1717 AD event is inferred to have ruptured a 300–500 km length of fault; the 1620 AD event ruptured 200–300 km; and the 1430 AD event ruptured 350–600 km. The geologically estimated moment magnitudes are 7.9 ± 0.3, 7.6 ± 0.3, and 7.9 ± 0.4, respectively. We conclude that large earthquakes ($M_w > 7$) on the Alpine fault will almost certainly occur in future, and it is realistic to expect some great earthquakes ($M_w \geq 8$).

INTRODUCTION

The Alpine fault is a mature dextral strike-slip fault that offsets basement rocks by ~470 km and offsets deposits of the last glaciation (18-80 ka) by 0.4–2.0 km [*Wellman and Willett*, 1942; *Wellman*, 1953; *Norris and Cooper*, 2001; *Sutherland et al.*, 2006]. The surface trace is continuous for at least 800 km (Figure 1), and has no separations of >5 km [*Wellman and Willett*, 1942; *Norris et al.*, 1990; *Berryman et al.*, 1992; *Barnes et al.*, 2005]. In central South Island, the fault is locally transpressive and the surface trace bounds the western edge of the mountains [*Norris et al.*, 1990]. Although the Alpine fault is identified as a laterally continuous crustal structure with an average late Quaternary surface displacement rate of 2–3 cm/yr [*Berryman et al.*, 1992; *Norris and Cooper*, 2001; *Sutherland et al.*, 2006], there have been no large earthquakes causing surface rupture on the fault

[1] GNS Science, Lower Hutt, New Zealand.
[2] GNS Science, Dunedin, New Zealand.
[3] U.S. Geological Survey, Menlo Park, California.
[4] Victoria University of Wellington, Wellington, New Zealand.
[5] University of Otago, Dunedin, New Zealand.
[6] University of Canterbury, Christchurch, New Zealand.
[7] Geotech Consulting, Lyttelton R.D., New Zealand.

A Continental Plate Boundary: Tectonics at South Island, New Zealand
Geophysical Monograph Series 175
10.1029/175GM12

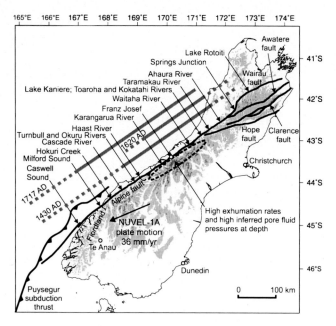

Figure 1. Location of the Alpine fault and places named in the main text. Topography >800 m is shown with a light shade and >1800 m with a dark shade. Bold grey lines indicate inferred extent of past Alpine fault ruptures. Bold arrow shows NUVEL-1A Pacific-Australia plate motion [*DeMets et al.*, 1994].

since European settlement in ca. 1800 AD. Since a national seismic network was installed and continuously improved (during the last 50 years), it has also become clear that small and moderate-sized earthquakes are less frequent near the Alpine fault in central South Island than at any other part of the plate boundary through New Zealand [*Evison*, 1971; *Eberhart-Phillips*, 1995]. The lack of moderate or large-magnitude earthquakes on the Alpine fault since reliable record keeping started means that the seismic potential of the Alpine fault must be inferred from indirect observations and theoretical considerations.

It has been argued that high heat-flow and a shallow brittle-ductile transition make it physically unlikely that elastic strains of sufficient size could accumulate to generate great earthquakes and that the measured geodetic strains can be explained if a significant proportion of plate boundary displacement is accommodated aseismically [*Walcott*, 1978; *Walcott*, 1998]. Moreover, an extensive region of low seismic wave speeds [*Smith et al.*, 1995; *Stern et al.*, 2001; *Eberhart-Phillips and Bannister*, 2002] and high electrical conductivity [*Wannamaker et al.*, 2002] within the Alpine fault zone suggests that high fluid pressures may be present, and supports the hypothesis that the Alpine fault is not able to sustain high shear stresses. This raises several key ques-

tions. Could some component of long-term displacement on the Alpine fault be accommodated by mechanisms other than large earthquakes [*Stern et al.*, 2001]? Continental strike-slip faults of similar dimensions to the Alpine fault typically fail in large earthquakes [*Wells and Coppersmith*, 1994; *Hanks and Bakun*, 2002; *Lin et al.*, 2002; *Eberhart-Phillips et al.*, 2003], but is the Alpine fault atypical? Is it possible that some component of long-term Alpine fault slip may be accommodated by slow ruptures, afterslip or during a succession of smaller earthquakes? Even if the Alpine fault does fail in large earthquakes, then it may not exclusively rupture in large earthquakes in every seismic cycle; by analogy, the Santa Cruz section of the San Andreas fault ruptured in earthquakes of different magnitude in 1906 and 1989.

One objective of the SIGHT project, and related projects, was to better determine Alpine fault geometry and physical properties at depth, so that its seismic potential in central South Island could be better understood. We review and synthesise the data that are now available and present new modeling to address what the seismic potential of the Alpine fault might be. Does the Alpine fault fail in large (M_w >7) earthquakes? Does the Alpine fault fail in great (M_w ~8) earthquakes? We explore these questions and, in the light of our conclusions, make recommendations for future data collection activities that could underpin improved quantitative seismic hazard models for central South Island.

SURFACE EXPRESSION

The Alpine fault varies significantly in character along its length (Figure 1). From its southern termination near the Puysegur subduction thrust to offshore from Caswell Sound, the Alpine fault scarp is up-to-the-northwest in most places and the trace is segmented into a series of right-stepping sections, small pull-apart basins, and minor transpressional ridges [*Barnes et al.*, 2001; *Barnes et al.*, 2005]. At Caswell Sound a right step in the fault trace of about 5 km occurs [*Barnes et al.*, 2005]. From Caswell Sound to the Cascade River (Figure 1), surface trace continuity is disrupted by minor (<200 m) step-overs and the scarp is up-to-the-northwest offshore and at most localities onshore. Slickensides and offset topography constrain the average surface slip vector onshore between Milford Sound and Cascade River to plunge at <5° [*Berryman et al.*, 1992; *Sutherland and Norris*, 1995; *Barnes et al.*, 2005; *Sutherland et al.*, 2006].

A zone of overlapping fault traces about 1 km wide near Cascade River is accompanied by a change in character of the fault trace. North of Cascade River to Haast River (Figure 1), the fault trace becomes up-to-the-southeast, but the displacement is still primarily dextral strike-slip and the fault trace remains about two kilometres northwest of the range front [*Berryman et al.*, 1992; *Cooper and Norris*, 1995]. The ca. 7° clockwise change in regional strike of the fault trace north of Cascade River should reduce the convergent component of plate motion across the fault, but reverse movement in fact becomes more pronounced; this observation suggests that regional changes in geology either side of the fault play a significant role in determining the nature of slip partitioning at the plate boundary.

North of Haast River to Taramakau River (Figure 1), the Alpine fault trace lies at the rangefront and is segmented at a scale of 1–5 km into a series of strike-slip and moderately southeast-dipping oblique-thrust sections [*Norris et al.*, 1990; *Berryman et al.*, 1992]; thrust sections are often difficult to recognise on air photos or in the field, except where river exposures reveal cataclasite thrust over Quaternary sediments.

Between Taramakau River and Lake Rotoiti (Figure 1), the Alpine fault surface trace is up-to-the-southeast and lies near to the range front [*Berryman et al.*, 1992; *Nathan et al.*, 2002]. Near Taramakau River, the Alpine fault and Hope fault intersect at an acute angle. There is a substantial restraining bend in the Alpine fault southwest of Lake Rotoiti, where the Awatere fault merges with the Alpine fault. Northeast of Lake Rotoiti, the Alpine fault is continuous with and becomes reclassified as the Wairau fault.

SURFACE DISPLACEMENT RATE

The Alpine fault surface displacement rate in the region between Milford Sound and Cascade River is 23 ± 2 mm/yr (95% confidence), as determined from multiple measured offsets that accumulated over a time interval of 18–79 kyr [*Sutherland et al.*, 2006]. The very similar character of the offshore Milford-Caswell section with that onshore, and a glacial moraine offset by a similar magnitude to those onshore, suggest that a similar displacement rate continues at least as far south as Caswell Sound [*Barnes et al.*, 2005]. The prominent surface trace and rapid rates of deformation in adjacent basins suggests that there is also a rapid (1–3 cm/yr) displacement rate south of Caswell Sound [*Barnes et al.*, 2001].

At Haast River, the surface strike-slip displacement rate is determined to be >21 mm/yr from an offset terrace of age <4.4 cal ka [*Cooper and Norris*, 1995] and 28 ± 4 mm/yr from a terrace surface offset 25 m since 0.74 cal ka [*Berryman et al.*, 1998]. Between Haast River and Taramakau River, numerous exposures of a discrete fault zone emplacing shattered mylonite over Quaternary sediments suggest rapid rates of Quaternary displacement; this is confirmed by estimates of strike-slip displacement rate of 27 ± 5 mm/yr [*Norris and Cooper*, 1997], >22 mm/yr [*Cooper and Norris*, 1994], and 29 ± 6 mm/yr [*Wright*, 1998], and estimated dip-

slip displacement rates of 7.8 ± 1, >6.7, 6 ± 1, >12, 8 ± 3, and 7.5 ± 1.5 mm/yr [*Norris and Cooper*, 2001].

Northeast of Taramakau River, where there exists a zone of intersection with the Hope fault, the Alpine fault displacement rate is determined to be $\geq 6.3 \pm 2$ mm/yr [*Yetton et al.*, 1998] and 10 ± 2 mm/yr [*Berryman et al.*, 1992]. The Alpine fault displacement rate diminishes to the northeast, as the intersection with the Awatere fault is passed, and is estimated to be 4 ± 1 mm/yr where the Alpine fault is reclassified as the Wairau fault [*Berryman et al.*, 1992; *Zachariasen et al.*, 2006].

EVIDENCE FOR ALPINE FAULT CREEP AT THE SURFACE

It has previously been suggested that shallow aseismic deformation may be occurring near or on the Alpine fault in central South Island [*Walcott*, 1978; *Walcott*, 1998]. Trenching, natural exposures, and geomorphic offsets indicate that long-term displacement on the shallowest part of the Alpine fault (<1 km depth) is localised in a narrow zone and has a rate of 2–3 cm/yr, when averaged over >1000 yr [*Norris and Cooper*, 2001].

The most famous experiment to investigate the existence of Alpine fault creep was the construction (under the supervision of F. Evison) in 1964 of a concrete monitoring wall near Springs Junction (Figure 1), which is north of the Alpine-Hope fault intersection [*Beanland*, 1987]. The "Evison wall" and sealed roads in ten locations are built across the Alpine fault, but show no signs of deformation over a period of >40 yr. Other man-made structures that cross the Alpine fault include delicate infrastructure items such as a petrol station at Franz Josef and several water races and tunnels associated with the Dilmans Hydroelectric scheme near Taramakau river (Figure 1), where even a very small displacement is likely to have been noticed. No such displacement has been reported.

In addition to human features, several natural river terraces that are >40 years old cross the Alpine fault in central South Island, but show no sign of deformation, even though a clear scarp exists on older terraces [*Adams*, 1980]. This observation is supported by trenches and exposures, where undeformed sediment is observed overlying faulted sediment [*Yetton et al.*, 1998].

If Alpine fault creep had occurred during the decades in which geodetic observations have been made, then higher strain rates would be expected on small-aperture survey networks that span the Alpine fault, as compared to larger-aperture networks. However, no significant difference is observed between networks with apertures of 1–5 km [*Wood and Blick*, 1986] and networks with apertures of ca. 30 km [*Walcott*, 1979; *Beavan et al.*, 1999].

By analogy with the creeping section of the San Andreas fault, spatial clustering of small-magnitude earthquakes would be expected close to the Alpine fault, if it were creeping near the surface. This is not observed [*Eberhart-Phillips*, 1995].

It is very difficult to rule out the possibility that some aseismic deformation may be occurring in the mountainous region immediately east of the Alpine fault trace, but the evidence described above precludes significant creep on the surface part of the Alpine fault during the last 40 yr and strongly suggests that none has occurred in >100 yr.

HISTORICAL SEISMICITY

Seismicity is distributed in a broad region near the Alpine fault and extending approximately 100 km to the southeast. Moderate magnitude earthquakes (M_w >5.0) recorded from 1928–1999 have occurred in the vicinity of the Alpine fault, but these historic epicentres are more poorly determined: it is not possible to say if they represent Alpine fault events.

The stress field near the Alpine fault was computed after the SAPSE experiment from focal mechanisms: the regional stress field has a subhorizontal axis of maximum compressive stress that trends 110–120° [*Leitner et al.*, 2001], a direction similar to the azimuth of maximum contractional strain rate [*Beavan et al.*, 1999] and at an angle that is ca. 60° oblique to the strike of the Alpine fault. At 80% confidence level, stress tensors for all regions in the Southern Alps are the same [*Leitner et al.*, 2001].

The maximum depth of seismicity provides an estimate of the thickness of the seismogenic zone and, therefore, relates to the depth of the brittle-ductile transition zone [*Scholz*, 2002]. The maximum depth of crustal seismicity is uniform over large parts of central South Island at ~12 km. The SAPSE experiment recorded 60 earthquakes in a swath extending from 5 km northwest to 15 km southeast of the surface trace of the Alpine fault [*Leitner et al.*, 2001]. Given an approximate fault dip of 45°, these earthquakes occurred on or close to the fault, delineating its seismogenic zone. Seismicity was highest just north of Milford Sound. In central South Island (near Franz Josef), seismicity near the Alpine fault was low during the SAPSE experiment and during the 8 years of New Zealand national seismic network (NZNSN) recordings. However, the occurrence of a few earthquakes with hypocentral depths of as much as 10 km suggests that crust in the vicinity of the Alpine fault is capable of storing and then releasing elastic strain.

When similar magnitude ranges are considered, the seismicity rate of the Alpine fault is comparable with seismicity rates along locked sections of the San Andreas fault [*Leitner et al.*, 2001]. The moment release rate, calculated from NZNSN seismicity between 1990 and 1997 across the region, is highest at the Alpine fault, decreases toward the east, and shows another small maximum at 75–95 km distance from the Alpine fault. The distribution of moment release rate across the Southern Alps has a similar pattern to model strain rates [*Koons et al.*, 1994; *Koons et al.*, 1998] and measured strain rates [*Beavan et al.*, 1999; *Wallace et al.*, submitted], but is 2–3 orders of magnitude smaller than predicted if all the plate convergence had been accommodated by earthquakes during that period. Only a small fraction of the plate boundary strain that accumulated during the last 150 years has been released during earthquakes.

EVIDENCE FOR ANCIENT ALPINE FAULT EARTHQUAKES

Direct Earthquake Indicators

Pseudotachylyte, which is interpreted to be quenched melt generated by friction during seismic slip, is well documented within fault rocks of the Alpine fault zone in central South Island [*Reed*, 1964; *Wallace*, 1976; *Sibson et al.*, 1979; *Bossiere*, 1991; *Warr and van der Pluijm*, 2005]. The direct association of friction melts with fault rocks of the Alpine fault supports the hypothesis that seismic slip occurs on the Alpine fault, but places little constraint on the magnitude of seismic events.

Trenches across the Alpine fault near Haast and Okuru (Figure 1) show extensive evidence for liquefaction of sand layers, sand dykes, and sand extrusion onto paleosols [*Berryman et al.*, 1998]. The evidence for extensive liquefaction is consistent with shaking of intensity MM>7. The coincidence of sand extrusion horizons with the bases of colluvial wedges adjacent to the fault scarp suggests a causal relationship between liquefaction and fault scarp formation. Liquefaction of sand adjacent to the Alpine fault scarp is also reported from a trench (Kokatahi-2) 20 km southwest of Taramakau River [*Yetton*, 1998; *Yetton et al.*, 1998].

Episodic Scarp Formation

In the Haast-Okuru-Turnbull region, five trenches across faulted fluvial terraces show clear evidence for episodic surface rupture [*Berryman et al.*, 1998]. Fault surfaces that are overlain by undeformed sediment layers indicate tectonic quiescence after the last movement at each fault location. Intercalation of colluvial wedges and paleosols on the downthrown side of the scarp indicate episodic scarp formation, followed by scarp degradation, and subsequent soil formation. The correspondence of the base of the colluvial wedges with strata containing sand extruded from liquefaction features, and the upper truncation surfaces of faults, strongly suggests a causal relationship between episodic scarp formation, faulting, and strong shaking. Radiocarbon dating of the three 'event horizons' limits the timing of the three most recent earthquakes in the Haast region to 700–800 AD, 1160–1410 AD, and post 1480 AD. Surface displacement of 25 m is measured from offset river channels on terrace surfaces at the Haast and Okuru sites, and hence it is inferred that the horizontal surface offset during each faulting event was 8–9 m, with about 1 m of vertical motion. Additional evidence for horizontal surface faulting of 8–9 m during each of the last two fault movement events comes from offset river channels at Hokuri Creek [*Hull and Berryman*, 1986; *Sutherland and Norris*, 1995].

Fault scarps along the Alpine fault between Haast River and Toaroha River have not been excavated for paleoseismic investigations. However, evidence for discrete uplift events comes from a flight of terraces adjacent to the Karangarua River (Figure 1) immediately upstream from the Alpine fault [*Adams*, 1980]. Based on the age of forest cover determined by cores in tree trunks, the youngest terrace that no longer floods was abandoned in 1710–1720 AD, the next highest terrace was abandoned in 1600–1620 AD, and the terrace above that was abandoned in 1405–1445 AD [*Yetton et al.*, 1998].

In the region between Ahaura River and Toaroha River, truncated fault planes in six trenches and variably-deformed or undeformed scarp collapse deposits provide evidence for two discrete faulting events during the last 500 years [*Yetton*, 1998; *Yetton et al.*, 1998]. Radiocarbon data indicate that the earlier event occurred at 1480–1645 AD and the most recent event occurred after 1660 AD, and probably during 1700–1750 AD [*Yetton*, 1998; *Yetton et al.*, 1998]. Trenching evidence shows that the most recent event ruptured the Alpine fault up to 25 km northeast of its intersection with the Hope fault.

Paleoseismic trenching between Ahaura River and Lake Rotoiti demonstrates that the most recent event along the northern section of the Alpine fault (1480–1645 AD) appears to correspond with the penultimate event (ca. 1620 AD) recognised in the more southern trenches [*Yetton*, 2002]. Paleoseismic trenching of the Wairau fault suggests it has not ruptured during the last ca. 2000 yr [*Zachariasen et al.*, 2006].

Age of Forest Damage and Landslides

Evidence for vegetation damage comes from: the age of dead trees, the age of regenerating forest in areas that have undergone a significant disturbance, and tree-ring evidence from mature living trees that have survived a significant period of growth disturbance. Forest damage close to the Alpine fault was reported on the north side of Milford Sound, where trees had lost their crowns; the forest damage event was dated by tree trunk circumference to have been during the period 1650–1725 AD [*Cooper and Norris*, 1990].

Regional studies of forest dynamics and tree growth suppression have revealed significant forest disturbance events at 1700–1730 AD, 1610–1640 AD, and 1410–1440 AD [*Wells et al.*, 1998; *Yetton et al.*, 1998; *Wells et al.*, 1999; *Wells et al.*, 2001; *Cullen et al.*, 2003]. Widespread slope instability during these forest disturbance events is implied by large stands of forest that have no trees older than a certain age and that are substantially younger than the maximum life expectancy of mature species [*Wells et al.*, 1998]. The most recent forest disturbance event is inferred to correlate with a significant growth suppression event recorded by cross-matched tree ring analysis to have occurred in 1717 AD [*Wells et al.*, 1999]. Another significant growth suppression event is recorded from the Waitaha River region at ca. 1620 AD [*Wright*, 1998].

Another study that investigated the rockfall and landslide history of the Southern Alps utilised the distribution of lichen sizes on exposed boulders; clusters in lichen size were used to infer intense regional shaking at 1738–1758 AD, 1479–1499 AD, and 1216–1236 AD [*Bull*, 1996]. The 1748 AD age is slightly older (33 ± 18 yr) and barely resolved from the 1700–1730 age of forest disturbance. The 1489 AD age is 64 ± 18 yr older than the 1410–1440 age of forest disturbance. Uncertainties in both approaches may be larger than the quoted formal errors, and so it could be that the true ages cluster after the two largest recent Alpine fault earthquakes. However, some caution must be applied when interpreting the lichen data, because the closest site to the Alpine fault is 18 km away, and most sites are >25 km distant. It is possible that the lichen study dates rockfall deposits that were caused by earthquakes in the central Southern Alps, rather than on the Alpine fault.

A Reappraisal of Radiocarbon Age Clustering

It has been suggested on the basis of radiocarbon dating of aggradation and mass-movement deposits younger than 1250 AD, that these deposits have ages that cluster at times around, or shortly after the times of substantial forest disturbance [*Adams*, 1980; *Yetton*, 1998; *Yetton et al.*, 1998]. We reconsider the statistical question of how significant the clustering of the 19 reported radiocarbon ages is, in the light of recent advances in calibration of the radiocarbon timescale (Figure 2).

We assume a null hypothesis that the true age of the wood samples is equally frequent during the period 1300-1750 AD. The oldest bracketing age was chosen on the basis of the

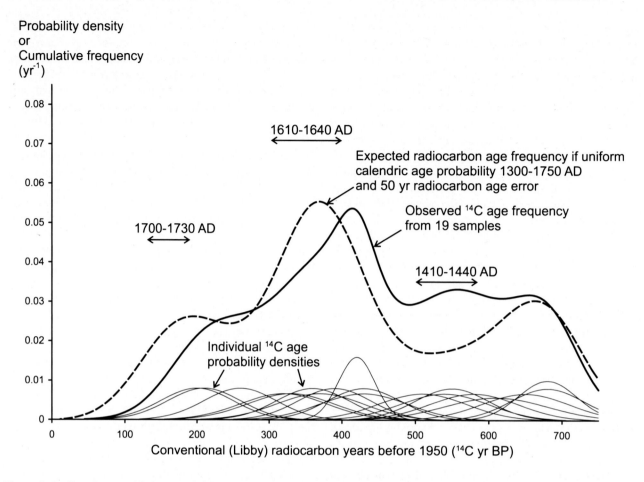

Figure 2. Radiocarbon ages from aggradation and mass-movement deposits inferred to be younger than 1250 AD [*Yetton*, 1998; *Yetton et al.*, 1998]. Thin lines show probability densities for individual samples. Bold line shows cumulative frequency, derived by summing the individual probability densities. The dashed line is the cumulative frequency expected from the SHCAL04 dataset [*McCormac et al.*, 2004], assuming the null hypothesis that the true calendric age of wood samples is equally likely during the interval 1300−1750 AD. The expected radiocarbon age ranges of inferred forest disturbance events at 1700-1730 AD, 1610−1640 AD, and 1410−1440 AD are also indicated [*Wells et al.*, 1998].

calibrated ages of the oldest radicarbon-dated samples. The youngest bracketing age was chosen to be slightly younger than the youngest period of substantial forest disturbance [*Wells et al.*, 1998; *Wells et al.*, 1999]. A model frequency distribution of radiocarbon ages was computed from the southern hemisphere calibration dataset [*McCormac et al.*, 2004], assuming a typical radiocarbon error of 50 14C yr, and then the data were compared with the model (Figure 2).

Although independent evidence provided by the forest disturbance ages do imply distinct episodes of enhanced mass-movement and aggradation, and these are a good match to radiocarbon ages of event horizons in paleoseismic trenches, it is apparent from Figure 2 that the random model (null hypothesis) and observed frequency distributions for radiocarbon ages of aggradation and mass movement deposits are

remarkably similar: the model peaks at 1700−1730 AD and 1610−1640 AD can be explained entirely from the distribution of radiocarbon ages within the calibration dataset, whereas a low frequency is predicted for the 1410−1440 AD period. We conclude that there is no definitive radiocarbon evidence for an increased frequency of aggradation events at either 1700−1730 AD or 1610−1640 AD. The observed frequency for the 1410−1440 AD event is nearly double the predicted frequency, but there are only five relevant samples and two of those are associated with the same (Toaroha river) terrace sequence. Although it is reasonable to suggest that times of enhanced mass-movement and terrace aggradation accompany earthquake events, we conclude that the existing mass movement and aggradation radiocarbon dataset is too small, and the ages too imprecise, to determine if there are significant clusters in the ages of mass movement and aggradation deposits.

ALPINE FAULT GEOMETRY AND PHYSICAL PROPERTIES AT DEPTH

Seismic reflection data have revealed reflectors associated with the Alpine fault zone at depth. Reflections with 9–15 s two-way-travel-time and reverse moveout were recorded east of the Alpine fault [*Davey et al.*, 1995; *Kleffman et al.*, 1998]; these reflections define a boundary, 20–30 km deep, dipping southeast at 40–60°, that, when projected to the surface, coincides with the surface trace of the Alpine fault. Stacked images reveal a bright reflective zone that can be traced from the mid-crust into a broad, sub-horizontal zone of diffuse, but strong, reflectivity at a depth of about 30 km [*Okaya et al.*, 2002].

Anomalous time delays on seismographs west of the Alpine fault were first reported after an experiment in 1995 that used large dynamite shots offshore of the east coast of South Island [*Smith et al.*, 1995]; the travel-time delays were consistent with a low-seismic-velocity region deep within the hanging wall of the Alpine fault. This preliminary model was tested with a higher resolution survey in 1998 [*Stern et al.*, 2001]: time delays from a series of 50–100 kg dynamite shots from the McKenzie basin were recorded on a 60 channel seismic array located in the lower part of the Karangarua river valley, just 6 km east of the Alpine fault. Maximum delays of 0.8 s were recorded. These data, in conjunction with observations of teleseismic P-wave delays permitted a roughly elliptical zone of low P-wave speeds to be delineated. The dimensions of the low-seismic-wave-speed zone are approximately 40 km by 25 km and the P-wave speed is 6-10% less than surrounding regions [*Stern et al.*, 2001]. The top of the low velocity volume is determined to lie at a depth of 6 ± 2 km from the differential delays of first arrival data [*Kleffman et al.*, 1998].

Tomography solutions that use all the SIGHT explosion data and natural earthquakes also show that rocks within the Alpine fault zone have unusually low seismic P-wave velocities. Southwest of Haast, a vertical seismic low-velocity zone is imaged to at least 15 km depth [*Eberhart-Phillips and Bannister*, 2002]. In central South Island, a localised 4–10% velocity reduction is consistent with a moderately southeast-dipping fault zone over a fault length of 50–100 km that penetrates to a depth of 20–30 km [*Stern et al.*, 2001; *Eberhart-Phillips and Bannister*, 2002; *Scherwath et al.*, 2003; *van Avendonk et al.*, 2004].

Alpine fault rocks may have low seismic velocity because of gouge, high crack density and high pore fluid pressure [*Eberhart-Phillips*, 1995]. Temperature and strain anomalies may be contributing factors, particularly in the fully ductile zone; laboratory measurements of metagreywacke show that the inferred temperature increase can cause 1–2 % velocity reduction at 15–30 km depths [*Christensen and Mooney*, 1995]. Active straining of rock produces small decreases in velocity, as observed in laboratory studies of dynamic moduli [*Winkler and Murphy*, 1995]. The combination of very low seismic velocity (6–10% reduction) and high electrical conductivity (30–300 ohm-m) at depths of 10–30 km suggests that pore fluid is both interconnected and at pressures approaching lithostatic [*Jones and Nur*, 1984; *Wannamaker et al.*, 2002]. The source of the pore fluids may be prograde metamorphism within the crustal root; large volumes of fluid may not be required because the rocks are likely to have low porosity.

HIGH HEAT FLOW AND GEOTHERMAL GRADIENT

Observational Evidence

A small number of direct heat-flow measurements from drillhole data [*Allis and Shi*, 1995] indicate a rise from 60 ± 5 mW/m^2 offshore to a maximum of 190 ± 50 mW/m^2 in the area of greatest erosion rate adjacent to the Alpine fault near Franz Josef (corresponding to an estimated geothermal gradient of $60° \pm 15°C/km$). The large uncertainty in heat-flow measurements is due to the effects of steep topography and near-surface water circulation near the fault. Further south, near Haast, measured heat-flow near the Alpine fault is 90 ± 25 mW/m^2, suggesting a lower geothermal gradient in regions with slower exhumation rates. West of the Alpine fault and north of Franz Josef, data from 24 drillholes allow a heat-flow estimate of 75 ± 25 mW/m^2 to be made [*Townend*, 1999].

Hot springs are abundant just east of the Alpine fault in central South Island. Although the water is thought to be predominantly meteoric, water in some of the hot springs contains a minor metamorphic component, suggesting fluids partly derived from depth [*Allis et al.*, 1979; *Koons and Craw*, 1991]. It has been suggested that elevated hydrothermal circulation near the Alpine fault may result from enhanced permeability owing to late stage brittle deformation and fracture [*Allis and Shi*, 1995].

Fluid inclusions trapped within quartz veins in high-grade schists exhumed next to the Alpine fault indicate formation at high temperature and low pressure [*Craw*, 1988; *Holm et al.*, 1989; *Craw et al.*, 1994; *Jenkin et al.*, 1994]; geothermal gradients of 70-90°C/km in the top 3-4 km of the crust are inferred for the central Southern Alps. The estimated depth range is dependent on an assumed lithostatic fluid pressure gradient; if fluid pressures were hydrostatic, the estimated temperature gradients would be lower (ca. 25–35°C/km).

Fluid inclusions may record transient, high-stress, high fluid pressure conditions associated with earthquakes, rather than typical conditions [*Allis and Shi*, 1995].

Thermochronological data indicate rapid cooling close to the Alpine fault. Fast erosion and exhumation have produced a pattern of decreasing metamorphic grade and increasing thermochronological age with increasing distance from the fault, consistent with the effects from rapid uplift and exhumation since ca. 5 Ma in central South Island [*Adams and Gabites*, 1985; *Tippett and Kamp*, 1993; *Batt and Braun*, 1999; *Batt et al.*, 2004; *Little et al.*, 2005].

Conclusions From Thermo-Mechanical Models

Thermal and thermo-mechanical models show that rapid exhumation along the Alpine fault inevitably perturbs isotherms to shallower depths, increasing the geothermal gradient in the vicinity of the fault and causing elevated heat-flow above it [*Allis et al.*, 1979; *Koons*, 1987; *Allis and Shi*, 1995; *Beaumont et al.*, 1996; *Batt and Braun*, 1999; *Gerbault et al.*, 2003]. An additional heat source (neglected in most models) is frictional shear heating [*Scholz et al.*, 1979], although the effect will only be significant if the fault has considerable shear strength. Assuming hydrostatic fluid pressure and a reasonable strength profile along the fault with depth, it has been shown that models without frictional shear heating fit heat-flow and thermochronological data at least as well as, or better than, models that include shear heating [*Shi et al.*, 1996].

Significant variations occur between different published thermo-mechanical models of the Southern Alps: some show a maximum upward deflection of 100–300°C isotherms east of the surface trace of the Alpine fault, matched by a downward perturbation of higher temperatures to the west and at greater depths as material is deflected downwards into a thickening crustal root [*Allis and Shi*, 1995; *Gerbault et al.*, 2003]; whereas others place the elevated 400°C isotherm directly above the Alpine fault at extremely shallow (<5 km deep) levels [*Koons*, 1987; *Batt and Braun*, 1999]. While differences exist, all the models predict elevated heat-flow above the Alpine fault is a consequence of rapid exhumation of lower crustal material. It follows that the highest exhumation rates, which are found near Franz Josef [*Tippett and Kamp*, 1993; *Batt et al.*, 2004; *Little et al.*, 2005], have led to a greater perturbation of isotherms and an increase in heat-flow in that region. Mechanical models of the Alpine fault demonstrate how elevated heat-flow and temperatures near the fault can cause a shallowing of the transition between brittle and ductile behaviour, so that the "locking depth" (i.e., the depth to which the fault remains essentially undeforming between earthquakes) is reduced [*Ellis and Stöckhert*, 2004b; *Ellis et al.*, 2006]. Mechanical models and geochemical observations also indicate that coupling between deformation and fluid flow may play an important role in allowing meteoric fluids to penetrate to depths where ductile rock deformation is occurring [*Upton et al.*, 1995].

ALPINE FAULT LOCKING INFERRED FROM GEODETIC DATA

Beavan et al. [1999] interpreted the velocity field from GPS surveys between 1994 and 1998 in a ~100-km wide swath across South Island through the central Southern Alps. They modeled the data using a two-fault model, one being the Alpine fault and the other dipping in the opposite direction with a deeper locking depth; this eastern fault was required to fit distributed surface deformation within the Alps and in reality is likely to be accommodated by numerous faults. Modeled Alpine fault strike-slip rates were lower than those inferred geologically, dip-slip rates were higher, and a shallow model locking depth of 5–8 km was obtained [*Beavan et al.*, 1999].

Pearson et al. [2000] modeled surface velocities from two GPS surveys in 1995 and 1998 that crossed the Southern Alps in a ~20-km wide transect near Haast, just north of where the fault is believed to be vertical and dip-slip motion is small. Modeling was done by grid search on slip rate and locking depth, with other parameters held fixed: first for a one-fault model, and then for a two-fault model similar to that used by *Beavan et al.* [1999]. The one-fault model gave a locking depth of ~20 km, which is greater than the 12 km maximum depth of recorded earthquakes. The two-fault model was best fit with a 10 ± 2 km Alpine fault locking depth, a strike slip rate slightly slower than geological estimates and a dip slip rate rather faster [*Pearson et al.*, 2000].

The previously described studies suffer from interpreting the velocity or strain-rate field projected onto a profile. In contrast, *Wallace et al.* [2007] interpreted the velocity field of South Island in a single model that solves for long-term rotations of crustal blocks, coupling distribution (the proportion of slip released episodically) along faults forming the block boundaries, and uniform strain fields within the blocks [*McCaffrey*, 2002]. For the ~200 km length of the Alpine fault in central South Island, they assumed a 45° dipping Alpine fault as one block boundary, and a distributed network of faults following the eastern foothills of the Alps as another. Geological slip rate and Australia-Pacific relative plate motion constraints were applied and both along-strike and down-dip variations in coupling were solved for, assuming the coupling coefficient decreases monotonically with depth. Wallace et al. [2007] were able to fit the GPS velocities with two possible models. (1) A model with a coupling coefficient averaging 70–85% down to 18 km depth, with slightly lower coupling values on the central Alpine fault. In this model, 70% locking to 18 km depth gives surface deformation roughly equivalent to 100% locking to about 15 km depth, which is still deeper than the maximum depth of small and moderate earthquakes. Alpine fault strike-slip rates for this model were at the high end of geological estimates (31 mm/yr) and dip-slip rates were similar to geological estimates [*Wallace et al.*, 2007]. (2) A model fitting the GPS velocities with an Alpine Fault strike-slip rate of 27 mm/yr (closer to the mid-point of geological estimates), and up to 5 mm/yr of distributed dextral deformation within the Southern Alps (< 50 km to the east of the Alpine Fault). In this model the coupling coefficients on the Alpine fault were smaller than in the first model, particularly in the central 80 km (reduced to 50% coupling) [*Wallace et al.*, 2007]. However, both cases require substantial interseismic coupling on the Alpine fault to match GPS observations.

DISCUSSION: ALPINE FAULT RUPTURE MECHANISMS

A long-term slip rate of 2–3 cm/yr on a discrete Alpine fault plane in the shallow surface (top 1 km) is demonstrated by geological observations [*Berryman et al.*, 1992; *Yetton et al.*, 1998; *Norris and Cooper*, 2001], but undeformed river terraces and man-made structures, and small aperture resurveyed networks convincingly show that no slip has occurred on the Alpine fault plane during the last 40-100 yr. Therefore, the Alpine fault must move episodically.

Globally, episodic movement on faults in the upper crust most commonly occurs during earthquakes [*Scholz*, 2002]. Seismological and geodetic observations suggest that the Alpine fault in central South Island is locked above a depth of 7–12 km and may fail during earthquakes [*Eberhart-Phillips*, 1995; *Beavan et al.*, 1999; *Leitner et al.*, 2001]. Direct evidence for previous Alpine fault earthquakes comes from the occurrence of pseudotachylyte within fault rocks and stratigraphic evidence for episodic scarp formation accompanied by liquefaction of adjacent sandy horizons.

However, in the past decade there has been increasing awareness that earthquakes only account for a fraction of plate boundary displacements, as predicted by models of present day plate velocities [*DeMets*, 1997]. The advent of continuous, coeval and collocated GPS and seismic measurements has revealed a suite of processes that may account, at least in part, for this slip deficit. These processes include slow-rupture earthquakes, silent earthquakes, aseismic creep and afterslip. For example, in both Japan [*Heki et al.*, 1997] and Kamchatka [*Burgmann et al.*, 2001], seismic events of around $M_w = 7.6$ were followed by afterslip for a period of ca. 12 months after the main event. In both cases, the afterslip events had a seismic moment greater than or equal to

the main seismic event. So called "tsunami earthquakes" are also types of slow earthquake in the sense that the magnitudes of such events appear deceptively small when measured at the standard 20 s period, but are much larger when estimated from long period (~250 s) surface waves [*Kanamori and Kikuchi*, 1993]. Slow earthquakes and other forms of aseismic fault slip are generally, but not solely, linked to evidence for high fluid pressures [*Kanamori and Kikuchi*, 1993; *Kennedy et al.*, 1997; *Dragert et al.*, 2001; *Obara*, 2002; *Douglas et al.*, 2005].

For models of rate and state dependent friction, high fluid pressures and a high geothermal gradient reduce the effective normal stress on the fault and suppress velocity weakening of the fault, and hence stable sliding, rather than stick-slip (earthquake) behaviour is promoted [*Scholz*, 1998; *Scholz*, 2002]. Both these characteristics are present in central South Island, so the suggestion that some earthquakes in the central region involve a transitional steady/unsteady style of frictional failure (i.e., slow earthquakes) remains plausible [*Stern et al.*, 2001]. However, petrological and structural evidence suggest that the 2–3 km thick Alpine fault mylonite zone is relatively dry above 5–10 km depth [*Vry et al.*, 2001]. Moreover, there is evidence from the geology and geometry of the electrical conductivity anomaly that water from the fault zone is released vertically ~8 km from the surface trace of the Alpine fault. Isotopic measurements show that a meteoric component is present in fluids that exist within the Alpine fault zone at depths of ~6–8 km, requiring some circulation of fluids from the surface to that depth [*Upton et al.*, 1995]. Hence, we suggest that, although the deepest parts of the Alpine fault may have near-lithostatic fluid pressures, the Alpine fault zone is not overpressured at depths <8 km, and will behave in a similar fashion to other shallow crustal faults and is likely to fail during earthquakes.

It may also be that the dynamics of the earthquake cycle are actually essential to produce the observed pattern of strain localisation in the mid and lower crust. Model experiments in which brittle slip on the Alpine fault occurs episodically show that an upper crustal earthquake may cause stress concentration in the mid-crust, resulting in a long-term strain pattern that is similar in geometry to the seismic velocity and electrical conductivity anomalies observed in central South Island [*Ellis and Stöckhert*, 2004a].

We suggest that earthquakes are likely to occur on the Alpine fault in central South Island, but important questions remain: what is the maximum earthquake magnitude, and what proportion of Alpine fault displacement is accommodated aseismically or in small or moderate magnitude earthquakes? Geological arguments of fault continuity, single event displacement, and coincidence of forest damage and landslides at widely-spaced localities near the Alpine fault have led numerous authors to conclude that the Alpine fault may fail in a great earthquake of M_w ~8.0 [*Adams*, 1980; *Berryman et al.*, 1992; *Sutherland and Norris*, 1995; *Bull*, 1996; *Berryman et al.*, 1998; *Wells et al.*, 1998; *Yetton*, 1998; *Wells et al.*, 1999; *Leitner et al.*, 2001]. However, great strike-slip earthquakes have rupture dimensions of several hundred km [*Wells and Coppersmith*, 1994; *Hanks and Bakun*, 2002]. If local regions of the Alpine fault in central South Island have had stress relieved at all but shallow depths because the fault has already slipped in that region, do these regions inhibit earthquake ruptures from propagating and hence limit the maximum earthquake magnitude that is possible? We constructed a simplified dynamic rupture model to test this hypothesis.

ALPINE FAULT DYNAMIC RUPTURE MODEL

The model fault is 100 km long and 12 km deep and has a rectangular patch at depth with zero initial shear stress (Figure 3). The patch simulates a region that has slipped due to elevated temperature and fluid pressure. Based on the observed pattern of seismicity, a 25 km long patch with an upper limit at 6 km depth was considered realistic, but other lengths and upper limits were also tested.

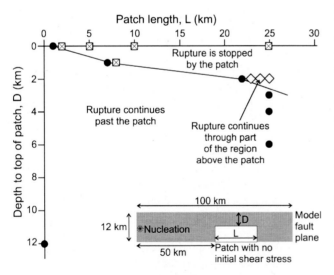

Figure 3. Rupture propagation model results showing the range of zero shear stress patch lengths, L, and shallow depth limits, D, that inhibit a dynamic rupture above the patch. The patch simulates a region of the fault that has slipped and had all shear stress relieved. Model runs are shown in which the rupture: successfully propagated past the patch (filled circles); propagated through part of the region above the patch (diamonds); and was completely stopped by the patch (checked squares).

We use a 3-D finite-difference computer program to simulate dynamic rupture of a strike-slip fault [*Harris and Day*, 1999]. The earthquake is artificially nucleated over a small area and then allowed to spontaneously propagate. The rupture velocity and the temporal and spatial patterns of rupture are not prescribed, but result from stress conditions on each sub-segment at each point in time. A slip-weakening fracture criterion is incorporated, so that the shear strength of the fault linearly decreases from a high static value to a low dynamic value after slipping a critical distance, d_0. Thus, it is much easier for an earthquake to continue to propagate once it is large enough and is only working against dynamic friction.

Taking parameter values calibrated to realistic models of the 1999 $M_w = 7.4$ Izmit, Turkey, earthquake [*Harris et al.*, 2002], we used critical slip (d_0) of 20 cm, static friction (μ_s) of 0.677, and dynamic friction (μ_d) of 0.525. Most model runs were for high-stress, below-failure conditions on the brittle fault segments with average initial stress (τ_0) of 70 MPa, and initial normal stress (σ_n) of 120 MPa. The fault segments are defined with zero cohesion so their static and dynamic yield strengths are 81 MPa and 63 MPa, respectively. The rest of the 3-D volume is assigned very high cohesion. To represent the decrease in stress toward the free surface, we include a linear taper over the upper 3 km for both τ_0 and σ_n. Actual shear stress is unlikely to be uniform, so a stochastic model is used to define τ_0 on the brittle fault (Figure 3). This is a 2-D self-similar random function with 2 km shortest wavelength, and variations up to ±6 MPa, giving the fault varied stress drop ranging from 1 to 13 MPa. The nucleation point is put near the end of the fault and in a region of the stochastic model that has a reasonably high τ_0.

The results of the numerical models are summarized in Figure 3, which shows the range of patch lengths and upper limits that were investigated. For a zero shear stress patch with length 25 km and an upper limit of 6 km, the rupture is able to continue to the far end of the fault, and propagates within the patch. With the upper limit of the patch raised to 3 km, the rupture still continues along the entire fault. Only when the upper limit is at <2 km does the rupture get suppressed and only part of the region above the patch is ruptured. For shorter patch lengths, shallower upper limits of the patch are required to stop the rupture. For a 7 km long patch, a >1 km depth brittle section is required to allow rupture to continue to the end of the fault. The rupture can jump over a 1 km long patch that reaches the surface, but rupture propagation is stopped by a 2 km long patch that reaches the surface.

The choice of nucleation point influences whether the earthquake turns into a full-length rupture. All models in Figure 3 have a distant nucleation point, but we also ran models with other nucleation points. When the rupture was nucleated at a point on the fault that was close to the patch, the rupture did not make it fully across a patch of width 25 km and upper limit 6 km. If the rupture was nucleated above the ductile patch, it was not able to rupture into the section with normal brittle behaviour to a depth of 12 km. Slip velocity and stress concentration at the leading edge of the rupture scales with fault depth, and thus it is more difficult for an initial 6 km depth rupture to grow into a 12 km depth rupture, than vice versa [*Harris and Day*, 1993].

The length of rupture along a fault with a varied depth of rupture is not dependent on whether it is high-stress or low-stress, but the low-stress case displayed some overshoot. We tested a series of fault models with low-stress conditions [*Harris et al.*, 2002]. The rupture patterns of the brittle fault were virtually the same as under high-stress conditions. However, there was an intriguing difference within the patch. For low-stress initial conditions, the rupture penetrates into the patch, even though no slip is required there. If such overshoot regularly occurs along an individual fault segment, it could have long-term effects, such as further localizing faulting on that segment.

Although the model is highly simplified, it clearly suggests that stress relief by creep or unstable sliding on a patch of the deeper part of the fault (>6 km depth) is insufficient to stop a fully developed earthquake rupture from propagating along the shallow part of the locked fault. However, a patch that has crept at depth may inhibit a large rupture from nucleating above it.

DISCUSSION: ALPINE FAULT RUPTURE SCENARIOS

We have established a strong case that the shallow part (<6–12 km depth) of the Alpine fault in central South Island is locked and capable of failing in an earthquake, but the maximum magnitude of any such earthquake depends upon how far that rupture propagates. Our model results indicate that patches of low shear stress, with dimensions that we consider to be realistic for a slow earthquake, are unlikely to stop the dynamic rupture of a large or great earthquake along the Alpine fault in central South Island; the precursory slip would have to reach depths of <1–4 km over a significant length of fault (>10–30 km) to inhibit a rupture.

Geometric irregularities of the fault surface are another factor that can influence rupture. Rupture simulations show that a strike-slip rupture can jump across gaps of <5 km between fault segments [*Harris and Day*, 1999]: moderate-sized earthquakes cascade into larger earthquakes, e.g., the 1992 M_w = 7.3 Landers, California, earthquake ruptured four fault segments. The 1999 M_w = 7.4 Izmit, Turkey, earthquake jumped several step-overs and propagated around a bend, but did not jump step-overs greater than 5 km [*Harris et al.*, 2002]. The 1855 M_w = 8.2 Wairarapa, New Zealand, earth-

quake ruptured across step-overs up to 6 km wide [*Schermer et al.*, 2004]. Ruptures may also change direction and transfer slip to adjacent faults with different strike: for example, the 2002 M_w = 7.9 Denali earthquake ruptured three segments, each of which differs in strike by 20–30° [*Eberhart-Phillips et al.*, 2003]. The differing-rake segments of the Alpine fault have no gap at the surface and this type of 'serial partitioning' of the surface trace is probably a response to erosional processes in operation at the range front [*Norris and Cooper*, 1997]. Thus, we suggest that the minor (<5 km) fault stepovers and changes in Alpine fault dip at the surface are not sufficient by themselves to inhibit a fully developed rupture that penetrates through the entire depth of the locked zone.

A number of scenarios remain for the possible rupture dimension of earthquakes on the central part of the Alpine fault (Figure 1). Theoretical considerations indicate that large or great earthquakes could occur, but cannot prove that they do. Does a combination of low shear stress and rough fault plane geometry inhibit ruptures in central South Island? Is the Alpine-Hope fault intersection a natural barrier to rupture? Could a rupture propagate between the Alpine and Hope faults? What is the significance of changes in hanging wall uplift rate, fault dip, and local segmentation style near Haast, and is this related to complexity at depth that could inhibit a rupture? Where might the southern limit of a fault rupture on the central Alpine fault be located: Haast River, Cascade River, Caswell Sound, Puysegur subduction thrust, or somewhere between? Where might the northern limit be?

The history of past earthquakes (Table 1; Figure 1) provides some guide as to what is possible in future. Radiocarbon data from trenches in the Haast-Okuru region indicate two coseismic displacement events of 8–9 m after ca. 1160 AD, and three since ca. 700–800 AD [*Berryman et al.*, 1998]. Estimated ages of forest disturbance at 1650–1725 AD [*Cooper and Norris*, 1990] and 1700–1730 AD [*Wells et al.*, 1998] near to the Alpine fault are correlated with terrace abandonment at Karangarua River at 1710–1720 AD and a major tree growth suppression event at 1717 AD [*Wells et al.*, 1999], which is inferred to be the year of the last Alpine fault earthquake. Evidence for the 1717 AD event spans a distance of ca. 400 km from Te Anau to Taramakau River and is associated with 8–9 m offsets at localities between Milford Sound and Haast. A trench at Crane Creek, a tributary of Ahaura River, demonstrates that surface rupture associated with this event died out 25–35 km northeast of the Taramakau River and the intersection with the Hope fault [*Yetton et al.*, 1998]. It is not known if the western segment of the Hope fault was also ruptured during this event [*Langridge and Berryman*, 2005].

At Karangarua River (Figure 1) there is evidence for an uplift and terrace abandonment event at 1600–1620 AD [*Yetton et al.*, 1998] and this corresponds to the time of a

Table 1. Summary of Alpine fault paleoseismic data (from north to south)

Region	Paleoseismic evidence
Wairau fault	No rupture during last ca. 2000 yr [*Zachariasen et al.*, 2006].
Lake Rotoiti to Springs Junction	Last slip event of 1–2 m at 1455–1700 AD; probably ca. 1620 AD, based on forest disturbance [*Yetton*, 2002]. No fault creep at roads or at the Evison monitoring wall constructed in 1964 near Springs Junction [*Beanland*, 1987].
Ahaura River	Trenching gives last slip event at 1480–1645 AD; probably ca. 1620 AD, based on forest disturbance [*Yetton*, 1998]. No fault creep at road.
Taramakau River	No fault creep at road.
Western Hope Fault	Single event dextral displacements of 3–4 m at recurrence interval of 310–490 yr, but timing of most recent surface ruptures is unknown [*Langridge and Berryman*, 2005].
Near Lake Kaniere	Trenching, forest disturbance, and tree ring data suggest earthquake events at 1717 and ca. 1620 AD [*Yetton*, 1998; *Wells et al.*, 1999].
Waitaha River	Tree ring and forest disturbance ages indicate earthquakes at 1717 AD and ca. 1620 AD [*Wright*, 1998; *Wells et al.*, 1999].
Near Franz Josef	Significant forest disturbance at 1717, ca. 1620 and ca. 1430–1450 AD [*Wells et al.*, 1999], and forest establishment on floodplains just after ca. 1720 AD and ca. 1620 AD [*Cullen et al.*, 2003]. No fault creep at road or within Franz Josef village.
Karangarua River	Significant forest disturbance at ca. 1710–20, 1610–20 and 1460 AD [*Wells et al.*, 2001] coincide with uplift and terrace abandonment events [*Adams*, 1980; *Yetton*, 1998].
Haast, Turnbull, and Okuru rivers	Dextral displacements of 8–9 m during last two events. Trenching indicates three events since 700–800 AD, with the penultimate event (see main text) at ca. 1430 AD [*Berryman et al.*, 1998]. Tree-ring suppression in a tree on the fault trace suggests the last event was at ca. 1720 AD [*Berryman et al.*, 1998]. No fault creep at road.
Hokuri Creek	Dextral displacements of 8–9 m during last two events, with last event at 1600–1800 AD [*Sutherland and Norris*, 1995].
Milford Sound	Tree damage near fault scarp indicates last displacement at 1650–1725 AD [*Cooper and Norris*, 1990] and regional tree-ring suppression at ca. 1717 AD [*Wells et al.*, 1999].

significant forest disturbance at 1610–1640 in the northern region [*Wells et al.*, 1998] and near Franz Josef [*Cullen et al.*, 2003], and of tree-ring suppression at ca. 1620 AD at Waitaha River [*Wright*, 1998]. The event is not recorded as far south as Haast [*Berryman et al.*, 1998; *Yetton et al.*, 1998], but is recorded north of the Hope fault intersection at Ahaura River [*Yetton et al.*, 1998] and 35 km southwest of Lake Rotoiti [*Yetton*, 2002].

The Karangarua River terrace abandonment event at 1405–1445 AD [*Yetton et al.*, 1998], the significant forest disturbance event at 1410–1440 AD [*Wells et al.*, 1998], and a radiocarbon date of 1420–1450 AD from a large rock avalanche near Lake Kaniere [*Yetton et al.*, 1998] all suggest that this was the time of a significant earthquake (Figure 1). The penultimate event identified from trenching in the Haast-Okuru area is constrained by radiocarbon evidence to have an age in the range 1160–1410 AD [*Berryman et al.*, 1998], though some of the dated material was clearly recycled from older sediment — leaving the likely possibil-

ity that the event is younger than stated. We propose two possibilities: either a single event at ca. 1430 AD ruptured the Alpine fault from south of Turnbull River to north of Ahaura River; or the rupture extent was only north of Haast, requiring an earlier rupture during the interval 1160–1410 AD to be the penultimate event in the Haast-Milford region. We favour the former scenario, because there is clear sedimentological evidence within the trench for reworking of the youngest soil into the the fault scarp colluvial wedge [*Berryman et al.*, 1998], and suggest that an earthquake in ca. 1430 AD ruptured the Alpine fault from north of the Hope fault intersection to at least as far south as Milford Sound.

DISCUSSION:
MAXIMUM EARTHQUAKE MAGNITUDE

Although paleoseismic data are relatively sparse and imprecise (Table 1), sufficient exist to evaluate the probable extent of the last three Alpine fault ruptures: the 1717 AD event is inferred to have ruptured a 300–500 km length of fault; the 1620 AD event ruptured 200–300 km; and the 1430 AD event ruptured 350–600 km (Table 1; Figure 1). If it is assumed that these events were single ruptures with an average down-dip rupture dimension of 12 km, then their respective magnitudes (M_w), as inferred from comparisons with similar global examples [*Hanks and Bakun*, 2002; *Lin et al.*, 2002; *Eberhart-Phillips et al.*, 2003], were 7.9 ± 0.3, 7.6 ± 0.3, and 7.9 ± 0.4.

Could the assumption that each paleoseismically recorded event corresponds to a single earthquake be wrong? The paleoseismic record allows us to determine: synchroneity to a precision of ca. 1–10 years for the last event (1717 AD) and ca. 10–50 years for the two previous events (ca. 1620 and 1430 AD); and that 8–9 m of slip occurred in the 1717 and 1430 AD events in the Milford-Haast region (Table 1; Figure 1). We accept that several earthquakes and some afterslip during a single interval of several years could combine to produce a pattern of fault displacement that is consistent with the paleoseismic evidence. If sections of Alpine fault of length 80–100 km ruptured during a sequence, then each event is inferred to have magnitude $M_w = 7.2 ± 0.3$ [*Wells and Coppersmith*, 1994]. While earthquake clustering in space and time can occur through various stress transfer mechanisms [*Freed*, 2005], the relative scarcity of global examples where a single continuous fault has ruptured in this manner, and the results of our dynamic rupture modeling, lead us to suggest that this is an unlikely scenario for the Alpine fault.

CONCLUSIONS

Geological observations require that episodic slip on the Alpine fault averages to a long-term displacement rate of 2–3 cm/yr. Patterns of natural seismicity and geodetic strain suggest that the Alpine fault in central South Island (between Haast and Taramakau River) is currently locked above a depth of 6–12 km and will probably fail in an earthquake.

In addition to earthquakes, aseismic slip at shallow crustal depths (6–12 km) may play a role in accommodating long-term fault displacement; this conclusion is made on the basis of high pore fluid pressures inferred from the very low seismic P-wave velocity and high electrical conductivity of the Alpine fault zone, and by comparison with other faults with high inferred fluid pressures and a record of aseismic slip.

Our modeling suggests that an episode of aseismic slip at depth may inhibit the nucleation of a large earthquake in the region above it, but is unlikely to inhibit the propagation of a fully developed rupture during a large or great earthquake. We did not investigate the stress loading of adjacent parts of the fault plane during a possible slow-slip event, and acknowledge that our model is highly simplified. However,

our conclusion is supported by a comparison with subduction zones, which are where most slow-slip earthquakes are documented, where high pore fluid pressures are inferred, but also where most great earthquakes occur.

A range of maximum earthquake magnitudes may be possible on the Alpine fault, depending on what length of fault ruptures. Based upon physical models that are calibrated using data from earthquakes elsewhere, we infer that: the observed Alpine fault geometry (Figure 1) is sufficiently smooth to allow ruptures to propagate; and low-stress patches on the fault are unlikely to inhibit a fully developed rupture from propagating. However, the combination of along-strike changes in Alpine fault geometry, thermal structure, fluid pressure, interactions with other faults, the possibility of episodic strain release at depth, and our inadequate knowledge of the spatially variable material properties and stress field means that our physical models are substantially oversimplified. It is, therefore, essential that any predictive model is consistent with past Alpine fault behaviour.

Past earthquakes on the central Alpine fault are inferred to have ruptured at least as far south as Fiordland (ca. 1717 AD; ca. 1430 AD) or to have been limited to the region north of Haast (ca. 1620 AD), and to have ruptured north to the Taramakau River (ca. 1717 AD) or to at least as far north as Ahaura River (ca. 1620 AD; ca. 1430 AD). Each of the three events documented is inferred to have different rupture dimensions (Table 1; Figure 1) and magnitudes: the 1620 AD event is inferred to have the most limited rupture extent (200-300 km); the 1430 AD event is inferred to have the largest rupture extent (350–600 km); and the 1717 AD event is intermediate (300–500 km), though it is not known if the western segment of the Hope fault was also ruptured during this event. Based upon the assumption that the paleoseismic events occurred in single earthquakes, the last three earthquakes most likely had magnitudes (M_w) in the range 7.6–7.9. We cannot rule out the possibility that some paleoseismically recorded events may be composed of a sequence of triggered earthquakes with smaller magnitude, but consider this scenario unlikely.

We conclude that large earthquakes ($M_w > 7$) on the Alpine fault will almost certainly occur in future. Great earthquakes ($M_w \geq 8$) with rupture dimensions of several hundred km are consistent with and provide a reasonable explanation for paleoseismic data, and hence remain a realistic future scenario. Better delineation of fault properties and structure at depth is required to allow more realistic models of Alpine fault earthquake rupture to be developed. Additional paleoseismic observations that improve the spatial extent of timing and displacement data, and sample a greater period of time (more past events) are required to test whether predictive models can accurately hindcast earthquakes.

Finally, we hypothesise that slow-slip events could occur at moderate depths (5–20 km) on the Alpine fault, and that a relationship may exist between the timing of such events and the nucleation of earthquakes. We suggest that continuous recording of GPS and seismicity data, and possibly some type of repeated stress measurement, around regions of high inferred fluid pressure (Figure 1) is required to test this hypothesis, record what actually happens during the next significant Alpine fault earthquake, and ultimately to provide a basis for time-varying seismic hazard estimates.

Acknowledgements. We thank Tom Brocher, Tom Hanks and Dick Walcott for comments on an early draft, Fred Davey as the scientific editor, and two anonymous reviewers. Funded by the New Zealand Foundation for Research Science and Technology.

REFERENCES

Adams, C. J., and J. E. Gabites (1985), Age of metamorphism and uplift in the Haast Schist Group at Haast Pass, Lake Wanaka and Lake Hawea, South Island, New Zealand, *N. Z. J. Geol. Geophys.*, *28*(1), 85–96.

Adams, J. (1980), Paleoseismicity of the Alpine Fault seismic gap, *Geology*, *8*, 72–76.

Allis, R. G., R. W. Henley, and A. F. Carman (1979), The thermal regime beneath the Southern Alps, *Bull. R. Soc. N. Z.*, (18), 79–85.

Allis, R. G., and Y. Shi (1995), New insights to temperature and pressure beneath the central Southern Alps, New Zealand, *N. Z. J. Geol. Geophys.*, *38*(4), 585–592.

Barnes, P. M., R. Sutherland, B. Davy, and J. Delteil (2001), Rapid creation and destruction of sedimentary basins on mature strike-slip faults: an example from the offshore Alpine Fault, New Zealand, *J. Struct. Geol.*, *23*, 1727–1739.

Barnes, P. M., R. Sutherland, and J. Delteil (2005), Strike-slip structure and sedimentary basins of the southern Alpine Fault, Fiordland, New Zealand, *Geol. Soc. Am. Bull.*, *117*, 411–435

Batt, G. E., S. L. Baldwin, M. A. Cottam, P. G. Fitzgerald, M. T. Brandon, and T. L. Spell (2004), Cenozoic plate boundary evolution in the South Island of New Zealand; new thermochronological constraints, *Tectonics*, *23*(4), 17.

Batt, G. E., and J. Braun (1999), The tectonic evolution of the Southern Alps, New Zealand: insights from fully thermally coupled dynamical modelling, *Geophys. J. Int.*, *136*, 403–420.

Beanland, S. (1987), Field guide to sites of active earth deformation, South Island, New Zealand, *N. Z. Geol. Surv. Rec.*, *19*, 1–104.

Beaumont, C., P. J. J. Kamp, J. Hamilton, and P. Fullsack (1996), The continental collision zone, South Island, New Zealand: comparison of geodynamical models and observations, *J. Geophys. Res.*, *101*(B2), 3333–3359.

Beavan, J., M. Moore, C. Pearson, M. Henderson, B. Parsons, S. Bourne, P. England, R. I. Walcott, G. Blick, D. Darby, and K. Hodgkinson (1999), Crustal deformation during 1994–98 due to oblique continental collision in the central Southern Alps, New Zealand, and implications for seismic potential of the Alpine fault, *J. Geophys. Res.*, *104* (B11), 25,233–25,255.

Berryman, K., A. F. Cooper, R. J. Norris, R. Sutherland, and P. Villamor (1998), Paleoseismic investigation of the Alpine Fault at Haast and Okuru, *Geol. Soc. N. Z. Misc. Publ.*, *101A*, 44.

Berryman, K. R., S. Beanland, A. F. Cooper, H. N. Cutten, R. J. Norris, and P. R. Wood (1992), The Alpine fault, New Zealand: variation in Quaternary structural style and geomorphic expression, *Ann. Tecton.*, *6*, 126–163.

Bossiere, G. (1991), Petrology of pseudotachylytes from the Alpine Fault of New Zealand, *Tectonophysics, 196*, 173–193.

Bull, W. B. (1996), Prehistorical earthquakes on the Alpine fault, New Zealand, *J. Geophys. Res.*, *101*, 6037–6050.

Burgmann, R., M. G. Kogan, V. E. Levin, C. H. Scholz, R. W. King, and G. M. Steblov (2001), Rapid aseismic moment release following the 5 December 1997 Kronotsky, Kamchatka, earthquake, *Geophys. Res. Lett.*, *28*(7), 1331–1334.

Christensen, N. I., and W. D. Mooney (1995), Seismic velocity structure and composition of the continental crust; a global view, *J. Geophys. Res., B, Solid Earth Planets*, *100*(6), 9761–9788.

Cooper, A. F., and R. J. Norris (1990), Estimates for the timing of the last coseismic displacement on the Alpine Fault, northern Fiordland, New Zealand, *N. Z. J. Geol. Geophys.*, *33*(2), 303–307.

Cooper, A. F., and R. J. Norris (1994), Anatomy, structural evolution, and slip rate of a plate-boundary thrust: The Alpine fault at Gaunt Creek, Westland, New Zealand, *Geol. Soc. Am. Bull.*, *106*, 627–633.

Cooper, A. F., and R. J. Norris (1995), Displacement on the Alpine Fault at Haast River, South Westland, New Zealand, *N. Z. J. Geol. Geophys.*, *38*(4), 509–514.

Craw, D. (1988), Shallow-level metamorphic fluids in a high uplift rate metamorphic belt; Alpine Schist, New Zealand, *J. Metamorphic Geol.*, *6*(1), 1–16.

Craw, D., M. S. Rattenbury, and R. D. Johnstone (1994), Structures within greenschist facies Alpine Schist, central Southern Alps, New Zealand, *N. Z. J. Geol. Geophys.*, *37*(1), 101–111.

Cullen, L. E., R. P. Duncan, A. Wells, and G. H. Stewart (2003), Floodplain and regional scale variation in earthquake effects on forests, Westland, New Zealand, *J. R. Soc. N. Z.*, *33*, 693–701.

Davey, F. J., T. Henyey, S. Kleffmann, A. Melhuish, D. Okaya, T. A. Stern, and D. J. Woodward (1995), Crustal reflections from the Alpine fault zone, South Island, New Zealand, *N. Z. J. Geol. Geophys.*, *38*(4), 601–604.

DeMets, C. (1997), Afterslip no longer an afterthought, *Nature*, *386*(6625), 549.

DeMets, C., R. G. Gordon, D. F. Argus, and S. Stein (1994), Effect of recent revisions to the geomagnetic time scale on estimates of current plate motions, *Geophys. Res. Lett.*, *21*, 2191–2194.

Douglas, A., J. Beavan, L. M. Wallace, and J. Townend (2005), Slow slip on the northern Hikurangi subduction interface, New Zealand, *Geophys. Res. Lett.*, *32*, L16305, doi:10.1029/2005GL023607.

Dragert, H., K. Wang, and T. S. James (2001), A silent slip event on the deeper Cascadia subduction interface, *Science*, *292*(5521), 1525–1528.

Eberhart-Phillips, D. (1995), Examination of seismicity in the central Alpine fault region, South Island, New Zealand, *N. Z. J. Geol. Geophys.*, *38*, 571–578.

Eberhart-Phillips, D., and S. Bannister (2002), Three-dimensional crustal structure in the Southern Alps region of New Zealand from inversion of local earthquake and active source data, *J. Geophys. Res., B, Solid Earth Planets*, *107*(10) (20.

Eberhart-Phillips, D., P. J. Haeussler, J. T. Freymueller, A. D. Frankel, C. M. Rubin, P. Craw, N. A. Ratchkovski, G. Anderson, G. A. Carver, A. J. Crone, T. E. Dawson, H. Fletcher, R. Hansen, E. L. Harp, R. A. Harris, D. P. Hill, S. Hreinsdottir, R. W. Jibson, L. M. Jones, R. Kayen, D. K. Keefer, C. F. Larsen, S. C. Moran, S. F. Personius, G. Plafker, B. Sherrod, K. Sieh, N. Sitar, and W. K. Wallace (2003), The 2002 Denali Fault earthquake, Alaska; a large magnitude, slip-partitioned event, *Science*, *300*(5622), 1113–1118.

Ellis, S., J. Beavan, D. Eberhart-Phillips, and B. Stöckhert (in press), Simplified models of the Alpine Fault seismic cycle: stress transfer in the mid-crust, *Geophys. J. Int.*

Ellis, S., and B. Stöckhert (2004a), Elevated stresses and creep rates beneath the brittle-ductile transition caused by seismic faulting in the upper crust, *J. Geophys. Res.*, *109*(B5), 10.

Ellis, S., and B. Stöckhert (2004b), Imposed strain localization in the lower crust on seismic timescales, *Earth Planets Space*, *56*, 1103–1109.

Evison, F. F. (1971), Seismicity of the Alpine Fault, *R. Soc. N. Z. Bull.*, *9*, 161–165.

Freed, A. M. (2005), Earthquake triggering by static, dynamic, and postseismic stress transfer, *Annu. Rev. Earth Planet. Sci.*, *33*, 335–367.

Gerbault, M., S. Henrys, and F. Davey (2003), Numerical models of lithospheric deformation forming the Southern Alps of New Zealand, *J. Geophys. Res., B, Solid Earth Planets*, *108*(7), 18.

Hanks, T. C., and W. H. Bakun (2002), A bilinear source-scaling model for M-log A observations of continental earthquakes, *Bull. Seismol. Soc. Am.*, *92*, 1841–1846.

Harris, R. A., and S. M. Day (1993), Dynamics of fault interaction; parallel strike-slip faults, *J. Geophys. Res.*, *98*(3), 4461–4472.

Harris, R. A., and S. M. Day (1999), Dynamic three-dimensional simulations of earthquakes on en echelon faults, *Geophys. Res. Lett.*, *26*(14), 2089–2092.

Harris, R. A., J. F. Dolan, R. D. Hartleb, and S. M. Day (2002), The 1999 Izmit, Turkey, earthquake; a 3D dynamic stress transfer model of intraearthquake triggering, *Bull. Seismol. Soc. Am.*, *92*(1), 245–255.

Heki, K., S.-i. Miyazaki, and H. Tsuji (1997), Silent fault slip following an interplate thrust earthquake at the Japan Trench, *Nature*, *386*(6625), 595–598.

Holm, D. K., R. J. Norris, and D. Craw (1989), Brittle and ductile deformation in a zone of rapid uplift; central Southern Alps, New Zealand, *Tectonics*, *8*(2), 153–168.

Hull, A.G., and K. Berryman (1986), Holocene tectonism in the region of the Alpine Fault at Lake McKerrow, Fiordland, New Zealand, *R. Soc. N. Z. Bull.*, *24*, 317–331.

Jenkin, G. R. T., D. Craw, and A. E. Fallick (1994), Stable isotopic and fluid inclusion evidence for meteoric fluid penetration into an active mountain belt; Alpine Schist, New Zealand, *J. Metamorphic Geol.*, *12*(4), 429–444.

Jones, T., and A. Nur (1984), Effect of temperature, pore fluids, and pressure on seismic wave velocity and attenuation in rock, *Geophysics*, *49*(5), 666.

Kanamori, H., and M. Kikuchi (1993), The 1992 Nicaragua earthquake; a slow tsunami earthquake associated with subducted sediments, *Nature*, *361*(6414), 714–716.

Kennedy, B. M., Y. K. Kharaka, W. C. Evans, A. Ellwood, D. J. De Paolo, J. J. Thordsen, G. Ambats, and R. H. Mariner (1997), Mantle fluids in the San Andreas fault system, California, *Science*, *278*(5341), 1278–1281.

Kleffman, S., F. Davey, A. Melhuish, D. Okaya, T. Stern, and the SIGHT team (1998), Crustal structure in the central South Island, New Zealand, from the Lake Pukaki seismic experiment, *N. Z. J. Geol. Geophys.*, *41*, 39–49.

Koons, P. O. (1987), Some thermal and mechanical consequences of rapid uplift; an example from the Southern Alps, New Zealand, *Earth Planet. Sci. Lett.*, *86*(2), 307–319.

Koons, P. O., and D. Craw (1991), Evolution of fluid driving forces and composition within collisional orogens, *Geophys. Res. Lett.*, *18*(5), 935–938.

Koons, P. O., D. Craw, S. C. Cox, P. Upton, A. S. Templeton, and C. P. Chamberlain (1998), Fluid flow during active oblique convergence; a Southern Alps model from mechanical and geochemical observations, *Geology*, *26*(2), 159–162.

Koons, P. O., R. J. Norris, M. R. Sutherland, D. Wilson, and A. F. Cooper (1994), Structure and activity of the Alpine Fault, north and south of Milford Sound, *Eos Trans. AGU*, *75*(44, Suppl.), 669.

Langridge, R. M., and K. Berryman (2005), Morphology and slip rate of the Hurunui section of the Hope Fault, South Island, New Zealand, *N. Z. J. Geol. Geophys.*, *48*, 43–57.

Leitner, B., D. Eberhart-Phillips, H. Anderson, and J. L. Nablek (2001), A focused look at the Alpine fault, New Zealand: seismicity, focal mechanisms and stress observations, *J. Geophys. Res.*, *106*, 2193–2220.

Lin, A., B. Fu, J. Guo, Q. Zeng, G. Dang, W. He, and Y. Zhao (2002), Co-seismic strike-slip and rupture length produced by the 2001 Ms 8.1 Central Kunlun earthquake, *Science*, *296*, 2015–2017.

Little, T. A., S. Cox, J. K. Vry, and G. Batt (2005), Variations in exhumation level and uplift rate along the oblique-slip Alpine Fault, central Southern Alps, New Zealand, *Geol. Soc. Am. Bull.*, *117*(5–6), 707–723.

McCaffrey, R. (2002), Crustal block rotations and plate coupling, *Geodyn. Ser.*, *30*, 101–122.

McCormac, F. G., A. G. Hogg, P. G. Blackwell, C. E. Buck, T. F. G. Higham, and P. J. Reimer (2004), SHCal04 Southern Hemisphere Calibration 0 − 11.0 cal kyr BP, *Radiocarbon*, *46*, 1087–1092.

Nathan, S., M. S. Rattenbury, and R. P. Suggate (2002), *Geology of the Greymouth area, 1:250 000 geological map 12*, 58 pp., Institute of Geological and Nuclear Sciences, Lower Hutt.

Norris, R. J., and A. F. Cooper (1997), Erosional control on the structural evolution of a transpressional thrust complex on the Alpine Fault, New Zealand, *J. Struct. Geol.*, *19*(10), 1323–1324.

Norris, R. J., and A. F. Cooper (2001), Late Quaternary slip rates and slip partitioning on the Alpine Fault, New Zealand, *J. Struct. Geol.*, *23*(2–3), 507–520.

Norris, R. J., P. O. Koons, and A. F. Cooper (1990), The obliquely-convergent plate boundary in the South Island of New Zealand; implications for ancient collision zones, *J. Struct. Geol.*, *12*(5–6), 715–725.

Obara, K. (2002), Nonvolcanic deep tremor associated with subduction in Southwest Japan, *Science*, *296*(5573), 1679–1681.

Okaya, D., S. Henrys, and T. Stern (2002), Double-sided onshore-offshore seismic imaging of a plate boundary; "super-gathers" across South Island, New Zealand, *Tectonophysics*, *355*(1–4), 247–263.

Pearson, C., P. Denys, and K. Hodgkinson (2000), Geodetic constraints on the kinematics of the Alpine Fault in the southern South Island of New Zealand, using results from the Hawea-Haast GPS transect, *Geophys. Res. Lett.*, *27*(9), 1319–1322.

Reed, J. J. (1964), Mylonites, cataclasites, and associated rocks along the Alpine Fault, South Island, New Zealand, *N. Z. J. Geol. Geophys.*, *7*, 654–684.

Schermer, E. R., R. J. Van Dissen, K. Berryman, H. M. Kelsey, and S. M. Cashman (2004), Active faults, paleoseismology, and historical fault rupture in northern Wairarapa, North Island, New Zealand, *N. Z. J. Geol. Geophys.*, *47*, 101–122.

Scherwath, M., T. Stern, F. Davey, D. Okaya, W. S. Holbrook, R. Davies, and S. Kleffmann (2003), Lithospheric structure across oblique continental collision in New Zealand from wide-angle P wave modeling, *J. Geophys. Res.*, *108*(12), 18.

Scholz, C. H. (1998), Earthquakes and friction laws, *Nature*, *391*, 37–42.

Scholz, C. H. (2002), The mechanics of earthquakes and faulting, pp. 471, Cambridge University Press.

Scholz, C. H., J. Beavan, and T. C. Hanks (1979), Frictional metamorphism, argon depletion and tectonic stress on the Alpine Fault, New Zealand, *J. Geophys. Res.*, *84*, 6770–6782.

Shi, Y., R. G. Allis, and F. Davey (1996), Thermal modelling of the Southern Alps, New Zealand, *Pure Appl. Geophys.*, *146*, 469–501.

Sibson, R. H., S. H. White, and B. K. Atkinson (1979), Fault rock distribution and structure within the Alpine Fault Zone: a preliminary account, *Bull. R. Soc. N. Z.*, *18*, 55–65.

Smith, E. G. C., T. Stern, and B. O'Brien (1995), A seismic velocity profile across the central South Island, New Zealand, from explosion data, *N. Z. J. Geol. Geophys.*, *38*, 565–570.

Stern, T., S. Kleffman, D. Okaya, M. Scherwath, and S. Bannister (2001), Low seismic-wave speeds and enhanced fluid pressure beneath the Southern Alps of New Zealand, *Geology*, *29*, 679–682.

Sutherland, R., K. Berryman, and R. J. Norris (2006), Quaternary slip rate and geomorphology of the Alpine fault: implications for kinematics and seismic hazard in southwest New Zealand, *Geol. Soc. Am. Bull.*, *118*, 464–474.

Sutherland, R., and R. J. Norris (1995), Late Quaternary displacement rate, paleoseismicity and geomorphic evolution of the Alpine fault: Evidence from near Hokuri Creek, south Westland, New Zealand, *N. Z. J. Geol. Geophys.*, *38*, 419–430.

Tippett, J. M., and P. J. J. Kamp (1993), Fission track analysis of the Late Cenozoic vertical kinematics of continental Pacific crust, South Island, New Zealand, *J. Geophys. Res.*, *98*, 16,119–16,148.

Townend, J. (1999), Heat flow through the West Coast, South Island, New Zealand, *N. Z. J. Geol. Geophys.*, *42*, 21–31.

Upton, P., P. O. Koons, and C. P. Chamberlain (1995), Penetration of deformation-driven meteoric water into ductile rocks; isotopic and model observations from the Southern Alps, New Zealand, *N. Z. J. Geol. Geophys.*, *38*(4), 535–543.

van Avendonk, H. J. A., W. S. Holbrook, D. Okaya, J. K. Austin, F. Davey, and T. Stern (2004), Continental crust under compression; a seismic refraction study of South Island geophysical transect; I, South Island, New Zealand, *J. Geophys. Res.*, *109* (B6), 16.

Vry, J. K., A. C. Storkey, and C. Harris (2001), Role of fluids in the metamorphism of the Alpine fault zone, *J. Metamorphic Geol.*, *19*, 1–11.

Walcott, R. I. (1978), Present tectonics and late Cenozoic evolution of New Zealand, *Geophys. J. R. Astron. Soc.*, *52*, 137–164.

Walcott, R. I. (1979), Plate motion and shear strain rates in the vicinity of the Southern Alps, *R. Soc. N. Z. Bull.*, *18*, 5–12.

Walcott, R. I. (1998), Modes of oblique compression: late Cenozoic tectonics of the South Island of New Zealand, *Rev. Geophys.*, *36*, 1–26.

Wallace, L. M., J. Beavan, R. McCaffrey, K. Berryman, and P. Denys (2007), Balancing the plate motion budget in the South Island, New Zealand, using GPS, geological and seismological data, *Geophys. J. Int.*, *168*(1), 332–352, doi:10.1111/j.1365-246X.2006.03183.x.

Wallace, R. C. (1976), Partial fusion along the Alpine Fault Zone, New Zealand, *Geol. Soc. Am. Bull.*, *87*, 1225–1228.

Wannamaker, P. E., G. R. Jiracek, J. A. Stodt, T. G. Caldwell, V. M. Gonzalez, J. D. McKnight, and A. D. Porter (2002), Fluid generation and pathways beneath an active compressional orogen, the New Zealand Southern Alps, inferred from magnetotelluric data, *J. Geophys. Res., B, Solid Earth Planets*, *107*(6), 22.

Warr, L. N., and B. A. van der Pluijm (2005), Crystal fractionation in the friction melts of seismic faults (Alpine Fault, New Zealand), *Tectonophysics*, *402*, 111–124.

Wellman, H. W. (1953), Data for the study of Recent and Late Pleistocene faulting in the South Island of New Zealand, *N. Z. J. Sci. Technol.*, (B34), 270–288.

Wellman, H. W., and R. W. Willett (1942), The geology of the west coast from Abut Head to Milford Sound, Part I, *Trans. R. Soc. N. Z.*, *71*, 282–306.

Wells, A., R. P. Duncan, and G. H. Stewart (2001), Forest dynamics in Westland, New Zealand: the importance of large, infrequent earthquake-induced disturbance, *J. Ecol.*, *89*, 1006–1018.

Wells, A., G. H. Stewart, and R. P. Duncan (1998), Evidence for widespread, synchronous, disturbance-initiated forest establishment in Westland, New Zealand, *J. R. Soc. N. Z.*, *28*, 333–345.

Wells, A., M. D. Yetton, R. P. Duncan, and G. H. Stewart (1999), Prehistoric dates of the most recent Alpine Fault earthquakes, New Zealand, *Geology*, *27*(11), 995–998.

Wells, D. L., and K. J. Coppersmith (1994), New empirical relationships among magnitude, rupture length, rupture width, rupture area, and surface displacement, *Bull. Seismol. Soc. Am.*, *84*(4), 974–1002.

Winkler, K. W., and W. F. Murphy III (1995), Acoustic velocity and attenuation in porous rocks, *AGU Ref. Shelf*, *3*, 20–34.

Wood, P. R., and G. H. Blick (1986), Some results of geodetic fault monitoring in South Island, New Zealand, *R. Soc. N. Z. Bull.*, *24*, 39–45.

Wright, C. A. (1998), Geology and paleoseismicity of the central Alpine Fault, New Zealand, MSc thesis, University of Otago, Dunedin, New Zealand.

Yetton, M. D. (1998), Progress in understanding the paleoseismicity of the central and northern Alpine Fault, Westland, New Zealand, *N. Z. J. Geol. Geophys.*, *41*, 475–483.

Yetton, M. D. (2002), Paleoseismic investigation of the north and west Wairau sections of the Alpine Fault, South Island, New Zealand, *EQC Res. Rep.*, *99/353*, 96.

Yetton, M. D., A. Wells, and N. J. Traylen (1998), The probabilities and consequences of the next Alpine Fault earthquake, *EQC Res. Rep.*, *95/193*.

Zachariasen, J., K. Berryman, R. M. Langridge, C. Prentice, M. Rymer, M. Stirling, and P. Villamor (in press), Timing of late Holocene surface rupture on the Wairau Fault, Marlborough, New Zealand, *N. Z. J. Geol. Geophys.*

S. Bannister, J. Beavan, K. R. Berryman, S. Ellis, S. Henrys, B. Leitner, M. Stirling, R. Sutherland, L. Wallace, GNS Science, 1 Fairway Drive, PO Box 30-368, Lower Hutt, NZ. (r.sutherland@gns.cri.nz)

A. F. Cooper, R. J. Norris, University of Otago, PO Box 56, Dunedin, NZ.

S. Cox, D. Eberhart-Phillips, GNS Science, Private Bag 1930, Dunedin, NZ.

R. A. Harris, U.S. Geological Survey, Menlo Park, CA 94025-3591, USA.

T. A. Little, T. Stern, J. Townend, Victoria University of Wellington, PO Box 600, Wellington, NZ.

J. Pettinga, University of Canterbury, Private Bag 4800, Christchurch 8020, NZ.

M. Yetton, Geotech Consulting, RD1 Charteris Bay, Lyttelton R.D., NZ.

Three-Dimensional Geodynamic Framework for the Central Southern Alps, New Zealand: Integrating Geology, Geophysics and Mechanical Observations

Phaedra Upton[1] and Peter O. Koons

Department of Earth Sciences, University of Maine, Orono, Maine, USA

Vertical and horizontal strain within oblique convergence reflects imposed tectonic and surficial boundary conditions as transmitted via the rheological structure of the deforming crust. We combine three-dimensional mechanical modeling with geological and geophysical observations to develop a geodynamic framework for the central Southern Alps of New Zealand. We propose that an along-strike variation in crustal thickness and strength exists within South Island, with Otago having a thicker and weaker crust. This rheological structure and the presence of elevated pore pressure beneath the Mt Cook-Aoraki region results in predictable distribution of strain and stress conditions which are compatible with observations of surface strain. The models predict that these two features lead to enhanced rates of rock uplift against the Alpine Fault adjacent to Mt Cook-Aoraki region. This results from two components of vertical stretching: the first caused by strong Canterbury lower crust pushing into weak Otago lower crust, causing vertical extrusion; and the second a result of elevated pore pressure in the mid-crust. Extensional strain predicted in the upper crust is confirmed by the presence of sub-vertical mineralized veins and earthquakes with a normal fault solution in the upper 5-8 kms of the Main Divide region.

1. INTRODUCTION

Two-dimensional models are generally insufficient to describe deformation fields resulting from oblique plate motions and/or spatially variable rheological behavior and boundary conditions. Due to the sensitivity of three-dimensional strain to rheological parameters and imposed boundary conditions, oblique deformation fields provide information unavailable from a purely two-dimensional system. By comparing crustal strain and displacement fields from three-dimensional mechanical models with surface and geophysical observations, we can impose bounds upon the dynamics compatible with real systems.

Using a three-dimensional mechanical framework in conjunction with geological and geophysical observations, we define characteristic strain regimes for an oblique orogen analogous to the Southern Alps of New Zealand. Within this framework we investigate the influence of driving forces and rheological variability on surface deformation and deep strain patterns within the Southern Alps orogen. The vertical and horizontal distribution of strain within a zone of oblique convergence reflects tectonic and surficial boundary conditions as transmitted through the rheological structure of the deforming lithosphere. Departures from rheological steady state are greatest through thermal perturbations and pore pressure fluctuations, both of which are present in the Southern Alps orogen.

[1]Now at: Geology Department, University of Otago, Dunedin, New Zealand.

A Continental Plate Boundary: Tectonics at South Island, New Zealand
Geophysical Monograph Series 175
10.1029/175GM13

2. THE SOUTHERN ALPS OF NEW ZEALAND

2.1. Tectonic and Geological Setting

Since ca. 45 Ma when a distinct boundary between the Pacific and Australian plates developed through the New Zealand region, 850 km of dextral shear is inferred to have occurred across South Island [*Sutherland*, 1999]. The Alpine Fault, an oblique reverse structure, dipping SE beneath the Southern Alps, is the present day plate boundary [*Norris et al.*, 1990]. 460 km of dextral strike slip offsetting the Dun Mountain Ophiolite belt is evident along the Alpine Fault. The remaining 400 km has been taken up by distributed shear across South Island, reflected in right lateral bending of the Dun Mountain Ophiolite belt and adjacent terranes [*Coombs et al.*, 1976; *Molnar et al.*, 1999; *Cox and Sutherland*, this volume]. Today, the relative plate vector driving fault movement and associated deformation is directed ENE-WSW at a rate of ca. 39 mm/yr [*DeMets et al.*, 1994]. Rock uplift rates are high (>8 mm/yr [*Wellman*, 1979; *Simpson*

et al., 1994]) in the central Southern Alps, immediately adjacent to the Alpine Fault and decrease eastward to ca. 1 mm/yr or less to the east of the Main Divide [*Blick et al.*, 1985].

To the north and south of central South Island the tectonics of the New Zealand micro-continent reflect the changing character of the Australian/Pacific plate boundary. To the north, the oceanic Pacific plate subducts beneath the Australian plate and to the south, the oceanic Australian plate subducts beneath the Pacific plate (Figure 1). Transition from continental collision to subduction occurs across Northern South Island. Here five major dextral-slip faults strike at a low angle to the present Australia-Pacific plate convergence vector (Figure 1). These are known as the Marlborough Fault Zone, and their strike-slip motion accommodates most of the plate motion with underthrusting of the Pacific plate and uplift and shortening within the Australian crust taking up most of the rest. The Marlborough faults young to the south, with the oldest and westernmost, the Wairau Fault, exhibiting the greatest total offset [*Little and Jones*, 1998].

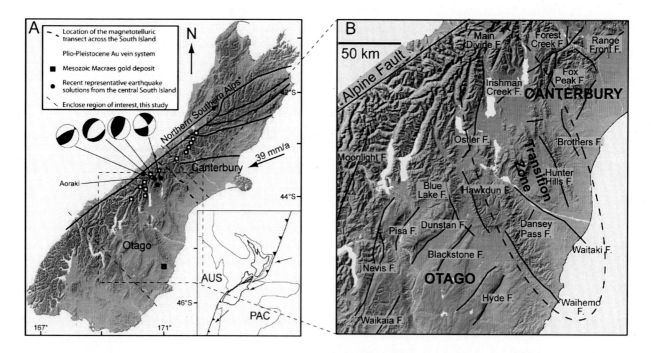

Figure 1. A: Topographic map of South Island of New Zealand constructed from a digital terrain image (Geographx.co.nz), showing the location of Plio-Pleistocene gold-bearing vein systems, the Mesozoic Macraes gold deposit [*Craw and Campbell*, 2004] and recent focal mechanisms for earthquakes in the central Southern Alps showing that normal, reverse and strike-slip events occur [after *Anderson et al.*, 1993; *Leitner et al.*, 2001]. The dashed line shows the location of the MT transect across South Island [*Wannamaker et al.*, 2002; *Jiracek et al.*, this volume]. Inset shows the plate tectonic setting with the Alpine Fault as the Pacific-Australian plate boundary and the relative plate motion vector [after *DeMets et al.*, 1994]. Light dashed lines outline the region of the Southern Alps that this manuscript addresses. B: Topographic map showing the transition from Canterbury to Otago. Active faults east of the Main Divide are shown. The dashed circle encloses the transition from Canterbury to Otago referred to in the text.

In central South Island, the eastward extent of modern deformation associated with oblique collision of the Australian and Pacific plates along the Alpine Fault can be broadly correlated with the width of elevated topography in South Island (Figure 1). In Canterbury, the Southern Alps are narrow, ca. 70 km, reflecting deformation concentrated close to the plate boundary and consequently high strain rates. In Otago, by contrast, deformation associated with the current tectonic regime extends ca. 200 km to the East Coast, including the Southern Alps and the Central Otago Ranges, in a broad zone of lower strain rates and distributed deformation (Figure 1). Combining geological and geophysical observations of variations along strike from Canterbury to Otago with three-dimensional mechanical models, as described in this manuscript, has led us to conclude that these regions have inherited distinct rheological profiles which are reflected in present day crustal strain.

The Southern Alps basement consists of Mesozoic metagreywacke that was differentially exhumed in the late Mesozoic as part of a convergent orogen along the Pacific margin of Gondwana [*Coombs et al.*, 1976; *MacKinnon*, 1983] Differential uplift is manifest in a significant change in the metamorphic grade from Canterbury to Otago across the Waihemo/Waitaki Fault Zones. Canterbury, at the surface, is largely made up of prehnite-pumpellyite grade greywacke with minor regions of actinolite-pumpellyite in South Canterbury [*MacKinnon*, 1983]. South of the Waihemo/Waitaki Fault Zones (Figure 1), Otago is largely made up of schists, varying in grade from pumpellyite-actilolite to garnet grade schists of textural zones IIIB and IV [*Bishop et al.*, 1985; *Mutch*, 1963; *Mortimer*, 1993]. The basement in both regions is truncated by the time-transgressive (Cretaceous-Oligocene) Waipounamu Erosion Surface [*LeMasurier and Landis*, 1996]. This relationship implies that relative motion between the two blocks had occurred prior to the late Cretaceous.

2.2. Along Strike Variation: Canterbury/Otago

Oblique collision between the Australian and Pacific plates in South Island of New Zealand has often been treated as varying only in the direction normal to the plate boundary [e.g., *Wellman*, 1979; *Koons*, 1990; *Beaumont et al.*, 1992; *Willett et al.*, 1993; *Upton*, 1998; *Koons et al.*, 2003]. Without the complexities described below and addressed later in this study, the standard model produces a two-sided, asymmetric surficial strain pattern typical of oblique collision influenced by the asymmetric orographic erosional regime, with a zone of deep exhumation adjacent to the indenting block and a broad zone of deformation to the east of the Main Divide backthrust structures [*Koons*, 1990; *Beaumont et al.*, 1992; *Koons et al.*, 1998, 2003]. In

reality, the width and character of the zone of active deformation associated with oblique collision varies significantly along strike of the plate boundary [e.g., *Norris et al.*, 1990; *Little et al.*, 2005; *Liu and Bird*, 2006]. We suggest that the orogen width and strain variations result from an along strike change in rheology from Canterbury to Otago associated with an increase in pre-collisional crustal thickness in Otago relative to Canterbury.

Present day crustal thickness varies both along and across South Island. Thicknesses along the East Coast, where deformation is at a minimum, are mostly likely to reflect pre-collisional values. Estimates of Moho depths in mid-Canterbury derived from seismic reflection, refraction and inversion of seismic velocities vary from 18 km in the Bounty Trough, ca. 25 km at the East Coast and ca. 30 km at the western edge of the Canterbury Basin [*van Avendonk et al.*, 2004]. 60 km to the southwest Moho depth is estimated at ca. 27km at the East Coast [*Scherwath et al.*, 2003]. The Moho deepens parallel to the coast to ~32 km east of Otago, and then deepens further to the northwest on a line normal to the plate boundary where velocity inversion suggests that lower crustal rocks extend an additional 10km downward in Otago relative to Canterbury [*Eberhart-Phillips and Bannister*, 2002].

3. THEORETICAL BACKGROUND

3.1. Thermal Evolution of Canterbury/Otago

The observed crustal thicknesses described above suggest that it is likely that the Otago crust was thickened from approximately 25 km to >30 km sometime prior to Cenozoic collision. Changes in crustal thickness affect both the thermal and rheological evolution of the lithosphere. Following thickening of the lithosphere, geothermal gradients are depressed as cool material replaces warmer (Figure 2A,B). Following relaxation over a period of at least 30Ma, the lower crust gradually warms due to both conduction and increased radiogenic heat production at lower crustal levels. The rheological consequence is to weaken the lithosphere as the weaker quartz- and feldspar-rich material that replaced stronger diabase within the lower crust warms up (Figure 2C). The opposite occurs following thinning of the lithosphere. Initially geothermal gradients are increased resulting in a weaker crust (Figure 2A,B). Following relaxation over a period of at least 30Ma, the resulting thermal profile is one with a depressed thermal gradient. The depressed profile develops by thinning the crust and replacing crustal material with mantle material which has a lower concentration of heat producing elements. The rheological consequence of this is to strengthen the lithosphere, especially the lower crust (Figure 2C).

3.2. Choice of Rheological Model

We have chosen a rheological model similar to the 'jelly sandwich' type [*Chen and Molnar*, 1983; *Burov and Watts*, 2006] consisting of a pressure-dependent upper layer on top of a temperature-dependent lower layer consisting of quartz- and feldspar-dominated crustal compositions [e.g., *Brace and Kohlstedt*, 1980; *Ranalli*, 1995]. Crustal rheology is unlikely to be as simple as the experimentally-derived flow laws would imply [*Brace and Kohlstedt*, 1980], but the sharp reduction in flow stress at temperatures above 300–350°C due to the exponential dependence upon temperature appears to be a robust feature of quartz-dominated crust [*Sibson*, 1982]. In this study, we employ thermally dependent rhe-

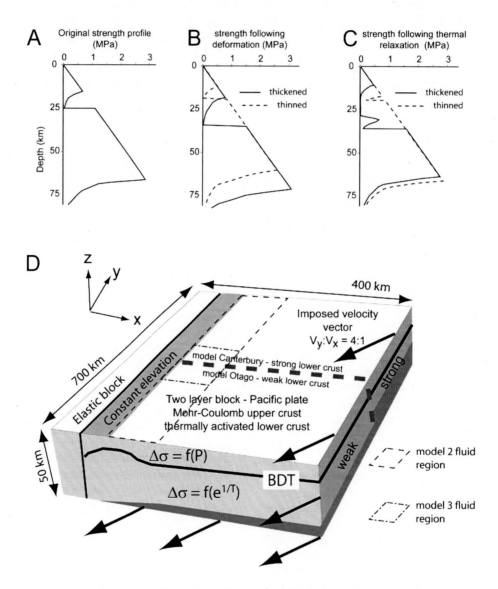

Figure 2. A–C: Rheological evolution of thickened (solid line) and thinned (dashed line) crust as a result of thermal relaxation. A: Original strength profile to 80 km for an average lithosphere and 25 km thick crust with a quartz-dominated upper crust, diabase-dominated lower crust and olivine-based mantle. B: Strength profiles following instantaneous crustal thickening to 35 km (solid line) and thinning to 20 km (dashed line). C: Strength profiles following 30 Ma of thermal relaxation. Note that the thinned crust is now much stronger than the thickened crust. D: Block modeling showing the geometry, boundary conditions and material properties used in the model. The Australian plate is modeled as a rigid elastic block. The Pacific plate is modeled as a two layered crust overlying a rigid (elastic) mantle. The upper crust has a Mohr-Coulomb rheology and the lower crust is modeled using a Von-Mises criterion which weakens as a function of exp(1/T). The rheological profiles are based on those shown in C, *Canterbury* is based on average crust and *Otago* is based on thickened crust. Boundary conditions are imposed on the far field boundaries as shown.

ologies to model the obliquely deforming continental crust of South Island. We incorporate the differing thermal histories of the two regions discussed above into the rheological models such that the lower crust of Canterbury is of average thickness and is strong whereas the lower crust of Otago is relatively thick and weak. Note that these conditions are based on our interpretation of crustal thickness before Southern Alps collision started and do not reflect the crustal thicknesses observed within the deforming region today.

3.3. Mechanical Models

We model the oblique deformation that occurs along South Island transpressional boundary using a two-layered crust overlying an elastic Pacific mantle (Figure 2D). The solution domain is a three-dimensional numerical region extending 400 km normal to the plate boundary (=x) by 700 km parallel to the plate boundary (=y) by 50 km depth (=z). The western edge of the model consists of an elastic block simulating the Australian plate which deforms minimally. The erosional regime of the Southern Alps is a result of a strongly asymmetric precipitation pattern that efficiently removes material from the western inboard slope [Koons, 1990; Beaumont et al., 1992; Willett et al., 1993]. We have simulated this orographic effect by maintaining the western slope at a constant elevation.

Our models are aimed at unraveling crustal deformation and the displacement of the mantle is input as a boundary condition. Two end member kinematic models have been used to explain how strain is accommodated in the mantle lithosphere beneath South Island with one evoking South Island-sub crustal mantle deformation over >200 km [Molnar et al., 1999; Moore et al., 2002; Baldock and Stern, 2005]. The other end member assumes little or no deformation in the uppermost mantle and resembles a subduction type boundary condition. [e.g., Beaumont et al., 1992; Koons et al., 2003; Ellis et al., 2006]. Models with a distributed shear in the mantle are the subject of an ongoing study, however, with a 'jelly sandwich' type rheology, initial models suggest that it is difficult to reconstruct the strain distribution in Canterbury using a distributed shear boundary condition. We have chosen the subduction type boundary condition where the Pacific plate is descending beneath the Australian plate without a large region of mantle deformation based on a comparison between mechanical models and observations [Ellis et al., 2006; Liu and Bird, 2006; Savage et al., this volume; Beavan et al., this volume]. These boundary conditions are similar to those used in previously published mechanical models of the Southern Alps [e.g., Beaumont et al., 1992; Koons et al., 1998, 2003]. The Pacific crust is dragged along its base, 25 km below sea level for Canterbury and 35 km below sea level for Otago, at a velocity of 40 mm/yr parallel to (along y axis) and 10 mm/yr normal to (along x axis) the plate boundary compatible with the calculated relative plate vectors from NUVEL 1A De Mets et al. [1994] (Figure 2D).

We represent crustal materials with an elastic/Mohr-Coulomb, pressure-dependent upper crust overlying an elastic/Von Mises plastic lower crust with flow parameters determined from the published creep laws at a reference strain rate of 10^{-14} s^{-1} (Table 1). We chose to define the lower crust in this way rather than using the power law rheology also available to us in FLAC^{3D} as we have found that our solutions are more stable when running to higher strains. The reference strain rate of 10^{-14} s^{-1} is likely to be low for material close to the Alpine Fault but the scale of our models is such that we cannot focus deformation to the extent that occurs in nature. As our analysis is concentrating on the distribution of strain across the whole orogen we feel that this simplifying assumption can be justified. Rheological properties are varied along strike of the boundary in a manner constrained by the geological and thermal arguments presented above. Note that throughout the remainder of this manuscript we refer to the model Canterbury, Otago, and Alpine Fault in italics to distinguish model features from nature. The northern half of the model represents the Canterbury region, the crust here has an initial thickness of 25 km [van Avendonk et al., 2004] and employs a relatively strong lower crust compatible with seismic evidence for a mafic dominated lower crustal layer [Mortimer et al., 2002]. The southern half of the model represents the Otago region, here the crust has an initial thickness of 35 km [Eberhart-Phillips and Bannister, 2002; Mortimer et al., 2002] and employs a relatively weak lower crust compatible with a thermally equilibrated quartzofeldspathic dominated lower crust (Figure 2D). Note that the Alpine Fault is not included 'a priori' in the model setup but develops as a through going shear as a result of the imposed boundary conditions [Koons et al., 2003].

3.4 Fluids Within the Models

Models (#2 and 3) were developed which include a region of interconnected fluid in the lower crust whose geometry was based upon the simplified U-shaped geometry of the region of high conductivity measured by magnetotellurics (MT) beneath the Southern Alps along SIGHT transect 2 [Wannamaker et al., 2002; Upton et al., 2002]. The MT transect shows the crust away from the U-shaped anomaly to be resistive and thus it is interpreted to be fluid-poor. A low-velocity region coincident with the MT anomaly observed from teleseismic waves on transect line 1 [Stern et al., 2001], suggests that pore fluid pressure is high and that it extends along strike for at least 80 km. Metamorphic dehydration reactions are proposed as the source of this fluid [Stern et

Table 1. Parameters, material properties and initial conditions used in the numerical models

Model #	1	2	3
Bulk Modulus (Pa)	1e10	1e10	1e10
Shear Modulus (Pa)	3e9	3e9	3e9
Density (kg m^{-3})	2800	2800	2800
Friction angle (°)	35	35	35
Geothermal gradient (°C km^{-1})	20 [elevated next to the Alpine Fault after *Koons et al.*, 2003]	20 [elevated next to the Alpine Fault after *Koons et al.*, 2003]	20 [elevated next to the Alpine Fault after *Koons et al.*, 2003]
Rheological Profile	*Canterbury* – strong lower crust *Otago* – weak lower crust	All one strength, based on *Canterbury* lower crust	*Canterbury* – strong lower crust *Otago* – weak lower crust
Kshear (strength parameter, Pa)*	*Canterbury*: 1e8 – 7.3e8 *Otago*: 5e6 – 7.3e8	Whole model: 1e8 – 7.3e8	*Canterbury*: 1e8 – 7.3e8 *Otago*: 5e6 – 7.3e8
Fluids	Fluid present, understaturated throughout model	U-shaped region of interconnected fluids at lithospheric pore pressure that extends the whole length of the model	U-shaped region of interconnected fluids at lithospheric pore pressure limited to 100 km along strike in the center of the model
Boundary conditions	Subduction boundary condition	Subduction boundary condition	Subduction boundary condition

*Creep flow law is $\dot{\varepsilon} = A\,\sigma^{n}\,e^{(-Q/RT)}$ A is a pre-exponential factor, σ is the differential stress (MPa), n is the power-law exponent, Q is the activation energy, R the universal gas constant and T is temperature (K). A, Q and n are determined from extrapolation of laboratory creep experiments on wet synthetic quartzite [*Paterson and Luan*, 1990]. A = 6.5 × 10^{-8} MPa^{-n}s^{-1}; Q = 135 KJ mol^{-1}; n = 3.1. In the lower crust the rheology is that of diabase, A = 2 × 10^{-4} MPa^{-n}s^{-1}; Q = 260 KJ mol^{-1}; n = 3.4 [*Shelton and Tullis*, 1981]. Kshear is determined from the calculated differential stress, kshear = 0.5σ.

al., 2001; *Wannamaker et al.*, 2002]. Although there are an infinite number of possible geometrical, rheological and kinematic conditions associated with this U-shaped anomaly, we concentrate on the following two in this manuscript where we can justify the conditions by observations external to the model. Model #2 has a U-shaped region of high pore pressure which runs the entire length of the model parallel to plate boundary. This model does not have any variation in lower crustal strength along strike and the strength profile is based on that of *Canterbury* in the original model. In model #3, the pore pressure anomaly is restricted to 100 km along strike. This model includes the *Otago/Canterbury* rheological variation and the pore pressure anomaly is centered on this transition. All models are initial wet, containing a fluid with saturation less than one to simulate unconnected fluids. The saturation and pore pressure in the U-shaped region were slowly increased during the runtime of the models, to simulate fluid pressure developing from strain-induced metamorphic reaction in the mid-crust, reaching lithostatic during the model runtime. Pore fluid pressure gradients are lithostatic. Note that we are not modeling fluid flow within the crust but are parameterizing fluid presence within the models.

3.5 Solution Method

Models were developed using the numerical code FLAC^{3D} a three-dimensional finite difference code, which we have modified to accommodate large strains and local erosion. (Version 3.0, Fast Lagrangian Analysis of Continua in Three Dimensions [*Cundall and Board*, 1988]). A large number of comparisons have been made between FLAC^{3D} results and exact solutions for plasticity problems [*ITASCA*, 1997]. The results show uniformly good agreement with theoretical solutions and in modeling geometrical instabilities well past the collapse or failure limit. Materials are represented by polyhedral elements within a three-dimensional grid that uses an explicit, time-marching solution scheme and a form of dynamic relaxation. Each element behaves according to a prescribed linear or nonlinear stress/strain law in response to applied forces or boundary restraints. The inertial terms in the equations of motion,

$$\frac{\partial \sigma_{ij}}{\partial x_j} + \rho\,b_i = \rho\,\frac{dv_i}{dt}$$

(Cauchy's equations of motion where σ_{ij} is the stress tensor, x_j, v_i are the vector components of position and velocity respectively ρ is the density of the material, b_i is the body force) are used as a numerical means to reach the equilibrium state of the system under consideration. The resulting system of ordinary differential equations is then solved numerically using an explicit finite difference approach in time. The drawbacks of the explicit formulation (i.e., small timestep limitation and the question of required damping) are overcome by automatic inertia scaling and automatic damping that does not influence the mode of failure. The governing differential equations are solved alternately, with the output for the solutions of the equations of motion used as input to the constitutive equations for a progressive calculation. Solution is achieved by approximating first-order space and time derivatives of a variable using finite differences, assuming linear variations of the variable over finite space and time intervals, respectively. The continuous medium is replaced by a discrete equivalent with all forces involved concentrated at the nodes of a three-dimensional mesh used in the medium representation.

4. RESULTS: KINEMATIC ANALYSIS OF MODELS

4.1 Kinematic Analysis of Canterbury/Otago (Model 1)

The three-dimensional mechanical models show that a pre-existing difference in rheology is sufficient to produce the observed differences in orogen-scale deformation style across the *Canterbury/Otago* transition (Figure 3). Deformation in *Canterbury* is restricted to a narrow region against the Alpine Fault, strain rates are high (closely spaced velocity contours) and there is little partitioning of the two horizontal velocity components (Figure 3A,B). In *Otago*, deformation is distributed over a wide region east of the Alpine Fault. *Otago* strain rates are considerably lower and there is partitioning between the two velocity horizontal components with the plate margin normal velocities extending further to the east (Figure 3A,B). Model vertical velocity (v_z) also reflects the along strike rheological variation. In *Canterbury*, uplift is concentrated into a ca. 60 km wide band adjacent to the *Alpine Fault*. The maximum vertical velocity is against

the fault as would be predicted. Vertical velocity in *Otago* falls into two zones. A region of moderate uplift exists adjacent to the *Alpine Fault*, along strike from the uplifted region in *Canterbury*. To the east, small but measurable vertical velocity extends to the edge of the continent. A robust feature of this model is a region of high uplift where the transition from weak to strong lower crustal rheology intersects the *Alpine Fault* (Figure 3C).

4.1.1 Plate margin normal deformation (Plate 1B,E,G). Plate-normal deformation is accommodated predominantly by contraction along the Alpine Fault zone and distributed to the east by shortening in the outboard fold and thrust belt. The component of shortening normal to the Alpine Fault is accommodated either by thrusting (represented by $-\partial v_x/\partial x$, Plate 1E) and upward velocities ($+v_z$, Plate 1C) in the upper crust or downward velocity ($-v_z$, Plate 1C) and crustal thickening in the lower crust. Structures taking up this movement develop in regions of strong rotation about the y-axis (Plate 1G). A region of basal shear has developed in the lower crust of Otago (Plate 1G).

4.1.2 Plate margin parallel deformation. Lateral (boundary parallel) strain is concentrated along the eastward dipping plate-bounding structure (Alpine Fault, Plate 1H). A small percentage of lateral strain is also taken up along an oppositely dipping zone to the east of the Main Divide. This secondary structure is stronger in Canterbury than in Otago.

4.1.3 Strain rate components; variation with depth. On map sections through the model (Figure 4; also on the CDROM which accompanies this volume) we can observe variations in three-dimensional strain rate across the models. Contractional strain rate normal to the plate boundary

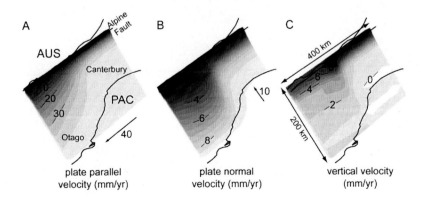

Figure 3. Model 1 results showing the three velocity components at sea level. A: plate margin parallel velocity in mm/yr. B: Plate margin parallel velocity in mm/yr. Note the differences between *Canterbury* and *Otago*. Strain rates in *Canterbury* are higher as shown by the more closely spaced velocity contours. C: Vertical velocity in mm/yr. The greatest uplift rates occur adjacent to the *Alpine Fault*, at the transition from *Canterbury* to *Otago*.

(directed NW-SE) varies along strike and with depth (Figure 4; also on the CDROM which accompanies this volume). At 25 km depth this component is concentrated adjacent to the plate boundary in *Canterbury*. It steps outward across the boundary from *Canterbury* to *Otago* and then reduces in magnitude across *Otago*. To the southwest of the transition zone, there is a region which is extending slightly, oriented NW-SE. Higher in the crust, at 15 km depth, strain rate normal to the plate boundary is contractional and is greatest in *Canterbury*. Contraction in y (parallel to the plate boundary) occurs at the transition from *Canterbury* to *Otago* at all depths. It is greatest at 25 km where the contraction is occurring entirely in *Otago*. At higher crustal levels it reduces in magnitude and begins to spread out across the transition into *Canterbury*. At 25 km depth, as the *Canterbury* lower crust is pushing into the *Otago* lower crust (resulting in the contraction in y described above), the *Otago* lower crust is being extruded upward as shown by a region of $+\partial v_z / \partial z$ on the *Otago* side of the transition zone.

4.2 Pore Pressure, no Along Strike Rheological Variation (Model 2)

Results of this model are illustrated by block diagrams of half the model showing a map view at sea level and a cross-section through the middle of the model cut normal to the plate boundary (Plate 2). The effect of the pore pressure on the mechanical solution is most obvious in the convergent velocity and the uplift pattern. The material moves through the orogen more rapidly at depths of 20–25 km, coincident with the pore pressure anomaly, than it does closer to the surface (Plate 2A). In fact, the material in the upper 5–10 km of the crust in the *Main Divide* region is virtually stationary, not moving towards the Australian plate at all, despite material beneath and west of it moving westward. This effect occurs because the mid-crust is effectively weakened by the elevated pore pressure, focusing deformation into a mid-crustal shear zone. This has the effect of creating a channel in the mid-crust beneath the *Main Divide* region. Note that we use the term channel to describe a narrow zone of material moving more rapidly than the material about or below it but the use of the term does not imply a driving force similar to that proposed by *Burchfield and Royden* [1985]. This velocity profile also creates a region of NW-SE directed stretching in the upper 5 km between the stationary material and the material moving westward along the *Alpine Fault*. The pattern of vertical velocity is also modified by the pore pressure anomaly (Plate 2C). Instead of vertical velocity being concentrated against the *Alpine Fault*; with the greatest amount of uplift against the fault, all the material above the

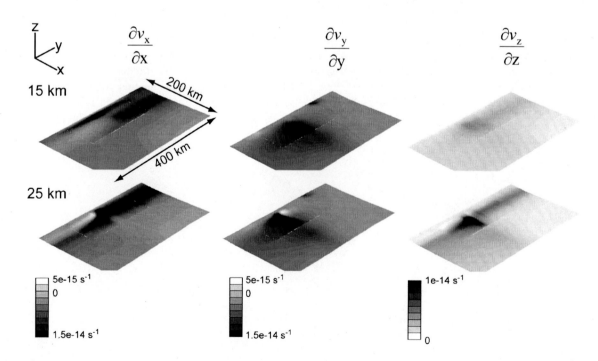

Figure 4. Contour map view slices at 15 and 25 km depth through the original model of the Southern Alps with a strong lower crust in Canterbury and a weak lower crust in Otago. At both levels, the three orthogonal strain rate components are shown, $\partial v_x / \partial x - \partial v_y / \partial y$, and $\partial v_z / \partial z$. +ve = expansion, −ve = contraction. Color version of this figure is on CDROM which accompanies this volume.

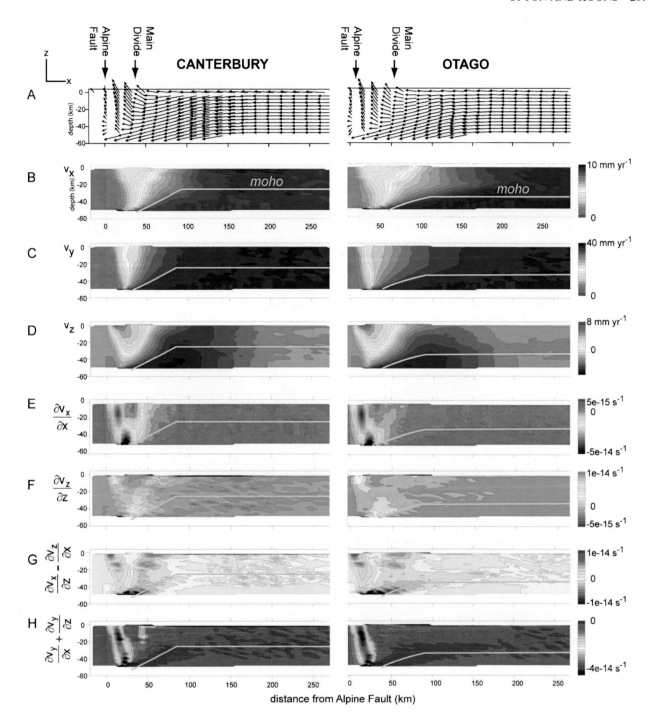

Plate 1. Velocity components and gradients for the *Canterbury* and *Otago* regions. A: Velocity vectors normal to the plate boundary, the maximum vector has a magnitude of 40 mm/yr; B: plate boundary normal velocity component (x-velocity); C: plate boundary parallel velocity component (y-velocity); D: vertical velocity component (z-velocity); E: Plate normal strain rate component, the change in the plate boundary normal velocity along x, $(\partial v_x / \partial x)$; F: vertical strain rate component, $(\partial v_z / \partial z)$; G: Convergent strain rate represented as rotation rates in the x-z plane about the y axis = $(\partial v_x / \partial z - \partial v_z / \partial x)$: Plate-normal deformation is accommodated predominantly by contraction along the Alpine Fault zone and distributed to the east by shortening in the outboard fold thrust belt. H: Lateral (boundary parallel) strain rate $(\partial v_y / \partial x - \partial v_y / \partial z)$: Lateral strain is concentrated along the eastward dipping plate-bounding structure and along an oppositely dipping zone to the east of the *Main Divide*, which is more developed in *Canterbury* than *Otago*.

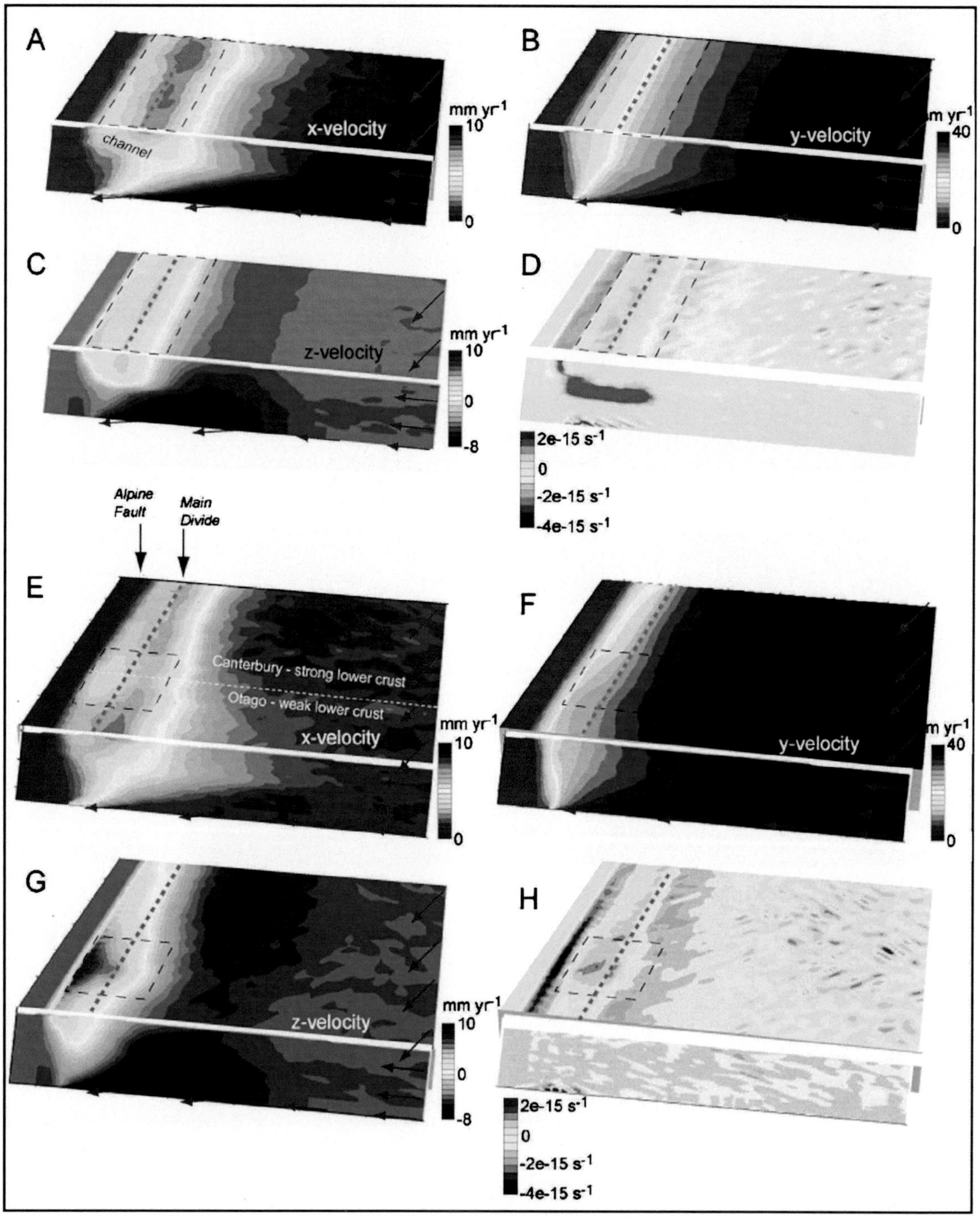

Plate 2. Block diagrams (map view and cross-section normal to the plate boundary) illustrating results from models 2 (A–D) and 3 (E–H) All figures are relative to fixed Australia. A, E: x-velocity (plate margin normal). B, F: y-velocity (plate margin parallel). C,G: z-velocity (rock uplift). D, H: Dilatation strain rate determined in two dimensions. +ve = expansion, –ve = contraction. The map view shows

$$\Delta_{xy} = \frac{1}{2}\left(\frac{\partial v_x}{\partial x} + \frac{\partial v_y}{\partial y}\right)$$ and the cross-section shows $$\Delta_{xz} = \frac{1}{2}\left(\frac{\partial v_x}{\partial x} + \frac{\partial v_z}{\partial z}\right).$$ See text for details. The black dashed boxes outline the

extent of the pore pressure anomaly in the models. The grey dashed lines shows the approximate location of the main divide. The location of the *Alpine Fault* and the *Main Divide* are noted on E.

pore pressure anomaly is being uplifted as a block. Plate 2D illustrates two-dimensional dilatation strain rate (Note positive = expansion, negative = contraction). In map view, a region of NW-SE directed positive dilatation along the *Main Divide* correlates to the region of diverging x velocity shown in Plate 2A. In cross-section, a large dilation strain is associated with the pore pressure anomaly particularly in the bottom of the U-shaped region and along the *Alpine Fault*.

4.3 Limited Extent Pore Pressure, Along Strike Rheological Variation (Model 3)

In this model there is a rheological variation along strike where the lower crust of the *Otago* region is weak and that of the *Canterbury* region is strong. The pore pressure anomaly is restricted to an along strike extent of 100 km centered about the transition from *Canterbury* to *Otago*. The model results are illustrated in a similar way to the model above except that $^2/_3$ of the model is shown with the cross-sections cutting through south of the *Canterbury/Otago* transition (Plate 2E–H).

The model velocity distributions reflect the presence of a high pore pressure region at depth within the crust (Plate 2). Material is brought more rapidly into the inboard region above the pore pressure anomaly by an increase in the v_x just east of the *Alpine Fault* (Plate 2E). Vertical velocity and positive dilatation are also greatest at this point adjacent to the *Alpine Fault*, coincident with the transition from *Canterbury* to *Otago* and the pore pressure anomaly. Positive dilatation (oriented NW-SE) occurs along the length of the *Main Divide* however it is greatest directly above the pore pressure anomaly (Plate 2H).

5. DISCUSSION AND IMPLICATIONS

Continental collision in South Island lies between two oppositely dipping subduction zones which dominate crustal dynamics at either end of the island. In this discussion we are concerned with the central part of South Island, enclosed within dashed lines on Figure 1. To the north and south of these dashed lines, the basal boundary conditions change significantly as the subducting slabs begins to affect the dynamics. Consequently, these regions require their own geodynamic description.

A simple variation in lower crustal rheology from strong to weak in an oblique orogen has a marked effect on the width of orogen and the strain rates both at the surface and at depth within the orogen. An interesting consequence of this change is that the greatest amount of uplift along the *Alpine Fault* is spatially coincident with the transition from a weak to a strong lower crust (Figure 3C). By looking at the Cartesian normal strain rate components derived from the models, we see at mid-crustal depths, the strong crust of *Canterbury* acts as an indentor into the weaker lower crust of *Otago* (Figure 4; also on the CDROM which accompanies this volume). Adjacent to the Alpine Fault and just southwest of the transition to *Canterbury* to *Otago*, two regions develop with distinctive strain ellipses (Figure 5). These regions are separated by the east dipping branch of the *Alpine Fault zone* [*Koons et al.*, 2003]. In both cases NE-SW convergence is occurring. However, vertical stretching is only occurring east of the *Alpine Fault*. West of the *Alpine Fault*, a combination of flattening and extension is predicted while east of the *Alpine Fault*, the predicted strain ellipse would resemble a cigar [*Koons et al.*, 2004]. Exhumation of this zone might present the structural geologist with a puzzle; two distinct strain regimes separated by a zone of high strain. Unraveling this puzzle would require recognition of a change in the regional lower crustal rheology and a dynamic understanding of how this rheological change is reflected in the local strain patterns.

In our models, the presence of elevated pore pressures at depth is noticeable in the surface strain (Plate 2) and is most significant for regions of positive dilatation within the orogen as discussed in the next section. The spatial coincidence of the pore pressure anomaly with the transition from *Canterbury* to *Otago* results in focusing of vertical uplift at the transition and against the plate boundary (Plate 2G), enhancing the effect that was observed in the original *Canterbury/Otago* model (Figure 4C; also on the CDROM which accompanies this volume). To fully explore the implications of this model, we need to consider the model and geological evidence for, and implications of, regional positive dilatation in the central Southern Alps.

5.1 Dilatation and the Main Divide Region: Model Results

Dynamically, minor regional extension is predicted to develop in the main divide region of an oblique orogen [*Koons and Henderson*, 1995; *Koons et al.*, 1998] In our initial fluid understaturated model (#1), a small component of NW-SE oriented dilatation is observed in the *Main Divide* region (Plate 1D). Addition of a U-shaped region of elevated pore pressure, beneath the Southern Alps in the models leads to a significant component of NW-SE oriented dilatation in the upper crust (Plate 2D,H). In model 2, the pore pressure anomaly runs the along strike length of the model and consequently the zone of dilatation extends along the full length of the model (Plate 2D). In model 3, the region of elevated pore pressure is restricted to 100 km along strike and in this case significant dilatation occurs only above the pore pressure anomaly (Plate 2H). The dilatation component of strain

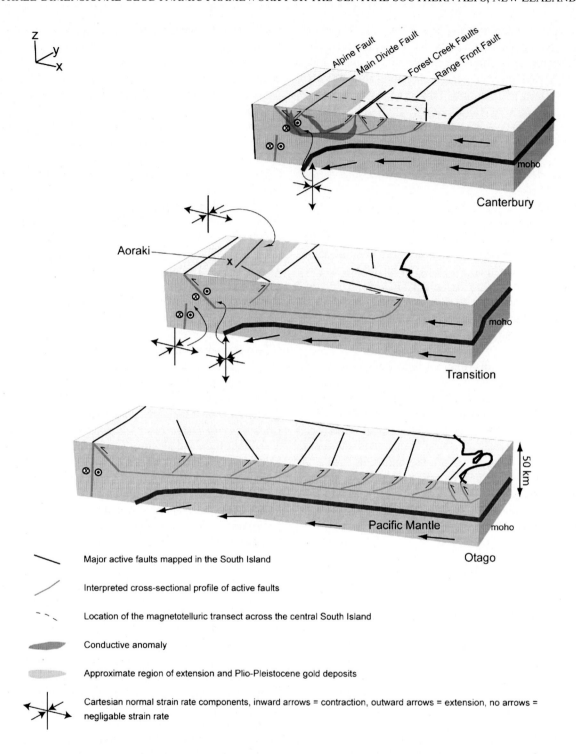

Figure 5. Block models of *Canterbury*, the transition between *Canterbury* and *Otago* and *Otago* showing the change in the nature of Southern Alps collision zone along strike. Active faults and the east coastline are shown as black lines on the surface. The faults are shown extending into the crust as grey lines. Detail of the Alpine Fault structure from *Koons et al.* [2003]. The moho (thick grey line) is at 25 km beneath east Canterbury and increases in depth to 35 km beneath east Otago. The magnetotelluric conductivity anomaly is shown on the Canterbury block but extends in some form southwest to at least past Mt Cook-Aoraki [*Wannamaker et al.*, 2002]. Cartesian normal strain rate components are shown for certain places within the orogen (see text for more discussion).

rate results from summing the symmetric components of the velocity gradient tensor, $\partial v_x / \partial x$, $\partial v_y / \partial y$ and $\partial v_z / \partial z$. Dilatation in the upper crust of our model is dominated by $\partial v_x / \partial x$, the component related to the plate margin normal velocity (Plate 2A). The material of the Pacific plate that is located at or just east of the *Main Divide* region is virtually stationary (in the reference frame of fixed Australia) while material to the west, east and at greater depths move toward the Australian plate. This has the effect of producing a region of NW-SE directed stretching between the stationary material and that to the west of it which is moving away toward the Australian plate. We would expect to see this deformation taken up along steeply dipping planes oriented approximately parallel to the *Alpine Fault*. At greater depths, a large dilatational strain rate is associated with the region of elevated pore pressure. Here the dilatation is dominated by the vertical velocity component and would thus be taken up on shallowly dipping structures.

5.2 Earthquake Distribution in the Southern Alps

Earthquake distribution varies both across and along strike from the Alpine Fault. South of the Hope and Porter's Pass faults earthquakes have a high thrust component and accommodate part of the dip slip convergence of the plate motion [*Leitner et al.*, 2001]. Solutions suggest distributed shear on NNE-SSW trending thrust faults [*Leitner et al.*, 2001]. East of the Main Divide, two sequences of shallow earthquakes have occurred whose solution suggests normal faulting with a NW-SE oriented extension axis [*Leitner et al.*, 2001]. *Leitner et al.* [2001] suggest that these may have been hydrologically triggered as they occurred following significant rainfall and spring snowmelt. Their occurrence and NW-SE oriented extension axis, however, are consistent with predicted dilatation occurring in the upper crust as a result of stretching in the direction normal to the plate boundary (Plate 2D, H). Thus while an increase in pore pressure due to significant rainfall and snowmelt may have triggered these minor earthquake swarms, extensional events in this region of the orogen are not unexpected. South of Mt Cook-Aoraki, seismicity is lower and more broadly distributed with no extensional solutions determined [*Leitner et al.*, 2001].

5.3 Gold in the Southern Alps

Vein material, including gold, forms in extensional sites within a deforming region [*Groves et al.*, 1998; *Craw et al.*, 2002]. Orogenic gold deposits are associated with continental collision [*Groves et al.*, 1998] and deposit most of their gold at temperatures of 200–400°C ("mesozone"), with associated weakly mineralized zones at shallow levels ("epizone"). Orogenic gold deposits in young collisional mountain belts, e.g., Southern Alps, Taiwan, the Alps, the Pyrenees and the Rocky Mountains in Canada and Alaska, are generally hosted in greenschist facies rocks of widely varying protoliths and fill extensional fractures or dilational sites in shear fractures [*Templeton et al.*, 1998a,b; *Koons et al.*, 1998; *Tan et al.*, 1991; *Shaw and Morton*, 1990; *Craw and Leckie*, 1996; *Diamond*, 1990; *Curti*, 1987; *Craw et al.*, 1993; *Ayora and Casa*, 1986; *McCaig et al.*, 1995; *Craw et al.*, 2002]. These gold-bearing vein systems are generally located at or near major topographic divides in the various mountain belts [*Craw et al.*, 2002]. Gold (along with arsenic and sulphur) dissolve in crustal fluids in contact with greenschist facies rocks near the brittle-ductile transition. Gold is then deposited from these fluids in fracture systems from below the brittle-ductile transition to the near-surface region [*Groves et al.*, 1998; *Craw et al.*, 2002].

Plio-Pleistocene mineralized veins have formed along the Main Divide of the Southern Alps associated with late Cenozoic continental collision [Figure 1; *Craw et al.*, 1999; *Craw et al.*, 2002]. For the purposes of this discussion we will focus only on the mineralized veins south of the Rakaia River (Figure 1). North of here the southernmost of the Marlborough Faults begins to interact with the Alpine Fault and Main Divide Fault Zone [*Becker et al.*, 2000]. There the intersection of the Alpine Fault and the Marlborough Faults is clearly important for the development of structural permeability and dilatation into which active mineralization is taking place [*Craw and Campbell*, 2004]. We will focus on a narrow band of gold occurrences centered on the Main Divide and extending from just south of Mt Cook-Aoraki north to the intersection of the Marlborough Faults (Figure 1). Gold mineralization of these veins has occurred over a range of temperatures, from >300 to ca. 100°C [*Templeton et al.*, 1999b] and geological arguments based on the amount of rock uplifted along the Alpine Fault suggest formation depths ranging from <1 to ca. 5 km [*Cox et al.*, 1997; *Templeton et al.*, 1999b]. *Craw et al.* [2002] suggest that these veins represent the shallow crustal levels above the mesozone of an orogenic gold-depositing system. The mineralized veins (assemblage quartz, carbonate, pyrite, arsenopyrite and gold) fill north to northeast striking, steeply dipping extensional fractures, including open tension gashes up to 1 m long and 50 cm wide [*Craw et al.*, 2002].

5.4 Combining Models and Observations

5.4.1 Main Divide Dilatation: Upper Crust. Our models predict a region of upper crustal, NW-SE oriented dilatation which is enhanced by the presence of elevated pore pressure beneath the Main Divide. We can further predict that we expect to see this deformation taken up on structures oriented approximately parallel to the plate boundary with a NW-SE extension direction. As described in the previous section, Plio-Pleistocene mineralized veins within the Southern Alps are found in just these types of structures. The isotopic signal of the mineralization in this region is dominated by a rock-exchanged signature, reflecting a deep source [*Cox et al.*, 1997; *Koons et al.*, 1998; *Becker et al.*, 2000]. The MT interpretation for the Main Divide region above 8 kms shows a resistive region and is therefore thought to contain a nonconductive or poorly connected fluid phase or possibly both [*Wannamaker et al.*, 2002; *Upton et al.*, 2003]. Fluid flow is likely to be episodic, with fluid being driven upward during short-term structural events [*Sibson*, 1996; *Upton et al.*, 2002] possibly associated with seismic events such as the sequences of shallow earthquakes with normal faulting solutions described above [*Leitner et al.*, 2001]. Models of elastic deformation in the central Southern Alps based on geodetic measurements also imply a NW-SE extension direction in the Main Divide region north of Lakes Pukaki and Tekapo [*Beavan et al.*, 1999].

Elevated pore pressure in the mid-lower crust beneath this zone is important to the development of the extension because it acts to weaken the mid-crust and results in the development of a mid-crustal channel. This has the effect that the material in the mid-crust is moving toward the Australian plate more rapidly than the upper crustal material overlying it, resulting in extension in the x-direction between the nearly stationary material and the material to the west. Elevated pore pressure is a very effective means of weakening the crust however we would also expect to see this effect if the lower crust was being weakened by some other mechanism, such as reaction weakening or the presence of fluid without a significantly high fluid pressure. For quartz dominated rocks undergoing diffusion creep, a small quantity of fluid (~0.5%) can significantly weaken the rock [*Tullis*, 1990; *Tullis and Yund*, 1980].

5.4.2 Main Divide Extension: Mid Crust. A component of dilatation is also predicted coincident with the region of elevated pore pressure at depth within the crust. In this convergent setting the regional stress orientations will be σ_3 (minimum principal compressive stress) sub-vertical and σ_1 (maximum principal compressive stress) sub-parallel to the compression direction. We are unable to observe this region directly to test our models but may look for possible analogs in the records of older, more deeply exposed mountain belts.

The Macraes gold deposit in east Otago developed in the Hyde-Macraes Shear Zone (HMSZ), a gently (15°) NE dipping Mesozoic shear zone [*Teagle et al.*, 1990]. The shear zone is parallel to the flat-lying foliation. It contains quartz

veins sub-parallel to the shear zone which formed under high fluid pressure during thrusting that was initiated below the brittle-ductile transition [*Teagle et al.*, 1990; *Begbie and Craw*, 2006]. These veins may well have formed in a shear zone setting similar to that we predict to exist beneath the Southern Alps where elevated pore pressures are allowing gently-dipping extension and extensional-shear vein systems to develop [*Craw and Campbell*, 2004].

5.4.3 Linkage of mid-crustal and upper-crustal dilatation and fluid flow. Our models predict that dilatation within the mid-crust is dominated by vertical stretching and sub-horizontal fracturing. Thus fluids generated by metamorphic reaction and responsible for the elevated pore pressure in this region would be expected to mostly flow horizontally within low angled shear zones which are taking up the dilatational strain. We predict that dilatation in the upper crust is dominated by near vertical NE-SW striking structures to depths of ~8–10 km, which is about the level of the brittle-ductile transition. Are these two regions of dilatation and fluid flow linked within the mid-upper crust? The $\delta^{18}O$ isotopic signature suggests that the fluid forming veins in the upper crustal extension fractures has a strongly rock-exchanged signature [*Cox et al.*, 1997; *Koons et al.*, 1998; *Becker et al.*, 2000; *Wightman and Little*, this volume]. The MT across this region shows a 'plume' moving upward from the U-shaped region in the mid-crust and this may well represent fluid from one extensional regime moving into another extensional regime.

5.4.4 Existence and extent of the elevated pore pressure along strike. Combining model results and observations from the Southern Alps allows us to make a reasonable guess at the along strike extent of the MT anomaly to the south of SIGHT transect 2. Earthquakes with an extensional solution and mineralized Plio-Pleistocene veins extend south from the MT transect line to approximately the head of Lake Pukaki, south of the *Canterbury/Otago* transition (Figure 1). None of these features of surface strain are observed south of this region in Otago. If a similar region of elevated pore pressure existed beneath the Otago region, we would expect to see it reflected in the surface strain. Thus we suggest that the region of elevated pore pressure that exists beneath the central Southern Alps has an along strike extent of ~60–100km. This result is consistent with observations that a low velocity region is observed on both transects 1 and 2 from the SIGHT program [*Stern et al.*, 2001; *Scherwath et al.*, 2003; *van Avendonk et al.*, 2004].

5.4.5 Higher rock uplift rates in the central section of the Southern Alps. Measured rock uplift rates, thermochronology, geological and geodetic observations suggest that the central section of the Southern Alps, including Mt Cook-Aoraki and the Fox and Franz Josef Glaciers, appears to be undergoing more rapid rock uplift than surrounding parts of the orogen [Simpson et al., 1994; *Henderson*, 2003; *Norris and Cooper*, 2001; *Little et al.*, 2005]. Various models have been put forward to explain this observation including the suggestion that there exists a restraining bend in the Alpine Fault at depth in this region [*Little et al.*, 2005]. We propose that the rheological transition from weak to strong lower crust from Otago to Canterbury, coincident with a region of interconnect pore pressure in the mid-crust, may contribute significantly to the development of this region of higher uplift rates. Note that the resolution of the models is such that while we can predict the deformation style at various regions within the orogen, we are not predicted the specific nature of the geological structures that might accommodate this deformation in nature. These structures will develop as a result of the factors discussed in this manuscript as well as local variability in rock properties and surficial boundary conditions.

Weakening of the mid-crust by a region of elevated pore pressure results in the development of a mid-crustal channel (Plate 2). As well as causing upper crustal dilatation (oriented NW-SE) as described above, it has the effect of bringing mid-crustal material rapidly to the surface along the Alpine Fault. The along strike extent of the pore pressure anomaly is postulated to be 60–100 km, consequently we would expect that the along strike extent of higher rock uplift associated with this anomaly to also be about 60–100 km. Alone, this feature accounts for an increase in rock uplift rates in the western central Southern Alps. We also observed that the rheological transition from a weak to strong lower crust is associated with higher rock uplift rates. This is due to the stronger lower crust of *Canterbury* acting as an indentor into the weaker lower crust of *Otago*, associated with this indention *Otago* material is extruded upward, resulting in a region of higher rock uplift rates centered on the *Otago* side of the transition (Figure 4; also on the CDROM which accompanies this volume). Combined, as they are in present day central region of the Southern Alps, the coincidence of change in lower crustal strength and a region of elevated pore pressure leads to a focused region of high rock uplift.

6. CONCLUSIONS

The three-dimensional pattern of surface deformation within the central Southern Alps contains a wealth of information on spatial variations of the rheological structure of the crust. When combined with geological observations and geophysical investigations of the deeper structure of the Southern Alps, three-dimensional mechanical models allow us to develop a geodynamic framework in which to characterize the deformation of the oblique orogen. A simple variation in lower crustal rheology from strong to weak has a marked effect on the pattern of stress, strain and strain rates both at the surface and at depth within the orogen. Changes in surface deformation patterns from Canterbury to Otago can be explained by this change in the lower crustal rheology. Against the Alpine Fault, the transition zone is characterized by higher rock uplift rates, due in part to the stronger lower crust of Canterbury indenting into the weaker lower crust of Otago. A region of elevated pore pressure beneath the central Southern Alps, constrained by crustal velocity and MT from the SIGHT experiments, has a predictable effect on the strain. A region of elevated pore pressure in the mid-crust enhances the convergent velocity component and leads to the development of a channel of material beneath the main divide that is moving more rapidly than that above it. By comparing model results with geological and geophysical observations, we suggest that the pore pressure anomaly has a lateral extent of 60–100 km along strike. The combination of a change in lower crustal rheology from Canterbury to Otago and a zone of elevated mid-crustal pore pressure results in a region of higher rock uplift coincident with the Mt Cook-Aoraki region.

Acknowledgments. Research presented in this manuscript was funded by the University of Otago, the University of Maine and the New Zealand Foundation of Research, Science and Technology (Contract UO0818). Discussions with Dave Craw, Mike Begbie, Mark Henderson and Scott Johnson helped to develop many of the ideas expressed herein. Thorough reviews by Tim Little, Susan Ellis and an anonymous reviewer have greatly improved the clarity of the manuscript.

REFERENCES

Anderson, H., T. Webb, and J. Jackson (1993), Focal mechanisms of large earthquakes in the South Island of New Zealand: Implications for the accommodation of Pacific-Australia plate motion, *Geophys. J. Int.*, *115*, 1032–1054.

Ayora, C., and J. M. Casas (1986), Stratabound As–Au mineralization in pre-Caradocian rocks from the Vall de Ribes, eastern Pyrenees, Spain. *Miner. Depos.*, *21*, 278–287.

Baldock, G., and T. Stern (2005), Width of mantle deformation across a continental transform: Evidence from upper mantle (Pn) seismic anisotropy measurements, *Geology*, *33*, 741–744.

Beaumont, C., P. Fullsack, and J. Hamilton (1992), Erosional control of active compressional orogens, in *Thrust Tectonics*, edited by K. R. McClay, pp. 1–18, Chapman & Hall, New York.

Beavan, J., M. Moore, C. Pearson, C. M. Henderson, B. Parsons, S. Bourne, P. England, R. I. Walcott, G. H. Blick, D. Darby, and K. Hodgkinson (1999), Crustal deformation during 1994–1998 due to oblique continental collision in the central Southern Alps, New Zealand, and implications for seismic potential of the Alpine Fault, *J. Geophys. Res.*, *104*, 25,233–25,255.

Beavan, J., S. Ellis, L. Wallace, and P. Denys (this volume), Kinematic constraints from GPS on oblique convergence of the Pacific and Australian plates, central South Island, New Zealand.

Becker, J. A., D. Craw, T. Horton, and C. P. Chamberlain (2000), Gold mineralization near the Main Divide, upper Wilberforce valley, Southern Alps, New Zealand, *N. Z. J. Geol. Geophys.*, *43*, 199–215.

Begbie, M., and D. Craw (2006), The geometry and petrography of stockwork vein swarms, Macraes Mine, Otago Schist, New Zealand, *N. Z. J. Geol. Geophys.*, *49*, 63–73.

Bishop, D. G., J. D. Bradshaw, and C. A. Landis (1985), Provisional terrane map of South Island, New Zealand, in *Tectonostratigraphic Terranes, Earth Sci. Ser.*, vol. 7, edited by D. G. Howell, pp. 515–521, Circumpacific Council for Energy and Mineral Resources, Houston, TX.

Blick, G. H., S. A. L. Read, and P. T. Hall, Ostler Fault zone (1985). Progress report on surveying results 1966–1985, levelling and tilt levelling, *New Zealand Geological Survey Report EDS 103*, 90 pp.

Brace, W. F., and D. Kohlstedt (1980), Limits on lithospheric stress imposed by laboratory experiments, *J. Geophys. Res.*, *89*, 6248–6252.

Burchfiel B. C., and L. H. Royden (1985), North-south extension within the convergent Himalayan region, *Geology*, *13*, 679–682.

Burov, E. B., and A. B. Watts (2006), The long-term strength of continental lithosphere: "jelly sandwich" or "crème brûlèe?, *GSA Today*, *16*, 4–10.

Chemenda, A., S. Lallemand, and A. Bokun (2000), Strain partitioning and interplate friction in oblique subduction zones: Constraints provided by experimental modeling, *J. Geophys. Res.*, *105*, 5567–5581.

Chen, W.-P., and P. Molnar (1983), Focal depths of intracontinental and intraplate earthquakes and their implications for the thermal and mechanical properties of the lithosphere, *J. Geophys. Res.*, *88*, 4183–4214.

Coombs, D. S., C. A. Landis, R. J. Norris, J. M. Sinton, D. J. Borns, and D. Craw (1976), The Dun Mountain ophiolite belt, New Zealand, its tectonic setting, construction, and origin, with special reference to the southern portion, *Am. J. Sci.*, *276*, 561–603.

Cox, S. C. (this volume), Regional geological framework: Constraints for the study of New Zealand plate boundary dynamics.

Cox, S. C., D. Craw, and C. P. Chamberlain (1997), Structure and fluid migration in a late Cenozoic duplex system forming the Main Divide in the central Southern Alps, New Zealand, *N. Z. J. Geol. Geophys.*, *40*, 359–373.

Craw, D., and J. R. Campbell (2004), Tectonic and structural setting for active mesothermal gold vein systems, Southern Alps, New Zealand, *J. Struct. Geol.*, *26*, 995–1005.

Craw, D., and D. A. Leckie (1996), Tectonic controls on dispersal of gold into a foreland basin: An example from the western Canada foreland basin, *J. Sediment. Res.*, *66*, 559–566.

Craw, D., D. A. H. Teagle, and R. Belocky (1993), Fluid immiscibility in late-Alpine gold-bearing veins, Eastern and Northwestern European Alps, *Miner. Depos.*, *28*, 3–28.

Craw, D., M. S. Rattenbury, and R. D. Johnstone (1999), Structural geology and vein mineralization in the Callery River headwaters, Southern Alps, New Zealand, *N. Z. J. Geol. Geophys.*, *30*, 273–286.

Craw, D., P. O. Koons, T. Horton, and C. P. Chamberlain (2002), Tectonically driven fluid flow and gold mineralization in active collisional orogenic belts: Comparison between New Zealand and western Himalaya, *Tectonophysics*, *348*, 135–153.

Cundall, P., and M. Board (1988), A microcomputer program for modeling of large-strain plasticity problems, in *Numerical Methods in Geomechanics. Proceedings of the 6th International Conference on Numerical Methods in Geomechanics*, vol. 2, edited by C. Swododa, pp. 101–108, Balkemapp, Innsbruck, Austria.

Curti, E. (1987), Lead and oxygen isotope evidence for the origin of the Monte Rosa gold lode deposits (western Alps, Italy): A comparison with Archean lode gold deposits, *Econ. Geol.*, *82*, 2115–2140.

DeMets, C., R. G. Gordon, D. F. Argus, and S. Stein (1994), Effect of recent revisions to the geomagnetic reversal time scale on estimates of current plate motions, *Geophys. Res. Lett.*, *21*, 2191–2194.

Diamond, L. W. (1990), Fluid inclusion evidence for P-V-T-X evolution of hydrothermal solutions in late-Alpine quartz veins at Brusson, Val d'Ayas, northwest Italian Alps, *Am. J. Sci.*, *290*, 912–958.

Eberhart-Phillips, D., and S. Bannister (2002), Three-dimensional crustal structure in the Southern Alps region of New Zealand from inversion of local earthquake and active source data, *J. Geophys. Res.*, *107*(B10), 2262, doi:10.1029/2001JB000567.

Ellis, S., J. Beavan, and D. Eberhard-Phillips (2006), Bounds on the width of mantle lithosphere flow derived from surface geodetic measurements: Application to the central Southern Alps, New Zealand, *Geophys. J. Int.*, doi:10.1111/j.1365–246X.2006.02918.x.

Enlow, R. L., and P. O. Koons (1998), Critical wedges in three dimensions: Analytical expressions from Mohr-Coulomb constrained perturbation analysis. *J. Geophys. Res.* *103*(B3), 4897–4914.

Henderson, C. M. (2003), The velocity field of the South Island of New Zealand derived from GPS and terrestrial measurements, PhD thesis, University of Otago, Otago, New Zealand.

Gerbault, M., F. Davey, and H. Stuart (2002), Three-dimensional lateral crustal thickening in continental oblique collision: An example from the Southern Alps, New Zealand, *Geophys. J. Int.*, *150*, 770–779.

Gerbault, M., S. Henrys, and F. Davey (2003), Numerical models of lithospheric deformation forming the Southern Alps of New Zealand, *J. Geophys. Res.*, *108*(B7), 2341, doi:10.1029/2001JB001716.

Groves, D. I., M. E. Barley, and S. E. Ho (1998), Nature, genesis and tectonic setting of mesothermal gold mineralization in the Yilgarn block, Western Australia, in *The Geology of Gold Deposits, Econ. Geol. Monogr.*, edited by R. Keays, R. Ramsay, and D. Groves, pp 71–85, Society of Economic Geologists, El Paso, TX.

ITASCA (1997), *FLAC³ᴰ (Fast Lagrangian Analysis of Continua in 3 Dimensions)*, ITASCA Consulting Group, Minneapolis, MN.

Jackson, J. A. (2002), Strength of the continental lithosphere: Time to abandon the jelly sandwich?, *GSA Today*, *12*, 4–10.

Jiracek, G. R., V. M. Gonzalez, T. G. Caldwell, P. E. Wannamaker, and D. Kilb (this volume), Seismogenic, electrically conductive, and fluid zones at continental plate boundaries in New Zealand, Himalaya, and California-USA.

Koons, P. O. (1990), Two-sided orogen: Collison and erosion from the sandbox to the Southern Alps, New Zealand, *Geology*, *18*, 679–682.

Koons, P. O., and C. M. Henderson (1995), Geodetic analysis of model oblique collision and comparison to the Southern Alps of New Zealand, *N. Z. J. Geol. Geophys.*, *38*, 545–552.

Koons, P. O., D. Craw, S. C. Cox, P. Upton, A. S. Templeton, and C. P. Chamberlain (1998), Fluid flow during active oblique convergence: A Southern Alps model from mechanical and geochemical observations, *Geology*, *26*, 159–162.

Koons, P. O., R. J. Norris, D. Craw, and A. F. Cooper (2003), Influence of exhumation on the structural evolution of transpressional plate boundaries: An example from the Southern Alps, New Zealand, *Geology*, *31*, 3–6.

Koons, P. O., P. Upton, S. E. Johnson, M. Jessell (2004), Model strain and model fabric development in 3D oblique orogens, *Geol. Soc. Am. Abstr. Programs*, *36*(5), 437.

Leitner, B., D. Eberhart-Philips, H. Anderson, and J. L. Nabelek (2001), A focused look at the Alpine Fault, New Zealand: Seismicity, focal mechanisms, and stress observations, *J. Geophys. Res.*, *106*, 2193–2220.

LeMasurier, W. E., and C. A. Landis (1996), Mantle-plume activity recorded by low-relief erosion surfaces in West Antarctica and New Zealand, *Geol. Soc. Am. Bull.*, *108*, 1450–1466.

Little, T. A., and A. Jones (1998), Seven million years of strike-slip and off-fault deformation on the Awatere Fault, South Island, New Zealand, *Tectonics*, *17*, 285–302.

Little, T. A., S. Cox, J. K. Vry, and G. Batt (2005), Variations in exhumation level and uplift rate along the oblique-slip Alpine fault, central Southern Alps, New Zealand, *Geol. Soc. Am. Bull.*, *117*, 707–723.

Liu, Z., and P. Bird (2006), Two-dimensional and three-dimensional finite element modeling of mantle processes beneath central South Island, New Zealand, *Geophys. J. Int.*, *165*, 1003–1028.

MacKinnon, T. C. (1983), Origin of the Torlesse terrane and coeval rocks, South Island, New Zealand, *Geol. Soc. Am. Bull. 94*, 967–985.

Maggi, A., J. A. Jackson, D. McKenzie, and K. Priestley (2000), Earthquake focal depths, effective elastic thickness, and the strength of the continental lithosphere, *Geology*, *28*, 495–498.

McCaig, A. M., D. M. Wayne, J. D. Marshall, D. Banks, and I. Henderson (1995), Isotopic and fluid inclusion studies of fluid movement along the Gavarnie Thrust, central Pyrenees: Reaction fronts in carbonate mylonites, *Am. J. Sci. 295*, 309– 343.

Molnar, P., H. J. Anderson, E. Audoine, D. Eberhart Phillips, K. Gledhill, E. Klosko, T. V. McEvilly, D. Okaya, M. K. Savage, T. Stern, and F. Wu (1999), Continuous deformation versus faulting through the continental lithosphere of New Zealand, *Science*, *286*, 516–619.

Moore, M., P. England, and B. Parsons (2002), Relation between surface velocity field and shear wave splitting in the South Island of New Zealand, *J. Geophys. Res.*, *107*, 2198, doi:10.1029/2000JB000093.

Mortimer, N. (1993), Geology of the Otago Schist and adjacent rocks. Scale 1:50000, *Institute of Geological and Nuclear Sciences Geological Map 7*, Lower Hutt, New Zealand.

Mortimer, N., F. J. Davey, A. Melhuish, J. Yu, and N. J. Godfrey (2002), Geological interpretation of a deep seismic reflection profile across the Eastern Province and Median Batholith, New Zealand: Crustal architecture of an extended Phanerozoic convergent orogen, *N. Z. J. Geol. Geophys.*, *45*, 349–363.

Mutch, A. R. (1963), Sheet 23-Oamaru. Geological Map of New Zealand, 1:250,000. New Zealand Geological Survey.

Norris, R. J., and A. F. Cooper (2001), Late quaternary slip rates and slip partitioning on the Alpine Fault, New Zealand, *J. Struct. Geol.*, *23*, 507–520.

Norris, R. J., P. O. Koons, and A. F. Cooper (1990), The oblique-convergent plate boundary in the South Island of New Zealand: Implications for ancient collision zones, *J. Struct. Geol.*, *12*, 715–725, 1990

Paterson, M., and F. Luan (1990), Quartzite rheology under geological conditions, in *Deformation Mechanisms, Rheology and Tectonics*, vol. 54, edited by R. J. Knipe and E. H. Rutter, pp. 299–307, *Geol. Soc. Lond. Spec. Publ.*, London.

Ranalli, G. (1995), *Rheology of the Earth*, 2nd ed., Chapman & Hall, London.

Savage, M. K., A. Tommasi, S. Ellis, and J. Chery (this volume), Modeling strain and anisotropy along the Alpine Fault, South Island, New Zealand.

Scherwath, M., T. Stern, F. Davey, D. Okaya, W. S. Holbrook, R. Davies, and S. Kleffmann (2003), Lithospheric structure across oblique continental collision in New Zealand from wide-angle P wave modeling, *J. Geophys. Res.*, *108*(B12), 2566, doi:10.1029/2002JB002286.

Shaw, R. P., and R. D. Morton (1990), Gold mineralization in Lower Cambrian McNaughton Formation, Athabasca Pass, Canadian Rocky Mountains: Structural, mineralogical, and temporal relationships, *Can. J. Earth Sci. 27*, 477–493.

Shelton, G., and J. A. Tullis (1980), Experimental flow laws for crustal rocks, *Trans. Am. Geophys. Union*, *62*, 396.

Sibson, R. H. (1982), Fault zone models, heat flow and the depth distribution of earthquakes in the continental crust of the United States, *Bull. Seismological Soc. Am. 72*, 151–163.

Sibson, R. H. (1996), Structural permeability of fluid driven fault-fracture meshes, *J. Struct. Geol.*, *18*, 1031–1042.

Simpson, G. D., A. F. Cooper, and R. J. Norris (1994), Late Quaternary evolution of the Alpine Fault at Paringa, South Westland, New Zealand, *N. Z. J. Geol. Geophys.*, *37*, 49–58.

Stern, T., S. Kleffmann, D. Okaya, M. Scherwath, and S. Bannister (2001), Low seismic-wave speeds and enhanced fluid pressure beneath the Southern Alps of New Zealand, *Geology*, *29*, 679–682.

Sutherland, R. (1999), Cenozoic bending of New Zealand basement terranes and Alpine Fault displacement: A brief review, *N. Z. J. Geol. Geophys.*, *42*, 295–301.

Tan, L. P., C. H. Chen, T. K. Yeh, and A. Takeuchi (1991), Tectonic and geochemical characteristics of Pleistocene gold deposits in Taiwan, in *Neotectonics and Resources*, edited by M. Jones and J. Cosgrove, pp. 290–305, Bellhaven Press, London.

Teagle, D. A. H., R. J. Norris, and D. Craw (1990). Structural controls on gold-bearing quartz mineralization in a duplex thrust system, Hyde-Macraes Shear Zone, Otago Schist, New Zealand, *Econ. Geol.*, *85*, 1711–1719.

Templeton, A. S., C. P. Chamberlain, P. O. Koons, and D. Craw (1998a), Stable isotopic evidence for mixing between metamorphic fluids and surface-derived waters during recent uplift of the Southern Alps, New Zealand, *Earth Plan. Sci. Lett.*, *154*, 73–92.

Templeton, A. S., D. Craw, P. O. Koons, and C. P. Chamberlain (1998b), Near-surface expression of a young mesothermal gold mineralizing system, Sealy Range, Southern Alps, New Zealand, *Miner. Depos.*, *34*, 163–172.

Tullis, J. (1990), Experimental studies of deformation mechanisms and microstructures in quartzo-feldspathic rocks, in *Deformation Processes in Minerals, Ceramics, and Rocks, Mineral. Soc. Ser.*, vol. 1, edited by D. Barbour and P. Meredith, pp. 190–227, Unwin & Hyman, London.

Tullis, J., and R. A. Yund (1980), Hydrolytic weakening of experimentally deformed Westerly granite and Hale albite rock, *J. Struct. Geol.*, *2*, 439–451.

Upton, P. (1998), Modelling localisation of deformation and fluid flow in a compressional orogen: Implications for the Southern Alps of New Zealand, *Am. J. Sci. 289*, 296–323.

Upton, P., D. Craw, T. G. Caldwell, P. O. Koons, Z. James, P. E. Wannamaker, G. J. Jiracek, and C. P. Chamberlain (2003), Upper crustal fluid flow in the outboard region of the Southern Alps, New Zealand, *Geofluids*, *3*, 1–12.

van Avendonk, H. J. A., W. S. Holbrook, D. Okaya, J. K. Austin, F. Davey, and T. Stern (2004), Continental crust under compression: A seismic refraction study of South Island Geophysical Transect: I, South Island, New Zealand, *J. Geophys. Res.*, *109*, doi:10.1029/2003JB002790.

Wannamaker, P. E., G. R. Jiracek, J. A. Stodt, T. G. Caldwell, V. M. Gonzalez, J. D. McKnight, and A. D. Porter (2002), Fluid generation and pathways beneath an active compressional orogen, the New Zealand Southern Alps, inferred from magnetotelluric data, *J. Geophys. Res.*, *107*, doi:10.1029/2001JB000186.

Wellman, H. W. (1979), An uplift map of the South Island of New Zealand, and a model for uplift in the Southern Alps, in *Origin of the Southern Alps, Bull. R. Soc. N. Z.*, vol. 18, edited by R. I., Walcott and M. M. Cresswell, pp. 13–20.

Wightman, R., and T. A. Little (this volume), Deformation of the Pacific Plate above the Alpine Fault ramp and its relationship to expulsion of metamorphic fluids: An array of backshears.

Willett, S. C., C. Beaumont, and P. Fullsack (1993), Mechanical model for the tectonics of doubly vergent compressional orogens, *Geology*, 21, 371–374.

_____

P. O. Koons, Department of Earth Sciences, University of Maine, Orono, ME 04469, USA.

P. Upton, Geology Department, University of Otago, PO Box 56, Dunedin, New Zealand.

Transpression Models and Ductile Deformation of the Lower Crust of the Pacific Plate in the Central Southern Alps, a Perspective From Structural Geology

Timothy Little[1], Ruth Wightman[1], Rodney J. Holcombe[2], and Matthew Hill[1]

[1]*School of Earth Sciences, Victoria University of Wellington, Wellington, New Zealand.*
[2]*Department of Earth Sciences, University of Queensland, Queensland, Australia.*

In contrast to common transpression models, oblique collision in the central Southern Alps is dominated by translation. Climatic conditions there have led to the development of a two-sided orogen across which the oblique plate motion components are, at most, weakly slip-partitioned. Rapid erosion on the western side of the range localizes deformation to the Alpine fault, and has resulted in deep exhumation along that oblique-reverse structure. Rocks rapidly migrate through the outboard part of the deforming zone, preventing large finite strains, and allowing older fabrics to be preserved through the veil of late Cenozoic deformation (this has an inferred shortening strain of ~30–50%). Lower crustal detachments are a feature of obliquely convergent (transpressive) orogens, including the Southern Alps. As metasediments in the lower crust of the Pacific Plate translate westward across South Island, they are transpressed and thickened. When they impinge upon the Alpine fault, they delaminate from deeper parts of the Pacific Plate along a subhorizontal or gently west-dipping interface, and undergo an oblique-slip backshearing process that tilts them onto the crystalline footwall ramp of that structure. This ramping has been associated with transient penetration of brittle deformation downward into the lower crust to depths of >20 km, and a ductile overprint that has constructively reinforced a pre-existing, vertical foliation while also deflecting it to a steeply SE-dipping attitude. Currently there is little evidence for "boosting" of the dip-slip velocity of the Pacific Plate rocks along the fault as a result of this shearing or as a result of any associated transpressional "extrusion." Final deformation has involved uplift of Pacific Plate crust by localized oblique-reverse ductile shear in the basal, high-strain mylonite zone, together with a more distributed "drag" extending ~2 km structurally higher into the non-mylonitic part of the Alpine Schist.

INTRODUCTION

In the central Southern Alps, the Alpine Schist is an eastward-tilted slab of the Pacific Plate crust that has been up-ramped from depths of 25–30 km in the hangingwall of the Alpine fault. Now exposed on the surface, these tectonites provide a rare opportunity to evaluate models of ductile strain accumulation in the lower crust of transpression zones

A Continental Plate Boundary: Tectonics at South Island, New Zealand
Geophysical Monograph Series 175
Copyright 2007 by the American Geophysical Union.
10.1029/175GM14

by examining rocks that have been naturally deformed within an actve and well-defined geodynamic context. In this paper, we will briefly review several conceptual transpressional strain models and consider their possible application to the San Andreas fault and especially the central Southern Alps of New Zealand. We will attempt to provide a concise (but by no means exhaustive) summary of the main findings and conclusions of recent published studies on the ductile structural geology of the Alpine Schist, especially in the Franz Josef and Fox Glaciers region. Our goal is to extract a geodynamic signal from these rocks that reveals something about the magnitude and nature of late Cenozoic ductile flow in the crust beneath the Southern Alps, to integrate these geological data with key geophysical observations of the SIGHT project, and to derive a new (but hopefully still simple) model for crustal transpression in obliquely convergent, two-sided orogens that is fundamentally underpinned by structural field data. The paper is intended as a brief review of recent geological studies addressing styles of ductile flow in the lower crust of the Pacific Plate in the Southern Alps, and their relationship to patterns of uplift and exhumation in the Southern Alps. It attempts to address the question, "how is transpression be accommodated in the lower crust of continent collision zones, and what does this ductile deformation *do* to the rocks?" It does not present new data, but attempts to synthesize a range of earlier studies under the unifying banner of "transpression." In an attempt to explore the generality of our results, and to understand what may be "different" about the Southern Alps collision zone, we will briefly compare the style of transpression in the crust and mantle of the Southern Alps to that in the much better known example of the central San Andreas fault.

Although independent in its focus and scope, the present paper provides a framework that may be useful to any who read the subsequent one by *Wightman and Little* [this volume]. The latter paper presents a quantitative data set from which a kinematic analysis is undertaken of a serial array of neotectonic shears exposed in a mid-crustal part of the Pacific Plate. These data allow estimation of magnitudes and directions of bulk finite shearing in the Pacific Plate, the number and volume of quartz veins emplaced syntectonically into these shears, and the corresponding volumes of fluid flow associated with this late Cenozoic deformation. They also analyse possible feedbacks between this brittle-ductile deformation, fluid flow in the central Southern Alps, and styles of transpression.

TECTONIC SETTING

The Alpine fault is a dominant structure in South Island, New Zealand, and represents the western edge of the ob-

liquely convergent Australian-Pacific plate boundary (Figure 1a, 1b). This active oblique-reverse fault is bounded to the east by a ~2 km-thick zone of mylonites [*Norris and Cooper*, this volume]. These are the strongly sheared equivalents of the structurally overlying package of non-mylonitic rocks, the Alpine Schist, that are chiefly derived from quartzofeldspathic rocks of the Torlesse Terrane. Metamorphic grade in the Alpine Schist decreases eastward from amphibolite facies near the Alpine fault to prehnite-pumpellyite facies near the Main Divide of the Southern Alps [*Grapes*, 1995]. The mean attitude of the mylonitic foliation near Franz Josef and Fox glaciers suggests that the Alpine fault in the central section of the Southern Alps dips ~40–60° SE [*Sibson et al.*, 1981; *Norris and Cooper*, 1995; *Little et al.*, 2002b], an observation that is consistent with seismic reflection imaging of that structure near Mt Cook from SIGHT [*Kleffman et al.*, 1998]. The relative plate motion across the central section of the Southern Alps is 37 ± 2 mm/yr at an azimuth of $071 \pm 2°$, calculated from the Nuvel-1A global plate model [*DeMets et al.*, 1990, 1994]. Estimates of late-Quaternary slip-rates indicate that two-thirds to three-quarters of the plate motion is accommodated by slip along the Alpine fault [*Norris and Cooper*, 2001]. The remainder of the plate motion must be accommodated by deformation on structures to the east of the Alpine fault. Rock uplift-rates along the fault appear to be in the range of 5–10 mm/yr [*Wellman*, 1979; *Tippet and Kamp*, 1993; *Bull and Cooper*, 1986; *Simpson et al.*, 1994; *Beavan et al.*, 2002; *Little et al.*, 2005], approximately equal to rates of erosion in the Southern Alps, suggesting an approximately steady-state topography in the range [*Adams*, 1981]. The structural field studies on the west flank of the Southern Alps that are reviewed in this paper were largely conducted in the Franz Josef and Fox Glaciers region between Whataroa River, to the north, the location of SIGHT Line 1, and Copland Valley, to the south, the location of SIGHT Line 2 (Figure 1a).

TRANSPRESSION MODELS

Transpression can be defined as a strike-slip deformation that deviates from simple shear by the addition of a coaxial shortening component that is orthogonal to the boundaries of the deformation zone [*Dewey et al.*, 1998]. The obliquity of motion of one block or plate relative to the other block can be specified by the angle (α) between the strike of the zone and the relative motion vector (Figure 2). Although inherently complex, natural and analogue modelling examples of upper crustal fault systems in transpressional settings are relatively well-understood [e.g., *Holdsworth et al.*, 1998, and references therein]. This complexity derives from the three-dimensional arrangement and shape of the fault surfaces, the

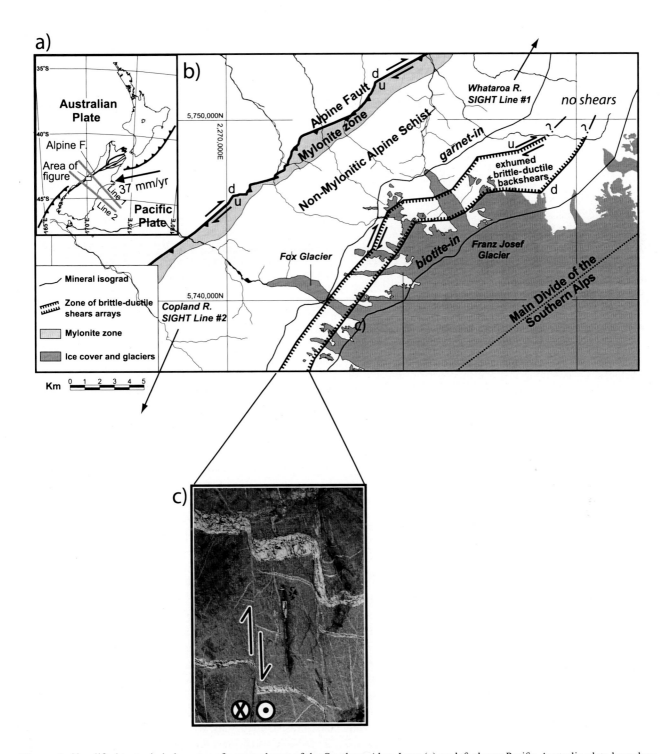

Figure 1. Simplified tectonic index map of a central part of the Southern Alps. Inset (a) on left shows Pacific-Australia plate boundary and relative motion vector and location of SIGHT lines 1 and 2. b) shows distribution of mylonitic and non-mylonitic parts of the "Alpine Schist," selected isograds, and distribution of exhumed brittle-ductile backshears. c) is outcrop photograph of backshears (SW-dipping outcrop face on east side, Franz Josef Glacier) showing their systematic spacing, and ductile to brittle, dextral-oblique shearing of pre-existing quartz veins. Pencil is 15 cm long.

obliquity and variability of slip on the faults, partitioning of slip components between spatially distinct faults of different type (e.g., strike-slip and reverse), and finally from rotation of fault blocks about vertical axes [e.g., *Bayasgalan et al.*, 1999; *Spotila and Sieh*, 2000; *Tavernelli et al.*, 2004]. By contrast, the deformation of the ductilely deforming mid to lower crust at depth beneath such fault systems may be simpler and more continuous [e.g., *Molnar*, 1992; *Teyssier et al.*, 2002]. Because the lower crust is concealed beneath active orogens, its flow pattern is less well understood than styles of upper crustal faulting. Of particular interest is the homogeneity of such flow; for example, to what degree is strike-slip deformation distributed evenly across a wide deforming zone [e.g., *Bourne et al.*, 1998; *Molnar et al.*, 1999]; or focused into narrow shear zones that are dynamically coupled to major upper crustal faults [e.g., *Ellis and Stockhert*, 2004].

Simple kinematic models have been widely used for visualizing and classifying potential patterns of ductile flow in transpression zones [e.g., *Tikoff and Fossen*, 1999]. Besides simplicity, an advantage of these is that they can be used to predict foliation and lineation orientation and development. Rock fabric is something that can be observed directly by a field geologist, or in some cases remotely (for example by seismic anisotropy experiments) so these models are in practice testable. The "vertical stretch" model (Figure 2a) has also been called "simple transpression" [*Sanderson and Marchini*, 1984] and "type B transpression" [*Fossen and Tikoff*, 1998]. Referring to a homogeneously deforming vertical deformation zone, it involves a combination of strike-slip (wrench) shear and a margin-transverse shortening that is volumetrically balanced by vertical thickening. In a "lateral stretch" model, the coaxial shortening component is balanced by elongation parallel to the strike of the deformation zone (Figure 2b). Of course combinations of these two end members are also possible. Like the vertical-stretch model, inclined transpression (Figure 2c) also involves an upwardly directed component of coaxial elongation, but the direction of simple shearing is allowed to be non-horizonal, and the deformation zone itself need not be vertical [*Holdsworth et al.*, 1998; *Lin et al.*, 1998]. Because the shear direction is plunging, whereas the principal axes of coaxial strain (in the simplest case) remain parallel and perpendicular to the zone's strike, the flow symmetry reduces from orthorhombic to triclinic, and the pattern of lineation development can be quite complex. The above models are homogeneous, so they require a free-slip condition at the margins of the zone to accommodate the strain discontinuities that exist against the less-deformed wall rocks. The "heterogeneous transpression" model (Figure 2d) dissipates this boundary shear across a marginal shear zone to obtain a complexly varying strain

field [*Robin and Cruden*, 1994; *Dutton*, 1997]. To a degree, the Alpine fault can be visualized as velocity discontinuity at the edge of the Southern Alps orogen (Figure 2f). In a "vertical stretch" zone, the marginal shearing will be dip-slip and will show opposite shear senses on either side (i.e., Figure 2d). Another type of heterogeneous transpression, widely discussed in the literature especially with respect to upper crustal faulting, is a "slip partitioned" style [e.g., *McCaffrey*, 1992; *Teyssier et al.*, 1995]. In this case, most of the margin-parallel motion is localized into narrow strike-slip simple shear zones (or faults), whereas the contraction, together with any residual strike-slip component, are distributed more or less uniformly across adjoining borderlands (Figure 2e).

STYLES OF LITHOSPHERIC TRANSPRESSION IN SAN ANDREAS FAULT SYSTEM AND SOUTHERN ALPS

Here we will briefly compare the style of transpression in the crust and mantle of the Southern Alps to that in the much better known example of the borderlands region of the central San Andreas fault. The San Andreas fault system in central California illustrates slip partitioning behaviour, at least in the upper crust [*Bloch et al.*, 1993; *Jones et al.*, 1994; *Tikoff and Teyssier*, 1994; *Page et al.*, 1998] (Figure 3a). Plate motion is oblique in central California, at a convergence angle, α, of only ~5°, whereas in the central Southern Alps that angle is ~11–18° [*DeMets et al.*, 1990, 1994; *Cande and Stock*, 2004]. Convergence began in central California at ~3.5 Ma [*Harbert*, 1991; *Page et al.*, 1998], whereas in the Southern Alps it began no later than 5–6.5 Ma [*Sutherland*, 1996; *Walcott*, 1998; *Batt et al.*, 2004], and possibly as early as ~20 Ma [*Cande and Stock*, 2004]. In both plate boundary zones, the relative velocities make an angle (α) of <20° to the margin, corresponding to the "wrench-dominated" case of *Tikoff and Teyssier* [1994]. For this situation, "vertical stretch" strain models predict that the instantaneous direction of maximum elongation will be horizontal, and that its azimuth will make an acute angle of <45° to the margin. The San Andreas is a strongly partitioned system, with 70–75% (~35 mm/yr) of the margin-parallel plate motion taken up as strike-slip on the San Andreas fault; whereas most of the remaining oblique motion, a crustal-scale (vertical-stretch) transpression, is accommodated in a distributed way across the thrust-faulted and wrench-folded borderlands of the Great Valley and Coast Ranges [e.g., *Page et al.*, 1998; *Teyssier and Tikoff*, 1998].

The narrower Southern Alps orogen (Figure 3b) is regarded as one of the world's best examples of an unpartitioned transpressional orogen [e.g., *Beaumont et al.*, 1996; *Koons et al.*, 2003]. Although the Alpine fault, like the San Andreas,

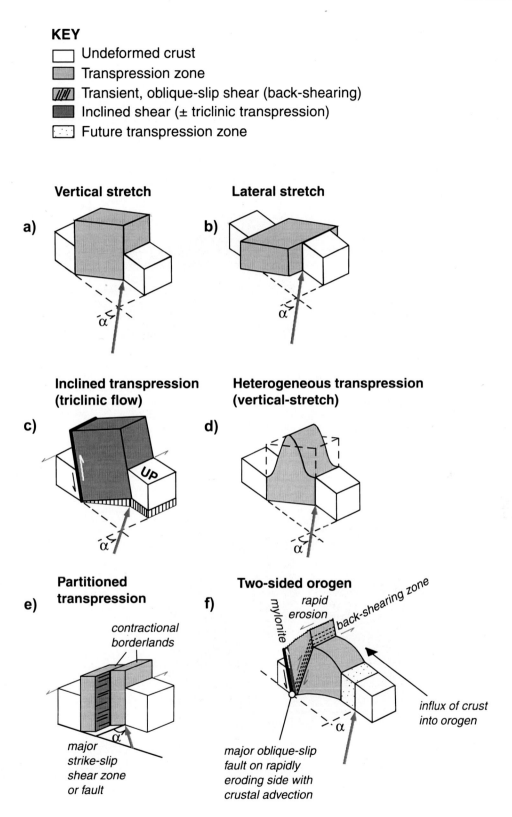

KEY
☐ Undeformed crust
▨ Transpression zone
▨ Transient, oblique-slip shear (back-shearing)
▨ Inclined shear (± triclinic transpression)
▨ Future transpression zone

a) Vertical stretch

b) Lateral stretch

c) Inclined transpression (triclinic flow)

UP

d) Heterogeneous transpression (vertical-stretch)

e) Partitioned transpression

contractional borderlands

major strike-slip shear zone or fault

f) Two-sided orogen

rapid erosion

mylonite

back-shearing zone

influx of crust into orogen

major oblique-slip fault on rapidly eroding side with crustal advection

Figure 2. Simplified kinematic models for bulk transpression [modified after *Dewey*, 1998]. See text for further discussion.

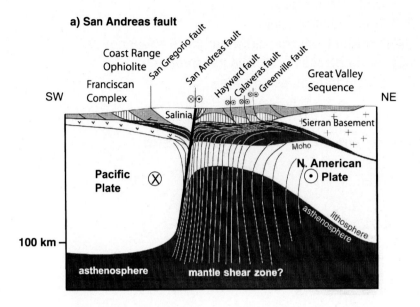

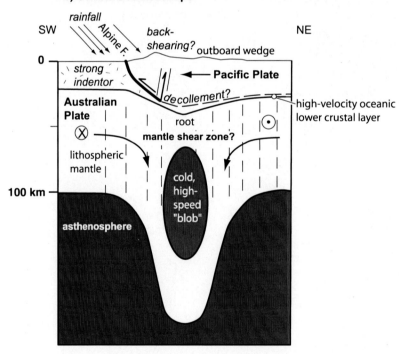

Figure 3. Highly simplified, interpretive cross-sections through the lithosphere across a) a central part of the San Andreas fault system [from *Tessyier and Tikoff,* 1998]; and b) across the central Southern Alps near SIGHT Line 2 [adapted from *Stern et al.,* 2002; *Scherwath et al.,* 2003].

accommodates 70–75% of the margin-parallel plate motion, the Alpine fault also moves with up to 8–12 mm/yr of dip-slip and is oblique-reverse [*Norris and Cooper*, 2001]. There is some uncertainty about whether the azimuth of slip on this fault is exactly parallel to plate motion, as might be expected for a 100% unpartitioned transpressional system, or whether it may be deflected slightly clockwise of that vector (and towards the fault's dip); for example, because of only a partial degree of slip-partitioning [*Sibson et al.*, 1981; *Norris and Cooper*, 1995, 1997; *Walcott*, 1998; *Beavan et al.*, 1999; *Little*, 2004]. Despite this lack of clarity, most would agree that the angle between the fault's slip azimuth and the plate motion vector is small (i.e., <20°). Said differently, the angles α and β in Figure 4 are similar to one another. One factor that has been suggested to contribute to the small degree of slip partitioning in the Southern Alps is the relatively high angle of plate convergence (α), as this geometry will tend to frictionally inhibit slip on vertical strike-slip faults, while favoring oblique-slip on dipping planes [*McCaffrey*, 1992; *Platt*, 1993; *Braun and Beaumont*, 1995; *Teyssier et al.*, 1995]. A more important reason for poorly developed slip partitioning in the Southern Alps, however, is probably the strong asymmetry in rainfall and erosion rates across South Island, New Zealand. In contrast to the more uniformly semiarid San Andreas system, this climatic asymmetry has led to the two-sided structure of the Southern Alps orogen [*Koons*, 1990; 1994; *Koons et al.*, 2003]. On the drier, outboard side of the Southern Alps, a contractional wedge can propagate eastward into previously undeformed crust of the Pacific Plate. By contrast, on the western, inboard side of the range, rapid erosion rates preclude such wedge propagation into the Australian Plate, and instead cause a strong focusing of deformation at or near the Alpine fault [*Norris and Cooper*, 1997; *Koons et al.*, 2003]. In the past 3–5 m.y., slip on this long-lived, SE-dipping structure has accumulated a finite dip-slip of 40–70 km [*Little et al.*, 2005] and has resulted in the exhumation of a 1–2 km-thick zone of amphibolite-facies mylonites [*Norris and Cooper*, 2003]. With time, the resultant upward advection of heat and increased crustal heat flow may have thermally weakened the Alpine fault (shear) zone to such an extent that it can now take up the strike- and dip-slip components of plate motion simultaneously [*Batt and Braun*, 1999; *Koons et al.*, 2003]. In a similar way, any other rheological process that weakens the Alpine fault zone relative to the surrounding crust of the Southern Alps crust, for example locally enhanced pore fluid pressures near the fault [e.g., *Stern et al.*, 2001], might also favor an "unpartitioned" style of oblique convergence [*McCaffrey*, 1992]. Conversely, any rheological process that preferentially weakens the Pacific Plate relative to the Alpine fault might have the opposite effect. If there were high

rates of strike-slip on steep shear planes in the Pacific Plate, the system must be at least partly slip partitioned [see *Wightman and Little*, this volume]. This might be expressed by an "excess" proportion of dip-slip on the Alpine fault (that is, $\beta > \alpha$ in Figure 4).

In transpressive systems, subhorizontal detachments can transfer plate motion between heterogeneously deforming crustal domains and accommodate velocity gradients and rheology contrasts between the lithospheric mantle and upper crust. Detachments in the mid to lower crust are inferred in many parts of the San Andreas system [e.g., *Namson and Davis*, 1988; *Brocher et al.*, 1994; *Jones et al.*, 1994; *Page et al.*, 1998]. Fault motions in the upper crust are complexly distributed across a >200 km width of California and Nevada, while the mantle may be deforming more uniformly and pervasively at depth across a much narrower width of zone (Figure 3a) [e.g., *Molnar*, 1992; *Teyssier and Tikoff*, 1998]. This difference can only be accommodated if there are one or more detachments in the lower crust. In New Zealand, patterns of crustal exhumation and the distribution and style of upper crustal faulting suggest that a detachment similarly exists at ~25–30 km depth in the lower crust of the Southern Alps near the Alpine fault, and that it may extend farther to the east as well (Figures 3b, 4) [*Wellman*, 1979; *Norris et al.*, 1990; *Grapes*, 1995]. This inference is supported by seismic images derived from SIGHT, which delineate a strongly reflective boundary that extends between a moderately SE dipping part of the Alpine fault in the mid crust, and a broad subhorizontal zone of strong but diffuse reflectivity at ~30 km depth in the lower crust [*Davey et al.*, 1998; *Okaya et al.*, 2002; *Stern et al.*, this volume]. The boundary overlies a ~2–10 km-thick, high-velocity (P-wave, ~7 km/s) layer at the base of the crust that has been interpreted to be a relict slab of old oceanic crust [e.g., *Stern et al.*, 2002; *Scherwath et al.*, 2003; *van Avendonk et al.*, 2004].

There appear to be important differences in mantle structure between the two orogens. SKS shear wave splitting data, commonly with delay times of 1–2 seconds, are quite complex (and regionally variable) in California, and have been used to derive several non-unique (often two-layered) models of mantle flow near the San Andreas [*Savage et al.*, 2004]. The interpretation in Figure 3a, based on the data of *Ozalaybey and Savage* [1995], is but one of these, showing a 50–100 km-wide zone of mantle shear exploiting a domain of unusually warm, thin (locally <50 km thick) and weak lithospheric mantle on the eastern side the east of the San Andreas fault [*Teyssier and Tikoff*, 1998]. By contrast, in the central Southern Alps, the plate boundary-deformed lithospheric mantle near the Alpine fault is inferred from the pattern of teleseismic P-wave delays to be unusually cold, strong, and thick (i.e., reaching depths of up to ~170 km)

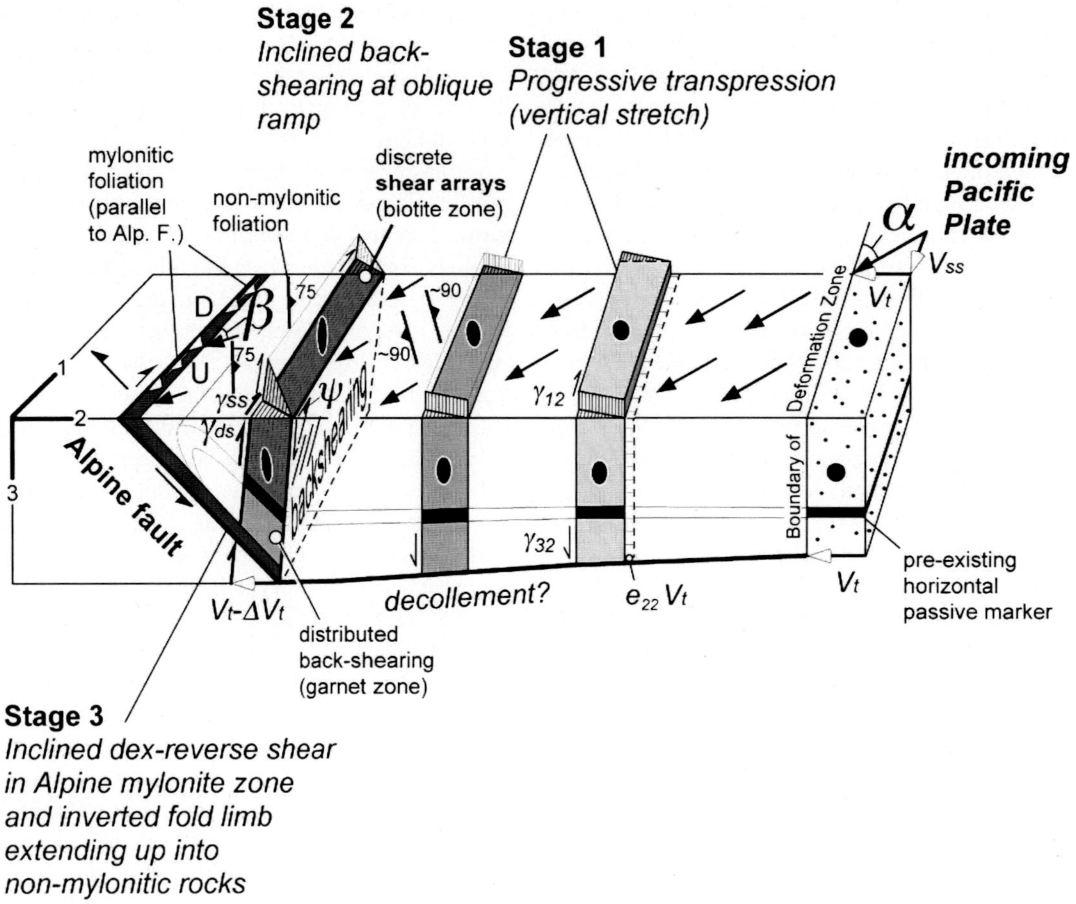

Figure 4. Cartoon illustrating style of obliquely convergent ductile deformation in the Pacific Plate crust of the central Southern Alps. Angle, α, is angle between plate motion vector and the margin; angle β, is angle between Alpine fault slip vector and strike of the Alpine fault. Vt is the convergent component of the relative plate velocity; this is reduced by ΔVt by the time a rock encounters the Alpine fault. e_{22} is the margin-perpendicular shortening rate. γ_{ss} is the strike-slip component of backshearing, γ_{ds} is the dip-slip component of backshearing, and ψ is the pitch angle of the backshearing vector.

and to define a vertically dipping "blob," up to ~80 km wide that is centered beneath the Alpine fault [*Stern et al.*, 2000] (Figure 3b). This "blob" has been interpreted to reflect a distributed lithospheric thickening resulting from >6 m.y. of plate convergence [*Stern et al.*, 2000]. In the Southern Alps, SKS delays are typically 2–3 seconds and define simpler and more regionally coherent azimuthal pattern of fast directions, making a 0–30° counterclockwise angle with respect to the Alpine fault [*Klosko et al.*, 1999; *Molnar et al.*, 1999; *Savage et al.*, 2004]. These SKS splitting results, together with local measurements of Pn anisotropy [*Scherwath et al.*, 2002], suggest that the mantle has a strong seismic anisotropy fabric on both sides of the Alpine fault across a span of South Island that is >200 km wide; there is continuing debate, however, about whether this necessarily reflects the

presence of a wide shear zone in the mantle of Cenozoic age [e.g., *Little et al.*, 2002].

ACCOMMODATING TRANSPRESSION IN THE LOWER CRUST OF THE CENTRAL SOUTHERN ALPS

We will now briefly summarize current knowledge about how bulk transpression has been accommodated in the lower to middle crust of the Pacific Plate (Figure 4). These interpretations are based on a combination of structural geological and (to a lesser extent) geodetic data. We will not attempt to describe or explain the considerable complexity of upper crustal faulting and folding that has been mapped in the central Southern Alps [e.g., *Cox and Findlay*, 1995; *Cox and Sutherland,* this volume], but focus instead on basic aspects of the ductile strain imprint acquired by deeper-level schists on their way to the surface, predominantly in garnet zone rocks cropping out within 7 km of the Alpine fault in the Franz Josef and Fox glaciers region. *Wightman and Little* [this volume] analyze late Cenozoic brittle-ductile structures exposed higher up in the Pacific Plate slab (biotite zone), and address some aspects of the how strain may be accommodated at higher structural levels. As we will see, several of the idealized strain models will have to be modified, or at least combined, for us to obtain a more realistic view of how bulk transpression is accomplished in natural two-sided orogens such as the Southern Alps.

In contrast to the San Andreas system, the oblique-reverse Alpine fault has exhumed a ductilely deformed sample of the lower crust of the Pacific Plate that is accessible for study by structural geologists. Such study is hindered, however, by: 1) inherited, pre-Cenozoic structural fabrics in the multiply deformed Alpine Schist [*Little et al.*, 2002a]; 2) difficulty in dating specific structures through the veil of young cooling ages that have "reset" the spectrum of commonly available thermochronometric systems in these rapidly exhumed rocks [*Batt et al.*, 2000; *Little et al.*, 2005], and 3) a late brittle overprint acquired by some rocks (especially the basal mylonites) during exhumation [*Norris and Cooper*, this volume].

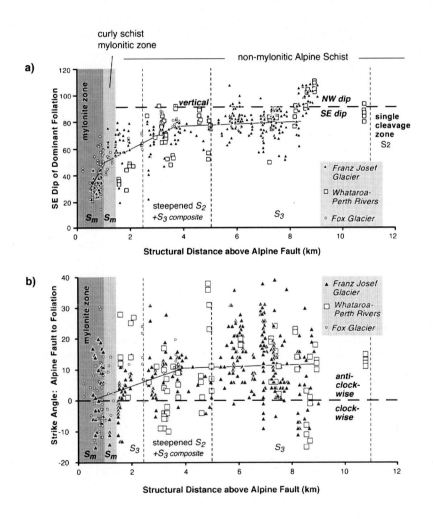

Figure 5. Scatter plots showing changing attitude of dominant foliations in the Alpine Schist as a function of structural distance orthogonal to the Alpine Fault. The "Alpine foliation" (S_3/S_2) is the dominant foliation in the non-mylonitic part of the Alpine Schist; whereas the mylonitic foliation (S_m) is the dominant foliation in the Alpine mylonite zone. a) foliation dip; b) foliation strike angle relative to Alpine fault. From *Little et al.* [2002a]. Fine lines show inferred mean trends. Note apparent kinks in both data sets at ~4 km from the Alpine Fault. The distance of structural data points from the Alpine fault plane were calculated in 3-D by vector algebra from map coordinates and elevations stored in a G.I.S. database.

Vry et al. [2004] dated the high-temperature (Barrovian) regional metamorphism at several localities in the central and northern part of the Alpine Schist as late Cretaceous (~86 Ma) using the Sm-Nd and Lt-Hf techniques on garnet; whereas *Mortimer and Cooper* [2004] obtained ages farther south near Haast of ~71–100 Ma using U-Pb on monazite and a Sm-Nd whole-rock isochron. Outside of the mylonite zone, the dominant (nonmylonitic) foliation in the Alpine Schist in the central Southern Alps strikes NNE at 025–040 at a counterclockwise acute angle to the ~055-striking Alpine Fault (Figure 5a). In garnet- and biotite- zone rocks between the Alpine fault and the Main Divide, the foliation typically dips steeply (70–80°) to the SE (Figure 5b), whereas in chlorite-zone rocks at and to the east of the divide, foliations (and fold axial surfaces) are consistently near-vertical [*Sporli and Lillie*, 1974; *Findlay and Sporli*, 1984; *Findlay*, 1987; *Cox and Sutherland*, this volume]. Textural relationships argue strongly that inception of this steeply dipping "Alpine" foliation coincided with the main phase of regional metamorphism in the Alpine Schist [*Grindley*, 1963; *Findlay*, 1987; *Little et al.*, 2002a], and thus we infer that this steep fabric was inherited into the late Cenozoic Southern Alps orogen [*Little et al.*, 2002a]. Microstructural data argue that a late Cenozoic increment of ductile deformation constructively overprinted and reinforced that pre-existing foliation, and that this overprint was accompanied by renewed growth of micas (e.g., Plate 1a and 1b). Late Cenozoic mineral growth is most obvious in the mylonite zone [see *Vry et al.*, 2004 for evidence of Cenozoic garnet growth there], but also took place in the structurally higher non-mylonitic schists. *Little* [2004] measured 3D grain-shape fabrics in polygonally recrystallised, monocrystalline quartz veins to quantify aspects of the late Cenozoic increments of ductile flow imprinted on these rocks (prior to the quenching in of these fabrics during their unroofing). The results of this detailed study indicate that in the central Southern Alps, the late-incremental ductile deformation caused additional flattening across the NNE-striking, steeply SE dipping foliation planes (e.g., Plates 1a and 1b).

As the Pacific Plate is transported westward into the Southern Alps orogen, ductile deformation in the lower crust initially accumulates as a vertical-stretch transpression (Stage 1 in Figure 4). We infer that the rocks experience both a dextral shearing and a vertical elongation. Evidence for this type of flow includes, first, the geodetically defined pattern of particle velocities on the Pacific Plate, all of which remain parallel to plate motion vector as they decelerate towards the Australian Plate across the >150 km width of the orogen [*Beavan and Haines*, 2001]. This velocity field is uniquely characteristic of "vertical-stretch" transpression; any horizontal elongation would cause a more complex pattern of velocities and their divergence away from parallelism with the plate motion vector [e.g., *Fossen and Tikoff*, 1998]. Kinematic modelling shows that exhumed fabric orientations in the Alpine Schist can be explained by a transpressive lower crustal flow that is just as homogenous and widely distributed as that implied by the present-day gradients in geodetic surface velocities [*Little*, 2004]. Second, is the gradual westward thickening of the crust revealed in seismic sections across South Island from ~27–28 km near the east coast, to 37–44 km near the Alpine fault, and also the gentle (4–5°) westward dip of lower crustal reflectors towards the Alpine fault [*Scherwath et al.*, 2003; *van Avendonk et al.*, 2004]. Although much of the crustal root of the Southern Alps probably lies below the decollement, this thickening trend implies a progressive westward accumulation of transpressional strain (Figure 3b). Third, is the progressive clockwise change in strike ("oroclinal bending") of near-vertical fabric elements (bedding, foliation, and faults) in the Torlesse Terrane and in the Dun Mountain ophiolite body (Maitai Terrane) across a ~50 km wide domain to the east of the current trace of the Alpine Fault [e.g., *Norris*, 1979; *Little et al.*, 2002; *Cox and Sutherland*, this volume]. This deflection from approximately due N to ~040 encompasses only the westernmost (and perhaps youngest) limb of the greater "New Zealand orocline." If these deflections reflect a late Cenozoic increment of dextral shearing along the Alpine Fault, then that deformation must have been dis-

---

Plate 1. (opposite) a) Photograph of foliation-parallel, steeply SE-dipping, quartz vein in biotite-zone Alpine Schist (Burster Range, east of Franz Josef Glacier). The vein shows an inherited (pre-Cenozoic) intersection lineation that was overprinted by extension gashes that are indicative of a steeply SW-pitching direction of late-incremental stretching. The cracks are healed by a new generation of biotite. Arrowed end of strike symbol is NNE. b) Photomicrograph of garnet zone schist that was cut perpendicular to the dominant foliation, and approximately parallel with the down-dip direction. Image (plain light) is arranged with the up-dip direction at the top and the down-dip direction at the bottom, as if looking towards the NE (from Whataroa River). On the left, note microboudinage of biotite porphyroblasts containing internal trails of graphite parallel to the exterior foliation. Late in their deformation history, these inherited elements were microboudinaged with a late growth phase of clear biotite healing between those separated, inclusion-rich fragments. On the right a similar porphyroblast, also with graphite inclusion trails, has been severely "bookshelfed" as a result of up-to-the NW shearing (This shear sense is also supported by oblique grain-shape fabrics in the quartz layers (not shown). Image is modified from *Little et al.* [2001a].

a)

late-incremental cracks

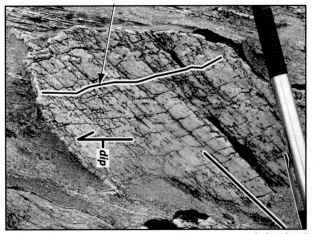

dip

**inherited
lineation**

b)

NW **ramp-related ductile shearing** *SE*

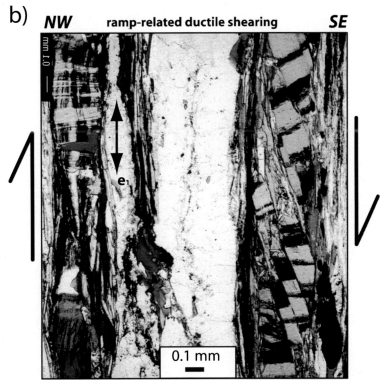

1.0 mm

e₁

0.1 mm

**down-dip extension
(boudinage) & renewed
biotite growth**

**"bookshelfing" of (inherited)
biotite porphyroblast**

tributed across the region to the east of the fault, and it must have been transpressive in nature, combining a dextral shear strain (γ) in the crust of ~0.1; with a fault-perpendicular contraction of ~25% [*Little et al.*, 2002]. Fourth is the consistently oblate (flattened) shape of late-incremental strain imprinted in the Alpine Schist as inferred from strain shadows adjacent to porphyroblasts [*Little et al.*, 2002b]; the grain-shape fabrics in deformed quartz veins [*Little*, 2004]; patterns of porphyroblast rotation [*Holcombe and Little*, 2001], the disposition of conjugate shear bands [*Holm et al.*, 1989; Little et al., 2002b]; near-orthohombic and quartz c-axis lattice preferred orientations [*Hill*, 2005]. Oblate strain shapes are a predictable consequence of vertical-stretch transpression [e.g., *Fossen and Tikoff*, 1998]. Fourth, the microstructures, especially the 3D pattern of rotation of biotite porphyroblasts relative to the foliation [*Holcombe and Little*, 2001], and the orientation of late-incremental extension fractures (Plate 1a) indicate that finite elongation in the biotite and garnet zone schists is at a maximum in a near down-dip direction.

The most significant (and also probably the shortest-lived) phase of strain accumulation in the metasedimentary crust of the Pacific Plate is inferred to have accompanied its upward ramping onto the crystalline footwall of the SE-dipping Alpine fault (Stage 2 in Figure 4). In the central Southern Alps, field observations reveal that an "escalator-like" shear failure process was the final phase of ductile deformation to affect these rocks. We refer to it as "back-shearing," while modelers [e.g., *Waschbusch et al.*, 1998] tend to refer to a similar (often 2D) process as "pro-step-up" shearing. Thermobarometry indicates that this shearing was accompanied by transient embrittlement and fluid pressure fluctuations at deep levels the crust (>20 km depth; ~450°C), including much of the biotite zone [*Wightman et al.*, in review]. At still deeper levels, in the garnet zone, evidence for a pervasively distributed, dextral-oblique back-shearing, in part accommodated by foliation-parallel slip, is confirmed by several types of microstructural shear indicators [*Little et al.*, 2002b] (Plate 1b). The pattern of variable rotation (and "bookshelf" sliding) of elongate biotite porphyroblasts (of different orientations and shapes) relative to the foliation plane records a bulk dip-slip finite shear strain of 0.6 [*Holcombe and Little*, 2001]. By itself, this amount of up-to-the-NW distributed shear would be sufficient to tilt any incoming horizontal markers to a NW dip of ~30°. Other predictable outcomes of the dextral-oblique, ramp-related shearing would be development (or strengthening) of any pre-existing foliations that strike NNE (oblique to the Alpine Fault) and also a change in their dip from originally vertical to steeply SE (Figure 5). The ramp deformation would also introduce a late-incremental stretching direction that pitches SW. This pitch reflects the

contribution of dextral-oblique shearing, as is supported by the microstructural kinematic indicators. All of these fabric attributes are well expressed in the Alpine Schist, both macroscopically (e.g., Plate 1a) and in its quartz grain-shape fabrics [*Little*, 2004]. Recent work by *Hill* [2005] on the lattice-preferred orientation of back-sheared quartz veins involved a vertical-stretch style of inclined transpression (with triclinic symmetry, e.g., Figure 2c).

The clearest and most informative record of back-shearing is preserved in systematically spaced arrays of near-vertical brittle-ductile shears, striking sub-parallel to the Alpine fault and dipping near Fox and Franz Josef Glaciers. These cut and displace older quartz veins embedded in biotite zone schists (Figure 1c). The serial offsets of these markers on well-exposed glaciated outcrops allow quantification of backshearing-related deformation gradients and strain in the Alpine fault's hangingwall [*Wightman and Little*, this volume]. These offsets and fault-surface lineations indicate that the average back-shearing slip-vector pitches ~40° to the SW (angle ψ in Figure 4). The shears thus accommodate significant dextral-slip. This pitch is in agreement with the SW pitch of the late-incremental ductile elongation direction defined by quartz grain-shape fabrics in the wall-rocks to the shears [*Little*, 2004]. *Little et al.* [2005] argue that the spatial restriction of backshearing in this central region may reflect the presence of a sharper ramp-angle on the Alpine fault at depth (Figure 6a). This footwall topology, the origin of which is unknown, may cause the hangingwall to plastically fail there during the up-ramping process (Figure 6b) rather than to undergo a longer-wavelength flexure, as may be happening outside of the central region of the Southern Alps (Figure 6c). Structural and thermochronometric evidence for enhanced vertical rates of uplift and exhumation in the central part of the Southern Alps, and the map-view thinning of bedrock markers in the tilted Pacific Plate slab across that region are both consistent with the Alpine fault steepening in its mean dip by at most ~10–20° across the central region [*Little et al.*, 2005]. There the fault may attain the ~50° mean dip imaged, for example, by *Kleffman et al.* [1998] near Mt Cook.

The final phase of deformation affecting rocks of the Pacific Plate involved their upward translation along the Alpine fault leading to erosional exhumation (Stage 3 in Figure 4). Most of this translation was taken up by brittle slip on the Alpine fault superposed on an older component of ductile shear in the high-strain mylonite zone [*Norris and Cooper*, 2003]. The Alpine mylonite zone can be sub-divided into a basal mylonite zone, which includes interlaced zones of ultramylonite; and a structurally higher zone of coarser-grained and less deformed protomylonitic rocks, locally referred to as the "curly schist" [*Sibson et al.*, 1981]. The late Cenozoic mylo-

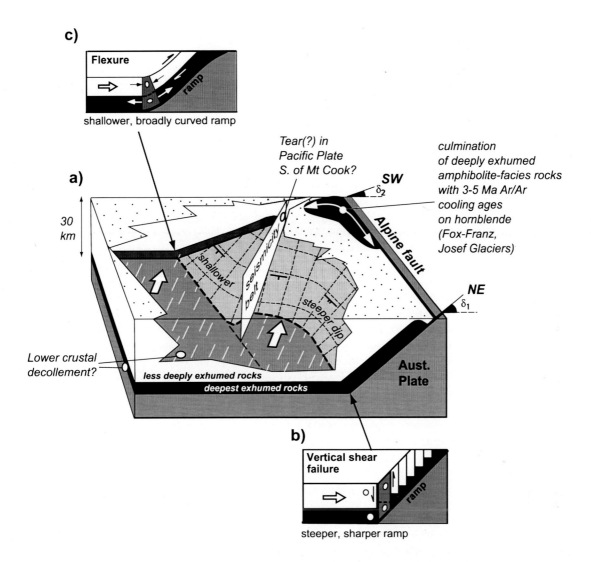

Figure 6. a) Schematic cut-away diagram showing curved topology of the Alpine fault as inferred for the central Southern Alps on the basis of structural geologic, thermochronometric, and seismicity data across that region [after *Little et al., 2005*]. Note inferred relationship between the restraining bend and a maximum in crustal exhumation level near Fox-Franz Josef glaciers and to an EW belt of seismicity near Mt Cook recognized by *Leitner et al.* [2001]. The dip angles δ_1 and δ_2 are inferred to differ by 10–20° and are mean dips at depth rather than surface dips; b) cartoon showing accommodation of up-ramping by shear failure (backshearing); c) cartoon showing accommodation of up-ramping by a broad flexure without shear failure.

nitic foliation (labelled "Sm" in Figure 4) strikes subparallel to the Alpine fault (Figure 5a) and, like that fault, dips at a gentle to moderate angle to the SE at the surface (Figure 5b). The locally variable dip of the mylonitic foliation is controlled by a structural segmentation of the fault, which alternates between oblique-thrust and strike-slip strands near the surface [*Norris and Cooper*, 1995]. Based on a progressive shallowing of foliation (Figure 5a), some ductile shear extends up to ~2 km upward into the non-mylonitic part of the Pacific Plate

[*Little et al.*, 2002b]. In the basal, amphibolite-facies part of the Pacific Plate, another expression of such distributed shear is a reversal in $^{40}Ar/^{39}Ar$ thermochrononometric age trends with respect to hornblende within 3 km of the Alpine fault, implying the presence of the inverted limb of a basal drag fold in the Pacific Plate [*Little et al.*, 2005].

While there is no doubt that the Alpine mylonites have been strongly deformed by dextral-reverse shearing, perhaps reaching finite shear strains of as high as 200–300 [*Norris and Cooper*, 2003], less obvious is the possible additional role of any coaxial strain component during this shearing. *Jiang et al.* [2001] suggested that the locally variable (or steep) NE-pitch of stretching lineations in some parts of the mylonite zone, as measured by *Sibson* [1981], are evidence, at least locally, for an inclined transpressional shearing (with triclinic symmetry). In apparent support of this idea, *Little et al.* [2002b] suggested that conjugate shear bands and poorly expressed stretching lineations that typify mylonite near Franz Josef Glacier indicate a transpressional type of shear. Countering this idea, *Koons et al.* [2003], present a numerical model that could explain the steeply plunging mylonitic lineations by structural overprinting of an original dip-slip simple shear zone by a younger oblique-slip simple shear during an expected thermo-mechanical evolution of the Alpine fault zone from a strongly partitioned (two-fault) system to an unpartitioned (single fault) system.

The role of ductile vertical extension (simple transpression, Figure 2a) in proximity to the Alpine fault is problematic. *Walcott* [1998] inferred that the basal ~5 km of the Pacific Plate (not just the mylonite zone) has been vertically "extruded" near Franz Josef Glacier, like a tectonic aneurysm, as a result of focused erosional exhumation there. The idea of significant vertical extrusion in the Pacific Plate is inconsistent, however, with the lack of any known normal faults on the SE flank of this proposed mass, the undeflected patterns of foliations and lineations across it, and lack of any evidence for higher erosion rates there [*Little et al., 2005; Cox and Sutherland*, this volume]. The trend of slip vectors inferred from in the orientations of S/C fabrics in the mylonite zone [*Little et al.*, 2002b] and from brittle slip-lineations on or near the Alpine fault [*Norris and Cooper*, 1995, 1997] typically overlap with, or are within error of, the contemporary Pacific-Australia plate motion vector; although some of these linear data are scattered towards the down-dip direction of the Alpine fault. This transport direction is not demonstrably rotated towards the dip of the fault, as would be expected on the margins of an extruded mass (in other words, the angles α and β in Figure 4 are similar). On the other hand, *Little* [2004] showed that a dip-parallel elongation is an expectation of the vertical-shear back-shearing process (Figure 4), as would be expected to increase dip-slip velocities on the Alpine fault. Such a deformationally induced "boost" might explain why the best-constrained late Quaternary slip rate along the central part of the Alpine Fault at Gaunt Creek, at >12 mm [*Norris and Cooper*, 2001], exceeds the total plate convergence velocity of 7–10 mm/yr [*DeMets et al.*, 1990, 1994; *Cande and Stock*, 2004], an observation that is at least consistent with geodetic data from the central Southern Alps [*Beavan et al.*, 1999]. Yet, as already stated, there is no direct structural evidence, for example in slip-striations, for

the expected clockwise steepening in the pitch of the slip-vector on the Alpine fault. Clearly more work is required to clarify these contradictory observations and conclusions.

DISCUSSION AND CONCLUSIONS

In contrast to commonly used kinematic models for transpression, oblique collision in the crust of the central Southern Alps is dominated by translation of the Pacific Plate (Figure 2f), and involves deep exhumation on one side. The style of transpression in the semiarid borderlands of San Andreas fault zone in central California, on the other hand, is widely distributed and does not involve exhumation of lower crustal rocks. It is also partitioned. Climatic conditions in New Zealand have led to the development of an asymmetric, two-sided orogen across which the oblique plate motion components are, at most, only weakly slip-partitioned. Rapid erosion rates on the wet, western side of the Southern Alps strongly localize deformation to the Alpine fault zone, and have resulted in deep exhumation along that oblique-reverse fault. Thermal weakening of the Alpine fault zone resulting from this crustal advection has probably contributed to the current poorly "slip-partitioned" nature of the orogen [*Koons et al.*, 2003]. Rocks rapidly migrate through the outboard part of the deforming zone, preventing the accumulation of large finite strains in any one place, and allowing older deformational fabrics to be preserved through the veil of the late Cenozoic deformation, while deforming, and in some cases reinforcing, those older, inherited elements (e.g., Plate 1a). Outside of the mylonite zone, late Cenozoic strain, which is oblate in shape, is inferred to involve only a modest foliation-orthogonal shortening (30–50%), based on the observed deformation of the youngest generation of pre- to syn-tectonic biotite grains [*Little et al.*, 2002b; *Little*, 2004], and is consistent with the predictions of 3D kinematic modelling [*Little*, 2004]. The somewhat higher shortening strains (i.e., 50–60%) measured by Holm et al., [1989] on folded veins, and of >75% by *Little et al.* [2002b] on several microstructural strain gauges dating back to the time of garnet growth are both inferred to include a portion of the older Mesozoic imprint and thus to overestimate the late Cenozoic contribution.

Lower crustal detachments are a typical feature of obliquely convergent orogens, including the better known San Andreas system, where they accommodate an inherently heterogeneous distribution of rheology and transpression styles through the lithosphere. As metasedimentary rocks in the Pacific Plate translate westward across South Island, they are progressively transpressed and thickened in the lower crust. When these rocks impinge upon the SE-dipping Alpine fault at the inboard edge of the orogen, they delaminate from more deep-seated parts of the Pacific Plate, and undergo an oblique-slip backshearing process that tilts them onto the crystalline footwall ramp of that structure. This is associated with a transient penetration of brittle deformation downward into the lower crust. Much of the late-incremental, ductile-to-brittle deformation in deep parts of the Alpine Schist was acquired during up-ramping of the Pacific Plate. The oblique-slip nature of this shearing contributed to non-vertical foliations and lineations in the (already deformed) Alpine Schist.

At present, there is little evidence for "boosting" of the dip-slip velocity of the Pacific Plate rocks along the fault as a result of this shearing or as a result of any transpressional "extrusion." The final phases of deformation involve uplift of Pacific Plate crust by localized oblique-reverse shear extending ~2 km beyond upper contact of high-strain Alpine mylonite zone to cause a shallowing of foliation dip towards parallelism with the Alpine fault and an inversion of the crustal section on the overturned limb of a "drag fold" defined by the $^{40}Ar/^{39}Ar$ isochron structure of hornblende.

ACKNOWLEDGEMENTS

This paper was improved by the reviews of John Dewey and Geoff Batt, and discussions over the years with Richard Norris, Susan Ellis, Tim Stern, Julie Vry, Simon Cox, Brad Ilg, Dick Walcott, and Basil Tikoff.

REFERENCES

Adams, C. J. (1981), Uplift rates and thermal structure in the Alpine fault zone and Alpine schists, Southern Alps, New Zealand, *Geol. Soc. Lond. Spec. Publ.*, *9*, 211–212.

Batt, G. E., S. L. Baldwin, M. Cottam, P. G. Fitzgerald, M. T. Brandon, and T. L. Spell (2004), Cenozoic plate boundary evolution in the South Island of New Zealand: new thermochronological constraints, *Tectonics*, *23*, TC4001, doi:10.1029/2003TC001527.

Batt, G. E., and J. Braun (1999), The tectonic evolution of the Southern Alps, New Zealand: insights from fully thermally coupled dynamic modelling, *Geophys. J. Int.*, *136*, 403–420.

Batt, G. E., J. Braun, B. P. Kohn, and I. McDougall (2000), Thermochronological analysis of the dynamics of the Southern Alps, New Zealand, *Geol. Soc. Am. Bull.*, *112*, 250–266.

Bayasgalan, A., J. Jackson, J.-F. Ritz, and S. Carretier (1999), "Forebergs," flower structures, and the development of large intracontinental strike-slip faults: the Gurvan Bogd fault system in Mongolia, *J. Struct. Geol.*, *21*, 1285–1302.

Beaumont, C., P. J. J. Kamp, J. Hamilton, and P. Fullsack (1996), The continental collision zone, South Island, New Zealand: comparison of geodynamical models and observations, *J. Geophys. Res.*, *101*, 3333–3359.

Beavan, J., Moore, M., Pearson, C., Henderson, M., Parsons, B., Blick, G., Bourne, S., England, P., Walcott, R. I., Darby, D., and Hodgkinson, K. (1999), Crustal deformation during 1994–1998 due to oblique continental collision in the central Southern Alps, New Zealand, and implications for seismic potential of the Alpine fault, *J. Geophys. Res.*, *104*(B11), 25,233–25,255.

Beavan, J., M. Denham, P. Denys, B. Hager, T. Herring, C. Kurnik, D. Matheson, P. Molnar, and C. Pearson (2002), A direct geodetic measurement of the uplift rate of the Southern Alps, *Eos Trans. AGU*, *83*(22), SE31D-01.

Beavan, J., and J. Haines (2001), Contemporary horizontal velocity and strain rate fields of the Pacific-Australian plate boundary through New Zealand, *J. Geophys. Res.*, *107*(B1), 741–770.

Beavan, J., M. Moore, C. Pearson, M. Henderson, B. Parsons, G. Blick, S. Bourne, P. England, R. I. Walcott, D. Darby, and K. Hodgkinson (1999), Crustal deformation during 1994–1998 due to oblique continental collision in the central Southern Alps, New Zealand, and implications for seismic potential of the Alpine fault, *J. Geophys. Res.*, *104*(B11), 25,233–25,255.

Bloch, R. B., R. V. Huene, P. E. Hart, and C. M. Wentworth (1993), Style and magnitude of tectonic shortening normal to the San Andreas fault across the Pyramid Hills and Kettleman hill South Dome, California, *Geol. Soc. Am. Bull.*, *105*, 464–478.

Bourne, S. J., P. C. England, and B. Parsons (1998), The motion of crustal blocks driven by flow of the lower lithosphere and implications for slip rates of continental strike-slip faults, *Nature*, *391*, 655–659.

Braun, J., and C. Beaumont (1995), Three-dimensional numerical experiments of strain partitioning at oblique plate boundaries: implications for contrasting tectonic styles in the southern Coast Ranges, California, and central South Island, New Zealand, *J. Geophys. Res.*, *100*(B9), 18,059–18,074.

Brocher, T. M., J. McCarthy, P. E. Hart, W. S. Holbrook, K. Furlong, T. V. McEvilly, J. A. Hole, and Klempereer (1994), Seismic evidence for a lower crustal detachment beneath San Francisco Bay, *Science*, *265*, 1436–1439.

Bull, W. B., and A. F. Cooper (1986), Uplifted marine terraces along the Alpine fault, New Zealand, *Science*, *234*, 1225–1228.

Cande, S. C., and J. M. Stock (2004), Pacific-Antarctic-Australia motion and the formation of the MacQuarie Plate, *Geophys. J. Int.*, *157*, 399–414.

Cox, S. C., and R. H. Findlay (1995), The Main Divide Fault Zone and its role in the formation of the Southern Alps, *N. Z. J. Geol. Geophys.*, *38*, 489–499.

Cox, S. C., and Sutherland, R. (this volume), Regional geological framework of South Island, New Zealand, and its significance for understanding the active plate boundary.

Davey, F. J., T. Henyey, W. S. Holbrook, D. Okaya, S. T. A., A. Melhuish, S. Henrys, H. Anderson, D. Eberhart-Phillips, T. McEvilly, R. Urhammer, F. Wu, G. R. Jiracek, P. E. Wannamaker, G. Caldwell, and N. Christensen (1998), Preliminary results from a geophysical study across a modern continent-continent collisional plate boundary— The Southern Alps, New Zealand, *Tectonophysics*, *288*, 221–235.

DeMets, C., R. G. Gordon, D. F. Argus, and S. Stein (1990), Current plate motions, *Geophys. J. Int.*, *101*, 425–478.

DeMets, C., R. G. Gordon, D. F. Argus, and S. Stein (1994), Effect of recent revisions to the geomagnetic reversal time scale on estimates of current plate motions, *Geophys. Res. Lett.*, *21*, 2191–2194.

Dewey, J. F., R. E. Holdsworth, and R. A. Strachan (1998), Transpression and transtension zones, in *Continental Transpressional and Transtensional Tectonics*, edited by R. E. Holdsworth, R. A. Strachan, and J. F. Dewey, pp. 1–14, Geological Society of London, London.

Dutton, B. J. (1997), Finite strains in transpression zones with no boundary slip, *J. Struct. Geol.*, *19*, 1189–1200.

Ellis, S., and B. Stockhert (2004), Imposed strain localisation in the lower crust on seismic timescales, *Earth Planets Space*, *56*, 1103–1109.

Findlay, R. H. (1987), Structure and interpretation of the Alpine schists in Copland and Cook River Valleys, South Island, New Zealand, *N. Z. J. Geol. Geophys.*, *30*, 117–138.

Findlay, R. H., and K. B. Sporli (1984), Structural Geology of the Mt. Cook Range and Main Divide, Hooker Valley region, New Zealand, *N. Z. J. Geol. Geophys.*, *27*, 257–276.

Fossen, H., and B. Tikoff (1998), Extended models of transpression and transtension, and application to tectonic settings, *Geol. Soc. Lond. Spec. Publ.*, *135*, 15–33.

Grapes, R. H. (1995), Uplift and exhumation of Alpine schist, Southern Alps, New Zealand: thermobarometric constraints, *N. Z. J. Geol. Geophys.*, *38*, 525–533.

Grindley, G. W. (1963), Structure of the Alpine Schists of South Westland, Southern Alps, New Zealand, *N. Z. J. Geol. Geophys.*, *6*, 872–930.

Harbert, W. (1991), Late Neogene relative motions of the Pacific and North American plates, *Tectonics*, *10*, 1–16.

Hill, M. P. (2005), Evolution of quartz and calcite microstructures exhumed from deep brittle-ductile shear zones in the Southern Alps of New Zealand, M.Sc. thesis, Victoria University of Wellington, Wellington, New Zealand.

Holcombe, R. J., and T. A. Little (2001), A sensitive vorticity gauge using rotated porphyroblasts, and its application to rocks adjacent to the Alpine Fault, New Zealand, *J. Struct. Geol.*, *23*, 979–990.

Holdsworth, R. E., R. A. Strachan, and J. F. Dewey (1998), *Continental Transpressional and Transtensional Tectonics*, Geological Society of London, London.

Holm, D. K., R. J. Norris, and D. Craw (1989), Brittle and ductile deformation in a zone of rapid uplift: central Southern Alps, New Zealand, *Tectonics*, *8*, 153–168.

Jiang, D., S. Lin, and P. F. Williams (2001), Deformation path in high-strain zones, with reference to slip-partitioning in transpressional plate-boundary regions, *J. Struct. Geol.*, *23*, 991–1005.

Jones, D., R. Graymer, M. wang, T. V., and A. Lomax (1994), Neogene transpressive evolution of the California Coast Ranges, *Tectonics*, *13*, 561–574.

Kleffman, S., F. Davey, A. Melhuish, D. Okaya, and T. Stern (1998), Crustal structure in the central South Island, New Zealand, from the Lake Pukaki seismic experiment, *N. Z. J. Geol. Geophys.*, *41*, 39–49.

Klosko, E., F. Wu, H. Anderson, D. Eberhart-Phillips, T. McEvilly, E. Audoine, M. K. Savage, and K. Gledhill (1999), Upper mantle anisotropy in the New Zealand region, *Geophys. Res. Lett.*, *26*, 1497–1500.

Koons, P. O. (1990), Two-sided orogen: collision and erosion from the sand-box to the Southern Alps of New Zealand, *Geology*, *18*, 679–682.

Koons, P. O. (1994), Three-dimensional critical wedges: tectonics and topography in oblique collisional orogens, *J. Geophys. Res.*, *99*, 12,301–12,315.

Koons, P. O., R. J. Norris, D. D. Craw, and A. F. Cooper (2003), Influence of exhumation on the structural evolution of transpressional plate boundaries: an example from the Southern Alps of New Zealand, *Geology*, *31*, 3–6.

Leitner, B., D. Eberhart-Phillips, H. Anderson, and J. L. Nabelek (2001), A focused look at the Alpine fault, New Zealand: seismicity, focal mechanisms, and stress observations, *J. Geophys. Res.*, *106*(B2), 2193–2220.

Lin, S., D. Jiang, and P. F. Williams (1998), Transpression (or transtension) zones of triclinic symmetry: natural example and theoretical modelling, *Geol. Soc. Lond. Spec. Publ.*, *135*, 41–57.

Little, T. A. (2004), Transpressive ductile flow and oblique ramping of lower crust in a two-sided orogen: insight from quartz grain-shape fabrics near the Alpine Fault, New Zealand, *Tectonics*, *23*, TC2013, doi:10.1029/2002TC0011456.

Little, T. A., S. Cox, J. K. Vry, and G. E. Batt (2005), Variations in exhumation level and uplift-rate related to oblique-slip ramp geometry, Alpine fault, central Southern Alps, New Zealand, *Geol. Soc. Am. Bull.*, *117*(4), 707–723.

Little, T. A., R. J. Holcombe, and B. R. Ilg (2002a), Ductile fabrics in the zone of active oblique convergence near the Alpine Fault, New Zealand: identifying the neotectonic overprint, *J. Struct. Geol.*, *24*, 193–217.

Little, T. A., R. J. Holcombe, and B. R. Ilg (2002b), Kinematics of oblique continental collision inferred from ductile microstructures and strain in mid-crustal Alpine Schist, central South Island, New Zealand, *J. Struct. Geol.*, *24*, 219–239.

Little, T. A., M. K. Savage, and B. Tikoff (2002), Relationship between crustal finite strain and seismic anisotropy in the mantle, Pacific-Australia plate boundary zone, South Island, New Zealand, *Geophys. J. Int.*, *151*, 106–116.

McCaffrey, R. (1992), Oblique plate convergence, slip vectors, and forearc deformation, *J. Geophys. Res.*, *97*, 8905–8915.

Molnar, P. (1992), Brace-Goetze strength profiles, the partitioning of strike-slip and thrust faulting at zones of oblique convergence, and the stress-heat flow paradox of the San Andreas fault, in *Fault Mechanics and Transport Properties of Rock*, edited by B. Evans and T. F. Wong, Academic Press, London, pp. 435–459.

Molnar, P., H. J. Anderson, E. Audoine, D. Eberhart-Phillips, K. R. Gledhill, E. R. Klosko, T. V. McEvilly, D. Okaya, M. K. Savage, T. Stern, and F. T. Wu (1999), Continuous deformation versus faulting through the continental lithosphere of New Zealand, *Science*, *286*, 516–519.

Mortimer, N., and A. F. Cooper (2004), U-Pg ages and Sm-Nd ages from the Alpine Schist, New Zealand, *N. Z. J. Geol. Geophys.*, *47*, 21–28.

Namson, J. S., and T. L. Davis (1988), Seismically active fold and thrust belt in the San Joaquin Valley, central California, *Geol. Soc. Am. Bull.*, *100*, 257–273.

Norris, R. J., A geometrical study of finite strain and bending in the South Island, in *The Origin of the Southern Alps*, vol. 18, edited by R. I. Walcott and M. M. Cresswell, pp. 21–28.

Norris, R. J., and A. F. Cooper (1995), Origin of small-scale segmentation and transpressional thrusting along the Alpine fault, New Zealand, *Geol. Soc. Am. Bull.*, *107*, 231–240.

Norris, R. J., and A. F. Cooper (1997), Erosional control on the structural evolution of a transpressional thrust complex on the Alpine Fault, New Zealand, *J. Struct. Geol.*, 19, 1323–1342.

Norris, R. J., and A. F. Cooper (2001), Late Quaternary slip rates and slip-partitioning on the Alpine fault, New Zealand, *J. Struct. Geol.*, *23*, 507–520.

Norris, R. J., and A. F. Cooper (2003), Very high strains recorded in mylonites along the Alpine fault, New Zealand: implication for the deep structure of plate boundary faults, *J. Struct. Geol.*, *25*, 2141–2158.

Norris, R. J., and A. F. Cooper (this volume), The Alpine Fault, New Zealand: surface geology and field relationships.

Norris, R. J., P. O. Koons, and A. F. Cooper (1990), The obliquely convergent plate boundary in the South Island of New Zealand: implications for ancient collision zones, *J. Struct. Geol.*, *12*, 715–726.

Okaya, D., S. N. Henrys, and T. Stern (2002), Double-sided onshore-offshore seismic imaging of a plate boundary: "supergathers" across, South Island, New Zealand, *Tectonophysics*, *355*(1–4), 247–263.

Ozalaybey, S., and M. K. Savage (1995), Shear-wave splitting beneath the western United States in relation to plate tectonics, *J. Geophys. Res.*, *100*, 18,135–18,149.

Page, B. M., G. A. Thompson, and R. G. Coleman (1998), Late Cenozoic tectonics of the central and southern Coast Ranges of California, *Geol. Soc. Am. Bull.*, *110*, 846–876.

Platt, J. P. (1993), Mechanics of oblique convergence, *J. Geophys. Res.*, *98*, 16,239–16,256.

Robin, P.-Y. F., and A. R. Cruden (1994), Strain and vorticity patterns in ideally ductile transpression zones, *J. Struct. Geol.*, *16*, 447–66.

Sanderson, D. J., and R. D. Marchini (1984), Transpression, *J. Struct. Geol.*, *6*, 449–458.

Savage, M. K., K. M. Fischer, and C. E. Hall (2004), Strain modelling seismic anisotropy and coupling at strike-slip boundaries: applications in New Zealand and the San Andreas Fault, in *Vertical Coupling and Decoupling in the Lithosphere*, edited by J. Grocott, B. Tikoff, K. J. W. McCaffrey, and G. Taylor, The Geological Society, London, pp. 9–40.

Scherwath, M., T. Stern, F. Davey, D. Okaya, W. S. Holbrook, R. Davies, and S. Kleffman (2003), Lithospheric structure across oblique continental collision in New Zealand from wide-angle P wave modelling, *J. Geophys. Res.*, *108*(B12), 2566, doi:10.1029/2002JB002286.

Scherwath, M. A., A. Melhuish, T. Stern, and P. Molnar (2002), Pn anisotropy and distributed upper mantle deformation associated with a continental transform fault, *Geophys. Res. Lett.*, *29*(8), 1175, doi:10.1029/2001GL014179.

Sibson, R. H., S. H. White, and B. K. Atkinson (1981), Structure and distribution of fault rocks in the Alpine Fault Zone, New Zealand, *Geol. Soc. Lond. Spec. Publ.*, *9*, 197–210.

Simpson, G. D., A. F. Cooper, R. J. Norris, and I. M. Turnbull (1994), Late Quaternary evolution of the Alpine Fault Zone at Paringa, South Westland, New Zealand, *N. Z. J. Geol. Geophys.*, *37*, 49–58.

Sporli, K. B., and A. R. Lillie (1974), Geology of the Torlesse Supergroup in the northern Ben Ohau Range, Canterbury, New Zealand, *N. Z. J. Geol. Geophys.*, *17*, 115–141.

Spotila, J. A., and K. Sieh (2000), Architecture of transpressional thrusting in the San Bernardino Mountains, southern California, from deformation of a deeply weathered surface, *Tectonics*, *19*, 589–615.

Stern, T., S. Kleffman, D. Okaya, M. Scherwath, and S. Bannister (2001), Low seismic-wave speeds and enhanced fluid pressure beneath Southern Alps of New Zealand, *Geology*, *29*, 679–682.

Stern, T., P. Molnar, D. Okaya, and D. Eberhart-Phillips (2000), Teleseismic P wave delays and modes of shortening the mantle lithosphere beneath South Island, New Zealand, *J. Geophys. Res.*, *105*(B9), 21,615–21,631.

Stern, T., D. Okaya, and M. Scherwath (2002), Structure and strength of a continental transform from onshore-offshore seismic profiling of South island, New Zealand, *Earth Planets Space*, *54*, 1011–1019.

Stern, T., D. Okaya, S. Kleffmann, M. Scherwath, S. Henrys, and F. Davey (this volume), Geophysical exploration and dynamics of the Alpine fault zone.

Sutherland, R. (1996), Transpressional development of the Australia-Pacific boundary through southern South Island, New Zealand: constraints from Miocene-Pliocene sediments, Waiho-1 borehole, South Westland, *N. Z. J. Geol. Geophys.*, *39*, 251–264.

Tavernelli, E., R. E. Holdsworth, P. Clegg, R. R. Jones, and K. J. W. McCaffrey (2004), The anatomy and evolution of a transpressional imbricate zone, Southern Uplands, Scotland, *J. Struct. Geol.*, *26*, 1341–1360.

Teyssier, C., and B. Tikoff (1998), Strike-slip partitioned transpression of the San Andreas fault system: a lithospheric-scale approach., in *Continental Transpression and Transtensional Tectonics*, edited by R. E. Holdsworth, R. A. Strachan, and J. F. Dewey, pp. 143–158.

Teyssier, C., B. Tikoff, and M. Markley (1995), Oblique plate motion and continental tectonics, *Geology*, *23*, 447–450.

Teyssier, C., B. Tikoff, and J. Weber (2002), Attachment between brittle and ductile crust at wrenching plate boundaries, *EUG Spec. Publ. Ser.*, *1*, 119–144.

Tikoff, B., and H. Fossen (1999), Three-dimensional reference deformations and strain facies, *J. Struct. Geol.*, *21*, 1497–1512.

Tikoff, B., and C. Teyssier (1994), Strain modeling of displacement-field partitioning in transpressional orogens, *J. Struct. Geol.*, *16*, 1575–1588.

Tippett, J. M., and Kamp, P. J. J. (1993), Fission track analysis of late Cenozoic vertical kinematics of continental Pacific crust, South Island, New Zealand, *J. Geophys. Res.*, *98*, 16,119–16,148.

van Avendonk, H. J. A., W. S. Holbrook, D. Okaya, J. K. Austin, F. Davey, and T. Stern (2004), Continental crust under compression: a seismic refraction study of South Island Geophysical Transect I, South Island, New Zealand, *J. Geophys. Res.*, *109*, B06302, doi:10.1029/2003JB002790.

Vry, J. K., R. Maas, T. A. Little, D. Phillips, R. Grapes, and M. Dixon (2004), Zoned (Cretaceous and Cenozoic) garnets and the timing of high grade metamorphism, Southern Alps, New Zealand, *J. Metamorph. Geol.*, *22*(3), 137–157, doi:10.1111/j.1525-1314.2004.00504.

Walcott, R. I. (1998), Modes of oblique compression: Late Cenozoic tectonics of the South Island of New Zealand, *Rev. Geophys.*, *36*, 1–26.

Waschbusch, P., G. Batt, and C. Beaumont (1998), Subduction zone retreat and recent tectonics of the South Island of New Zealand, *Tectonics*, *17*, 267–284.

Wellman, H. W. (1979), An uplift map for the South Island of New Zealand, and a model for the uplift of the Southern Alps, *R. Soc. N. Z. Bull.*, *18*, 13–20.

Wightman, R., and T. A. Little (this volume), Deformation of the Pacific Plate above the Alpine Fault ramp and its relationship to expulsion of metamorphic fluids: an array of backshears.

Wightman, R., T. A. Little, S. B. Baldwin, and J. W. Valley (in review), Stress, fluid pressure cycling and transient deep embrittlement of the lower crust recorded in a paleo-brittle-ductile transition zone, Southern Alps, New Zealand, *J. Geophys. Res.*

M. Hill, T. Little, and R. Wightman, School of Earth Sciences, Victoria University of Wellington. (timothy.little@vuw.ac.nz)

R. J. Holcombe, Department of Earth Sciences, University of Queensland, Queensland, Australia 4072.

Modeling Strain and Anisotropy Along the Alpine Fault, South Island, New Zealand

M. K. Savage[1], A. Tommasi,[2] S. Ellis[3], and J. Chery[2]

Near-surface deformation between the Australian and Pacific plates in South Island, New Zealand is concentrated in a narrow zone marked by the Alpine Fault, but strong and widespread anisotropy inferred across New Zealand from shear wave splitting suggests diffuse deformation at mantle depths. To constrain this interpretation, we calculate temperature- and stress-dependent strain fields, crystal preferred orientation, anisotropy and resultant shear-wave splitting beneath a lithospheric fault deforming by either pure strike-slip or transpression. In pure strike-slip experiments, strain localizes in the high temperature regions under the thick continental landmass. Under the oceanic regions, which have thin crust, anisotropy is weak (delay times (dt) <0.5 s for a 100 km thick lithosphere). Under the continents, dt increases to 1.0 s for 100 km and 2.3 s for 200 km thick lithosphere. Dt saturates before 15 My (525 km displacement), at which point the polarizations of the first arriving shear waves (ϕ) are 25° from fault-parallel, similar to measurements in southern South Island. Further strain does not increase dt, but ϕ rotates to 15° from fault-parallel for displacements > 1400 km. In transpression experiments that have initial structure based on estimates of the present crustal thickness and temperature, the cold root inhibits strain beneath the fault and leads to anisotropy patterns that do not explain the observations. This suggests that (a) a weak zone in the mantle lithosphere is needed to explain present-day deformation; and (b) most of the anisotropy measured may be "frozen in" from strike-slip deformation that occurred before the present mantle root developed.

1. INTRODUCTION

In central South Island, New Zealand, over 70% of the relative motion between the Pacific and Australian plates is currently accommodated along the Alpine Fault [e.g., *Norris*

[1]Institute of Geophysics, School of Earth Sciences, Victoria University of Wellington, New Zealand
[2]University de Montpellier II, France
[3]GNS Science, New Zealand

A Continental Plate Boundary: Tectonics at South Island, New Zealand
Geophysical Monograph Series 175
10.1029/175GM15

and Cooper, 2001, and references therein]. Yet strong anisotropy inferred across the width of New Zealand from shear wave splitting [*Molnar et al.*, 1999] and *Pn* phases [*Baldock and Stern*, 2005] has been cited as evidence that the mantle undergoes diffuse deformation over length-scales of hundreds of kilometers (Figure 1). The anisotropy is assumed to be generated in the lithospheric mantle. Modeling of the deformation of a thin elastic lid above a broad shear zone embedded in an elastic half space [*Moore et al.*, 2002] can fit both the surface deformation measured by GPS and the shear wave splitting data, although the fit is quite non-unique [e.g., *Ellis et al.*, 2006]. Furthermore, crustal deformation has been measured away from the fault, and the close similarity between the fast polarizations from shear-wave splitting and the maximum extension determined structurally has

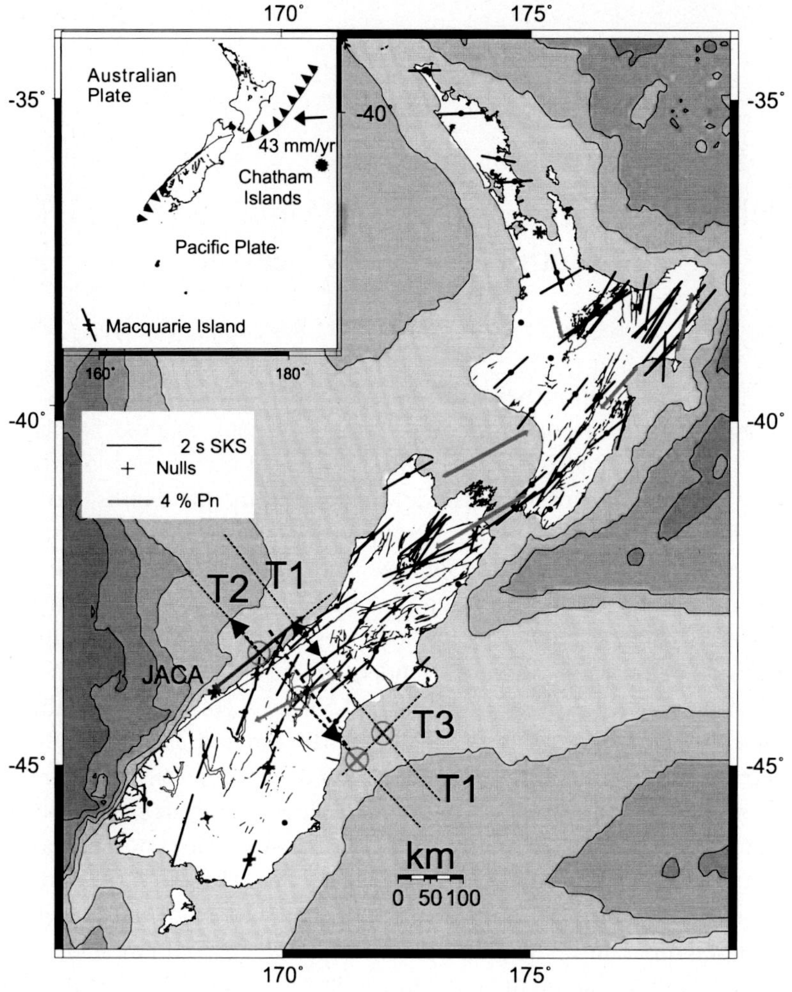

Figure 1. Seismic anisotropy measurements in New Zealand from Pn and SKS splitting, with bathymetry in 1000 m contours. Small filled circles: seismic stations on which SKS splitting has been analyzed. Bars: Positive measurements of splitting. Orientation of the line represents average fast polarization measured at the station, and length of the line is proportional to the average delay time as given in the scale. Crosses: null measurements (in which no splitting was observed), with the possible orientations that could yield no splitting. From *Duclos* [2005] and references therein. Gray arrows: Pn anisotropy, oriented along the line in which the fastest speed was measured. Length is proportional to anisotropy percentage as in the scale. Gray circles enclose the intersections of the lines used for the Pn measurements reported by *Baldock and Stern* [2005]; *Bourguignon et al.* [2007] and *Scherwath et al.* [2002]. Cross within the circle represents the lack of anisotropy measured by *Baldock and Stern* [2005] at the intersection of Lines T1 and T3. Other Pn measurements from regional earthquakes according to *Smith and Ekstrom* [1999]. Large arrows are the inferred limits of Pn anisotropy determined by the change in Pn along the SIGHT lines T1 and T2 [*Scherwath et al.*, 2002; *Baldock and Stern*, 2005] Inset: Regional plate tectonic setting of NZ, and splitting measurements at nearby oceanic islands. The line on South Island is the Alpine Fault. Arrow represents present Pacific plate motion vector relative to the Australian Plate [*Kreemer et al.*, 2003]. Chatham Islands splitting measurements were all null, suggesting no anisotropy in the horizontal plan. Color version of this figure is on CDROM which accompanies this volume.

been used to suggest that there is strong coupling between the crust and the mantle, with coherent deformation between the two [*Little et al.*, 2002]. Worldwide studies also suggest that strong crust-mantle coupling may be widespread beneath major strike-slip faults [*Silver*, 1996; *Tommasi et al.*, 1999].

Most measurements of seismic anisotropy come from shear-wave splitting of SKS phases. Shear wave splitting, or birefringence, occurs when shear waves impinge on an anisotropic medium. One component travels faster than the other, nearly orthogonal component. The polarization of the first shear arrival yields the fast orientation, and the time delay between the two components gives information about the strength of anisotropy combined with the length of the anisotropic path. SKS phases are difficult to use to constrain the depth extent of anisotropy, since the measured splitting may result from anisotropy anywhere along the path between the core mantle boundary and the surface [e.g., *Savage*, 1999]. The strength of anisotropy may vary with depth. Variations in fast orientations based on shear wave splitting measurements from local earthquakes in northern South Island, suggest that the uppermost mantle anisotropy is more localized than the SKS results suggest [*Audoine et al.*, 2000]. Also, coherence of splitting measurements between North and South Islands hints that some of the anisotropy beneath New Zealand results from asthenospheric flow rather than lithospheric strain.

To better understand the origin of the anisotropy, strain can be calculated using geodynamic models, and resultant shear wave splitting computed to compare to observations [*Blackman et al.*, 1996]. In a previous study, we used two-dimensional geodynamic models with simple linear rheologies and kinematic boundary conditions to calculate strain; we then assumed simplified relations between the strain and anisotropy to evaluate the effects of strike-slip deformation on splitting measurements in California and New Zealand [*Savage et al.*, 2004]. The New Zealand seismic anisotropy could be explained by strong coupling between the crust and mantle. These models therefore supported the suggestions of *Moore et al.*, [2002] and *Little et al.*, [2002] that the change in fast polarisations from between ca.30° to fault-parallel in southern South Island to fault-parallel in central South Island are caused by smaller strain to the south. However, the relationship between strain and delay time assumed in the modeling did not explain the constant delay times across the island, unless the delay times saturated at smaller strains than assumed.

Here we use more sophisticated models of both strain and anisotropy. Strain is calculated using an elasto-viscoplastic finite element model with a temperature- and stress-dependent viscosity [*Chéry et al.*, 2004], in which full mechanical evolution is allowed. We then calculate the crystal preferred orientations formed in olivine-pyroxene aggregates in response to this strain field using a polycrystal plasticity approach [*Tommasi et al.*, 2000] and the resulting elastic constants [*Mainprice*, 1990]. Ray-tracing through these models allows evaluation of the resulting shear wave splitting. We will show that these models better predict the anisotropy in South Island, in the sense that the delay times saturate at smaller strains than previously expected. We will further show that the formation of a cold lithospheric root, which has been interpreted from thermal modeling [*Shi et al.*, 1996] and from measurements of early arrivals from teleseismic earthquakes with paths traveling through the root

region [*Stern et al.*, 2000; *Kohler and Eberhart-Phillips*, 2002], inhibits strain and anisotropy formation beneath the fault. Therefore, it is likely that the anisotropy measured in the vicinity of the Alpine Fault was formed during the strike-slip deformation period, before compressional deformation formed a cool root.

2. METHOD

Geodynamic Modeling

We use the 3D finite element (FE) code Adeli3D (version 3d4, [*Chéry et al.*, 2004]) to investigate both the long-term evolution of the Southern Alps and the present deformation. Adeli3D solves for the quasi-static mechanical behavior of the lithosphere using an explicit scheme based on the Dynamic Relaxation Method [*Cundall*, 1988]. The model is not thermally coupled, but an initial temperature field is prescribed, which is then advected along with material points. We approximate the 3D plate boundary deformation in central South Island by prescribing a thin cross-sectional slice perpendicular to the Alpine Fault as an out-of-plane model, so that velocities are assumed not to vary along-strike, even though they change with distance and depth across-strike. This approximation allows us to use an unstructured mesh with an average element length-scale of 5 km.

Both crust and mantle lithosphere are assumed to have an elasto-visco-plastic rheology with constant elastic strength, pressure-dependent yield, and thermally-activated viscous creep with either linear or non-linear viscous flow laws, (parameter values described in Tables 1 and 2). Frictional behavior follows a standard Drucker-Prager yield relationship:

$$f(\sigma) = J_2(\sigma) - \alpha[\overline{\sigma} + \frac{c}{\tan\phi_0}] < 0$$

$$J_2(\sigma) = \sqrt{\frac{3}{2}}\sqrt{dev\sigma : dev\sigma}$$

$$\alpha = \frac{6\sin\phi_0}{3 - \sin\phi_0} \qquad (1)$$

$$\overline{\sigma} = -\frac{1}{3}tr(\sigma)$$

where σ is the stress tensor, $\overline{\sigma}$ is the mean stress, *dev* is the deviatoric part of a tensor, and the colon represents the contracted product. ϕ_0 is internal angle of friction, and c is cohesion [*Leroy and Ortiz*, 1989]. The dilatancy angle is set to zero, therefore leading to a non-associated plastic flow

Table 1. Material properties used in the experiments

Material	Density (kg m^{-3})	ELASTIC: Young's modulus (Pa), Poisson's ratio	FRICTIONAL: ϕ_0, C (MPa)	VISCOUS: γ (Pa^{-n} s^{-1}), E_a (KJ mol^{-1}), n
Crust	2800	1×10^{11}, 0.25	15°, 1	3.28×10^{-19}, 44, 1 or 1.63×10^{-26}, 135,3.1 (nonlin)
Fault	2800	1×10^{11}, 0.25	3°, 1	3.28×10^{-19}, 44, 1 or 1.63×10^{-26}, 135,3.1 (nonlin)
Strong Mantle	3300 or eqn (3) (dens)	1×10^{11}, 0.25	15°, 1	8.2×10^{-18}, 111, 1 or 7.24×10^{-18}, 535,3.5 (nonlin)
Weak Mantle	3300 or eqn (3) (dens)	1×10^{11}, 0.25	15°, 1	16.4×10^{-18}, 111, 1

(the plastic potential is simply equal to $J_2 (\sigma)$). For most of the materials in the numerical experiments, we use $\phi_0 = 15°$. We also include a frictionally weak zone with $\phi_0 = 3°$ to represent the strained, weakened brittle part of the Alpine Fault, dipping at 45° in most experiments but vertical in some of the experiments.

Thermally-activated creep obeys the following relationship [e.g., *Turcotte and Schubert*, 1982]:

$$\dot{\varepsilon}_{ij} = \gamma \tau_{ij}^{\ n} \exp\left(-E_a\big/ RT\right) \qquad (2)$$

where γ is the fluidity in Pa^{-n} s^{-1}, $\dot{\varepsilon}_{ij}$, τ_{ij} are the deviatoric i-jth component of the strain-rate and stress tensors, respectively, n is the power-law exponent, E_a the activation energy, R the gas constant, and T is temperature in Kelvin. Creep parameters in the crust and mantle are chosen to approximate the behavior of hydrous quartz and dry olivine in high-temperature laboratory experiments [*Chopra and Paterson*, 1984; *Paterson and Luan*, 1990]. The strong anisotropy measured in New Zealand implies that dislocation creep, which is characterized by a non-linear stress-strain rate relation, is the dominant deformation mechanism in the shallow mantle beneath New Zealand, since diffusion creep does not produce crystal preferred orientations and hence no large-scale anisotropy develops. However, some numerical experiments were run with a linear relation between stress and strain (Newtonian behavior; $n = 1$) because the results were numerically more stable than for nonlinear rheologies. The linear viscous parameter values were chosen to approximate the experimentally observed non-linear behavior for a representative temperature gradient and strain-rate. Linearized rheologies thus have E_a and γ recalculated in order to produce the same viscosity of the actual dislocation creep law at similar temperatures for a constant strain-rate of 4×10^{-15} s^{-1}.

In some experiments we also applied a ductile strain-softening in the mantle, where the fluidity is multiplied by $(0.2)^{-n}$ for finite strains larger than 1. These experiments yielded little differences in the final strain distribution and shear-wave splitting values from the cases without strain-softening.

In some experiments (those with suffix dens in Table 3), we also tested the effect of a temperature-dependent density in the mantle. In this case, mantle density obeys the following function [*Turcotte and Schubert*, 1982]:

$$\rho = \left(\frac{\rho_0}{1 - \rho_0 gz\beta_a}\right)\left(1 - \Delta T \alpha_V\right) \qquad (3)$$

where ρ_0 is the initial density (3300 kg/m^3 for the mantle), g is the acceleration of gravity (assumed -10 ms^{-2}), z is the depth below sea level, β_a is the adiabatic compressibility, (assumed 8.7×10^{-12} Pa^{-1}), ΔT is the temperature difference between the true temperature and a reference temperature (chosen as 1223 °K, the temperature at 50 km depth), and α_V is the coefficient of thermal expansion at constant volume (assumed to be 3×10^{-5} K^{-1}).

Lattice Preferred Orientation From Polycrystal Plasticity Models

The development of olivine and orthopyroxene lattice preferred orientations (LPO) as a function of strain is calculated using a viscoplastic self-consistent model [*Lebensohn and Tomé*, 1993]. This model simulates the LPO evolution due to plastic deformation by dislocation glide in a bimodal (60% olivine – 40% enstatite) polycrystalline material. In polycrystal plasticity models, LPO evolution is essentially controlled by the imposed deformation, the initial texture (crystal preferred and, to a lesser extent, shape preferred orientation), and the active slip systems. The latter depend

Table 2. Viscoplastic model slip systems' data

Mineral	Slip system	CRSS	n
olivine	(010)[100]	1	3
	(001)[100]	1	3
	(010)[001]	2	3
	(100)[001]	3	3
enstatite	(100)[001]	2	3
	(010)[001]	5	3

on the mineral structure, but also on the temperature and pressure conditions, which control their relative strength or critical resolved shear stress. In order to simulate olivine and pyroxene LPO evolution in the lithospheric mantle and asthenosphere, we used, as in previous viscoplastic self-consistent simulations for olivine [*Wenk et al.*, 1991; *Tommasi et al.*, 2000], slip system data (Table 2) derived from high-temperature, low-pressure single-crystal deformation experiments [*Bai et al.*, 1991]. Higher critical resolved shear

stresses for enstatite account for its higher strength relative to olivine.

The deformation history is derived directly from the final velocity gradient field in the mechanical model by assuming that for times > 100 ky, i.e., after the relaxation of elastic stresses, the flow pattern in the mantle does not change significantly. This allows us to calculate the anisotropy for times that are longer than those used in the geodynamic models, which become unstable when they become too deformed. To reduce calculation times, we avoid calculating the LPO evolution in neighboring elements with similar deformation histories. Instead, we search over the entire finite-element model and create velocity gradient tensor "classes". Within each class, variations for each of the tensor components are less than 5×10^{-16} s^{-1}. Olivine and enstatite LPO evolution are then calculated for each velocity gradient tensor "class". These LPO are used to compute the aggregate seismic properties based on a Voigt-Reuss-Hill average of the elastic constants of olivine and enstatite at ambient conditions [*Kumazawa and Anderson*, 1969; *Mainprice*, 1990]. The calculated LPO and elastic constant tensor are then tracked

Table 3. Summary

Experiment Name run	Description	Extent of Weak mantle (km from fault) at 0 My	Temp	Motion
PASTTR, PASTTRnonlin used in paper for Plate 1a,c,e, and Figure 2	Pre-root continent Alpine Fault dips 45° No strain-softening Mantle extends to 100 km.	none	Linear Gradient	35 mm/yr along-fault 10 mm/yr across-fault
PRESTR, PRESTRnonlin used in paper for Plate 1b,d,f and Figure 2	Root [*Scherwath et al.*, 2003] Alpine Fault dips 45°. No strain-softening Mantle extends to 100 km.	none	*Shi et al.*, 1996	same
PASTSS, PASTSSnonlin Used in paper for Plate 1h, Figures 2 and 3	Pre-root continent. Alpine Fault vertical Strain-softening used. Mantle extends to 100 km.	none	Linear Gradient	35 mm/yr along-fault 0 across-fault
PASTSSdeep, PASTSSdeepnonlin Used in paper for Figure 3	Same as above but mantle extends to 200 km.	none	Linear Gradient	same as above
PRESTRdensdeep, PRESTRdensdeepnonlin Used in paper in Plate 2 and Figure 4.	Root [*Scherwath et al.*, 2003]. Alpine Fault dips 45°. No strain-softening Mantle extends to 200 km.	−195 km to 150 km	*Shi et al.*, 1996	35 mm/yr along-fault 10 mm/yr across-fault
PASTTRdens, PASTTRdensnonlin Used in paper in Plate 2 and Figure 4	Pre-root continent. Alpine Fault vertical No strain-softening Mantle extends to 100 km. No erosion	−20 km to 35 km	linear	35 mm/yr along-fault 10 mm/yr across-fault

back to the final positions of the elements' baricenters in the finite-element model.

Splitting Calculation

To calculate shear-wave splitting, we modify a method outlined previously [*Fischer et al.*, 2000]. Seismic ray paths are traced through the AK135 average global velocity model [*Kennett et al.*, 1995] to determine which blocks of the model the ray passes through. A linear particle motion with a period of 10 s is passed through the first block and split using the Christoffel equation with the tensor of elastic coefficients calculated from the LPO models. The Christoffel equation gives the wave speeds and polarizations of the quasi P and two quasi S waves propagating through the medium with the given direction [*Babuska and Cara*, 1991]. The resultant particle motion is split again in the next block, and so on until the top of the model is reached. The net effects of splitting observed at the surface are measured using the same type of particle motion analysis [*Silver and Chan*, 1991] that is typically applied to real data.

3. GEODYNAMIC EXPERIMENTS

We consider two experimental setups (Table 3). The first, with labels starting by PAST, has initial conditions representing the time just before motion on the central Alpine Fault changed from nearly pure strike-slip to transpressional, at either 6.4 Ma [*Walcott*, 1998] or 11 Ma [*Cande and Stock*, 2004]. It uses a flat topography on the continental region, with crustal thickness of 27 km (see arguments in *Stern et al.*, [2002]) from a distance of 200 km west of the Alpine fault to 230 km east of the fault [*Scherwath et al.*, 2003]. East and west of the continental material, we use an oceanic Moho depth of 14.5 and 17 km, respectively. Experiments in which we apply transpression are labeled PASTTR, and when we apply strike-slip, they are called PASTSS. PASTTR experiments impose a weak zone dipping at 45° in the crust (to 27 km depth), representing the proto-Alpine Fault, and PASTSS experiments impose a vertical weak zone through the crust. We use a laterally uniform geotherm, based on the background temperature profile of *Shi et al.*, [1996] for shallow depths (≤50 km) and a steeper thermal gradient in the deep lithosphere (Plate 1).

The second experiment setup, labeled PRESTR (Table 3), simulates the present situation. It uses a recent crustal structure model [*Scherwath et al.*, 2003]. Note that this model has a slightly thicker oceanic crust on the east, while a seismic line about 50 km to the north reveals a slightly thicker oceanic crust to the west [*van Avendonk et al.*, 2004]. The initial temperature profile is based on the temperature field obtained by finite element modeling of the compressional component of Alpine Fault motion [*Shi et al.*, 1996], which is extrapolated to 100 km depth. Other initial temperature structures, like the one presented in Figure 6 of *Stern et al.*, [2000], were also tested. However, due to space limitations, the results of these additional models are not shown here, since they do not yield more insight than the PRES experiments.

Most experiments were run with a model thickness of only 100 km. However, several experiments with thicknesses of 200 km (suffix deep, Table 3) were run to test that model thickness did not significantly affect results.

In all experiments, we apply a lithostatic pressure condition at the base of the model. The upper boundary is free and gravity forces are fully taken into account. Both PAST and PRES model experiments start from geostatic equilibrium, attained by applying an initial one-dimensional lithostatic stress state and then allowing the experiments to equilibrate stresses over a short time interval. Velocity conditions mimicking the Pacific plate motion relative to Australia are applied on the Pacific side boundary of the model. The Australian side boundary is kept fixed. The actual plate motion has changed over time [*Cande and Stock*, 2004]. Thus two types of lateral boundary conditions were implemented: pure strike-slip experiments, in which we impose a 35 mm/yr displacement of the southeastern boundary parallel to the fault (models with prefix PASTSS), and transpressional experiments, which have both a strike-slip component of 35 mm/yr and a compressional component of 10 mm/yr (experiments with prefix PRESTR and PASTTR).

4. RESULTS

Plate 1 shows the temperature and strain profiles for transpressional experiments PASTTR, which starts from the continental crust and mantle temperature profile assumed to occur before compression, and PRESTR, which uses the present temperature field and crustal structure (Table 3). For experiment PASTTR at 0 My, before any compression has taken place, the temperature is slightly higher under the continental than the oceanic mass. This is consistent with the expected warmer temperatures due to the crustal heat production and blanketing effect in the continental lithosphere. For experiment PRESTR, the temperature is colder under the Alpine Fault due to the localized crustal and mantle thickening (root). After 5 My of transpressive deformation, the temperatures under the continental mass in experiment PASTTR have decreased relative to those under the ocean, because the compressional component of motion induced thickening leading to displacement of cooler shallow material to deeper levels. In the crust, the deformation is localized within the weak (low friction) Alpine Fault block

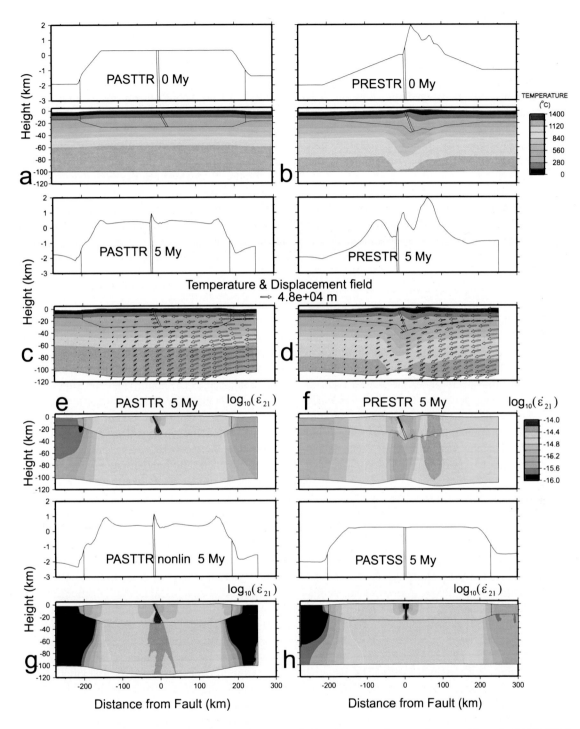

Plate 1. Results of geodynamic modeling. (a). Starting temperature field and structure for experiments with prefix PAST (Table 3). Top panel is topography; bottom panel is a cross section, with x = 0 at the surface intersection of the Alpine Fault. (b) Starting temperature field and structure for experiments with prefix PRESTR. (c) Temperature, topography and displacement fields after 5 My of transpression for experiment PASTTR. (d) Temperature, topography and displacement fields after 5 My of transpression for experiment PRESTR. (e) Shear strain rate $\log_{10}(\dot{\varepsilon}_{21})$ where 1 and 2 are horizontal and vertical Cartesian coordinates, respectively, for experiment PASTTR after 5 My of transpression (f) Same for experiment PRESTR after 5 My. (g) Shear strain rate and topography for experiment PASTTRnonlin after 5 My (h) Same for experiment PASTSS after 5 My.

or directly beneath it. Strain localization is much weaker in the mantle. Strain is enhanced in the mantle underneath the continent compared to the adjacent oceanic mantle, but it is nearly constant across the width of the continental region. Non-linear models display slightly higher shear strains in a 50km wide domain beneath the Alpine fault. However, only small mountains are being formed near the Alpine Fault, and no localized cold mantle root has developed, despite deformation having occurred for 5 My. The shear strain for pure strike-slip deformation (model PASTSS) is similar to that for transpression (Plate 1).

The temperature distribution after 5 My of deformation in the present geometry experiments (PRESTR) shares similar features to PASTTR; isotherms under the continent have been broadly depressed compared to the temperature field at 0 My (Plate 1). The strain field, however, is markedly different from the strain field for experiment PASTTR. The cold root remains relatively undeformed due to the strong temperature dependence of viscosity and the major deformation is localized in the continental lithosphere that borders the thickened fault domain. This causes mountains as high or higher than the Southern Alps to develop on both coasts and thickened lithosphere to form beneath these mountains. In contrast, the cool lithosphere under the Alpine Fault in experiment PASTTR does not deform or thicken as much as that at the edges. This experiment is therefore not consistent with present deformation. *Liu and Bird* [2006] find similar growth of mountains east & west of South Island when a lithospheric root is present without an embedded weak zone.

Including nonlinear dependence of stress on strain localizes the deformation for experiment PASTTRnonlin more strongly beneath the continental landmass (Plate 1), but still not enough to explain the formation of the present crustal root beneath the Alpine Fault, for which a strong localization of the horizontal shortening along the fault is necessary [e.g., *Norris and Cooper*, 2003; *Beavan et al.*, 2006]. Splitting patterns computed for these experiments are shown in Figure 2. For all PAST experiments, the delay times in the oceanic regions are small (<0.3 s), and, where they are determinable, the fast polarizations are close to −45° from the Alpine Fault strike (which is 0°). The patterns are as expected for small shear strains: olivine [100] axes and hence the fast polarization of S-waves align with the extension direction, but the anisotropy is very weak. Under the continent, the delay times are over 0.5 s, and the larger strain aligns fast orientations closer to fault parallel (−40° to −25°), but not as close as the measured values in the central Southern Alps (Figure 1). Yet all the PAST experiments show relatively consistent fast polarizations and delay times across the continental region, in agreement with the SKS measurements (Figure 1) [e.g., *Klosko et al.*, 1999; *Duclos et al.*, 2005]). Nonlinear models show somewhat stronger anisotropy localization than the linear models, as evidenced by fast polarizations more strongly fault-parallel and larger delay times toward the Alpine Fault.

The cold root in both linear and nonlinear experiments PRESTR and PRESTRnonlin inhibits strain beneath the fault, so that splitting varies strongly across the continental region, with fast polarizations at about −45° to the fault plane and small delay times in the centre of the region above the root (Figure 2). Away from the root, delay times approach those of the PASTTR experiments.

The splitting predictions in the experiments starting from the present structure (PRES) are very different from the measured splitting values. In contrast, splitting predictions from the PAST experiments are quite similar to measured values, suggesting that the present anisotropy can be fully explained by olivine LPO formed in the lithosphere during an earlier phase of deformation. Because the relative plate motion during the period from ~45 to 10 Ma was nearly parallel to the Alpine Fault, we examine the strain and the shear wave splitting expected from such a deformation (Figure 3). We show only the nonlinear cases here. We examine both a 100 km thick (PASTSSnonlin) and 200 km thick (PASTSS-deepnonlin) model. The strain is similar to that for the experiment PASTTRnonlin (Plate 1). The splitting is also similar, showing a rotation over time of the fast polarizations above the continental region to becoming progressively more fault-parallel (Figure 3). Both depth ranges show similar fast polarizations at similar times (compare 6 My values). Linear models yield nearly constant polarizations across most of the continental regions, while nonlinear models are closer to fault-parallel over the fault (Figures 2,3). By 40 My, the fast polarizations are closer than 15° to the fault-plane. The delay times increase until about 15 My, at which point they saturate, yielding the same delay time regardless of increasing strain, even though the fast polarizations continue to rotate. For the 200 km thick experiment, the delay times are about twice as large because they travel through more anisotropic material, reaching 2.3 s. The fast polarizations take longer to become fault-parallel in the deep (200 km thick) model experiment compared to the 100 km thick experiment (for example, the fast polarizations at 20 My in the deep experiment are similar to those at 15 My in the 100 km thick experiment), because the strain is spread across a thicker region and is therefore somewhat weaker.

The experiments discussed so far have included constant densities of 2800 kg/m³ in the crust and 3300 kg/m³ in the mantle. However, these simplified experiments were not able to produce the present lithospheric structure by starting from the inferred past deformation. For example, experi-

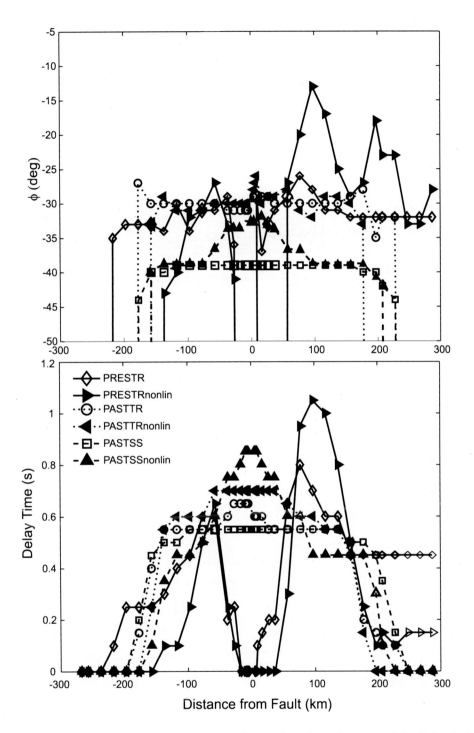

Figure 2. Modeled splitting parameters as a function of horizontal distance from the surface trace of the Alpine Fault, for model times of 6 My. Top: Fast polarization, measured relative to the strike of the Alpine Fault at 0°; e.g., −30° means that the fast polarization is 30° clockwise from the strike of the Alpine Fault. Bottom: Delay time. Experiment names (Table 3) are given in the key. Color version of this figure is on CDROM which accompanies this volume.

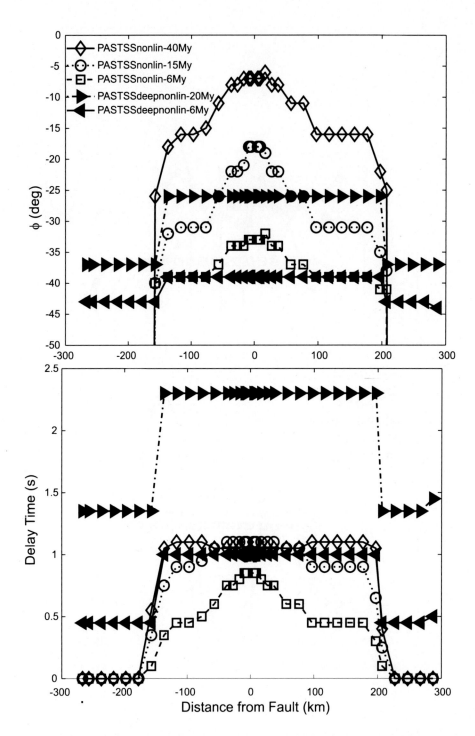

Figure 3. Splitting measurements as a function of distance from the fault for nonlinear experiments PASTSSnonlin and PASTSSdeepnonlin for various time intervals. Note that the delay times saturate by 15 My for the experiment with lithosphere thickness of 100 km. Color version of this figure is on CDROM which accompanies this volume.

ments PASTTR and PASTRnonlin yielded a broad down warp of the mantle, but not a concentrated root, despite running for 5 My (Plate 1). As temperature-dependent density is expected to play a role in the formation of such crustal roots via Rayleigh-Taylor instabilities [e.g., *Houseman and Molnar*, 1997; *Molnar et al.*, 1998]), we modified the program to test whether temperature- and depth-dependence of the density alters the solution, as described in the methods section. The experiments shown here do not include erosion, and also include a weak zone under the continents, because the weak zone resulted in a stronger response to the density structure.

As expected, the cold lithospheric root in the present (PRES) experiment deepens when temperature-dependent density is included (Plate 2, experiment PRESTRdensdeep). Because erosion is not included, the compressional component of motion causes the crustal root to thicken further. When erosion is included (not shown), the crustal root maintains a more constant thickness, with material loss through the surface erosion. There is also more upward flow of material near the surface when erosion is included. The surface elevation in both experiments is brought down by the pull of the cold lithospheric root, and both experiments have shear deformation concentrated away from the cold root region. Nonlinear versions of the deep models became unstable before the linear models, and so are not shown in Plate 2 although they were used to calculate splitting in Figure 4.

The PAST model experiment that most closely evolves into the present-day configuration of central South Island when transpressional boundary conditions were applied, was when we included a narrow weak zone in the mantle beneath the fault (Plate 2, experiment PASTTRdens). The lithosphere in the weak region thickened more than the surrounding regions, making it closer to the present lithospheric structure inferred from P delays [*Stern et al.*, 2000]. Weakening in this domain may result from a large variety of strain dependent processes, such as development of mechanical anisotropy due to the progressive orientation of olivine crystals in easy glide positions [*Tommasi et al.*, 2000] or reduction in grain size due to dynamic recrystalization. Erosion was not included in this case, but is expected to have little effect because the surface remains low in elevation. Predicted splitting from the PASTTRdens experiments shows a variation of delay time and fast polarization with distance from the fault (Figure 4). The strong variation in the fast polarization in the centre of the PRESTRdensdeep experiment is caused by the plunging of the axes of symmetry of olivine when the motion in the "drip" becomes vertical. The observed ϕ and dt do not vary as a function of distance from the Alpine Fault (Figure 1), and are thus inconsistent with this experiment.

5. DISCUSSION

All the experiments show the highest strain rates underneath the continental landmass. The rates are not highest at the border between continental and oceanic crust (the two types of crust are not distinguished compositionally in our models). Instead the rates are highest where the temperatures are hottest and therefore the viscosities are lowest; in these models the hottest temperatures are under the continental crust. While most measurements of anisotropy in New Zealand are on continental material and hence cannot test the prediction that higher strain-rates occur beneath continental cf. oceanic crust, there are a few measurements close to the margin that do suggest a decrease in anisotropy where the continental crust decreases in thickness. The limits of anisotropy determined by Pn measurements along SIGHT line T2 [*Scherwath et al.*, 2002; *Baldock and Stern*, 2005], occur above the 1000 m bathymetric contour (Figure 1). These authors found that the wave speeds along line T2 were small under the continent compared to the much larger values along perpendicular lines, which led them to infer high anisotropy in South Island upper mantle just east of the Alpine Fault. The wave speeds increased to more average mantle speeds at the points marked by large arrows in Figure 1, which led them to suggest that the anisotropy decreased at these points. Similar arguments suggested a smaller region of anisotropy along line T1 [*van Avendonk et al.*, 2004]. In addition, station JACA, the southernmost station on the Australian Plate in the SAPSE deployment, is close to oceanic crust, and yielded very little anisotropy (Figure 1). Single events at JACA yielded only null measurements, and the stacked values yielded 0.6 s delay time, the smallest of all South Island stations [*Klosko et al.*, 1999]. The Fresnel zone radius at the surface for SKS phases is about 40 km, and includes oceanic material deeper than 1000 m. This suggests the 1000 m bathymetry line may represent the edge of the continental crust and therefore the edge of the region of high anisotropy.

In all the experiments with the present temperature structure, shear strain maxima occurred away from the cold root. Calculated shear-wave splitting for the strain from these experiments yielded both fast polarizations and delay times that varied substantially with distance from the fault, contrary to the measured values. The models simulate a simplified time evolution of the plate boundary motion, and a model could be devised in which rheology weakens in a direct tradeoff with thickening, such that the strain and resulting anisotropy would be more consistent across the continent with the present temperature and structure. However, there was a long period (at least 20 My) of strike-slip motion, with significant strain accumulation, before the onset of compressional motion

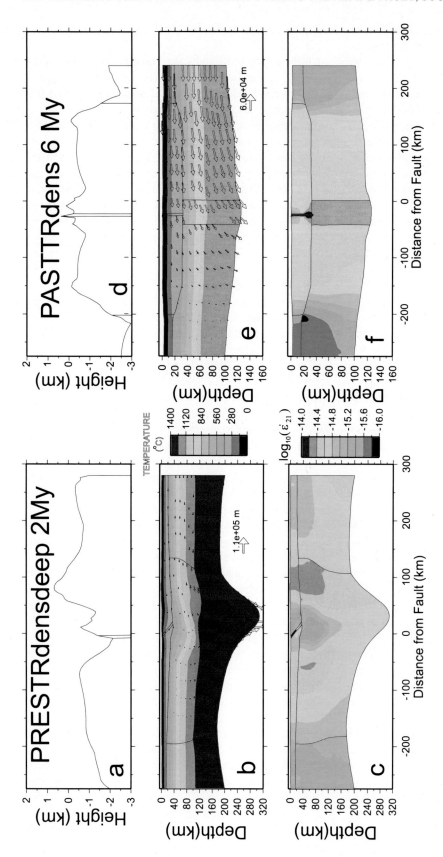

Plate 2. Effect of temperature-dependent density at the end of the model runs. a,d) topography. b,e) temperature (contours) and displacement fields (arrows). c,f) strain rate, as measured by $\log_{10}(\dot{\varepsilon}_{21})$ at the end of the model run. a,b,c) transpression was run for 2 My applied to experiment PRESTRdensdeep (Table 3). In this experiment, the mantle below the continent has a factor of two larger fluidity (i.e., a lower effective viscosity), to simulate the possibility of strain softening in the most highly strained region during the previous strike-slip episode. Temperature and depth dependent density is included, but erosion is not included. d,e,f) experiment PASTTRdens; the postulated pre-compressional temperature structure was used with transpression applied for 6 My, for a total of 60 km of shortening. The large-fluidity region is confined to a narrow region of width 55 km.

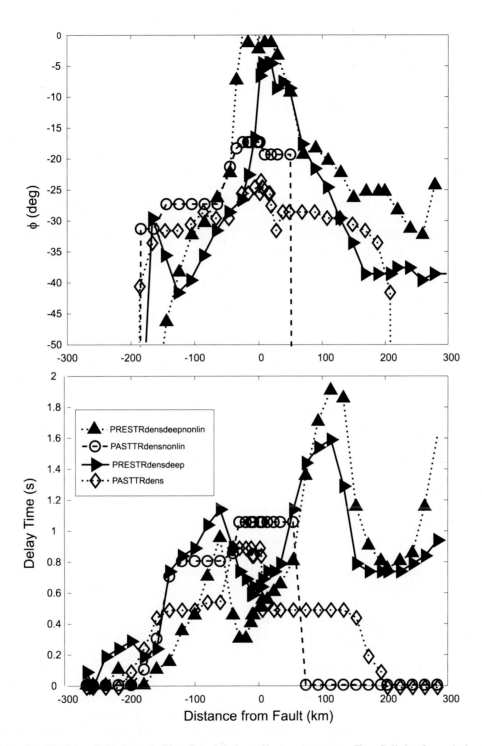

Figure 4. Splitting values for the models shown in Plate 2, and their nonlinear counterparts. Top: Splitting fast polarization as a function of distance from the fault. Bottom: Delay time as a function of distance from the fault. Experiment names (Table 3) are given in the key. The experiments all predict patterns of fast polarizations and delay times that vary with distance from the fault. Color version of this figure is on CDROM which accompanies this volume.

that most likely formed the present temperature structure [*Walcott*, 1998; *Cande and Stock*, 2004]. Our modeling shows that such significant strike-slip strain can produce anisotropy splitting parameters that can match the present-day shear-wave splitting measurements. This is consistent with the view that at least a large part of the anisotropy measured today is due to past deformation during the previous pure strike-slip episode, before the cold root was formed.

The LPO models we calculate start with isotropic, unstrained aggregates and build up strain and LPO over time. In the real earth, deformations are built on top of previous ones. This means that an LPO developed during previous strike-slip deformation could dominate the present-day LPO. This is particularly the case for the cool lithospheric root that is thought to have formed during the later, transpressional phase. We expect superposition of strain during the two phases, with smaller strain in the lithospheric root in the transpressional phase once it cooled and became stronger than surrounding material, would preserve the strike-slip olivine LPO in the root. Possibly the material further from the fault would strain more strongly, and become more fault-parallel over time than the material directly under the fault. However, the error bars in the splitting measurements available now are too large to distinguish between these possibilities.

If anisotropy formed in a previous episode of strike-slip deformation, then SKS measurements cannot be used to distinguish between end-member models for mantle deformation of central South Island, such as inferred subduction of Pacific or Australian mantle along a localized "antifault" [e.g., *Wellman*, 1979] vs. distributed thickening of mantle lithosphere over hundreds of kilometers [e.g., *Molnar et al.*, 1999]. Moreover, the unrealistic topography developed for the symmetric cold root model both herein, and in *Liu and Bird* [2006], suggests that if such a root is present, it is likely to have been formed in a previously weakened region. Previous models that have investigated evolution of the lithosphere beneath South Island [*Pysklywec*, 2002; *Gerbault*, 2003] have shown that an embedded weak zone can initiate deformation in the upper mantle along a dipping shear plane, which acts to localize crustal deformation and can explain both the topographic uplift of the Southern Alps, and development of a crustal root. The results here, including the "negative" result in which we were unable to reproduce current South Island dynamics with a homogeneous mantle rheology, also suggest that some variation in mantle strength is necessary. More localized mantle deformation at present is also supported by geodetic measurements of strain currently accumulating across central South Island [e.g., *Beavan et al.*, 2006; *Ellis et al.*, 2006; *Wallace et al.*, 2007].

For the strike-slip motion without a root, the delay times increase until about 15 My, at which point they saturate,

yielding the same delay time regardless of increasing motion (Figure 3). The average delay time measured for central South Island stations is 1.4 s [*Duclos et al.*, 2005]. This is higher than the 1.0 s to which the delay times over the continental regions saturate with our 100 km thick experiments after 15 My, but smaller than the 2.3 s delay time to which our 200 km thick lithosphere experiments saturate. This suggests either that the lithospheric thickness is between 100 and 200 km, or that another mechanism has allowed anisotropy to be smaller or larger than that modeled.

Although the delay times saturate, the fast polarizations continue to rotate to become more fault-parallel with time, i.e., increasing strain. This may explain the variations observed between southern and central South Island, in which the fast polarization measured in the southern region is consistent with the extension direction determined by the surface strain measurements [*Little et al.*, 2002], with an orientation further away from fault-parallel in the south than in the central region, but with delay times nearly constant despite the smaller strain. The measured splitting along central South Island is closer to fault-parallel than the calculated splitting, even after 40 My of movement of the input strain, which yields 1400 km of displacement, greater than the 850 km measured [*Sutherland*, 1999]. 850 km displacement equates to 23 My in our experiments, yielding φ about 25 degrees from fault-parallel (Figure 2 and 3). Thus, to explain the nearly fault-parallel fast polarizations observed along the central Alpine Fault, we need to consider that additional deformation mechanisms, such as dynamic recrystallization, accelerate the development of olivine LPO, leading olivine [100] axes to align with the shear direction at shear strains of 1, as observed in simple shear deformation experiments[*Zhang and Karato*, 1995; *Bystricky et al.*, 2000]. Nevertheless the splitting from the present experiments matches better to the measurements than those determined by simpler 2D modeling [*Savage et al.*, 2004]. This is because the delay times in the experiments shown here saturate with fairly small strains, less than 525 km displacement or slightly over 1, taking the strain rate of $10^{-14.5}$ s^{-1} (Plate 1) over 15 My. Thus, delay times are similar across the continental region despite differences in strain and fast polarization. Another major difference between the present experiments and the earlier study is in the boundary conditions. In the earlier study, the plate motion was driven from above, i.e., a prescribed velocity was set at the surface in opposite directions on the two sides of a fault, and the bottom boundary was fixed to zero slip at great depth. In that case, the strain decreases with distance from the fault, both in depth and laterally. The integrated effect of the anisotropy with depth was to weaken the splitting that would have been observed if the surface strain extended to all depths. In the models shown here, plate motion was driven from the side, and the bottom boundary was free-slip. This allowed the strain to be similar at all depths (e.g., Plates 1,2). The seismic waveforms therefore encounter similar anisotropy through their entire path, allowing large delay times and consistent polarizations to be measured. Lateral variations in *Pn* anisotropy [*Baldock and Stern*, 2005] and shear wave splitting from local earthquakes [*Audoine et al.*, 2000] both suggest that the deformation might be smaller in the uppermost mantle than in the region sampled by the SKS phases, suggesting that the true conditions in the Earth might be in between the boundary conditions examined in these two studies.

The coherence of SKS between the southern North Island and South Island suggests a common cause. If so, there are two possibilities. Both regions may be affected by current, trench-parallel asthenospheric flow or by past strike-slip strain such as modeled here. We discuss the possibilities more fully in a review paper in this volume [*Savage et al.*, 2007].

6. SUMMARY

a) The out-of-plane geodynamic experiment results that start from pre-collisional conditions (PAST) predict that thick continental crust localizes the strain beneath the continental compared to the oceanic regions, but the strain is distributed throughout the continental region. Shear-wave splitting parameters for pure strike-slip deformation through such a region yield delay times that saturate at a maximum value of 1.0 s to 2.3 s for lithospheric thicknesses of 100 and 200 km, respectively, across most of the continental region. The fast polarizations are constant across the region but rotate with increasing strain to become more fault-parallel, reaching 25° at the maximum expected strains. The measured delay time average of 1.4 s suggests lithosphere either greater than 100 km thick, or more anisotropic than modeled. Dynamic recrystallization, leading to faster orientation of olivine [100] parallel to the shear direction, may be needed to explain the fault-parallel splitting fast polarizations in the central Alpine Fault, and could also cause larger anisotropy than modeled, helping to fit the delay times.

b) Experiments starting from the present-day configuration (PRES) and no weak zone in the mantle predict that a cold root should inhibit strain and therefore the formation of anisotropy. Such experiments have difficulty reproducing observed shear wave splitting and topography. A weak zone within the mantle that localizes deformation there is necessary to reproduce current deformation patterns, in which case mantle anisotropy accumulates in a narrow region. Therefore, it is likely that the present wide-spread anisotropy was formed during the previous episode of strike-slip deformation, before the cold root was formed.

c) We predict that the upcoming ocean bottom seismometer deployment off the coasts of South Island will find anisotropy changes at the edge of the continental crust (i.e., the regions where the bathymetry deepens in Figure 1), similar to what has already been inferred for the Pn anisotropy along SIGHT line T2.

Acknowledgements. Karen Fischer allowed us to modify her codes for the splitting calculations. Financial support was supplied by the NZ Marsden Fund and the NZ Foundation for Research, Science and Technology. The research was begun while M. Savage was on sabbatical leave at the University of Montpellier II, with support from Victoria University and from the Centre National de la Recherche Scientifique (France) through a Chercheur étranger associé position. Thorough, constructive reviews by Peter Molnar and Zhen Liu are much appreciated.

REFERENCES

Audoine, E., M. K. Savage, and K. Gledhill (2000), Seismic anisotropy from local earthquakes in the transition region from a subduction to a strike-slip plate boundary, New Zealand, *J. Geophys. Res.*, *105*, 8013–8033.

Babuska, V., and M. Cara (1991), *Seismic Anisotropy in the Earth*, 217 pp., Kluwer Academic Publishers, Dordrecht/Boston/London.

Bai, Q., S. J. Mackwell, and D. L. Kohlstedt (1991), High-temperature creep of olivine single crystals,1, Mechanical results for buffered samples, *J. Geophys. Res.*, *96*, 2441–2463.

Baldock, G., and T. Stern (2005), Width of mantle deformation across a continental transform: Evidence from upper mantle (Pn) seismic anisotropy measurements, *Geology*, *33*, 741–744.

Beavan, J., S. Ellis, L. Wallace, and P. Denys (Eds.) (2006), *Kinematic Constraints from GPS on Oblique Convergence of the Pacific and Australian Plates, Central South Island, New Zealand*, in revision pp.

Blackman, D. K., J.-M. Kendall, P. R. Dawson, H.-R. Wenk, D. Boyce, and J. P. Morgan (1996), Teleseismic imaging of subaxial flow at mid-ocean ridges: travel-time effects of anisotropic mineral texture in the mantle, *Geophys. J. Int.*, *127*, 415–426.

Bourguignon, S., T. A. Stern, and M.K. Savage (2007), Crust and mantle thickening beneath the southern portion of the Southern Alps, New Zealand, *Geophys. J. Int.*, *168*, 681–690

Bystricky, M., K. Kunze, L. Burlini, and J.-P. Burg (2000), High shear strain of olivine aggregates: rheological and seismic consequences, *Science*, *290*, 1564–1567.

Cande, S. C., and J. M. Stock (2004), Pacific-Antarctic-Australia motion and the formation of the Macquarie Plate, *Geophys. J. Int.*, *157*, 399–414 doi:310.1111/j.1365-1246X.2004.02224.x.

Chéry, J., M. D. Zoback, and S. Hickman (2004), A mechanical model of the San Andreas fault and SAFOD pilot hole stress measurements, *Geophys. Res. Lett.*, *31*, doi:10.1029/2004GL019521.

Chopra, P. N., and M. S. Paterson (1984), The role of water in the deformation of dunite, *J. Geophys. Res.*, *89*, 7861–7876.

Cundall, P. A. (1988), Formulation of a three-dimensional distinct element model; Part I, A scheme to detect and represent contacts in a system composed of many polyhedral blocks *International Journal of Rock Mechanics and Mining Sciences & Geomechanics Abstracts*, *25*, 107–116.

Duclos, M. (2005), Insights on Plate Boundary Deformation from Seismic Anisotropy in the New Zealand Upper Mantle, 210 pp., Victoria University of Wellington, Wellington, New Zealand.

Duclos, M., M. K. Savage, A. Tommasi, and K. R. Gledhill (2005), Mantle Tectonics beneath New Zealand Inferred from SKS Splitting and Petrophysics, *Geophys. J. Int.*, *163*, 760–774, doi:710.1111/j.1365-1246X.2005.02793.x.

Ellis, S., J. Beavan, and D. Eberhart-Phillips (2006), Bounds on the width of mantle lithosphere flow derived from surface geodetic measurements: application to the central Southern Alps, New Zealand, *Geophysical Journal International*, doi: 10.1111/j.1365-1246X.2006.02918.x.

Fischer, K. M., E. M. Parmentier, A. R. Stine, and E. R. Wolf (2000), Modeling anisotropy and plate-driven flow in the Tonga subduction zone back arc, *J. Geophys. Res.*, *105*, 16,181–16,192.

Gerbault, M., Henrys, S., and Davey, F.J. (2003), Numerical models of lithospheric deformation forming the Southern Alps of New Zealand, *J. Geophys. Res.*, *108*, 2341: doi:2310.1029/2001JB001716.

Houseman, G. A., and P. Molnar (1997), Gravitational (Rayleigh-Taylor) instability of a layer with non-linear viscosity and convective thinning of continental lithosphere, *Geophys. J. Int.*, *128*, 125–150.

Kennett, B. L. N., E. R. Engdahl, and R. Buland (1995), Constraints on seismic velocities in the Earth from traveltimes, *Geophys. J. Int.*, *122*, 108–124.

Klosko, E. R., F. T. Wu, H. J. Anderson, D. Eberhardt-Phillips, T. V. McEvilly, E. Audoine, M. K. Savage, and K. R. Gledhill (1999), Upper mantle anisotropy in the New Zealand region, *Geophys. Res. Lett.*, *26*, 1497–1500.

Kohler, M. D., and D. Eberhart-Phillips (2002), Three-dimensional lithospheric structure below the New Zealand Southern Alps, *J. Geophys. Res.*, *107*, 2225, doi:2210.1029/2001JB000182.

Kreemer, C., W.E. Holt, and A. J. Haines (2003), An integrated global model of present-day plate motions and plate boundary, *Geophys. J. Int.*, *154*, 8–34.

Kumazawa, M., and D. L. Anderson (1969), Elastic moduli, pressure derivatives, and temperature derivatives of single-crystal olivine and single-crystal forsterite, *J. Geophys. Res.*, *74*, 5961–5972.

Lebensohn, R. A., and C. N. Tomé (1993), A self-consistent anisotropic approach for the simulation of plastic deformation and texture development of polycrystals: application to zerconion alloys, *Acta Metall. Mater.*, *41*, 2611–2624.

Leroy, Y., and M. Ortiz (1989), Finite element analysis of strain localization in frictional materials, *Int. J. Numer. Anal. Methods Geomech.*, *13*, 53–74.

Little, T. A., M. K. Savage, and B. Tikoff (2002), Relationship between crustal finite strain and seismic anisotropy in the mantle, Pacific-Australia plate boundary zone, South Island, New Zealand, *Geophys. J. Int.*, *151*, 106–116.

Liu, Z., and P. Bird (2006), 2-D and 3-D finite element modeling of mantle processes beneath central South Island, New Zealand, *Geophys. J. Int.*, *165*, 1003–1028.

Mainprice, D. (1990), A FORTRAN program to calculate seismic anisotropy from lattice preferred orientation of minerals, *Comput. Geosci.*, *16*, 385-393.

Molnar, P., H. Anderson, E. Audoine, D. Eberhart-Phillips, K. Gledhill, E. Klosko, T. McEvilly, D. Okaya, M. Savage, T. Stern, and F. Wu (1999), Continuous deformation versus faulting through the continental lithosphere of New Zealand, *Science*, *286*, 516–519.

Molnar, P., G. A. Houseman, and C. P. Conrad (1998), Rayleigh-Taylor instability and convective thinning of mechanically thickened lithosphere: effects of non-linear viscosity decreasing exponentially with depth and of horizontal shortening of the layer, *Geophys. J. Int.*, *133*, 568–584.

Moore, M., P. England, and B. Parsons (2002), Relation between surface velocity field and shear-wave splitting in the South Island of New Zealand, *J. Geophys. Res.*, *107*, 2198, doi:2110.1029/2000JB000093.

Norris, R. J., and A. F. Cooper (2001), Late Quaternary slip rates and slip partitioning on the Alpine Fault, New Zealand *J. Structural Geol.*, *23*, 507–520.

Norris, R. J., and A. F. Cooper (2003), Very high strains recorded in mylonites along the Alpine Fault, New Zealand; implications for the deep structure of plate boundary faults., *Journal of Structural Geology*, *25*, 2141–2157.

Paterson, M. S., and F. C. Luan (1990), Quartzite rheology under geological conditions, in *Deformation mechanisms, rheology and tectonics* edited by R. J. R. Knipe, E H pp. 299-307, Geological Society Special Publications, Leeds, United Kingdom.

Pysklywec, R. N., Beaumont, C., and Fullsack, P. (2002), Lithospheric deformation during the early stages of continental collision: Numerical experiments and comparison with South Island, New Zealand, *J. Geophys. Res.*, *107*, doi:10.1029/2001JB000252.

Savage, M. K. (1999), Seismic anisotropy and mantle deformation: what have we learned from shear wave splitting?, *Rev. Geophys.*, *37*, 65–106.

Savage, M. K., M. Duclos, and K. A. Marson-Pidgeon (2007), Seismic anisotropy in the South Island, New Zealand, this volume.

Savage, M. K., K. M. Fischer, and C. E. Hall (2004), Strain modelling, seismic anisotropy and coupling at strike-slip boundaries: Applications in New Zealand and the San Andreas Fault, in *Vertical Coupling and Decoupling in the Lithosphere*, edited by J. Grocott, Tikoff, B., McCaffrey, K. J. W. & Taylor, G., pp. 9-40, Geological Society of London, Special Publication, London.

Scherwath, M., T. Stern, F. Davey, D. Okaya, W. S. Holbrook, R. Davies, and S. Kleffmann (2003), Lithospheric structure across oblique continental collision in New Zealand from wide-angle P wave modeling *J. Geophys. Res.*, *108*, 18 pp., doi:10.1029/2002JB002286.

Scherwath, M., T. Stern, A. Melhuish, and P. Molnar (2002), Pn anisotropy and distributed upper mantle deformation associated with a continental transform fault, *Geophys. Res. Lett.*, *29*, 16-11–16-14, doi:10.1029/2001GL014179.

Shi, Y., R. Allis, and F. Davey (1996), Thermal modelling of the Southern Alps, *PAGEOPH*, *146*, 469–501.

Silver, P. G. (1996), Seismic anisotropy beneath the continents: probing the depths of geology, *Ann. Rev. Earth Planet. Sci.*, *24*, 385–432.

Silver, P. G., and W. W. Chan (1991), Shear wave splitting and subcontinental mantle deformation, *J. Geophys. Res.*, *96*, 16,429–16,454.

Smith, G. P., and G. Ekstrom (1999), A global study of Pn anisotropy beneath continents, *J. Geophys. Res.*, *104*, 963–980.

Stern, T., D. A. Okaya, and M. Scherwath (2002), Structure and strength of a continental transform from onshore-offshore seismic profiling of South Island, New Zealand, *Earth, Planets, Space*, *54*, 1011–1019.

Stern, T. A., P. Molnar, D. Okaya, and D. Eberhart-Phillips (2000), Teleseismic *P* wave delays and modes of shortening the mantle lithosphere beneath South Island, New Zealand, *J. Geophys. Res.*, *105*, 21,615–21,631.

Sutherland, R. (1999), Cenozoic bending of New Zealand basement terranes and Alpine Fault displacement: a brief review, *N.Z. Joul. Geol. & Geophys.*, *42*, 295–302.

Tommasi, A., D. Mainprice, G. Canova, and Y. Chastel (2000), Viscoplastic self-consistent and equilibrium-based modeling of olivine lattice preferred orientations: Implications for the upper mantle seismic anisotropy, *J. Geophys. Res.*, *105*, 7893–7908.

Tommasi, A., B. Tikoff, and A. Vauchez (1999), Upper mantle tectonics: three-dimensional deformation, olivine crystalographic fabrics and seismic properties, *Earth and Planet. Science Lett*, *168*, 173–186.

Turcotte, D. L., and G. Schubert (1982), *Geodynamics: Applications of Continuum Physics to Geological Problems*, 450 pp., John Wiley & Sons, New York.

van Avendonk, H. J. A., W. S. Holbrook, D. A. Okaya, J. K. Austin, F. Davey, and T. Stern (2004), Continental crust under compression: A seismic refraction study of South Island Geophysical Transect I, South Island, New Zealand, *J. Geophys. Res.*, *109*, doi:10.1029/2003JB002790.

Walcott, R. I. (1998), Modes of oblique compression: late Cenozoic tectonics of the South Island of New Zealand., *Rev. Geophys.*, *36*, 1–26.

Wallace, L. M., J. Beavan, R. McCaffrey, and K. Berryman (2007), Balancing the plate motion budget in the South Island, New Zealand, using GPS, geological and seismological data, *Geophys. J. Int.*, *168*, 332–52; doi:10.111/j-1365.

Wellman, H. W. (1979), An uplift map for the South Island of New Zealand, *Bull. Roy. Soc. N.Z.*, *18*, 13–20.

Wenk, H.-R., K. Bennett, G. R. Canova, and A. Molinari (1991), Modelling plastic deformation of peridotite with the self-consistent theory, *J. Geophys. Res.*, *96*, 8337–8349.

Zhang, S., and S.-I. Karato (1995), Lattice preferred orientation of olivine aggregates deformed in simple shear, *Nature*, *375*, 774–777.

J. Chery and A. Tommasi, University de Montpellier II, France.

S. Ellis, GNS Science, New Zealand.

M. K. Savage, Institute of Geophysics, School of Earth Sciences, Victoria University of Wellington, New Zealand. (Martha .Savage@vuw.ac.nz)

A Comparison Between the Transpressional Plate Boundaries of South Island, New Zealand, and Southern California, USA: The Alpine and San Andreas Fault Systems

Gary S. Fuis[1], Monica D. Kohler[2], Martin Scherwath[3,6], Uri ten Brink[4], Harm J.A. Van Avendonk[5], and Janice M. Murphy[1]

There are clear similarities in structure and tectonics between the Alpine Fault system (AF) of New Zealand's South Island and the San Andreas Fault system (SAF) of southern California, USA. Both systems are transpressional, with similar right slip and convergence rates, similar onset ages (for the current traces), and similar total offsets. There are also notable differences, including the dips of the faults and their plate-tectonic histories. The crustal structure surrounding the AF and SAF was investigated with active and passive seismic sources along transects known as South Island Geophysical Transect (SIGHT) and Los Angeles Region Seismic Experiment (LARSE), respectively. Along the SIGHT transects, the AF appears to dip moderately southeastward (~50 deg.), toward the Pacific plate (PAC), but along the LARSE transects, the SAF dips vertically to steeply northeastward toward the North American plate (NAM). Away from the LARSE transects, the dip of the SAF changes significantly. In both locations, a midcrustal decollement is observed that connects the plate-boundary fault to thrust faults farther south in the PAC. This decollement allows upper crust to escape collision laterally and vertically, but forces the lower crust to form crustal roots, reaching maximum depths of 44 km (South Island) and 36 km (southern California). In both locations, upper-mantle bodies of high P velocity are observed extending from near the Moho to more than 200-km depth. These bodies appear to be confined to the PAC and to represent oblique downwelling of PAC mantle lithosphere along the plate boundaries.

[1]Earthquake Hazards Team, U.S. Geological Survey, Menlo Park, California.

[2]Center for Embedded Networked Sensing, University of California, Los Angeles, California.

[3]School of Earth Sciences, Victoria University of Wellington, Wellington, New Zealand.

[4]Coastal and Marine Geology Team, U.S. Geological Survey, Woods Hole, Massachusetts.

[5]Institute of Geophysics, University of Texas, Austin, Texas.

[6]Now at Leibniz Institute of Marine Sciences, IFM-GEOMAR, Kiel, Germany.

A Continental Plate Boundary: Tectonics at South Island, New Zealand
Geophysical Monograph Series 175
Copyright 2007 by the American Geophysical Union.
10.1029/175GM16

INTRODUCTION

The South Island Geophysical Transect (SIGHT), on the South Island of New Zealand, and the Los Angeles Region Seismic Experiment (LARSE), in southern California, USA, were carried out to image and understand the lithospheric structure across two large and comparable right-lateral, transpressional fault systems, the Alpine and San Andreas Faults (AF and SAF, respectively; Figure 1). Although there are clear similarities between the two fault systems and their structural and tectonic settings, there are also notable differences. It was hoped at the outset of both experiments that similarities and differences between the two fault systems would manifest themselves and reveal the nature of the

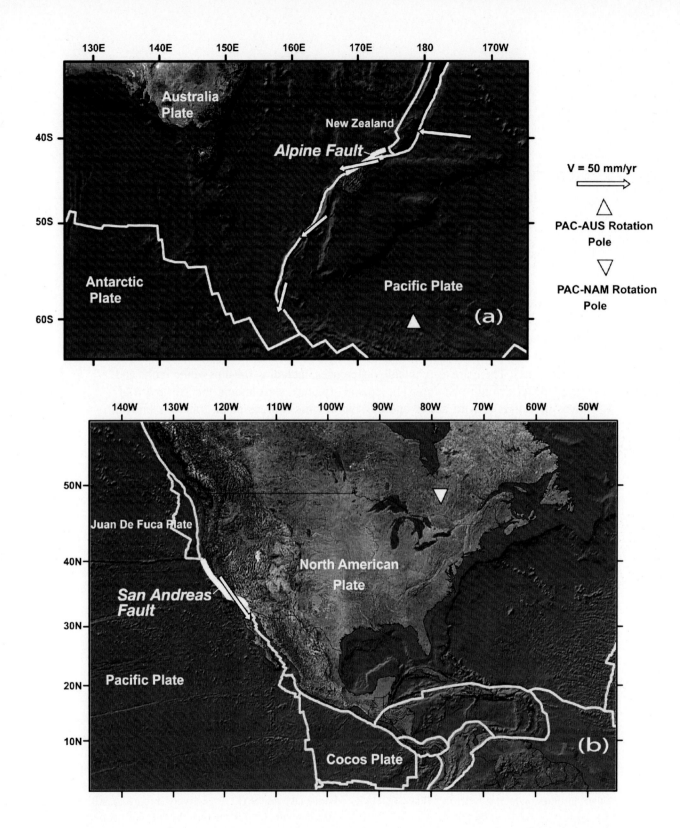

Figure 1. Shaded relief maps of (a) southwest Pacific Ocean and (b) North America, showing plate boundaries (white lines), poles of rotation (white triangles) for Pacific-Australian (PAC-AUS) plates and Pacific-North American (PAC-NAM) plates, and plate-motion vectors (white arrows) near Alpine and San Andreas Faults (heavy white lines). Ocean, dark gray; regions above sealevel, various shades of light gray. Color version of this figure is on CDROM which accompanies this volume.

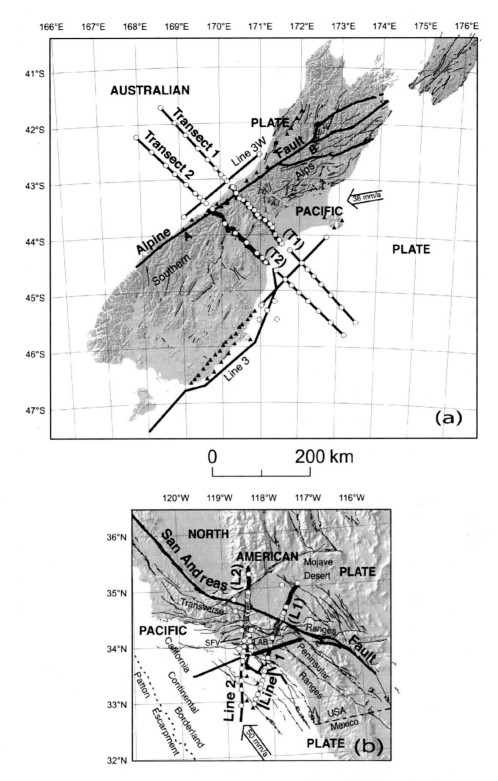

Figure 2. Shaded relief maps of (a) South Island of New Zealand and (b) southern California, USA, showing transects of South Island Geophysical Transect (SIGHT) and Los Angeles Region Seismic Experiment (LARSE). Black triangles, portable onshore seismographs; white-filled circles, shotpoints (onshore) and ocean-bottom seismographs (offshore); thin black lines, locations of airgun sources and towed streamers; heavy black lines, plate boundary faults (including Alpine Fault and its branches and San Andreas Fault); thin black lines, other Cenozoic faults. A,B outline central part of Alpine Fault studied by *Sibson et al.* [1979]. Pacific plate motion vectors shown in both (a) and (b) relative to Australian and North American plates, respectively, LAB, Los Angeles basin, SFV, San Fernando Valley. Color version of this figure is on CDROM which accompanies this volume.

underlying geological processes in both locations. This hope has, to a large extent, been realized.

On South Island and in southern California, relatively detailed seismic profiles and complementary geophysical data were collected across the fault systems (Figure 2). SIGHT and LARSE, collected in the middle to late 1990's, included two transects each separated by distances of 50–70 km, in order to obtain information on the three-dimensional nature of both fault systems. Both had offshore, offshore-onshore, and onshore active-source components [e.g., *Davey et al.*, 1998, this volume; *ten Brink et al.*, 2000; *Fuis et al.*, 2001, 2003], as well as passive components [e.g., *Kohler and Eberhart-Phillips*, 2002; *Kohler et al.*, 2003]. Both experiments produced models of seismic velocity and reflectivity that permit detailed comparison. These two fault systems may be the best studied transpressional plate boundaries to date.

GENERAL SIMILARITIES BETWEEN THE ALPINE AND SAN ANDREAS FAULT SYSTEMS

The AF and onshore part of the SAF systems originated more than 20 Ma ago [22–26 Ma for the AF and 22 Ma (onshore) to 28 Ma (offshore) for the SAF; see *Molnar et al.*, 1975; *Carter and Norris*, 1976; *Kamp*, 1986, for the AF, and *Atwater*, 1989; *Atwater and Stock*, 1998; *Powell*, 1993, for the SAF]. Both fault systems have similar total fault offsets of 450–480 km, although the the onshore SAF has less offset in southern California [315 km; *Powell*, 1993]. The plate boundary regions in New Zealand and the western U.S. have both experienced additional deformation. In New Zealand, oroclinal bending, beginning as early as the Eocene, may account for 300 km of additional right-lateral deformation, although much debate has centered around the origin and timing of this oroclinal bending [see summary in *Sutherland*, 1999; *Cox and Sutherland*, this volume]. Extension and strike slip faulting inboard of the SAF, in the Basin and Range province, plus offset along older faults of the SAF system, beginning as early as 28 Ma, account for 600+ km of additional right-lateral deformation [*Atwater and Stock*, 1998]. Both fault systems changed from pure strike-slip or transtensional regimes to transpressional regimes at approximately similar times: 6.4 Ma for the AF and ~5–6 Ma for the SAF [see *Walcott*, 1998, for a summary of New Zealand; *Powell*, 1993; *Oskin et al.*, 2001, for southern California]. Both the current traces of the AF and SAF have somewhat similar geodetic right-lateral slip rates of 35–37 mm/yr (AF) to 20–30+ mm/yr (SAF) [see *Beavan et al.*, 1999; this volume, for AF; *Savage and Lisowski*, 1998 (30+ mm/yr); *Argus et al.*, 2005 (20 mm/yr), for SAF]. The geodetic compression rates across the regions are also similar, 10–11 mm/yr [*Beavan et al.*, 1999, this volume, for AF; *Argus et al.*, 2005, for SAF].

Total relative plate motion rates at the latitudes of South Island and southern California are ~38 mm/yr for PAC-AUS, and ~50 mm/yr for PAC-NAM. Note that the AF takes up the majority (60–80 percent) of the PAC-AUS plate motion, whereas the SAF takes up a smaller fraction (40–70 percent) of the PAC-NAM plate motion.

GENERAL DISSIMILARITIES BETWEEN THE ALPINE AND SAN ANDREAS FAULT SYSTEMS.

The plate-tectonic histories differ for the AF and SAF. The AF originated as the transform boundary between the PAC and AUS at 22–26 Ma. The AF has had a single primary strand until the last few Ma and currently connects two oppositely dipping subduction zones [see summary in *Walcott*, 1998; *Cox and Sutherland*, this volume]. The current pole of rotation between the two plates is a little over 15 degrees away from the AF. It has migrated southward to a position where it puts the two plates into transpression across the AF (Figure 1a). The nearness of the current pole of rotation leads to compression across the fault that changes along strike. In contrast, the history of the SAF system has involved the interactions of three plates, the Farallon, Pacific, and North American plates (Figure 1b). The Farallon plate was subducting beneath most of the western USA until 28 Ma, when a segment of the PAC-Farallon spreading ridge approached the NAM trench in the region of northern Mexico, and a transform fault was created connecting two oppositely migrating triple junctions [*Atwater*, 1989]. (The largest remnants of the Farallon plate are the Juan de Fuca and Cocos plates (Figure 1b).) This transform fault migrated landward, gradually adding fragments of NAM to PAC, until it jumped ashore creating the onshore SAF system. Baja California was the most recent addition of NAM to PAC, which occurred between 6.4 and 4.7 Ma [*Oskin et al.*, 2001]. When the PAC-NAM plate boundary jumped inland in southern California to the current Gulf of California, the "Big Bend" in the SAF was created, and transpression replaced transtension.

The dips of the two faults appear to differ. Based on foliations in outcrops within the AF zone, *Sibson et al.* [1979] provide evidence that, in its central part (A to B, Figure 2a), including the vicinity of the SIGHT transects, the fault dips consistently 40 to 50 degrees southeastward. Gravity driven nappes flatten the fault zone to gentle dips in many places, but these places were avoided by *Sibson et al.* [1979] in determining the consistent moderate southeast dip. A steeper dip of 50–70 degrees was determined by *Woodward* [1979] from gravity, with the AF connecting to the steep northwest side of a crustal root (see below). In contrast, the SAF has a steep northeast to vertical dip in the vicinity of the LARSE lines, based on seismic results and potential-field studies (see below). To the northwest, however, it reverses dip to moderately southwestward [*Griscom and Jachens*, 1990]. To the southeast, it dips moderately northeastward [*Jones et al.*, 1986; *Griscom and Jachens*, 1990; *Fuis et al.*, 2007], although local dip reversals with depth ("wedge tectonics") have been modeled in one place from potential-field data [*Langenheim et al.*, 2005].

Rainfall amounts and erosion patterns differ strongly between South Island and southern California. On South Island rainfall and erosion are highly asymmetric, with the most rainfall (3 m/yr) and erosion occurring on the northwest side of the Southern Alps. The southeast side of the Southern Alps gets 0.65 m/yr rainfall, by contrast. Erosion removes much of the rock being uplifted on the northwest side of the Southern Alps, along the Alpine fault [see *Koons*, 1990], keeping elevations low and exposing rocks uplifted from mid-crustal levels [see *Sibson et al.*, 1979]. In contrast, the rainfall/erosion pattern in southern California is only moderately asymmetric, but the rainfall is six to ten times lower than on South Island. The Transverse Ranges still contain preserved sedimentary basins dating from the Miocene and earlier, indicating much lesser uplift and erosion than on South Island.

COMPARISON BETWEEN CRUSTAL STRUCTURES ACROSS THE ALPINE AND SAN ANDREAS FAULTS

SIGHT, which includes transects T1 and T2, and LARSE, which includes Lines 1 and 2 (L1 and L2), involved offshore and onshore recording of airguns and onshore recording of explosions (Figure 2). The length of each SIGHT transect was ~600 km (onshore ~150 km; offshore ~450 km), and the length of the LARSE transects ranged from 350–400 (onshore ~150 km; offshore ~200–250 km). Onshore shotpoint and station spacing for SIGHT was ~10–20 km and 300–400 m, respectively, vs 1–25 km and 100-500 m for LARSE. Offshore recordings were much more extensive for SIGHT, because they could be collected on both sides of South Island. On all four transects, structures associated with the modern Alpine and San Andreas Fault systems are seen as well as structures associated with older continent-building events.

Water Depth and Sedimentary Basins

Water depths on the continental shelves offshore of South Island and southern California are up to ~1.5 km. Late Cretaceous to Cenozoic sedimentary basins offshore of South Island vary in thickness between AUS and PAC, with a common thickness range and a maximum thickness of 1–2 km and 4 km, respectively, for AUS, and 2–3 km and 6 km,

respectively for PAC [*Scherwath et al.*, 2003; *van Avendonk et al.*, 2004; Plates 1a, 1b]. Cenozoic sedimentary basins offshore of southern California are typically shallow along the LARSE lines [0.5 km; *Baher et al.*, 2005] but deepen toward the shore, reaching variable depths of 1.5 to 4 km [*Wright*, 1991] (Plates 1c, 1d). Onshore sedimentary basins, on the other hand, are shallower on South Island compared to southern California, at least in the vicinity of the transects discussed here. For example, onshore along T1 on South Island, Late Cretaceous to Cenozoic sedimentary and volcanic rocks reach maximum depths of 2.5–3 km, on the southeast side of South Island (Plate 1a). Onshore along Line 1 in southern California, Cenozoic sedimentary and volcanic rocks thicken abruptly to as much as 10-km depth in the Los Angeles basin [*Fuis et al.*, 2001; Plate 1d]. Onshore along Line 2 in southern California, these basins reach various maximum depths, ranging from ~5 km in the San Fernando Valley to ~3 km farther north [*Fuis et al.*, 2005; Plate 1c].

Basement and Lower Crust

South Island. "Basement" is used here to refer to upper-crustal rocks having velocities in the approximate range of 5.0–6.5 km/s. "Lower crust" here refers to rocks having velocities in the approximate range of 6.5–7.5 km/s. On both transects T1 and T2 offshore of South Island, basement rocks reach depths ranging between 15 and 19 km on both AUS and PAC, but these depths increase to more than 20 km in the vicinity of the coast [Scherwath et al., 2003; van Avendonk et al., 2004; Plates 1a, 1b]. Onshore they deepen further to form the upper part of an asymmetrical crustal root, which has a steep side on the northwest and a gentle side on the southeast (Plates 1a, 1b). On both transects, basement velocity reaches slightly higher values on the PAC (6.2–6.3 km/s) than on AUS (6.0–6.1 km/s). On both transects, a prominent zone of relatively low basement velocity (5.8–6.0 km/s) is seen in the vicinity of the crustal root from near the surface to the lower crust (see discussion below). The base of basement rocks is constrained by intracrustal reflections (PiP) from the top of lower crust.

The AF is contained within the upper-crustal low-velocity zone, but unfortunately, its dip through the crust is uncertain. Surface outcrops of the fault near the SIGHT lines indicate a dip of 40–50 degrees southeastward [*Sibson et al.*, 1979]. On T2, moderately southeast-dipping reflections are seen at mid-crustal depths that are possibly related to the AF (Plate 1b). On a supplementary reflection profile along or near T2, Stern et al. (this volume) find steeply southeast-dipping reflections, which when migrated, suggest a 60-degree dip for the AF at ~15–20 km depth and a listric shape, whereby the fault becomes subhorizontal between 30- to 35-km depth. Thus, the AF appears to flatten into a mid- to lower-crustal decollement (see below).

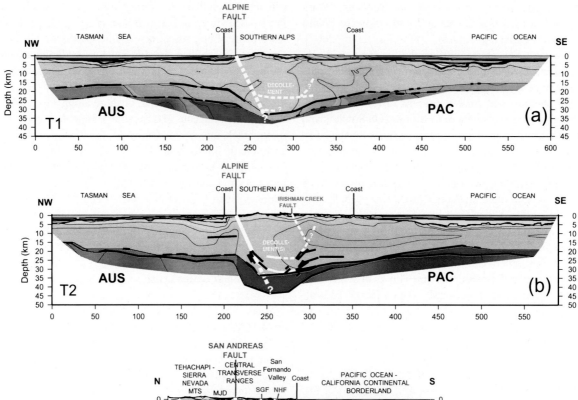

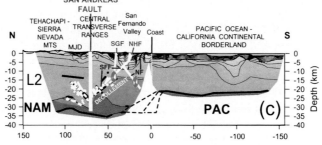

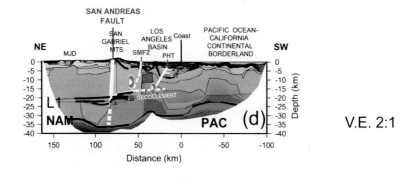

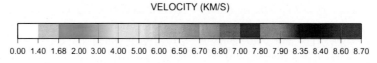

VELOCITY (KM/S)

0.00 1.40 1.68 2.00 3.00 4.00 5.00 6.00 6.50 6.70 6.80 7.00 7.80 7.90 8.35 8.40 8.60 8.70

Offshore of South Island, AUS lower crust is thicker on T1 (5–7 km) than on T2 (0.5–2 km), but PAC lower crust is similar on both transects (1–5 km) [*Scherwath et al.*, 2003; *van Avendonk et al.*, 2004; Plates 1a, 1b]. Onshore lower crust on T1 thins to 2–4 km at the base of the crustal root, but on T2, it thickens to 8–10 km within the root (Plates 1a, 1b). Velocity of the lower crust and its base (Moho) are constrained by Moho reflections (PmP). Moho depth (crustal thickness) is in the range of 25–28 km at both coasts on both T1 and T2. It is interesting to note that this same coastal crustal thickness is observed on both transects in southern California (see below; Plates 1c, 1d). Moho depth shoals seaward of the coast and deepens landward into the crustal root discussed above. The maximum crustal thickness at the root is 37 km on T1 and 44 km on T2; its relief is ~10 km on T1 and 18 km on T2. On both T1 and T2, the maximum crustal thickness is located ~15 km southeastward of the main topographic divide of the Southern Alps and 40 km southeastward of the AF.

Based on seismic velocities and crustal thicknesses, both of the two transects across South Island outside of the Southern Alps represent thin continental crust [see *Meissner*, 1986, for continental thicknesses; *Christensen and Mooney*, 1995, for continental velocities]. Beneath the Southern Alps, crustal thicknesses are in the normal range. At the northwest end of T2 (offshore), crustal thinning and an increase of sedimentary basin thickness probably represents a transition toward oceanic crust [*Scherwath et al.*, 2003].

Southern California. Offshore of southern California, ten Brink et al. [2000] modeled basement on Line 1 extending to the Moho at ~ 20-km depth. Baher et al. [2004, 2005] have modeled base- ment on this line extending to ~17–18-km depth, with a thin (~5-km thick) lower crust and a deeper Moho (~23 km). Unlike South Island, there are no intracrustal reflections (PiP) to directly constrain the top of the lower crust. As Moho depth is directly correlated with thickness and velocity of the lower crust (or, alternatively, with average crustal velocity), the models of ten Brink et al. [2000] and Baher et al. [2004, 2005] can be regarded as alternate models. The model of Baher et al., however, is constrained by additional data, namely, onshore-offshore data, and we have chosen to show their model in Plate 1d. For Line 2, we have only the model of ten Brink et al. [2000], which we regard as preliminary, as onshore-offshore data can be used to better constrain this model now that the onshore model has been finalized [Fuis et al., 2005]. In any case, a large step or slope in Moho is required near the coast on this line [Plate 1c; see ten Brink et al., 2000].

Onshore in southern California, Line 1 is underlain by a three-part crust south of the San Gabriel Mountains (part of the Central Transverse Ranges), a two-part crust in the San Gabriel Mountains, and a one-part crust in the Mojave Desert, north of the SAF [Plate 1d; *Fuis et al.*, 2001]. From the coast to the San Gabriel Mountains, the crust contains deep (5- to 10-km-deep) Cenozoic sedimentary basins, basement, and lower crust with a prominent intrusion or horst-like feature projecting upward to near the base of the sedimentary rocks in the northern Los Angeles basin (Plate 1d; intrusion/ horst-like feature is at 35–45 km in the model coordinate system). The San Gabriel Mountains are underlain by basement to ~20-km depth and lower crust to 36-km depth. Basement consists of relatively high velocities (6.2 km/s) in the upper 5 km overlying relatively low velocities (5.8–6.0 km/s) that extend to 20-km depth. At the base of basement rocks, strong wide-angle reflections and a marked zone of bright near-vertical-incidence reflections are observed that appear to project gently upward and southward from the SAF to the 1987 M 5.9 Whittier Narrows earthquake hypocenter [features A and B, Plate 1d; *Ryberg and Fuis*, 1998; see discussion below]. North of the SAF, basement velocities (< 6.45 km/s) occur down to Moho depths of 32–36 km. Strong wide-angle reflections are seen at 20–24-km depth in the Mojave Desert. A steeply north-dipping SAF (85 degrees) is interpreted from an offset in wide-angle reflectors (termination of reflective layer A) at a point a few km north of the surface trace [*Fuis et al.*, 2001]. Potential-field studies, however, indicate a vertical dip [*Langenheim*, 1999]. The SAF is not associated with a wide upper-crustal low-velocity zone as on South Island, although on Line 1, a faint, poorly resolved, narrow (~5 km) low-velocity zone extends to midcrustal levels (Plate 1d).

Plate 1. (opposite) Models of crust and uppermost mantle for (a) SIGHT T1 [*van Avendonk et al.*, 2004], (b) SIGHT T2 [*Scherwath et al.*, 2003], (c) LARSE Line 2 (L2) [*ten Brink et al.*, 2000; *Fuis et al.*, 2003, 2005], and (d) LARSE Line 1 (L1) [*Fuis et al.*, 2001; *Baher et al.*, 2004, 2005]. Heavy black lines, wide-angle reflectors; thin, dashed black lines in (c), alternate Moho geometries from *ten Brink et al.* [2000]. Heavy white lines, plate boundary faults, dashed where not clearly observed; thin white-line segments in (c), near-vertical-incidence reflectors. Blue beach balls in (c) and (d), focal mechanisms for M 5.8–6.7 earthquakes within 10 km of transects; far hemispheres displayed; vertically exaggerated along with rest of figure. Blue bodies A and B in (d), bright-layer reflections. AUS, Australian plate; PAC, Pacific plate; NAM, North American plate. Abbreviations in (c) and (d) MJD, Mojave Desert; NF, Northridge fault; NHF, Northridge Hills fault; PHT, Puente Hills thrust fault; SFF, San Fernando fault; SGF, San Gabriel fault; SMFZ, Sierra Madre fault zone. V.E. 2:1.

Beneath Line 1, the Moho has several steps. It steps downward from 24 to 28 km near the coast, and from 28 to 36 km near the boundary between the Los Angeles basin and the San Gabriel Mountains. It steps upward again, from 36 to 32 km in the southern Mojave Desert. The latter two steps bound a crustal root that is approximately centered on the SAF, ~6 km north of the topographic high of the San Gabriel Mountains [*Fuis et al.*, 2001; *Baher et al.*, 2004]. The root has a relief of 4–8 km, smaller than the root on South Island. *Kohler and Davis* [1997] found that a crustal root with vertical relief of ~10 km, also centered below the SAF, explained LARSE teleseismic P-wave travel-time residual observations.

Similar to onshore Line 1, onshore Line 2 [Plate 1c; *Fuis et al.*, 2003, 2005] is underlain by a three-part crust south of the Central Transverse Ranges and a two-part crust in the Central Transverse Ranges. In the western Mojave Desert north of the SAF, it is underlain by a three-part crust, as opposed to the one-part crust of Line 1. From the coast to the Central Transverse Ranges, the crust contains moderately deep (3- to 5-km-deep) Cenozoic sedimentary basins, basement, and lower crust. The Central Transverse Ranges are underlain by basement down to a north-dipping interface extending from the 1971 M 6.7 San Fernando earthquake hypocenter, at ~13-km depth, to the SAF at 25-km depth. This interface and a fainter interface a few km above it bracket a near-vertical-incidence to wide-angle reflective zone. The western Mojave Desert is underlain by a moderately deep sedimentary basin (3+ km), basement, and a relatively thin lower crust. Basement contains a south-dipping, near-vertical-incidence reflective zone that is symmetric to the reflective zone south of the SAF. A subvertical SAF appears to offset all layers above the Moho. Lower crust is much thicker south of the SAF and has a poorly constrained velocity. The Moho ranges in depth from 22 km just offshore [*ten Brink et al.*, 2000] to 36 km beneath the Transverse Ranges, forming a crustal root with a relief of 5 km (north) to 14 km (south). The SAF is located toward the north side of this root.

Based on seismic velocities and thicknesses, both of the transects across southern California represent thin continental crust southwest of the crustal roots and normal continental crust elsewhere. The transition to PAC oceanic crust occurs approximately 150 km southwest of the offshore ends of the LARSE lines, at the Patton Escarpment (Figure 2b).

UPPER-MANTLE VELOCITIES AND STRUCTURE

Observations and Models

South Island. Owing to recordings at long offsets on the T1 and T2 transects and to recordings along cross lines, upper-mantle Pn velocities are relatively well resolved on South Island. Along T1 and T2, mantle velocities are above 8.0 km/s (as high as 8.2–8.3 km/s) everywhere except in a region beginning at the bottom of the crustal root (T1) or the north side of the crustal root (T2; no Pn coverage below the root) and extending northwestward ~100 km, where the velocity is as low as 7.8 km/s [*Scherwath et al.*, 2003; *van Avendonk*, 2004; Plates 1a, 1b]. Crossline results near the northwest and southeast coasts reveal Pn velocities of 8.2–8.6 km/s [*Scherwath et al.*, 2002; *Melhuish et al.*, 2005] and 8.2 [*Godfrey et al.*, 2001], respectively. The crossline near the northwest coast (Figure 2a, Line 3W) traverses the region of 7.8-km/s velocity on transects T1 and T2 and clearly indicates horizontal anisotropy in the upper mantle [*Scherwath et al.*, 2002]. Thus, high Pn velocities are parallel to the AF. The cross line near the southeast coast (Figure 2a, Line 3), more than 200 km away from the plate boundary, agrees approximately with transects T1 and T2 in upper mantle velocity and indicates no significant horizontal anisotropy there [*Godfrey et al.*, 2001; cf. *Baldock and Stern*, 2005].

Steeply traveling P waves from ~40 teleseisms recorded on South Island reveal a body of relatively high upper-mantle velocity (up to 3% higher than the average for South Island) that underlies most of South Island, except for a couple of regions on the southeast side of the island [*Kohler and Eberhart-Phillips*, 2002; Plate 2a]. In three dimensions, this body of relatively high velocity is elongate (nearly 600 km long), ~100 km across, appears to cross the AF obliquely, and extends to depths of more than 200 km. It extends the length of the AF and into the subduction zones on either end (Plate 2a). A cross section along T1 shows that this body is tabular and dips moderately northwestward (~55 degrees; Plate 3a, top); however, resolution tests for this region indicate that at least some of this northwestward dip is likely an artifact of raypaths used in the inversion (for example, a vertical test cylinder dips ~80 degrees north after inversion). At 45-km depth, its northwest edge is located near the center of the crustal root. Similarly shaped tabular regions of relatively low velocity are seen both northwest and southeast of this high-velocity body. A cross section along T2 shows more complexity, with the high-velocity body being much wider and having both northwest- and southeast-dipping legs (Plate 3a, bottom). At 50-km depth, the northwest boundary of the feature is located near the northwest edge of the crustal root.

In contrast to the results of *Kohler and Eberhart-Phillips* [2002], *Stern et al.* [2000] found from modeling three teleseisms that occurred inline with portable recorders on T1 and T2 that the upper-mantle high-velocity body is constrained to be nearly vertical, approximately 70 km wide, limited in depth extent (from ~60- to 160-km depth), having a maximum velocity of 8.5-8.6 km/s, and centered approximately

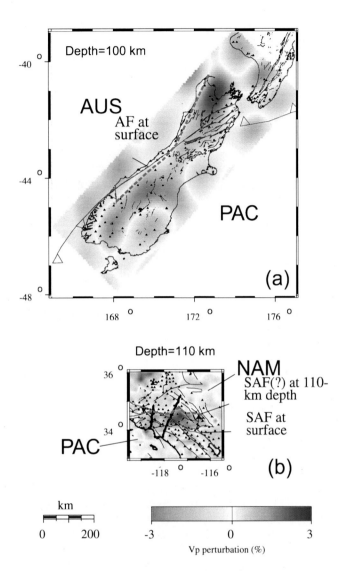

Plate 2. Horizontal slices of mantle for (a) South Island (100-km depth) and (b) southern California (110-km depth). Black triangles, seismographs that recorded teleseisms for mantle tomography; solid red lines, plate boundary faults (including Alpine Fault (AF) and its branches and San Andreas Fault (SAF)); red lines with teeth, subduction zones (teeth on upper plate); dashed cyan lines, northwest and north sides of high-velocity bodies (blue) at 100-, 110-km depth for (a), (b), respectively; thin black lines, other Cenozoic faults. (Note that all surface features, including AF and SAF, are projected downward 100 to 110 km.) Zero on color scale is 8.3 km/s at 100-km depth for (a) and 8.1 km/s at 110-km depth for (b). In southern California, geometric relationship can be established between SAF (red line) and north side of high-velocity body (dashed cyan line) that gives rise to apparent counterclockwise rotation of latter away from former (black rotation arrows; see text). On South Island, origin of similar apparent counterclockwise rotation is not as clear.

beneath the crustal root. Instrument density was greater in the two-dimensional study of *Stern et al.* [2000; 2- to 4-km spacing] than in the three-dimensional study of *Kohler and Eberhart-Phillips* [2002; 20- to 100-km spacing], but only three inline earthquakes were used in the former study versus ~40 earthquakes from many azimuths in the latter. Differences between the two studies also stem in part from the fact that the three-dimensional study was an inverse model and, therefore, necessarily smooth, whereas the two-dimensional study was a forward model, with relatively sharper boundaries. In our illustrations (Plate 3a), we show both models of the high-velocity body.

Southern California. In southern California, average Pn velocities determined from local earthquake sources are generally higher north and south of the Transverse Ranges (8.0–8.2 km/s) than within the Transverse Ranges (7.7–7.8 km/s) [*Hearn*, 1984], whether one considers anisotropy or not. From active sources on the LARSE transects, Pn is poorly resolved but seems to be less than or equal to about 8.0 km/s everywhere. The azimuthal variation in Pn for the central part of southern California (Mojave Desert, Transverse Ranges, and Los Angeles basin) is quite striking [*Hearn*, 1984]. Hearn calculates that the fast direction is N75W, approximately parallel to the strike of the SAF through the Transverse Ranges, and that anisotropy is ~3–4%. (A prior study of Pn anisotropy for approximately the same region [*Vetter and Minster*, 1981] found approximately the same value for anisotropy but a fast direction of N50W, approximately parallel to the direction of relative motion between PAC and NAM.)

A number of authors have imaged or modeled a prominent high-velocity body in the upper mantle of southern California [*Hadley and Kanamori*, 1977; *Raikes*, 1980; *Bird and Rosenstock*, 1984; *Humphreys et al.*, 1984; *Sheffels and McNutt*, 1986; *Humphreys and Clayton*, 1990; *Humphreys and Hager*, 1990; *Kohler*, 1999; *Kohler et al.*, 2003]. The study of *Kohler et al.* [2003], which includes data from several hundred teleseisms recorded on portable deployments (2-km spacing) along the LARSE lines and on the permanent southern California array, provides the highest resolution image of this body to date (Plates 2b, 3b). This body extends more than 250 km east-west through the Transverse Ranges and is somewhat narrower in plan view than the South Island body (Plate 2). Similar to the body on South Island, it appears to cross the plate boundary obliquely and extends to more than 200-km depth. This body is ~70–80 km wide at or near the Moho but broadens to 160- to 240-km width at a depth of 200 km (compare Plate 3b with Plate 3a). At 40-km depth on Line 1, the northeast edge of the feature is located at or near the projection of the north-dipping SAF to the Moho. It is irregular and split below 160-km depth (similar to the fea-

ture on T2), and the dips of the two components below 160 km are ~80 degrees and 55 degrees northward. The more steeply dipping component of the feature has approximately the same dip as the deep projection of the SAF. At 40-km depth on Line 2, the northeast edge of the body is also at or near intersection of the SAF (here vertical) and the Moho. The north side of this feature also appears to be vertical.

In the cross sections along the LARSE lines (Plate 3b) and in cross sections across the SAF farther northwest and southeast [*Fuis et al.*, 2006, 2007], the north side of the high-velocity body appears coplanar or at least continuous with the SAF. Both the SAF and the north side of the high-velocity body change dip along strike: in the northwest, the dip is moderately southwest; at the LARSE lines, it is steep; and in the southeast it is moderately northeast. Thus, the overall shape of a surface fitted to the SAF and the north side of the high-velocity body appears to be crudely propeller shaped. This geometry gives rise to the appearance in plan view that the north side of the high-velocity body crosses and is rotated counterclockwise from the surface trace of the SAF (Plate 2b, dashed cyan line). Although one sees a similar apparent plan-view geometry on South Island for the northwest side of the high-velocity body (Plate 2a, dashed cyan line), the deep structure of the plate boundary beyond the vicinity of the SIGHT lines is less clear.

Resolution

Formal lateral resolution limits differ between South Island and southern California for the upper-mantle high-velocity bodies because of different station spacing and different inversion parameterizations. In southern California, station spacing ranges from 15 to 40 km in the central part of the network, although LARSE portable arrays were also used, where spacing was 2 km. The mantle was gridded in 20- by 20-km blocks and the formal lateral resolution length-scale limit for the whole region is ~ 40 km. In checkerboard and vertical-cylinder tests, however, one observes near the LARSE lines that the inverted boundaries of the input checkerboard or vertical cylinders differ from input locations by generally less than 15–20 km (at depths of 40 to 200 km), indicating good resolution at a length scale of 15–20 km. On South Island, station spacing for the permanent network averages ~100 km, but several portable arrays with denser station spacing were used. A two-dimensional Fourier series parameterization was used (on 40-km-thick layers) and the formal lateral resolution length-scale limit for the whole region is a bit over 100 km. As for southern California, checkerboard and vertical-cylinder tests near the SIGHT lines indicate boundary displacements by the inversion of less than ~20 km (at depths of 100 to 200 km). As

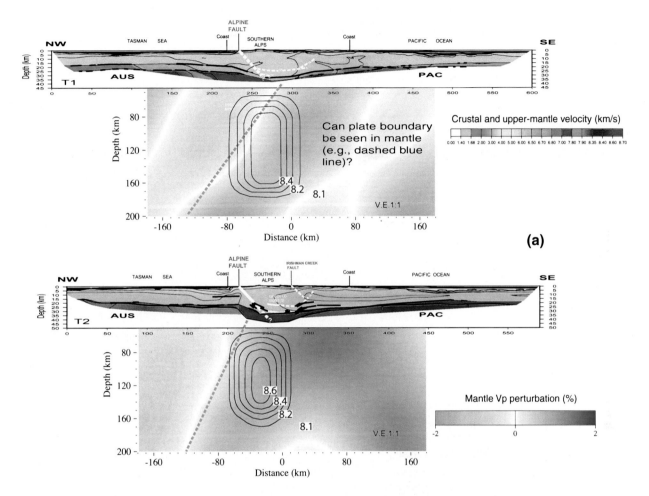

Plate 3. (a) Crustal models from Plates 1a, 1b combined with vertical slices of mantle below each line. See Plate 1 for explanation of symbols in crust. In mantle, color scale is from *Kohler and Eberhart-Phillips* [2002]. Zero on color scale is 8.2 km/s at 40-km depth and 8.5 km/s at 200-km depth. [For intermediate depths, refer to *Kohler and Eberhart-Phillips*, 2002]. Thin purple lines with P velocities (in km/s), velocity contours from *Stern et al.* [2000]; heavy cyan and purple dashed lines, our interpretations of northwest boundaries of high-velocity bodies in mantle from both studies above: cyan, *Kohler and Eberhart-Phillips* [2002], purple, *Stern et al.* [2000]. These dashed lines may or may not represent traces of plate boundary in mantle (see text). V.E. 1:1. (b) Crustal models from Plates 1c, 1d combined with vertical slices of mantle below each line. See Plate 1 for explanation of symbols and abbreviations in crust. In mantle, color scale is from *Kohler et al.* [2003]. Zero on color scale is 7.8 km/s at 40-km depth and 8.2 km/s at 200-km depth. [For intermediate depths, refer to *Kohler et al.*, 2003]. Heavy dashed cyan lines, our interpretations of north boundaries of high-velocity body; thin dashed white line in (b), projection of SAF below Moho. As in Plate 3a, heavy dashed cyan lines may or may not represent traces of plate boundary in mantle (see text). Abbreviations: CI, Santa Catalina Island; MJD, Mojave Desert, SMM, Santa Monica Mts.; SCI, San Clemente Island; SCR, San Clemente Ridge. V.E. 1:1.

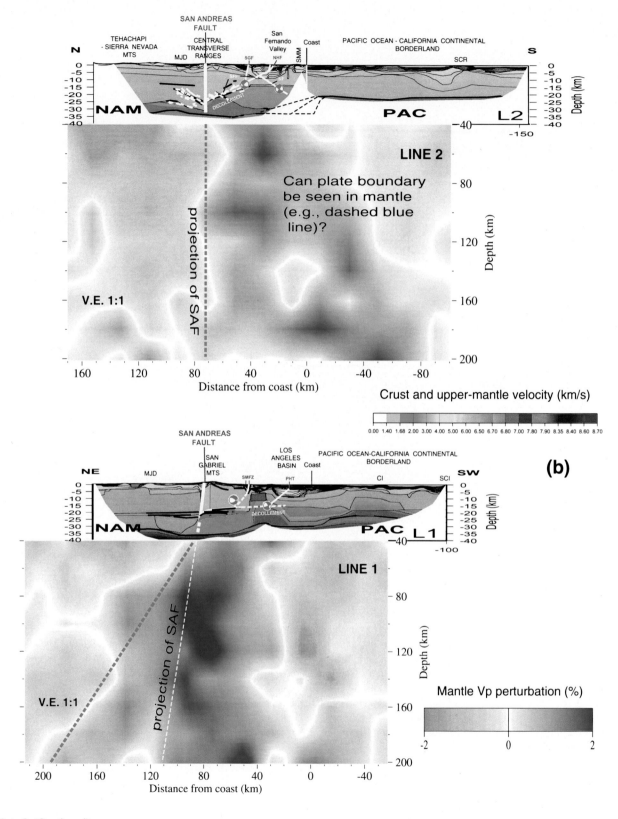

Plate 3. (Continued)

noted above, however, the inversion tends to add a small northward dip to vertical features.

Vertical resolution limits for features imaged on South Island and in southern California are poorly known and are the subject of ongoing investigation. Steeply dipping imaging rays would be expected to smear features vertically. The fact that we see features dipping moderately in different directions, suggests that vertical smearing is not extreme.

Interpretations of the Upper-Mantle High-Velocity Bodies

Based on surface studies of *Sibson et al.* [1979], the AF dips moderately southeastward and would project to approximately the center of the crustal root (Plate 3a). A line fitted to the northwest side of the high-velocity upper-mantle body also projects to within 20–30 km of the center of the crustal root, but dips moderately northwestward. This geometry is suggestive of indentation of the PAC by the AUS (see below). Alternatively, the inferred plate boundary in the mantle may not connect with the crustal plate boundary, and there may be a zone of decoupling in the upper mantle to lower crust. In a third alternative, the inferred plate boundary in the mantle is vertical and lies within the high-velocity body of *Stern et al.* [2000] (Plate 3a).

The cross sections shown here for southern California (Plate 3b), as well as cross sections across the SAF farther northwest and southeast [*Fuis et al.*, 2006, 2007], show that the high-velocity upper-mantle body is south of the SAF where the SAF intersects or is projected to intersect the Moho. The high-velocity body is, therefore, on the PAC and receives no clear contribution from the NAM near the Moho. This body may, thus, have a different thermal structure or composition from lithosphere of the NAM. The north side of the body is continuous with the crustal plate boundary (SAF) everywhere, and is coplanar with it everywhere except on LARSE Line 1. Given this geometry, a likely interpretation of this body is that it is downwelling of PAC mantle lithosphere beneath the Transverse Ranges. Specifically, this body is downwelling of lithosphere of the Peninsular Ranges and Continental Borderland, which constitute the PAC south of the Transverse Ranges (see Figure 2). This downwelling also involves a significant component of right-lateral movement of the lithospheric mantle. The crust does not descend into the mantle with the lithosphere (as in oceanic subduction) but becomes decoupled at some point and remains behind in a crustal root.

Upper-mantle high-velocity bodies on South Island and in southern California may arise from differences between these bodies and the surrounding mantle in density or in orientation of velocity components in an anisotropic medium. Density differences may arise, in turn, from thermal or compositional differences [see e.g., *Humphreys and*

Hager, 1990]. At this point, it is not possible to eliminate any of these possibilities with certainty, although an origin from anisotropy alone seems unlikely. In both locations, it appears likely that lithospheric mantle of the PAC is downwelling obliquely along the plate boundary. In both places, downwelling of lithospheric mantle is likely characterized by a slight density difference between lithospheric mantle of the PAC and the opposing plates caused by thermal or compositional differences, or both.

Models of high-velocity bodies in the upper mantle of continental orogens can be divided into at least 2 classes, one class based on Rayleigh-Taylor density instability, and one class based on subduction of one slab of lithosphere beneath another [see *Pysklywec et al.*, 2000]. In Rayleigh-Taylor instability, a denser layer of upper-mantle lithosphere downwells unstably ("drips") into a less dense asthenosphere. Two characteristics of Rayleigh-Taylor models are that the drips they produce in symmetric starting models are symmetric, contributed from the plates on either side of the point of initiation, and they are visible downward from the base of the lithosphere but not from the base of the crust [see e.g., *Houseman et al.*, 1981, 2000]. Subduction involves thrusting of one plate of lithosphere beneath another, followed by gravitational instability that results from penetration of the underthrust plate into the less dense asthenosphere. Two characteristics of subduction models are that the subducted slab is contributed primarily by one plate, and the geometry is asymmetric with respect to the point of initiation [see e.g., *Beaumont and Quinlan*, 1994; *Pysklywec et al.*, 2000]. These two classes of models can represent end members, given the broad spectrum of choices for convergence rates and rheological parameters for the various layers involved [crust, mantle lithosphere, asthenosphere; *Pysklywec et al.*, 2000].

Rayleigh-Taylor models. The models of *Stern et al.* [2000] for South Island and *Houseman et al.* [2000] for southern California are Rayleigh-Taylor models. In the model of *Stern et al.* [2000], plate convergence is postulated to cause depression of isotherms beneath some point of initiation. A high-velocity upper-mantle body develops gradually with depth as colder material in the center of the downwelling isotherms acquires a velocity contrast with hotter material on either side. *Houseman et al.* [2000] numerically simulate viscous flow in a lithospheric drip with the characteristics of the southern California body. Although they show a drip appearing only from the base of the lithosphere, in reality, the single layer they model as lithospheric mantle would likely have a temperature gradient, and downward flow of isotherms would create a drip that would become gradually visible with depth beneath the Moho. One might calculate that a 3% contrast

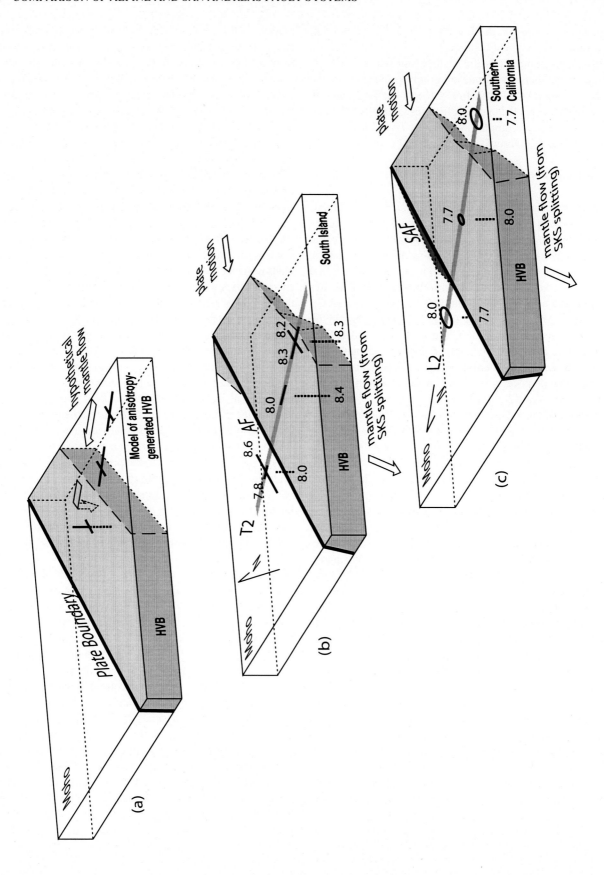

in velocity, such as shown in Plates 2, 3, would be acquired at approximately 50 km below the Moho, assuming an average mantle geotherm of 13 deg. C./km [*Williams*, 1996; for southern California], an adiabatic temperature gradient of 3 deg. C./km [*Lachenbruch and Sass*, 1980], and a P- velocity/temperature relationship of -0.5 km/s/deg. C. [*Creager and Jordan*, 1986]. One important feature of the Houseman et al. model is return upflow on the side of the drip, or convection, that serves to thin the lithospheric layer above and to enhance the thermal/density contrast of the drip to a somewhat shallower depth. In an attempt to simulate the asymmetric topography in southern California (high mountains only on the PAC), *Billen and Houseman* [2004] introduced a density contrast across the SAF and moved the point of drip initiation (shear weakening caused by strike-slip faulting) to a point south of the SAF, reasoning that older strands of the SAF are located in this region. They succeed in producing a drip south of the SAF, as we observe, but they start with equal contributions from the PAC and NAM.

Subduction models. The models of Pysklywec et al. [2000] demonstrate the varying influence of convergence rate and gravitational instability in the process of subduction. These models involve a relatively weak plastic crust, a relatively stiff plastic upper sublayer within the lithospheric mantle, a viscous lower sublayer within the lithospheric mantle, and an asthenosphere with a lower viscosity than the lower lithospheric layer. Some of their models [Figures 2, 4, and 5 of Pysklywec et al., 2000] match qualitatively the observations/inferences one might make from our Plate 3b, namely, an upper-mantle body of relatively high density that begins near the Mohos of the PAC and NAM, extends to 250+ km, and is chiefly on the PAC (with no to minimal contribution from the NAM). In these models, no significant amount of crust descends into the mantle, although a crustal root is formed. On the other hand, a model dominated by gravitational instability (where convergence is reduced to zero after initial underthrusting; their Figure 3), does not match the

features of our Plate 3b as closely—for example, the drip is relatively too thin.

Anisotropic models. Anisotropy may play a part in the observation of upper-mantle high-velocity bodies. For example, one might imagine the fast axis of olivine grains aligned with flow in the mantle [see summary in Fuchs, 1977], which might be horizontal and perpendicular to an orogen on either side of the orogen but which might rotate to vertical in a downwelling beneath the orogen (Figure 3a). Thus, steeply traveling imaging rays from teleseisms would image a high-velocity body beneath the orogen that owed its existence to nothing more than an anisotropy rotation. What is actually observed both on South Island and in southern California, however, is that, where a fast direction is defined in the mantle from Pn studies, it is parallel to the plate boundaries (and parallel to the orogens) and cannot be simply flipped downward by a component of mantle flow perpendicular to the plate boundary (Figures 3b, 3c). (Note that velocities in Figures 3b, 3c do not necessarily represent the true velocity components of an anisotropic medium; these velocities were not determined at exactly the same points and are the results of independent studies. For example, we do not know for certain that the vertical velocities from the tomographic studies apply to the uppermost mantle where the Pn results apply, given the likelihood of vertical smearing in the tomographic modeling. At best, the velocities shown in Figures 3b, 3c are estimates of the three components of an anisotropic velocity.) Nevertheless, both high-velocity bodies and the mantle on either side appear to be at least in part anisotropic.

Splitting of SKS phases results from anisotropy (for horizontal vibration directions) at some unknown location(s) along the long path of SKS from the Earth's core to the surface [see e.g., *Savage*, 1999]. For both South Island and southern California, the fast direction of SKS is on average approximately parallel to the long axes of the high-velocity bodies: NE-SW on South Island and E-W in southern California [for South Island, see *Savage et al.*, this volume, and discussion in *Scherwath et al.*, 2002; for southern

Figure 3. (opposite) Schematic diagrams for (a) hypothetical generation of a high-velocity upper-mantle body (HVB) from simple rotation of anisotropic velocity components, (b) observed anisotropy on South Island compared to boundaries of observed HVB, plate-motion vector (with respect to AUS), and mantle-flow vector (inferred from SKS splitting), (c) observed anisotropy in southern California similar to (b); plate-motion vector with respect to NAM. Note change in north arrow from (b) to (c). Top of blocks, Moho; heavy lines, plate boundaries; moderately heavy lines for (b), schematic Pn velocity components with attached values in km/s from *Scherwath et al.* [2003], van *Avendonk et al.* [2004], and *Godfrey et al.*, [2001]; moderately heavy ellipses for (c), schematic Pn velocity components from *Hearn* [1984] with long axes labeled with values of fast component, in km/s; moderately heavy dotted lines in (b) and (c), teleseismic mantle velocities from *Kohler and Eberhart-Phillips* [2002] (b) and *Kohler et al.* [2003] (c). Note that vertical components of velocity in (b) and (c) are not connected to horizontal, or Pn, components; it is not clear that these components were measured at same points as Pn components, namely in uppermost mantle (see text).

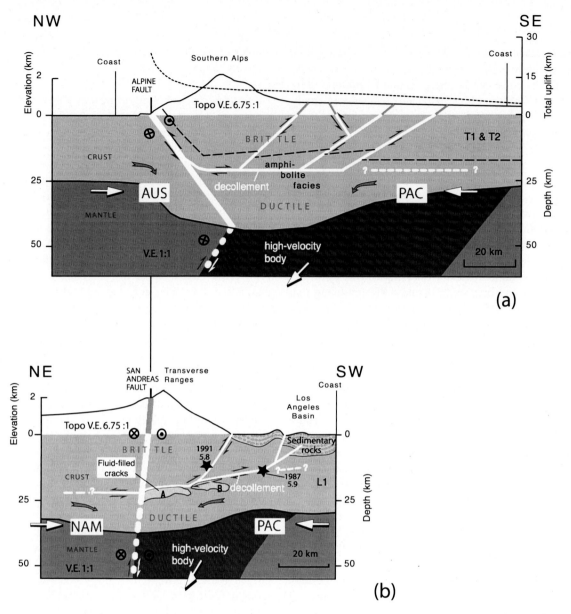

Figure 4. (a) Schematic diagram of tectonics of South Island along transects T1 and T2, adapted from *Norris et al.* [1990]. Light gray, crust; medium gray, relatively low-velocity mantle; dark gray, relatively high-velocity mantle. Heavy white line, Alpine Fault; heavy dotted line, interpreted plate boundary in mantle; thin white lines, other active faults (with relative-motion arrows); dashed black line, top of amphibolite-facies rocks (datum for fault offsets); curved arrows, rock motions for lower crust; large white arrows, relative plate-motion components in plane of section. (Note that motion along plate boundary in crust and mantle is primarily strike slip (circle with x, motion away from viewer; circle with dot, motion toward viewer), with a component of compression (large white arrows—see above)). AUS is interpreted to indent PAC along AF and shear zone in mantle. Compression leads to formation of decollement separating brittle upper crust from ductile lower crust. Above decollement, rock travels through orogen, undergoing deformation on both sides of Southern Alps, with erosion removing most of uplifted rock, especially on northwest (see uplift scale on right). Below decollement, ductile rock has no escape and accumulates in crustal root. Note that decollement on South Island might be a broad zone, based on thickness of mylonite zone exposed along AF [*Sibson et al.*, 1979] and width of observed reflective zone [*Stern et al.*, this volume]. Topography exaggerated by 6.75; region below sealevel not vertically exaggerated. (b) Schematic diagram of tectonics of southern California along Line 1 (L1), adapted from *Fuis et al.* [2001]. For symbols and shading see caption for Figure 4a. Black stars, large earthquakes (magnitudes and dates attached); very light gray patches with black outlines, bright reflective zones A and B (see Plate 1d). Fluid-lubricated decollement is interpreted to ascend gently southwestward from bright reflective zone A near SAF. Above decollement, brittle upper crust imbricates along thrust and reverse faults. Below decollement, lower crust flows toward SAF (curved arrows in lower crust) and forms crustal root. Color version of this figure is on CDROM which accompanies this volume.

California, see *Polet and Kanamori*, 2002]. The fast directions and delays between slow and fast directions are independent of whether the observation is made inside or outside the footprints of the high-velocity bodies. From this last observation, it would appear that the high-velocity bodies do not significantly perturb the overall S-wave anisotropy of the mantle for horizontal vibration directions, and thus do not owe their visibility to anisotropy, at least in those directions.

Other models. Bird and Rosenstock's [1984] analysis of the kinematics of surface fault movement in southern California requires downwelling of the mantle lithosphere along the oblique segment of the SAF. In their Figure 5, they infer a propeller-shaped lithospheric mantle downwelling that is remarkably similar to that observed. They also predict a decollement at the Moho in the vicinity of the SAF, but their model allows no decollements within the crust, as we actually observe.

Sheffels and McNutt [1986] modeled the gravity and topography of the vicinity of the Transverse Ranges by using two elastic lithospheric plates of different thicknesses and rigidities. The southern plate (Peninsular Ranges, Los Angeles basin, and Continental Borderland) is modeled as a thin, weak plate subducting beneath a thicker, stronger plate beneath the Transverse Ranges and Mojave Desert. The high-velocity upper-mantle body imparts a terminal force and bending moment to the southern plate, but its northern boundary is traced to a subduction zone (plate break) at the southern margin of the Transverse Ranges rather than to the SAF, as observed. The SAF is located within their strong elastic plate in the Central and Western Transverse Ranges and is not a primary discontinuity in the mechanics of the modeled subduction.

A few authors have suggested that the high-velocity upper-mantle body of southern California may be a fragment of oceanic lithosphere, such as the Monterey subplate of the Farallon plate [*Humphreys*, 1995; *Atwater and Stock*, 1998], and we also entertained such an idea [*Fuis et al.*, 2006]. The Monterey plate was captured by the PAC at ~19 Ma, and, as it moved northward with the PAC, its position beneath the western Peninsular Ranges caused break-away and clockwise rotation of an outboard fragment of the Peninsular Ranges, which is currently found in the western Transverse Ranges [*Nicholson et al.*, 1994]. The plate boundary jumped landward at this time, presumably to the eastern side of the Monterey plate, or to some deep juncture in that plate. The plate boundary inferred at ~19 Ma is the San Francisquito-Fenner-Clemens Well fault [*Powell*, 1993]. This fault is deformed and offset by younger traces of the SAF, including the modern trace. The geometric link between the upper-mantle

high-velocity body and the current trace of the SAF strongly suggests that this body formed no earlier than ~6 Ma.

TECTONICS

A schematic cross section of central South Island by *Norris et al.* [1990; their Figure 4b], was constructed prior to the deep seismic imaging of SIGHT but remains a very useful interpretation (Figure 4a). The fact that the highest-grade metamorphic rocks exposed along the AF represent a crustal depth of no more than 20–25 km, in spite of the large component of convergence across the AF (now known to be ~90 km) led them to include a decollement at 20-km depth beneath the Southern Alps, merging with the AF to the northwest. Above the decollement and AF, rock in the actively deforming Southern Alps is interpreted to move upward and northward to become exposed and eroded along the AF. The models of critical-wedge mechanics [e.g., *Davis et al.*, 1984] led them to connect active thrust faulting southeast of the AF with this decollement. Using the gravity results of *Woodward* [1979], they hypothesized that lower crustal rock was accumulating in a crustal root centered southeast of the crest of the Southern Alps, as is seen most dramatically on T2 (Plate 1b). Stern et al. (this volume) provide some support for this tectonic model. They observe a curved reflective zone beneath the Southern Alps that they interpret as a decollement connecting northwestward with a (listric) AF. The subhorizontal part of this reflective zone is, however, deeper (30–35 km deep) than the decollement interpreted by *Norris et al.* [1990].

A very similar schematic diagram of deformation of the crust in southern California was constructed by *Fuis et al.* [2001; their Figure 3]. In this diagram, the SAF, unlike the AF, dips steeply, but it is connected to a mid-crustal decollement similar to the inferred decollement on South Island, that, in turn, connects upward to active thrust faults located southeastward in the Los Angeles region (Figure 4b). This decollement is clearly imaged on the LARSE lines as highly reflective zones, including a near-vertical-incidence "bright reflective layer" on Line 1 [*Ryberg and Fuis*, 1998; Plate 1d, blue bodies A and B]. Because of the highly reflective character of the southern California decollement, it is interpreted to be lubricated by fluids that are perhaps injected cyclicly into the brittle fault zones above the decollement in the manner described by *Sibson* [1992].

On both South Island and in southern California, crustal deformation is largely one-sided and largely within the PAC. In both cases, the upper crust behaves as a collection of brittle blocks that escape the collision zone upwardly and laterally. The lower crust, on the other hand, with no place to escape (at least in the central areas of both regions), thickens to produce a crustal root. In both cases, volumes of the crustal roots appear consistent with estimated total convergence [*Godfrey et al.*, 2002; *Henrys et al.*, 2004].

On both South Island and in southern California, mantle lithosphere of the PAC appears to be obliquely downwelling along the plate boundary. On South Island, the northwest side of this downwelling dips northwest and may be a zone of oblique shear that connects at or near the Moho to the crustal plate boundary (AF) which dips in the opposite direction. This geometry is suggestive of indentation of the PAC by the AUS (Figure 4a). Alternatives are also possible (see above). In southern California, the north side of this downwelling is approximately co-planar with, or at least continuous with, the crustal plate boundary (SAF), and would, thus, appear to constitute a zone of oblique shear between the PAC and NAM (Figure 4b). On both South Island and in southern California, downwelling may be modeled by strike slip plus a combination of thrusting of PAC mantle lithosphere beneath the opposing plates (AUS and NAM) and gravitational instability, although other models are possible.

SUMMARY OF LITHOSPHERIC STRUCTURE AND TECTONICS

The seismic images across the AF on South Island and the SAF in southern California reveal features of the plate boundary in both locations.

1. On South Island, the upper crust (5.4–6.3 km/s) contains a prominent, wide (50 km) zone of low velocity (5.8–6.0 km/s) beneath the Southern Alps that extends from near the surface to mid-crustal levels. The AF is contained within this zone. Surface outcrops of the fault near T1 and T2 indicate a dip ~50 degrees southeastward. On or near T2, migrated reflections are interpreted to represent a listric AF dipping ~60 degrees southeastward at 15–20 km depth and flattening into a decollement at 30- to 35-km depth. In southern California, in contrast, the SAF near Lines 1 and 2 dips steeply northward (85–90 degrees), based on seismic results and is associated with low velocities only near the surface, although on Line 1, a faint, poorly resolved, narrow (~5 km) LVZ extends to midcrustal levels. The SAF is indirectly more visible than the AF because it terminates strong mid-crustal reflective zones associated with interpreted mid-crustal decollements.

2. On South Island, lower crust (6.5–7.0 km/s) is relatively thin (1–10 km). On T1 it is thicker on AUS (~7 km) than on PAC (<5 km) and it thins within a crustal root beneath the Southern Alps to 2- to 4-km thickness. On T2 the reverse is true, where it is thicker on PAC (1–5 km) than on AUS (0.5–2 km) and it thickens within a crustal root to 10-km thickness. The top of lower crust on South Island is in most places marked by wide-angle reflections. In southern Califor-

nia, lower crust (6.6–6.9 km/s) is moderately thick on PAC (5–20 km) but non-existent to thin on NAM (0 km, Line 1; 5 km, Line 2). Beneath the Transverse Ranges, the top of the lower crust is marked by strong near-vertical-incidence to wide-angle reflections, that originate at the SAF and extend southward and upward in the PAC, terminating at, or projecting to, the hypocenters of significant recent earthquakes (1971 M 6.7 San Fernando and 1987 M 5.9 Whittier Narrows earthquakes). These reflections are believed to represent one or more fluid-lubricated decollements that connect upward to active thrust faults along the southern margin of the Transverse Ranges.

3. On South Island, the Moho is depressed beneath the Southern Alps to maximum depths ranging from 37 to 44 km. This crustal root ranges in width from 85–95 km and is asymmetric, with its steepest side on the northwest side of the island. Northwest of the root, the Moho is 25 km deep, and on the southeast side of the root, ~30 km deep. The maximum relief on this root ranges from 10 km (T1) to 18 km (T2). The surface trace of the AF is above the northwest edge of the crustal root, but if the fault dips moderately southeastward, it may project to the center of the crustal root at its maximum depth. In southern California, the Moho is depressed beneath the Transverse Ranges to a maximum depth on both Lines 1 and 2 of 36 km. These roots are somewhat narrower than the South Island roots and have somewhat lesser reliefs (Line 1, maximum of 8 km; Line 2, maximum of 14 km). Like the South Island roots, they are approximately centered on the deep projection of the plate-boundary fault.

4. On South Island and in southern California, a mid-crustal decollement is inferred from reflectivity that connects the plate-boundary fault to thrust faults farther south in the PAC. This decollement allows upper crust to escape the collision zone laterally and vertically, but forces the lower crust to form a crustal root.

5. On South Island and in southern California, upper-mantle bodies of high P velocity are observed extending from near the Moho to more than 200-km depth. On South Island the north side of the upper-mantle body dips moderately northwestward, although other dips are allowable. The AF is inferred to dip oppositely (see above), and may (or may not) join the north side of the mantle body at the Moho. If the AF and north side of the mantle body represent the plate boundary, then the AUS would appear to indent the PAC. In southern California, the north side of the body is coplanar or at least continuous with the SAF or its projection to the Moho. The SAF and the north side of the body change dip along strike and have a crude propeller shape.

6. Both on South Island and in southern California, the mantle bodies are inferred to represent oblique downwelling

of PAC mantle lithosphere along the plate boundaries. This process may be modeled by strike slip plus a combination of underthrusting of the PAC mantle lithosphere beneath opposing plates (AUS and NAM) and ensuing gravitational instability, although other models are possible.

Acknowledgments. This paper benefited greatly from discussions with many people, although the interpretations we present are not necessarily consensus interpretations. These persons include (alphabetically) Peter Bird, Glenn Biasi, Paul Davis, Kevin Furlong, Greg Houseman, Gene Humphreys, Ray Ingersoll, Craig Jones, Art Lachenbruch, Vicki Langenheim, Pat McCrory, Craig Nicholson, Mike Oskin, Tom Parsons, Fred Pollitz, Bob Powell, Dan Scheirer, Tim Stern, Joann Stock, Bill Stuart, Colin Williams, Charley Wilson, Doug Wilson, Bob Yeats, and George Zandt. We thank John Hole, Simon Cox, Fred Davey, Fred Pollitz, and Donna Eberhart-Phillips for helpful reviews of earlier drafts of this paper. J. Luke Blair and Marco Ticci are primarily responsible for Figures 1 and 2. Funding for SIGHT and LARSE came mostly from the U.S. National Science Foundation with contributions from the New Zealand Foundation for Research Science and Technology, the U.S. Geological Survey, and the Southern California Earthquake Center.

REFERENCES

Argus, D. F., M. B. Heflin, G. Peltzer, F. Crampe, and F. H. Webb (2005), Interseismic strain accumulation and anthropogenic motion in metropolitan Los Angeles, *J. Geophys. Res., 110,* B04401.

Atwater, T. (1989), Plate tectonic history of the northeast Pacific and western North America, in *The Geology of North America, Vol. N: The Eastern Pacific Ocean and Hawaii,* edited by E. L. Winterer, D. M. Hussong, and R. W. Decker, Geological Society of America, Boulder, CO, pp. 21–72.

Atwater, T., and J. Stock (1998), Pacific-North America plate tectonics of the Neogene southwestern United States: an update, *Int. Geol. Rev., 40,* 375–402.

Baher, S., G. Fuis, C. Wilson, V. Langenheim, and J. Murphy (2004), The onshore-offshore LARSE I transect: San Clemente Island to the Mojave Desert—crustal blocks and the Moho: *EOS Trans. AGU, 85,* F1416.

Baher, S., G. Fuis, R, Sliter, and W. Normark (2005), Upper-crustal structure of the Inner Continental Borderland near Long Beach, California, *Bull. Seismol. Soc. Am., 95,* 1957–1969.

Baldock, G., and T. Stern (2005), Width of mantle deformation across a continental transform: evidence from upper mantle (Pn) seismic anisotropy measurements, *Geology, 33,* 741–744.

Beaumont, C., and G. Quinlan (1994), A geodynamic framework for interpreting crustal scale seismic reflectivity patterns in compressional orogens, *Geophys. J. Int., 116,* 754–783.

Beavan, J., M. Moore, C. Pearson, M. Henderson, B. Parsons, G. Blick, S. Bourne, P. England, R. I. Walcott, and K. Hodgkinson (1999), Crustal deformation during 1994–1998 due to oblique continental collision in the central Southern Alps, New Zealand, and implications for seismic potential of the Alpine fault, *J. Geophys. Res., 104,* 25,233–25,255.

Beavan, J., S. Ellis, L. Wallace, and P. Denys (this volume), Kinematic constraints from GPS on oblique convergence of the Pacific and Australian Plates, Central South Island, New Zealand.

Billen, M. I., and G. A. Houseman (2004), Lithospheric instability in obliquely convergent margins: San Gabriel Mountains, southern California, *J. Geophys. Res., 109,* B01404.

Bird, P., and R. W., Rosenstock (1984), Kinematics of present crust and mantle flow in southern California, *Geol. Soc. Am. Bull., 95,* 946–957.

Carter, R. M., and R. J. Norris (1976), Cainozoic history of southern New Zealand; an accord between geological observations and plate tectonic predictions, *Earth Planet. Sci. Lett., 31,* 85–94.

Christensen, N. I., and W. Mooney (1995), Seismic velocity structure and composition of the continental crust: a global review, *J. Geophys. Res., 100,* 9761–9788.

Cox, S. C., and R. Sutherland (this volume), Regional geological framework of South Island, New Zealand, and its significance for understanding the active plate boundary.

Creager, K. C., and T. H. Jordan (1986), Slab penetration into the lower mantle beneath the Mariana and other island arcs of the Northwest Pacific, *J. Geophys. Res., 91,* 3573–3589.

Davis, D., J. Suppe, and F. A. Dahlen (1984), Mechanics of fold-and-thrust belts and accretionary prisms, *J. Geophys. Res., 88,* 1153–1172.

Davey, F. J., D. Eberhart-Phillips, M. Kohler, S. Bannister, G. Caldwell, S. Henrys, M. Scherwath, and H. van Avendonk (this volume), Geophysical structure of the Southern Alps orogen, South Island, New Zealand.

Davey, F. J., et al. (1998), Preliminary results from a geophysical study across a modern, continent-continent collisional plate boundary—the Southern Alps, *Tectonophysics, 288,* 221–235.

Fuchs, K. (1977), Seismic anisotropy of the subcrustal lithosphere as evidence for dynamical processes in the upper mantle, *Geophys. J. R. Astron. Soc., 49,* 167–179.

Fuis, G. S., R. W. Clayton, P. M. Davis, T. Ryberg, W. J. Lutter, D. A. Okaya, E. Hauksson, C. Prodehl, J. M. Murphy, M. L. Benthien, S. A. Baher, M. D. Kohler, K Thygesen, G.Simila, and G. R. Keller (2003), Fault systems of the 1971 San Fernando and 1994 Northridge earthquakes, southern California: relocated aftershocks and seismic images from LARSE II, *Geology, 31,* 171–174.

Fuis, G., M. Kohler, M. Scherwath, U. ten Brink, and H. van Avendonk (2006), A comparison between the transpressional plate boundaries of the South Island, New Zealand, and southern California, USA, *Seismol. Soc. Am., Res. Lett., 77,* 200.

Fuis, G. S., V. E. Langenheim, D. S. Scheirer, and M. D. Kohler (2007), The San Andreas Fault in southern California is almost nowhere vertical—Implications for tectonics, shaking hazards, and geodetic modeling, EarthScope National Meeting, Monterey, California, p. 18.

Fuis, G. S., J. M. Murphy, S. Baher, T. Ryberg, M. D. Kohler, and D. A. Okaya (2005), Lithospheric refraction/reflection/teleseismic model of LARSE Line 2: thrusting of the Santa Monica Mountains-San Fernando Valley block beneath the central Transverse Ranges, southern California, *EOS Trans. AGU, 86*(52), Fall Meet. Suppl., Abstract S41A-0971.

Fuis, G. S., T. Ryberg, N. J. Godfrey, D. A. Okaya, and J. M. Murphy (2001), Crustal structure and tectonics from the Los Angeles basin to the Mojave Desert, southern California, *Geology, 29*, 15–18.

Godfrey, N. J., F. Davey, T. A. Stern, and D.Okaya (2001), Crustal structure and thermal anomalies of the Dunedin Region, South Island, New Zealand, *J. Geophys. Res., 106*, 30,835–30,848.

Godfrey, N. J., G. S. Fuis, V. Langenheim, D. A. Okaya, and T. M. Brocher (2002), Lower crustal deformation beneath the central Transverse Ranges, southern California: results from the Los Angeles Region Seismic Experiment, *J. Geophys. Res., 107, B7*, ETG 8-1-8-19.

Griscom, A., and R. C. Jachens (1990), Crustal and lithospheric structure from gravity and magnetic studies, in *The San Andreas Fault System,* edited by R. E. Wallace, *U.S. Geological Professional Survey Paper 1515*, 239–260.

Hadley, D. and H. Kanamori (1977), Recent seismicity in the San Fernando region and tectonics in the west-central Transverse Ranges, California, *Bull. Seismol. Soc. Am.*, 68, 1449–1457.

Hearn, T. (1984), Pn travel times in southern California, *J. Geophys. Res., 89*, (B3), 1843–1855.

Henrys, S. A., D. J. Woodward, D. Okaya, and J. Yu (2004), Mapping the Moho beneath the Southern Alps continent-continent collision, New Zealand, using wide-angle reflections, *Geophys. Res. Lett., 31*(17), L17602.

Houseman, G. A., D. P. McKenzie, and P. Molnar (1981), Convective instability of a thickened boundary layer and its relevance for the thermal evolution of continental convergent belts, *J. Geophys. Res., 86*, 6115–6132.

Houseman, G. A., E. A. Neil, and M. D. Kohler (2000), Lithospheric instability beneath the Transverse Ranges of California, *J. Geophys. Res., 105*(B7), 16,237–16,250.

Humphreys, E. D. (1995), Post-Laramide removal of the Farallon slab, western United States, *Geology, 23*, 987–990.

Humphreys, E. D., and Clayton, R. W. (1990), Tomographic image of the southern California mantle, *J. Geophys. Res., 95, B11*, 19,725–19,746.

Humphreys, E. D., R. W. Clayton, and B. H. Hager (1984), A tomographic image of mantle structure beneath southern California, *Geophys. Res. Lett., 11*, 7, 625–627.

Humphreys, E. D., and B. H. Hager (1990), A kinematic model for the recent development of southern California crust and upper mantle, *J. Geophys. Res., 95*(B11), 19,747–19,762.

Jones, L. M., L. K. Hutton, D. D. Given, and C. R. Allen (1986), The North Palm Springs, California, earthquake sequence of July 1986, *Bull. Seismol. Soc. Am.*, 76, 1830–1837.

Kamp, P. J. J. (1986), The mid-Cenozoic Challenger Rift System of western New Zealand and its implications for the age of Alpine fault inception, *Geol. Soc. Am. Bull.*, 97, 255–281.

Kohler, M. D. (1999), Lithospheric deformation beneath the San Gabriel Mountains in the southern California Transverse Ranges, *J. Geophys. Res., 104*, 15,025–15,041.

Kohler, M. D., and P. M. Davis (1997), Crustal thickness variations in southern California from Los Angeles Region Seismic Experi-

ment passive phase teleseismic travel times, *Bull. Seismol. Soc. Am.*, *87*, 1330–1344.

Kohler, M. D., and D. Eberhart-Phillips (2002), Three-dimensional lithospheric structure below the New Zealand Southern Alps, *J. Geophys. Res., 107*(B10), ESE 6-1-6-15.

Kohler, M. D., H. Magistrale, and R. W. Clayton (2003), Mantle heterogeneities and the SCEC Reference Three-Dimensional Seismic Velocity Model Version 3, *Bull. Seismol. Soc. Am.*, *93*, 2, 757–774.

Koons, P. O. (1990), Two-sided orogen: collision and erosion from the sandbox to the Southern Alps, New Zealand, *Geology, 18*, 679–682.

Lachenbruch, A. H., and J. H. Sass (1980), Heat flow and energetics of the San Andreas fault zone, *J. Geophys. Res., 85*, 6185–6222.

Langenheim, V. E. (1999), Gravity and aeromagnetic models along the Los Angeles Region Seismic Experiment (Line 1), California, U.S. Geological Survey Open-File Report 99-388, 22 pp.

Langenheim, V. E., R. C. Jachens, J. C. Matti, E. Hauksson, D. M. Morton, and A Christensen (2005), Geophysical evidence for wedging in the San Gorgonio Pass structural knot, southern San Andreas fault zone, southern California, *Geol. Soc. Am. Bull.*, *117*, 1554–1572.

Meissner, R. (1986), *The Continental Crust—A Geophysical Approach*, Academic Press, San Diego, CA.

Melhuish, A., W. S. Holbrook, F. Davey, D. A. Okaya, and T. Stern (2005), Crustal and upper mantle seismic structure of the Australian plate, South Island, New Zealand, *Tectonophysics, 395*, 113–135.

Molnar, P., T. Atwater, J. Mammerickx, and S. M. Smith (1975), Magnetic anomalies, bathymetry and the tectonic evolution of the South Pacific since the late Cretaceous, *Geophys. J. R. Astron. Soc., 40*, 383–420.

Nicholson, C., C. C. Sorlien, T. Atwater, J. C. Crowell, and B. P. Luyendyk (1994), Microplate capture, rotation of the western Transverse Ranges, and initiation of the San Andreas transform as a low-angle fault system, *Geology, 22*, 491–495.

Norris, R. J., P. O. Koons, and A. F. Cooper (1990), The obliquely-convergent plate boundary in the South Island of New Zealand: implications for ancient collision zones, *J. Struct. Geol., 12*, 5/6, 715–725.

Oskin, M., J. Stock, and A. Martin-Barajas (2001), Rapid localization of the Pacific-North America plate motion in the Gulf of California, *Geology, 29*, 459–462.

Polet, J., and H. Kanamori (2002), Anisotropy beneath California: shear wave splitting measurements using a dense broadband array, *Geophys. J. Int., 149*, 313–327.

Powell, R. E. (1993), Balanced palinspastic resonstruction of pre-late Cenozoic paleogeology, southern California: geologic and kinematic constraints on evolution of the San Andreas fault system, Boulder Colo., *Geol. Soc. Am. Mem.*, *178*, 1–106.

Pysklywec, R. N., C. Beaumont, and P. Fullsack (2000), Modeling the behavior of the continental mantle lithosphere during plate convergence, *Geology, 28*, 655–658.

Raikes, S. A. (1980), Regional variations in upper mantle structure beneath southern California, *Geophys. J. R. Astron. Soc., 63*, 187–216.

Ryberg, T., and G. S. Fuis (1998), The San Gabriel Mountains bright reflective zone: possible evidence of young mid-crustal thrust faulting in southern California, *Tectonophysics*, *286*, 31–46.

Savage, J. C., and M. Lisowski (1998), Viscoelastic coupling model of the San Andreas Fault along the big bend, southern California, *J. Geophys. Res.*, *103*, 7281–7292.

Savage, M. K. (1999), Seismic anisotropy and mantle deformation: what have we learned from shear wave splitting?, *Rev. Geophys.*, *37*, 65–106.

Savage, M., M. Duclos, and K. Marson-Pidgeon (this volume), Seismic Anisotropy in the South Island, New Zealand.

Scherwath, M., T. Stern, F. Davey, and D. Okaya (2003), Lithospheric structure across oblique continental collision in New Zealand from wide-angle P wave modeling, *J. Geophys. Res.*, *108*, B12, 2566.

Scherwath, M., T. Stern, A. Melhuish, and P. Molnar (2002), Pn anisotropy and distributed upper mantle deformation associated with a continental transform fault, *Geophys. Res. Lett.*, *29*, 8.

Sheffels, B., and M. McNutt (1986), Role of subsurface loads and regional compensation in the isostatic balance of the Transverse Ranges, California: evidence for intracontinental subduction, *J. Geophys. Res.*, *91*, 6419–6431.

Sibson, R. H. (1992), Implications of fault valve behavior for rupture nucleation and recurrence, *Tectonophysics*, *211*, 283–293.

Sibson, R. H., S. H. White, and B. K. Atkinson (1979), Fault rock distribution and structure within the Alpine Fault Zone: a preliminary account, in *The Origin of the Southern Alps*, *R. Soc. N. Z. Bull.*, vol. 18, edited by R. I. Walcott and M. M. Creswell, pp. 55–65.

Stern, T., P. Molnar, D. Okaya, and D. Eberhart-Phillips (2000), Teleseismic P wave delays and modes of shortening the mantle lithosphere beneath South Island, New Zealand, *J. Geophys. Res.*, *105*(B9), 21,615–21,631.

Sutherland, R. (1999), Cenozoic bending of New Zealand basement terranes and Alpine fault displacement: a brief review, *N. Z. Geol. Geophys.*, *42*, 295–301.

ten Brink, U. S., J.Zhang, T. M. Brocher, D. A. Okaya, K. D. Klitgord, and G. S. Fuis (2000), Geophysical evidence for the evolution of the California Inner Continental Borderland as a metamorphic core complex, *J. Geophys. Res.*, *105*, 5835–5857.

Van Avendonk, H. J. A., W. S. Holbrook, D. Okaya, J. K. Austin, F. Davey, and T. Stern (2004), Continental crust under compression: a seismic refraction study of South Island Geophysical Transect I, South Island, New Zealand, *J. Geophys. Res.*, *109*, B06302.

Vetter, U., and J.-B. Minster (1981), Pn velocity anisotropy in southern California, *Bull. Seismol. Soc. Am.*, *71*(5), 1511–1530.

Walcott, R. I. (1998), Modes of oblique compression: late Cenozoic tectonics of the South Island of New Zealand, *Rev. Geophys.*, *36*(1), 1–26.

Williams, C. F. (1996), Temperature and the seismic/aseismic transition; observations from the 1992 Landers earthquake, *Geophys. Res. Lett.*, *23*, 2029–2032.

Woodward, D. J. (1979), The crustal structure of the Southern Alps, New Zealand, as determined by gravity, in *The Origin of the Southern Alps*, *R. Soc. N. Z. Bull.*, vol. 18, edited by R. I. Walcott and M. M. Creswell, pp. 95–98.

Wright, T. L. (1991), Structural geology and tectonic evolution of the Los Angeles basin, California, in *Active Margin Basins*, *Am. Assoc. Petrol. Geol. Mem.*, vol. 52, edited by K. T. Biddle, pp. 35–134.

G. S. Fuis and J. M. Murphy, Earthquake Hazards Team, U.S. Geological Survey, Menlo Park, CA, USA. (fuis@usgs.gov)

M. D. Kohler, Center for Embedded Networked Sensing, University of California, Los Angeles, Calif., USA

M. Scherwath, Leibniz Institute of Marine Sciences, IFM-GEOMAR, Kiel, Germany.

U. ten Brink, Coastal and Marine Geology Team, U.S. Geological Survey, Woods Hole, MA, USA.

H. J. A. van Avendonk, Institute of Geophysics, University of Texas, Austin, TX, USA.

Taiwan and South Island, New Zealand: A Comparison of Continental Collisional Orogenies

Francis T. Wu[1], Fred J. Davey[2], David A. Okaya[3]

The young, active collision zones in Taiwan and South Island, New Zealand, are bracketed by two subduction zones of opposite polarities. Whereas in South Island two fragments of Gondwanaland are colliding with each other, in Taiwan the Luzon Arc collides with the Eurasian continental margin. The relative plate motion between the Philippine Sea and the Eurasian plates in the Taiwan area is about 3 convergence to 1 transcurrent (80 mm/yr to 27 mm/yr), and in South Island the Pacific to Australian plate motion is 1 convergence to 4 transcurrent (10 mm/yr to 38 mm/yr). Geologically, northern and central Taiwan are underlain by continental crust whereas the southern part corresponds to the transition zone between continent and ocean lithospheres. The eastern Coastal Range is separated from the Central Range by the Longitudinal Valley and is formed by a telescoped arc-forearc sequence of the Luzon arc. In South Island, schists dominate the main part of the orogen; rocks from the Australian plate appear mostly in the northwestern part of South Island and in the central part of the collision zone they form a narrow strip west of the plate boundary, the Alpine fault. The aerial strain measured by GPS in Taiwan is about one order of magnitude greater than that in South Island. In the central part of the two orogens seismicity tends to be shallower and less frequent in the high mountain regions than in the lower ranges. In Taiwan, a portion of the Central Range is nearly aseismic. Substantial roots have been built under both orogens and exhibit an asymmetry mirroring the asymmetry of the mountains. In both orogens the fast-direction of the S-splitting measurements follow the geological trends and in both cases the known crustal delays are too small to account for the observed total delays, arguing for a mantle source of anisotropy and coherent deformation of the crust and upper mantle. The along-strike lithospheric variations in both regions make the orogens strongly three-dimensional.

[1]Department of Geological Sciences and Environmental Studies, State University of New York, Binghamton, New York

[2]GNS Science, Gracefield, Lower Hutt, New Zealand

[3]Department of Earth Sciences, University of Southern California, Los Angeles, California

A Continental Plate Boundary: Tectonics at South Island, New Zealand
Geophysical Monograph Series 175
Copyright 2007 by the American Geophysical Union.
10.1029/175GM17

INTRODUCTION

Arc-continent and continent-continent collisions are major processes in the accretion and development of the continental lithosphere, and their study is important for understanding natural hazard and for resource accumulation. Taiwan and South Island, New Zealand, provide complementary examples of such collisions and a comparative study may allow us to understand the processes more fully. The two regions can be viewed as two end member environments of continental collision –

329

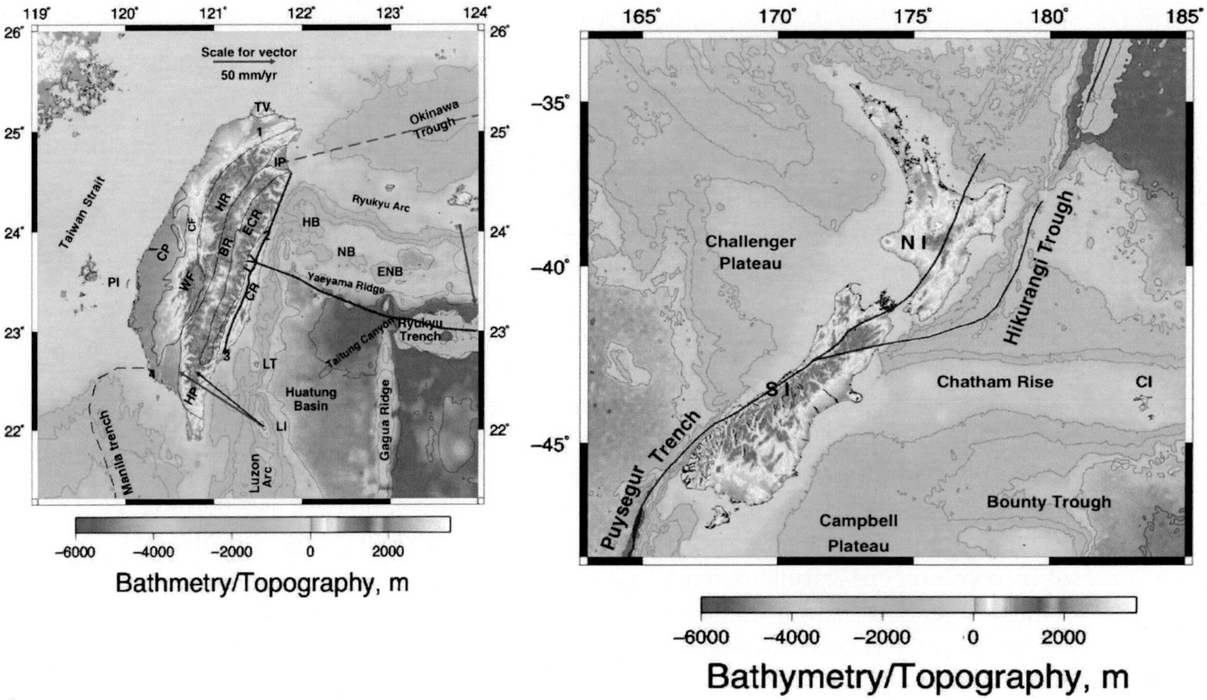

Plate 1. Plate configurations of Taiwan (left) and South Island, New Zealand (right). On the left: The physiographic-geologic units on land are: CR=Coastal Range, ECR=Eastern Central Range, BR=Backbone Range, HR=Hsueshan Range, WF=Foothills, CP=Coastal Plain, IP=Ilan Plain. CF=Chelungpu fault; TV=Tatun Volcano. LV=Longitudinal Valley. Offshore, HB=Hoping Basin, NB=Nanao Basin, ENB=East Nanao Basin, LI=Lanhsu Island, LT=Lutao Island, and PI=Penghu Islands. Cities, 1=Taipei, 2=Hualian, 3=Taitung. The blue arrows are GPS velocity vectors and the red arrow the NUVEL-1 predicted plate velocity. On the right, CI=Chatham Island, NI=North Island and SI=South Island.

broadly, the convergence of island arc lithosphere with a continental margin in the case of Taiwan and the convergence of two established continental crustal plates in the case of South Island (Plate 1). Although both involve oblique convergence, the rates and proportions differ significantly. The strain rate, seismicity and overall deformation in these two tectonic environments collectively reflect these differences.

Since this volume contains updated and comprehensive papers on the structure and tectonics of South Island we shall refer to them for most of the basic materials and discussions. This paper will provide materials on Taiwan, some previously published and some new, in order to bring out the contrast and compare the tectonics of the two regions.

In South Island, as detailed by *Davey et al.* [this volume] and *Cox and Sutherland* [this volume] two Gondwanaland continental lithospheric blocks of pre-Cretaceous age but different lithology are colliding. As a result of this collision, the Pacific Plate (PAC) east of the Alpine fault deformed to form the Southern Alps. The Australian Plate (AUS) side, west of the main range, shows little deformation apart from downwarping due to loading by the overthrust Pacific plate. The height of the Southern Alps tapers off slowly toward the north (Marlborough region) and south (Fiordland). The main end member models in this situation relate to how the plate boundary zone evolves - in dip and width - and how the shortening and thickening of the crust and lithospheric mantle are accommodated [*Walcott*, 1998; *Molnar et al.*, 1999].

In Taiwan, the oceanic Luzon island arc embedded in the Philippine Sea plate (PSP) is colliding with the passive continental margin of the Eurasian plate (EUR) (Plate 1). This collision has resulted in the construction of the Central Range on the EUR side and the Coastal Range on the PSP side. Although the main deformation occurs on the island, there have been infrequent M7 or greater events on the Taiwan Strait side to the west and much more frequent M7 or greater events on the PSP side, indicating that the tectonic stresses are high not only under the island but also in its vicinity.

Collisional orogenies such as those in Taiwan and South Island are the primary mechanisms for growth of continents. Much of the Archean and early Proterozoic cratonal consolidation involved the accretion of arc terranes [*Condie*, 2000]. The Paleozoic and early Mesozoic collision of arc complexes results in the tangled record of orogenesis along the Appalachian and Cordilleran margins (e.g., the Taconic, Antler, Sonoma orogenies). Despite the ubiquity of arc-continent collision in the Earth record of continental assembly, most ancient arc-continent collisional events are obscured by later tectonic events, and the plate geometry, plate kinematics, and rates of synorogenic processes are largely unknown. Taiwan and South Island are young, actively forming mountain belts, and the tectonic processes that have been

in operation can be readily imaged by geophysical means. While the current seismicity provides information about the stress conditions, the crustal and mantle structures are the integrated effect of orogeny.

Studying the similarities and differences in the plate tectonics that drive these orogenies and the resultant responses of the crust and upper mantle may allow us to better understand how orogenic processes operate. Many questions arise from our study: what are the roles of pure shear and simple deformations in the orogen? How do the different convergence rates affect the style of deformation? Do the orogenies exhibit vertical coherence in the crust and upper mantle? Does the continental plate in the collision subduct? These are among the key questions to be answered not only for these two but for all orogenies. Although we do not yet have complete answers, the seismicity, GPS-based surface deformation, crustal structures and SKS studies of these orogens point toward some possibilities.

GEOLOGIC FRAMEWORK

Taiwan

The Taiwan orogen is built on pre-Tertiary basement with most of the exposed edifice made of Miocene and older rocks. Pliocene and earlier Pleistocene rocks formed deep sedimentary troughs in western Taiwan and some have been folded and faulted to form the western Foothills.

In general, there are seven distinct physiographic and tectonostratigraphic provinces in Taiwan [*Ho*, 1986]. They are shown in the simplified geologic map (Figure 1):

1) the Coastal Plain – the foreland basin,
2) the Western Foothills – the foreland fold-thrust belt in Pliocene-Pleistocene rocks,
3) the Hsueshan Range – a mountain range built from Eocene-Miocene rocks on the continental shelf,
4) the Central Range – including the metamorphosed Tertiary shelf-slope sediments and the pre-Tertiary metamorphic core,
5) the Longitudinal Valley – a basin that overlies the boundary between continental and arc crust,
6) the Coastal Range – the telescoped Luzon arc complex, trench-forearc complex capped by Miocene andesites, and
7) the Hengchun Peninsula in the south, overlying an active subduction zone, is located in a zone where collision is just beginning [*Liu et al.*, 1997].

Recent detailed mapping of the offshore regions north and south of Taiwan suggests a continuation of many of the structural and geologic patterns recognized on land [*Huang et al.*, 1992; *Liu et al.*, 1997; *Sibuet and Hsu*, 2004]. North and northeast of Taiwan, the collision zone appears to have

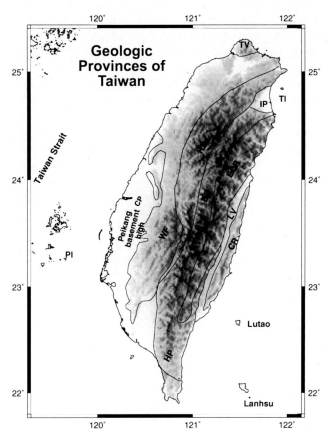

Figure 1. Simplified geologic provinces of Taiwan. Refer to Plate 1 for abbreviations. The additional one is TI=Turtle Island. BR and ECR are often combined and referred to as the Central Range.

subsided below sea level owing to the subduction of the PSP along the Ryukyu Trench and to the backarc spreading in the Okinawa Trough (Plate 1). In the region south of Taiwan, subduction of the South China Sea rather than continental lithosphere results in a submarine accretionary prism [*Reed et al.*, 1992] and an active, rather than inactive, magmatic arc.

South Island

The geology of South Island is generally split into two provinces corresponding largely to the eastern margin of the Australian plate and to the adjacent PAC crust, separated by the Alpine Fault [see Plate 2 in *Cox and Sutherland*, this volume]. West of the Alpine fault, Western Province rocks comprised of Paleozoic metasedimentary and igneous rocks intruded by Cretaceous granitoids form the basement. Western Province rocks are considered to be a fragment of Gondwanaland continental crust [*Cooper*, 1989]. Limited sampling by drillholes shows that similar rocks occur on the offshore Campbell and Challenger plateaus [*Beggs et al.*,

1990; *Wood*, 1991] and at Lord Howe Rise. Eastern Province rocks are a sequence of Mesozoic island arc and meta-sediments that have been accreted to the Gondwanaland crust. Western Province terranes are separated from those of Eastern Province by the Median Tectonic Zone or the Median Batholith [*Landis and Coombs*, 1967; *Bradshaw*, 1993; *Kimbrough et al.*, 1994; *Mortimer et al.*, 1999; *Tulloch and Kimbrough*, 2003], which records discrete pulses of subduction-related magmatism from about 360 Ma to 105 Ma.

Comparison

In Taiwan the age and metamorphic grade increase eastward towards the Longitudinal Valley, which divides the arc suites from the continental rocks, i.e., the suture; in South Island the grade of metamorphism increases westward toward the plate boundary (the Alpine fault). The folded and faulted Tertiary sediments found in the western Foothills of Taiwan have no corresponding sections in South Island. In Taiwan, thrust faults apparently separate rocks into discrete blocks of common age and grade, but in South Island it is more gradational. The general pattern of foliation changes from west-vergent to east vergent going from west to east across the Central Range of Taiwan as observed by *Lee* [1997]. This pattern is most probably related to the current mode of vertical deformation of the orogen in Taiwan (see later discussion).

The Taiwan orogenic belt is often referred to as a classic example of an arc-continent collision [*Suppe*, 1985; *Twiss and Moores*, 1992; *Davis and Reynolds*, 1996]. Based on mainly surface geology and exploration data down to a few kilometers, a thin-skinned model of mountain building has been proposed to explain the fold-and-thrust belt of the western Foothills [*Davis*, 1983; *Suppe*, 1987] and the detachment was later extended to underlie the whole island [*Carena et al.*, 2002]. In the simplest form of this model a shallow dipping detachment isolates the top layer, where the mountain building is taking place, from the lower layer below, the subducting EUR. Similar models have been used to explain other orogens [*Tozer et al.*, 2002]. But such models are not unique and thick-skinned interpretations of the same orogen have been expounded [e.g., *Coward*, 1996]. In Taiwan, models involving the basement [e.g., *Fuller et al.*, 2006] or lithospheric deformation [*Wu et al.*, 1997] have both been proposed. In South Island, the dominance of the relatively steep Alpine fault in the construction of Southern Alps seems to have ruled out a thin-skinned interpretation. Instead of altered products from processes that have long ceased their actions, here products of ongoing processes can be studied directly, both at the surface and at depth, because Taiwan and South Island are both young and active orogens.

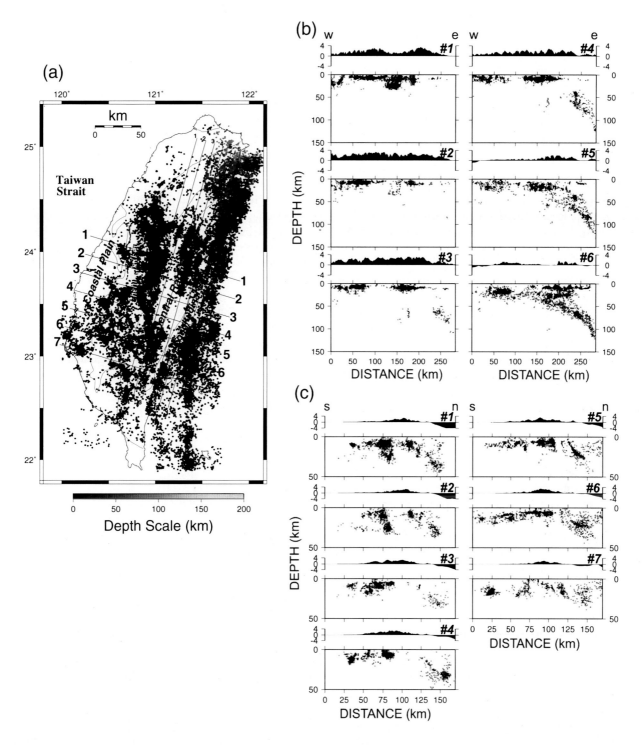

Figure 2. Seismicity in Taiwan (1994–1999). (a) Map view of seismicity and locations of two sets (NNW or N16°E and ESE or S74°E) of profiles. The NNE profiles are nearly in the dip-direction of the Benioff zone. (b) NNE profiles along the trend of the island. (c) ESE profiles perpendicular to the trend of Taiwan. See Plate A1 on the CD-ROM which accompanies this volume, where the map (a) is shown in color. Additionally Figures A1 and A2 on the CD-ROM show the same profiles as in (b) and (c) with the addition of focal mechanisms.

PLATE TECTONIC ENVIRONMENTS

Both Taiwan and South Island lie on a transpressive plate boundary between oppositely dipping subduction zones – NW-dipping subduction in the north and east dipping subduction in the south (Plate 1). However, there are several significant differences. South Island is about four times the size of Taiwan. The geometry of the colliding plates and rates of convergence are significantly different. The relative plate motion between PSP and EUR in the Taiwan area is about 3 convergence to 1 transcurrent (80 mm/yr to 27 mm/yr), whereas in South Island the PAC to AUS plate motion is 1 convergence to 4 transcurrent (10 mm/yr to 38 mm/yr). In Taiwan, the Ryuku Island Arc and backarc lie at +90° to the continental collision suture (Longitudinal Valley or LV) whereas in South Island the Hikurangi Trench is at +20° to the Alpine fault trend. The Luzon subduction is at −15° to the LV compared to the Puysegur/Fiordland subduction at −30° to Alpine Fault. In other words, subduction zones in South Island can be considered subparallel to the strike of the orogen but the northern subduction zone in Taiwan is perpendicular to the trend of the structures.

The seismicity in these regions illustrate the plate structures quite well because the subduction and collision are very active. We have relocated the seismicity in Taiwan using the double difference technique [*Waldhauser*, 2001]; maps and sections are shown in Figure 2 and also a map of seismicity in color in Plate A1 on the CD-ROM which accompanies this volume. The geometry of the PSP under northern Taiwan can be constructed from the sections in Figure 2b. For comparison with South Island, 3-D animation movies (Animations A1 and A2) and java interactive displays (Animation A3 "Live3D") are included on the CD-ROM which accompanies this volume. In the Live3D display a comparison of the two can be viewed at any angle and can be expanded to show details.

Wu et al. [1997, 2004], among others, interpreted the seismicity in terms of the tectonics of Taiwan; that of South Island has been summarized by *Davey et al.* [this volume]. In the rest of this section we will compare the seismicity of the two regions as related to subduction and discuss the shallow seismicity in central Taiwan later.

Taiwan

Figure 2b shows clearly that under Taiwan the Ryukyu Benioff zone is a well-defined northward dipping zone (Figure 2b, profile #6; see Figure 2a for the location of profiles). However 20 km inland the deeper (>50 km) seismicity is disconnected from the shallow crustal seismicity (Figure 2b, #4). Further inland the deeper seismicity disappears altogether leaving only crustal events (Figure 2b, #2 and #1). Relative to Taiwan, the PSP is moving northwestward at a rate of more than 8 cm/yr and its NNE component causes the PSP to subduct along the westward extension of the Ryukyu Trench (Plate 1). The configuration of the dipping subduction/collision zone implies that the collision boundary of the PSP with EUR varies along the strike of the collisional boundary. North of the junction of the Ryukyu Arc with Taiwan the collision occurs at increasing depth. Although the Ryukyu Trench bathymetric low near Taiwan (Plate 1) is absent, the location of the subduction boundary of the PSP can best be determined using the shape of Benioff zone. As shown in Figure 2b (#6) the Benioff zone begins to bend at about km 130 along the distance axis, corresponding to about 23.7°N. Thus south of 23.7°N the PSP and the EUR are in collision from the surface through the base of the lithosphere but north of this latitude the collision occurs at increasing depths, following the increasing depth of the Benioff zone [*Wu et al.*, 1997, 2007]. With the spreading the Okinawa Trough [*Sibuet et al.*, 1995] the Ryukyu arc moves southward, at a rate of ~60 mm/yr with respect to Taiwan (Plate 1) and the junction of Ryukyu arc with Taiwan will therefore migrate at that rate as well.

The disconnected Benioff zone inland (profile #3 in Figure 2b) probably indicates that as the PSP enters the asthenosphere it can overcome the resistance and move westward. The geometry of the plate edge can be seen more directly in Animations A1, A2 or A3 (Live3D display) on the CD-ROM which accompanies this volume. In Figure A1 and A2 on the CD-ROM we also show the focal mechanisms of M>4.7 events derived from the broadband seismic network (BATS) in Taiwan [*Kao and Jian*, 2001]. For the deeper part of the Benioff zone (>50 km), the down-dip T-axes and horizontal and slab-parallel P-axes dominate, the latter probably related to the continuous push by the PSP westward and the resistance provided by the viscous mantle. The Ryukyu subduction zone has also been imaged as a high velocity anomaly [*Rau and Wu*, 1995; *Kim et al.*, 2005].

The east-dipping seismic zone under the southern tip of Taiwan can be recognized easily from catalog seismicity [*Wu et al.*, 1997]. The double-difference relocation of hypocenters is not effective in the south because the spatial coverage of the land-based network is poor for most of the events, especially the deeper ones to the east of the Hengchun Peninsula. However, the active subduction tectonics of southernmost Taiwan are illustrated by a series of events offshore of western Hengchun at the end of 2006 and the beginning of 2007. Based on world-wide data (Harvard CMT Catalog) the first mainshock of the series has a M_w of 7.0 and centroid depth of 22.5 km (21.83°N, 120.39°E); the normal faulting mechanism has NNW-trending planes. Judging from the crustal structure of the area determined by *McIntosh et al.* [2005] the hypocenter of the earthquake (relocated by *C.S. Chang*, CWB, 2007) places it inside the subducting PSP; similar earthquakes have been observed in many other active subduction zones [*Mikumo et al.*, 2002].

As the Luzon Arc is at an angle with the EUR margin and its collision with the margin created the orogen, a dia-chronous closing of the ocean basin apparently took place [*Suppe*, 1987]. The Hengchun Peninsula is often considered to be the continuation of the southward propagating Central Range. However, the Coastal Range compressed together the Luzon arc volcanics, the forearc sediments, trench sediments and the volcanic arc (Lutao and Lanhsu – see Figure 1) while just south of the Coastal Range the Manila Trench (the plate boundary west of Hengchun in Plate 1) and the Luzon arc (Lanhsu and Lutao, see Figure 1) are separated by a distance of over 100 km. This implies a discontinuity of the southward propagation of orogeny. That is to say, that south of the lati-tude of Lutao the collision has barely started while to the north it is a mature collision zone.

Comparison

In South Island, the plate tectonic environment is not as complex, yet is similarly dynamic. Here the plate bounda-ries and the trend of the orogen are sub-parallel. In addition, the angle between the relative plate motions and the plate boundaries is small however, similar to Taiwan, the main zone of collision is located between the ends of the two subduction zones. Since the southern end of the Hikurangi Trench is known to have been further north than it is at present, the collision section extended farther north than at present. Thus the Marlborough region was within the con-tinental collision zone a few million years ago [*Cox and Sutherland*, this volume].

For both Taiwan and South Island noticeable topography or a sedimentary basin, or both, are found near the junction of the subduction and the collision zones. These tectono-physiological and sedimentological features are related to the geometry and dynamics of the plates. We refer to the steep NE coast of Taiwan and to the west coast of Fiordland and the Seaward Kaikoura Ranges of southwest and northeast South Island, respectively. *Eberhart-Phillips and Reyners* [1997] present 3D velocity images for the latter region de-rived from the tomographic inversion of earthquake data. In the uppermost mantle they image a low-velocity zone as-sociated with seismicity. They infer this to be related to the continental nature of the subducted plate with the increase in the amplitude of the velocity anomaly to the southwest as-sociated with the increasing thickness of the continental crust being subducted. They suggest that subduction is relatively minor and the plate interface may be locked, leading to the

intense deformation at the coastal region. As a result of this attempt to subduct buoyant continental crust, the overriding plate has been compressed, leading to the development of the Kaikoura Ranges (Inland and Seaward) with their very sharp onset at the coast. The very steep topography and the pres-ence of a deep offshore sedimentary basin in Taiwan are ap-parently the result the fact that only the deeper part of EUR is in collision with PSP, causing the upper portion of EUR to be subjected to EW tension and normal faulting (Plate 1).

Volcanism has not played a very prominent role in either Taiwan or South Island. For Taiwan, in addition to the vol-canic arc rocks southeast of Taiwan, the Tatun volcanoes in northern Taiwan overlie the tip of the PSP are possibly still active [*Lin et al.*, 2005]. But we note that no andesitic volcan-ism is found on South Island, either in the Marlborough or in the Fiordland area. The Solander Island, 30 km south of Fiordland, is calc-andesitic dated at 1 m.y. old. In the north, andesitic volcanism starts half way up North Island (Mt. Ruapehu) and continues along the back-arc basin to offshore of northern North Island. Offshore of northern Taiwan, the Okinawa Trough - an active back-arc basin, extends south-westward into the Ilan Plain area of northern Taiwan, with the dormant Turtle Island volcano offshore (Figure 1). The spreading rate of the Okinawa Trough has been estimated to be about 15 mm/yr for the last two million years [*Sibuet and Hsu*, 1995] but the contemporary rate of southward migration of the southwestern Ryukyu Islands from GPS measurements is about 60 mm/yr (Plate 1) [*Nakamura*, 2004], equivalent to 30 mm/yr of bilateral extension. The rate of extension of the back-arc basin north of North Island is about 20 mm/yr.

Taiwan and South Island share several common tectonic environments, such as a proximity to subduction zones of opposite polarities, a transpressive regime and differences in lithospheric properties across the collision boundaries. There are sufficient differences in boundary conditions and in rates that we should see distinct effects in deformation rates mea-sured at the surface, the seismic response and the resulting crus-tal structures. We shall now turn towards these observations.

TOPOGRAPHY, SURFACE DEFORMATION AND EROSION

Topography

Central South Island comprises five main zones. From the west, the Australian plate side is a narrow coastal plain that is followed by sharp linear slope change that corresponds to the Alpine fault – the expression of the plate boundary at the surface. Moving eastward, a steep slope up the range front to the main divide of the Southern Alps, a broad intermontane plateau and a less steep slope down to the broad outwash

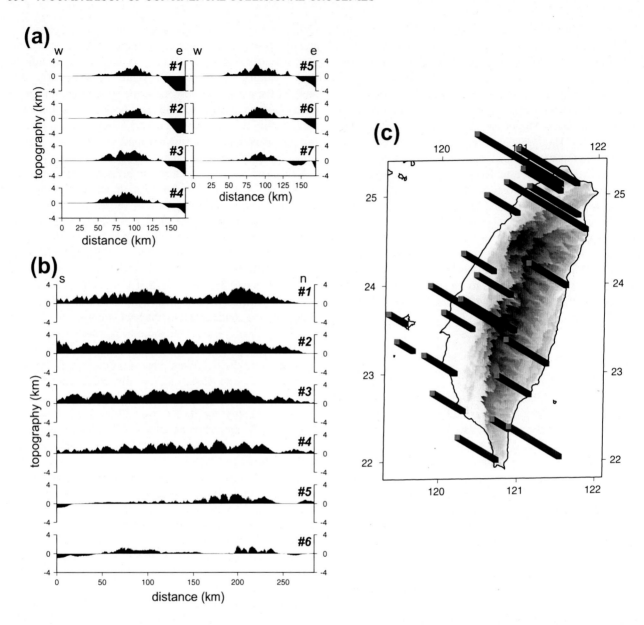

Figure 3. Topography of Taiwan. (a) E-W Profiles perpendicular to the structural trend. Locations same as seismicity profiles in Figure 2b. In southern Taiwan (profiles #6–7), where subduction is taking place, the topography is largely symmetric and in central Taiwan (profiles #1–2), in the collision section, a noted asymmetry is observed. (b) Parallel to the structural trend, locations same as seismicity profiles in Figure 2c. In profile #1, the northern peak is the Hsueshan Range. Profile #3 shows the topography along the axis of the Central Range. (c) 1943–1996 rainfall recorded at cities around the island; the rainfall is proportional to the length of the bar.

Plate 2. (Opposite) Results from the continuous GPS network in Taiwan (L.C. Kuo, IES, Academia Sinica). Arrows show horizontal GPS velocities relative to Penghu Islands and the areal strain rate computed from the velocities using the STRAINSIMPRO326 program from Rick Allmendinger [personal communication, 2007]. The color scale indicates the magnitude of areal strain rate: compression is negative and dilatation is positive. The islands to the southeast of Taiwan sits on the Philippine Sea plate and its velocity can be viewed as the relative velocity between the Philippine Sea plate and the Eurasian plate.

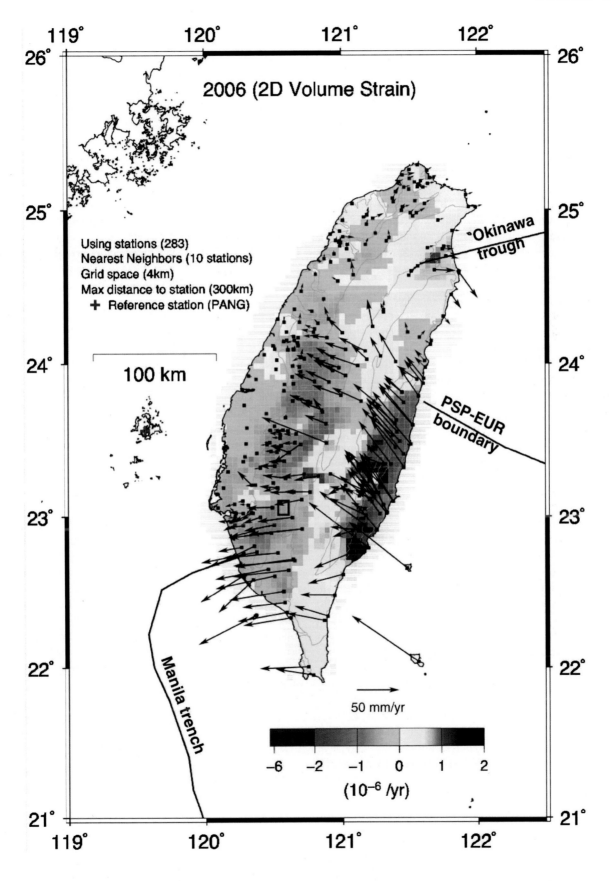

2006 (2D Volume Strain)

Using stations (283)
Nearest Neighbors (10 stations)
Grid space (4km)
Max distance to station (300km)
+ Reference station (PANG)

100 km

Okinawa trough

PSP-EUR boundary

Manila trench

50 mm/yr

-6   -2   -1   0   1   2

$(10^{-6} / yr)$

plain form the eastern coastal plain (Figure 2, profile (b) in *Davey et al.* [this volume]). In the south, a clear orocline turns the trend from NE in central South Island to SE in the Otago region [see *Cox and Sutherland*, this volume].

In central Taiwan, from the PSP side and going west, first the Coastal Range is encountered, reaching a maximum height of approximately 1000 m in the south and 300 m in the north (Figure 3a). The maximum width of the Coastal Range is about 13 km (Figure 1). Across the ~2–3 km wide Longitudinal Valley is the Eastern Central Range. The Coastal Range forms a barrier with only one river from the Central Range flowing across it. The Longitudinal Valley is a long, near sea-level, linear valley that receives sediments while both sides are rising. There is no range-front equivalent of the Alpine fault at the foot of the Central Range side of the Valley, and the rise to the high peaks is not as rapid as the western side of the Southern Alps.

In Taiwan, referring to the physiographic/tectonic provinces mentioned earlier, the highest topography coincides with the Tertiary slates in the Central Range and descends more rapidly on the east side (toward the Longitudinal Valley) than on the west side. In the southwest, the Foothills, formed mainly of sediments deposited on the continental slope and beyond, taper off into the Plain. To the northwest another mountain range, the Hsueshan Range, forms a second topographic high (Figure 3a Profile #3 and Figure 3b Profile #1) and is surrounded by its Foothills and an apron of terraces.

The landforms of both Taiwan and South Island are largely the result of dramatic tectonic deformation and sculpting by erosion. While the deformation process is complex and the subject of current studies, the main agent of erosion is clearly rainfall. As shown in Figure 2 of *Davey et al.* [this volume] the asymmetry of the trend-normal topographic profiles of South Island is very pronounced. It is clear that the orographic rainfall concentrates on the AUS side of the Southern Alps and is the main agent responsible for the erosion [*Batt and Braun*, 1999]. In contrast, Taiwan experiences three types of rainfall: the monsoon from the southwest, the northeastern seasonal rain from the northeast and typhoons that could approach Taiwan from the east, southeast or south. Typhoons are evidently the most powerful erosion mechanism. Rotating moist air masses may hit the rising topography of Taiwan from the east or west, depending on each typhoon's path and may cause heavy rainfall on the windward side of the island. Examples of the rainfall patterns are shown in Plate A2 on the CD-ROM which accompanies this volume. A fair assessment of the cumulative rainfall from typhoons in a few decades is not yet possible. The average rainfall in cities around the island between 1943 and 1996 (Figure 3c) does not show a clear asymmetry. These considerations may point to the fact that an important part of the asymmetry is created by the collisional deformation.

Continuous GPS Measurements and Deformation

The present rate of deformation is best assessed by campaign or continuous GPS measurements. *Beavan et al.* [this volume] reviewed the GPS data of South Island thoroughly. In Taiwan campaign GPS surveys started in the late 1980's, but a dramatic increase of continuous GPS stations began after the September 20, 1999, Chi-Chi earthquake. *Wu et al.* [2007] have calculated the velocities and strains using the dataset from 2000–2007. Plate 2 shows the velocity vectors, the areal volume strain and the shear strain for 2006. The overall patterns of the year-to-year results do not change but the magnitude of stresses and the rate of deformation rose slowly following the 1999 Chi-Chi earthquake. The large scale patterns of velocity vectors are best explained in terms of plate boundaries, plate motion and deformation. For example, in the central Taiwan collision zone, the velocity vectors gradually decrease from the east coast of Taiwan across the Central Range and the Foothills to very small amplitudes in the Coastal Plain indicating the shortening across the orogen. In the south, the Hengchun Peninsula as a whole is moving at ~40 mm/yr westward with PSP. In contrast, for northern and northeastern Taiwan north of the point where the PSP begins to subduct, a counterclockwise rotation of the vectors, in the area centered around 24°N and 121°E, signifies the end of collision at least in the shallow part of the lithosphere [e.g., *Hu et al.*, 2001] and a decrease in compression. On the other hand, the areal volume strain (Plate 2) shows the relatively high compression in the Coastal Range and the Foothills but a dilatation in most parts of the Central Range. The noise on the vertical component is much higher (~3 mm/yr), and the effect of the 1999 Chi-Chi earthquake prevents us from deciphering the long term trend of the vertical rate at present. In comparison with the results of *Beavan et al.* [this volume], the volume strain is approximately one order of magnitude higher, and the vertical rate is about twice as high. *Lee et al.* [2006] argued that Taiwan was subjected to a two-staged uplift history. From 6 mybp to 1 mybp it rose at a rate of about 1mm/yr and in the last million years it rose at 4–10 mm/yr. If this is the case, then the GPS measurements may be indicating the accelerated deformation.

CRUSTAL AND UPPER MANTLE STRUCTURES

Crustal Structures

While detailed crustal seismic data may image deformational structures at a range of scales (e.g., 1–10 km) in an orogen, configuration of the Moho records the cumulative results of the long term deformation, especially if an

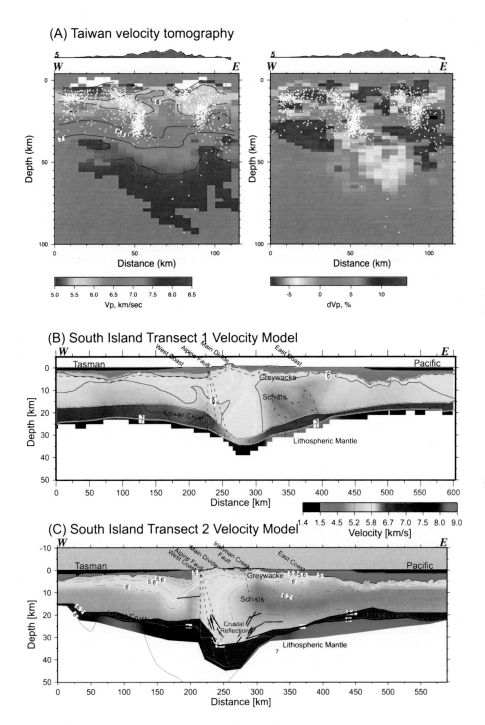

(A) Taiwan velocity tomography

(B) South Island Transect 1 Velocity Model

(C) South Island Transect 2 Velocity Model

Plate 3. Tomographic section across (a) central Taiwan and (b,c) two transects in central South Island. For (a), cross-sections of Vp (left) and dVp (right), the difference between the final result and the initial model. The white dots are hypocenters of earthquakes. Notice the asymmetry of the crustal root reflecting the topography. If we use the 7.5 km/sec contour as a proxy for Moho, the crustal thickness may exceed 55 km at a distance of about 70 km from the left side of the profile. South Island velocity structures are for Transect 1 (b) and Transect 2 (c) [from *Stern et al.*, this volume].

initial configuration can be assessed. Two key active source transects in South Island (Plate 3b and 3c and also see *Stern et al.* [this volume] and *Davey et al.* [this volume]) show several key features of the crust. Foremost is the presence of a root under the high Southern Alps, noticeably asymmetric with a steeper dipping Moho on the AUS side. We also see the contrast of a gradually thickening crust on the PAC side from the Chatham Rise – Campbell Plateau toward the Southern Alps in both profiles compared to the nearly unchanging Moho depth on the AUS side along Transect 1 (Plate 3b).

Although only limited results from active source work have been obtained in Taiwan so far, crustal images based on local earthquake tomography [*Rau and Wu*, 1995; *Kim et al.*, 2007; *Wu*, 2007] provide a base for comparison. General crustal properties are similar in these studies, represented by a profile across the northern portion of the Central Range (Plate 3a). The profile location is close to Profile #1 in Figure 2b. The earthquakes shown in the profile include the deeper Chi-Chi aftershocks to be discussed in the following seismicity section. The most recently published tomographic result is that of *Wu et al.* [2007], who incorporated a large number of S-P times in the tomography and improved the resolution of Vp/Vs structures.

Using Vp of 7.5 km/s as a proxy for the base of the crust, the crustal thickness of central Taiwan is approximately 55 km, ~10 km thicker than that shown in South Islands's transect 2 (Plate 3c). However, the asymmetry is similar with the steep increase of thickness on one side of the colliding plate (PSP in Taiwan and the Australian plate in South Island). A N-S profile in *Rau and Wu* [1995] shows that under the northern tip of the island and in western Taiwan the crust is about 30 km thick; along the spine of the Central Range the base of the crust increases its depth to a maximum depth of 55 km or so as the elevation of the range increases to the north, and then its depth decreases gradually further toward the south.

Mantle Structures

The teleseismic delay times recorded during the SIGHT transect project [*Stern et al.*, 2000, and this volume] were used to identify a high P-wave velocity zone in the upper mantle under South Island. Later, *Kohler and Eberhart-Phillips* [2002] used SAPSE teleseismic delay times for a three-dimensional velocity inversion. They added significant constraints for the exploration of upper mantle structures under South Island. *Davey et al.* [this volume] discuss these results extensively.

Using teleseismic delay times at Taiwan stations [*Chen et al.*, 2004], teleseismic tomography [*Lallemand et al.*, 2001] and combined local earthquake and teleseismic tomography

[*Wang et al.*, 2007] a steeply dipping upper mantle high velocity anomaly has been detected. The resolution and velocity contrasts vary in these studies. These anomalies were interpreted to be subducted EUR lithosphere. The zone is not clearly associated with seismicity and this prompted *Lallemand et al.* [2001] to hypothesize a detached lithosphere. Since the collision sections lack significant deep seismicity, the use of high velocity anomalies to identify possible active or remnant subduction zones is a very important problem. In Taiwan an experiment is now underway (the TAIGER project, see Conclusion) and one of the targets is the upper mantle anomalies.

SEISMICITY

We have already discussed the seismicity associated with the subduction zones in Taiwan and in South Island. The collision zones between the two subductions in these two areas bear interesting comparisons. The earthquake foci in this zone in South Island are concentrated above 25 km [*Davey et al.*, this volume], and probably above 12 km under the high Southern Alps and deeper under lesser topography. The seismicity in central Taiwan, roughly between 23°–24°N – where the collision is most active, is higher than that of South Island, consistent with the higher rate of convergence. In all the profiles of Figure 2c, an overall arch is seen in the seismicity from the Coastal Plain to the high peaks region though there are some gaps in some profiles under the Central Range. In the Coastal Range area, on the PSP side, events down to about 50 km are common but in the Foothills they extend only down to about 35 km. In Figure 2c the Chi-Chi aftershocks are included; *Wu et al.* [2004] compared pre-Chi-Chi and post-Chi-Chi seismicity in detail. Whether pre- or post-Chi-Chi, the boundary between the Foothills and the Central Range is an important one for the seismicity distribution. While the aseismic Central Range is defined by this boundary, so is a steeply west-dipping seismic zone after Chi-Chi (Figure 2c #1 and #2) at depths between 15 and 35 km. A large aftershock occurred in this zone on June 20, 2000. Its own aftershock sequence within 10 hours define a narrow zone and its orientation agrees with the steep plane of it focal mechanism

On the eastern side of the southern part of Central Taiwan (Figure 2c, #5–7) a very clear zone of east-dipping zone of seismicity projects into the LV on land. The larger events along the zone show clear thrust motions (Figure A2 on the CD-ROM which accompanies this volume), providing a mechanism for the Coastal Range to form.

The aftershock seismicity of the M7.6 Chi-Chi earthquake remained high for several years and only began to return to the pre-Chi-Chi level in 2006. The majority of events occurred under the Foothills but around the latitude of 24°N shallow aftershocks continue across the Central Range. The group of deeper events located from 15–35 km under the boundary between the Foothills and the Central Range mentioned above was unprecedented. As shown by the focal mechanism of a large foreshock in the zone on June 10, 2000, and its own aftershocks within 10 hours of the origin time, the dipping zone is most probably associated with a reverse fault with the Foothills side up and the Central Range side down. This counterintuitive result can be interpreted as a part of the processes that built the Central Range root

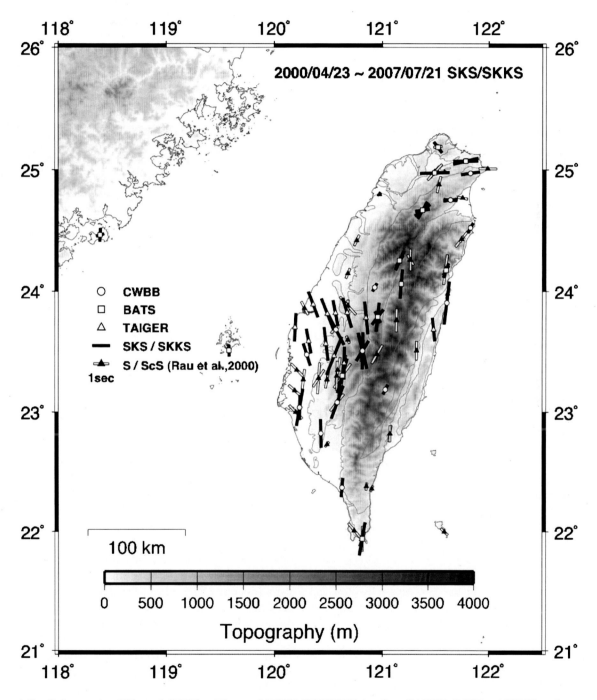

Figure 4. S-splitting results of *Wu et al.* [2007] and *Rau et al.* [2000]. SKS/SKKS data from TAIGER, BATS and CWBB stations are used by *Wu et al.* and short period CWBSN and BATS data are used by *Rau et al.* The length of the bar shows the delay time between the fast and slow S-wave and the direction of the bar shows the direction of the fast S-wave.

[*Wu et al.*, 2004]. A full account of the Chi-Chi aftershocks can be found in *Wu et al.* [2004].

S-SPLITTING MEASUREMENTS

Crustal S-Splitting

In South Island the schistose crustal rocks are highly anisotropic in laboratory measurements but S-splitting measurements yield delays less than 0.1 sec [see *Savage et al.*, this volume]. In Taiwan a few crustal S-splitting results show similar results [*Kuo et al.*, 1994; *Rau R.J.*, personal communication, 2006]. Recent measurements of schistose rocks in Taiwan indicate large anisotropy of more than 15% [*Christensen*, personal communication, 2007], with the anisotropic axes aligning to the foliation.

Mantle S-Splitting

The SKS splitting measurements on South Island were among some of the earlier datasets showing large delay times and the concordance of the fast directions and the surface geological trends [*Klosko et al.*, 1999; *Savage et al.*, this volume: Figure 1]. *Savage et al.* [this volume] summarized the S-splitting results of South Island as well as the North Island. For Taiwan, *Rau et al.* [2002] used S and ScS phases recorded on short-period CWBSN and some broadband BATS stations to determine the splitting parameters. They obtained delay times ranging from 0.5 to 2.1 seconds and found the fast directions to be generally parallel to the trends of local structures (see Figure 4). In northern Taiwan, where the structure turns from NNE to nearly EW, the fast direction of SKS follows it. Using SKS data from one Mozambique event recorded at a line of temporary broadband stations across southern Taiwan and several BATS stations, *Huang et al.* [2006] show a very dramatic increase of delay times in the Central Range – from less than 1 second to more than 2 seconds, with the high values within a distance of about 15 km. Recently *Hao et al.* [2007] added more SKS/SKKS data from several events recorded at BATS as well as temporary stations (the "TAIGER" network) and found a more pervasive trend-parallel pattern. The data around the "Peikang basement high" (Figure 1) is especially clear and the new data also confirmed the fast-direction measurements of *Rau et al.* [2000] in northern Taiwan (Figure 4).

DISCUSSION

The tectonics of South Island and Taiwan are both dominated by a young transpressional collision in a zone between two subduction systems. The history of the evolutionary orogeny in these two regions began more than 5 millions ago. In Taiwan however, an acceleration of the process may have taken place in the past million years. Currently both orogens have about the same physical stature although the strain rate and the erosion mechanisms are different. Taiwan is experiencing much more shortening than South Island although South Island is subjected to more shear. The strain rate calculated from GPS measurements reflects these differences; the areal strain rate of Taiwan is about one order of magnitude greater than that of South Island at 1.5 ppm. The high strain rate in Taiwan may have led to a different mode of deformation, at least at shallow depth: the high Central Range is in a region of dilation in Taiwan but across South Island only compression is observed.

One of the obvious differences between the two orogens is the degree of contrast in the deformation on the two sides of the plate boundary. In South Island the AUS is apparently so rheologically strong that little deformation is found on that side of the Alpine fault. On the other hand, there is considerable shortening of the PSP side to form the Coastal Range in Taiwan. Furthermore, frequent large earthquakes (M6–M7.5) occur on the PSP side and they are known even on the Taiwan Strait side. These earthquakes can be explained by the higher rate of convergence near Taiwan, implying higher stress and strain rates.

The geometry of the colliding and subducting PSP with the EUR east of central and northern Taiwan should affect the deformation on the island. South of the intersection of the Ryukyu Arc with Taiwan, full collision – from surface down to depth – is taking place. But to its north, collision occurs at increasingly greater depth; there the top part of the orogen is no longer in collision and the compression from the collision should be lessened. The expected changes can be seen clearly in Plate 2; north of the Ryukyu intersection the magnitude of the velocities decrease quickly. In southern Taiwan, across the transition from collision to subduction near 23°N, the area should be coupled to the PSP and move westward with it. In Plate 2, a very rapid increase of velocity toward the west and southwest in the Hengchun Peninsula is shown, consistent with the westward motion of PSP over the subducting EUR. In South Island, although the transitions from subduction to collision in the north and the south are sharp in terms of seismicity, the changes in surface deformation in terms of GPS measurements [*Beavan et al.*, this volume: Figures 3 and 4] are much subtler.

One of the most striking similarities of South Island to Taiwan is in the relation between S-splitting parameters and the structural trend of the orogen. The concordance of fast directions with structures is illustrated in South Island and is further affirmed in Taiwan. The wrap-around in the Peikang Basement High in central western Taiwan and the large clockwise rotation to EW in northern Taiwan are clear with the recent addition of data. The usual argument that the source of the splitting is anisotropy in the upper mantle [*Silver*, 1996] can be applied here and equally applicable is the conclusion that shear strain parallel to the orogen may responsible for the lattice-preferred orientation (LPO) of minerals in the upper mantle. Thus both South Island and Taiwan orogenies may involve vertically coherent deformation through the crust and upper mantle. The EW orienta-

tion of fast directions in northern Taiwan is curious in that the region overlies the PSP subduction and the orientation of the PSP fast direction is not known. At the same time, we discussed earlier the westward movement of PSP under Taiwan; it is possible that the flow around the plate or the end of the subduction zone flow may be the cause. We note that away from South Island or Taiwan delay times become small. In the case of Taiwan, the delay times in the Penghu Islands and near the shore of SE China the delays are less

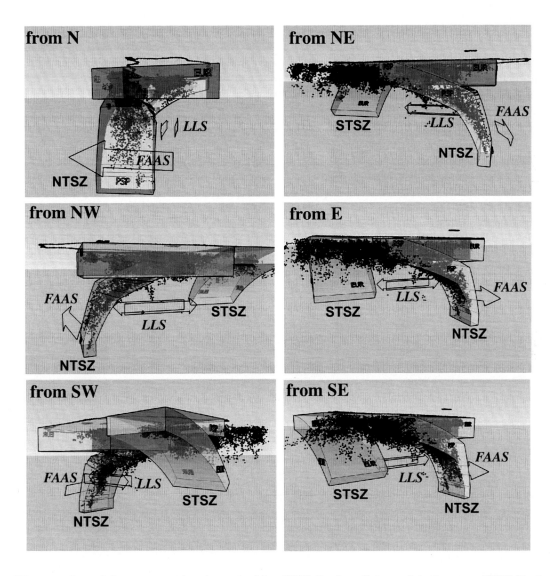

Figure 5. 3D perspectives of plate structures based on seismicity. STSZ=Southern Taiwan subduction zone; NTSZ=Northern Taiwan subduction zone; LLS=Left-lateral shear, shown by a pair of oppositely directed arrows; FAAS=possible upper mantle flow around the westward advancing slab, shown by a single arrow north of the northward subducting PSP. For the different panels: "From N" shows the N-directed subduction zone in frontal view "From NW" shows a side view of the NTSZ and the backside view of STSZ. "From SW" indicates that PSP is overriding EUR as it advances westward. "From NE" and "From E" and "From SE" indicate the subduction of the PSP and a kink in the PSP.

than 0.5 seconds (Figure 4). It would be difficult to explain the large delays on Taiwan if EUR subducts under Taiwan, unless it has an undetected vertical anisotropy or the subducted lithosphere has been subjected to large shear.

Both orogens are already quite mature, having developed a mountain range and a substantial root. The shallowness or absence of seismicity in both ranges argues for relatively high temperature in comparison to surrounding areas. The relatively high velocity in the Central Range [*Rau and Wu*, 1995] (Plate 3a) indicates that mid-crustal rock under the Central Range may have been uplifted as a result of shortening. In South Island the uplift is achieved mainly through thrusting along the Alpine fault [*Cox and Sutherland*, this volume]. In Taiwan it seems to be accomplished in a broader zone. There is now evidence in Taiwan to show that a high angle reverse fault at depth between the Foothills and the Central Range may be the mechanism for root formation. In other words, pure shear deformation forms the Central Range of Taiwan while simple shear controls the uplift of the Southern Alps.

A summary of the tectonics of Taiwan is presented in Figure 5. The block structures are constructed to surround the seismicity and represent the plate structure. The northern Taiwan subduction zone (NTSZ) is the most prominent structure in the region. Implied in the model is that the subducted PSP is connected to the PSP as a whole and the continued motion of the plate pushes the subducted PSP westward into the mantle. At shallow depth, PSP is shortened to form the Coastal Range (and partially eroded away). Once it enters the region below the lithosphere it starts to move westward and created the curved edge of the NTSZ in the "From N" view. The advancing slab moves the materials in front of it out of the way to form the "flow around advancing slab" or FAAS (Figure 5). This flow could create LPO and may explain the orientation of the fast direction of the S-splitting measurements. In the south, EUR subducts under PSP. However, considering the absolute plate motion (in the no-net-rotation or hot-spot frameworks [*Stein and Wysession*, 2003]), PSP has a rate of 43 cm/yr in the N47°W direction while EUR has a rate of 26 cm/yr in the direction of S70°E, i.e., in the absolute motion framework PSP is overriding EUR. The plate structure seen from the east (in "from NE", "from E" and "from SE" views of Figure 5) indicates the bending of the PSP as it subducts under EUR. The bending of the PSP could be related to the opening of the Okinawa Trough pushing the Ryukyu accretionary wedge southward. The oblique convergence may have also created a shear in the mantle as indicated by the pair of arrows below Taiwan, marked "LLS" or left-lateral shear. In this case, shear is the mechanism that creates the LPO that causes the SKS-splitting.

CONCLUSION

Although the plate tectonic frameworks for both South Island and Taiwan have many similarities, the rates of plate motion and the materials involved in the orogenic processes are different enough to create two different mountain ranges. Under the Central Range of Taiwan pure shear deformation, involving the simultaneous construction of the mountain and the root, is taking place. Dilatation at the earth surface, shown by strain calculations, point to "flowering" of the high mountain range. In South Island simple shear focussed about the Alpine fault built the Southern Alps.

The common observation of large S-splitting delays and the parallelism of the fast-directions with the structural trend are very well illustrated in both regimes. That the source of the splitting resides in the upper mantle seems to be the best hypothesis at present. If the implied vertical coherence [*Molnar et al.*, 1999] is the correct interpretation, then the mountain building processes must involve the crust and upper mantle. Splitting measurements in both Taiwan and South Island are so numerous and robust they provide important constraints to the modeling of orogeny.

The orogenies of Taiwan and South Island have drawn wide attention in the global geosciences community. The SIGHT and SAPSE project in the late 1990's advanced the state of knowledge concerning South Island. In terms of the knowledge regarding crustal and upper mantle structures, South Island is at a much more advanced state than it is in Taiwan. In fact these experiments served as examples for an ongoing project in Taiwan. The Taiwan Integrated Research in Geodynamics (TAIGER) will conduct a series of experiments, including land active source and sea-land transects, and broadband seismic deployment not only on land but also under the ocean. One of the unanswered questions in South Island is the width of the anomalous SKS zone. The TAIGER project will address this problem in Taiwan.

Finally, the accumulated data for both orogens are now quite extensive. Geodynamic modeling constrained by observational data in these two regions will signifcantly advance our understanding of mountain building processes.

Acknowledgements. The authors would like to acknowledge contributions from their colleagues on the SAPSE team. FTW and DO are supported by NSF Continental Dynamics program (NSF EAR-0410227). FD is supported by the NZ Foundation for Research Science and Technology. They would like to dedicate this paper to the memory of Tom McEvilly who was instrumental in organizing SAPSE as a cooperative effort with Helen Anderson. FTW thanks Dr. L.C. Kuo of IES, Academia Sinica for his effort in building an open GPS database. FTW also thanks H. Kuo-Chen for his assistance and Rick Allmendinger for his strain calculation program. Editorial effort by Leon Teng of USC and Eleanor Sonley of SUNY/Binghamton are greatly appreciated.

REFERENCES

Batt, G. E., and J. Braun (1999), The tectonic evolution of the Southern Alps, New Zealand: insights from fully thermally coupled dynamical modeling, *Geophys. J. Int., 136,* 403–420.

Beavan, J., S. Ellis, L. Wallace, P. Denys (this volume), Kinematic constraints from GPS on oblique convergence of the Pacific and Australian Plates, central South Island, New Zealand.

Beggs, J. M.; G. A. Challis, and R. A. Cook (1990), Basement geology of the Campbell Plateau; implications for correlation of the Campbell magnetic anomaly system, *N. Z. J. Geol. Geophys., 33,* 401–404.

Carena S., J. Suppe, and H. Kao (2002) - The active detachment of Taiwan illuminated by small earthquakes and its control of first-order topography, *Geology, 30,* 935–938.

Chen, P. F., B. S. Huang, and W. T. Liang (2004), Evidence of a slab of subducted lithosphere beneath central Taiwan from seismic waveforms and travel times, Earth Planet, *Earth Planet. Sci. Lett., 229,* 61–71.

Condie, K. C. (2000), Episodic continental growth models; after-thoughts and extensions, *Tectonophysics, 322,* 153–162.

Cooper, M. R. (1989), The Gondwanic bivalve Pisotrigonia (Family Trigoniidae), with description of a new species, *Palaeontol. Z., 63,* 241–250.

Coward, M. P. (1983), Thrust tectonics, thin skinned or thick skinned, and the continuation of thrusts to deep in the crust, *J. Struct. Geol., 5,* 113–123.

Coward, M. P. (1996), Balancing sections through inverted basins, in *Modern Developments in Structural Interpretation, Validation and Modelling, Geol. Soc. Spec. Publ.,* vol. 99, edited by P. G. Buchanan and D. A. Niewland, pp. 51–77.

Cox, S. C., and R. Sutherland (this volume), Regional geological framework of South Island, New Zealand, and its significance for understanding the active plate boundary.

Davey, F., D. Eberhart-Phillips, M. D. Kohler, S. Bannister, G. Caldwell, S. Henrys, M. Scherwath, T. Stern, and H. van Avendonk (this volume), Geophysical structure of the Southern Alps Orogen, South Island, New Zealand.

Davis, G. H., and S. J. Reynolds (1996), *Structural Geology of Rocks and Regions,* John Wiley and Sons, New York.

Davis, D., J. Suppe, and F. A. Dahlen (1983), Mechanics of fold-and-thrust belts and accretionary wedges, *J. Geophys. Res., 88*(B2), 1153–1172.

Eberhart-Phillips, D., and M. E. Reyners (1997), Continental subduction and three-dimensional crustal structure: the northern South Island, New Zealand. *J. Geophys. Res., 102,* 11,843–11,861.

Fuller, C. W., S. D. Willett, D. Fisher, and C. Y. Lu (2006), A thermomechanical wedge model of Taiwan constrained by fission-track thermochronometry, *Tectonophysics, 425,* 1–24.

Ho, C. S. (1986), A synthesis of the geologic evolution of Taiwan, *Tectonophysics, 125,* 1–285.

Hu, J.-C., S.-B. Yu, J. Angelier, and H.-T. Chu (2001), Active deformation of Taiwan from GPS measurements and numerical simulations, *J. Geophys. Res., 106,* 2265–2280, doi:10.1029/2000JB900196.

Huang, B.-S., W.-G. Huang, W.-T. Liang, R.-J. Rau, and N. Hirata (2006), Anisotropy beneath an active collision orogen of Taiwan: results from across islands array observations, *Geophys. Res. Lett., 33,* L24302, doi:10.1029/2006GL027844.

Huang, C. Y., C. T. Shyu, S. B. Lin, T. Q. Lee, and D. D. Sheu (1992), Marine geology in the arc-continent collision zone off southeastern Taiwan; implications for late Neogene evolution of the Coastal Range, *Mar. Geol., 107,* 183–212.

Kimbrough, D. L., A. J. Tulloch, D. S. Coombs, C. A. Landis, M. R. Johnston, and J. M. Mattinson (1994), Uranium-lead zircon ages from the Median Tectonic Zone, South Island, New Zealand, *N. Z. J. Geol. Geophys., 37,* 393–419.

Kao, H., and P. R. Jian (2001), Seismogenic patterns in the Taiwan region; insights from source parameter inversion of BATS data, *Tectonophysics, 333,* 179–198.

Kim, K. H., J. M. Chiu, J. Pujol, K. C. Chen, B. S. Huang, Y. H. Yeh, and P. Shen (2005), Three-dimensional VP and VS structural models associated with the active subduction and collision tectonics in the Taiwan region, *Geophys. J. Int., 162*(1), 204–220. doi:10.1111/j.1365-246X.2005.02657.x

Kuo, B. Y., C. C. Chen, C. C., and Shin, T. C. (1994), Split S waveforms observed in northern Taiwan; implications for crustal anisotropy, *Geophys. Res. Lett., 21,* 1491–1494.

Kuo-chen, H., F. T. Wu, D. A. Okaya, B. S. Huang, and W. T. Liang (2007), *S-Splitting Measurements and Taiwan Orogeny,* AGU Fall Meeting, San Francisco, CA.

Lallemand, S., Y. Font, H. Bijiwaard, and H. Kao (2001), New insights on 3-D plates interaction near Taiwan from tomography and tectonic implications, *Tectonophysics,* 335, 229–253.

Landis, C. A., and D. S. Coombs (1967), Metamorphic belts and orogenesis in southern New Zealand, *Tectonophysics, 4,* 501–518.

Lee, Y. H. (1997), Structural Evolution of Middle Central Range during the Penglai Orogeny, Taiwan, Ph.D. thesis, Institute of Geology, National Taiwan University, Taipei, Taiwan.

Lee, Y. H., C. C. Chen, T. K. Liu, H. C. Ho, H. Y. Lu, and W. Lo (2004), Mountain building mechanisms in the southern Central Range of the Taiwan orogenic belt; from accretionary wedge deformation to arc-continental collision, *Earth Planet. Sci. Letts., 229,* 61–71.

Lee, Y. H., and C. C. Chen (2006), Mountain building mechanisms in the southern Central Range of the Taiwan orogenic belt; from accretionary wedge deformation to arc-continental collision, *Earth Planet. Sci. Letts., 252,* 413–422.

Lin C. H., K. I. Konstantinou, W. T. Liang, H. C. Pu, Y. M. Lin, S. H. You, and Y. P. Huang (2006), Preliminary analysis of volcanoseismic signals recorded at the Tatun Volcano Group, northern Taiwan, *Geophys. Res. Lett.,* 32, L10313. doi:10.1029/2005GL022861

Liu, Char-Shine, I. L. Huang, and L. S. Teng (1997), Structural features off southwestern Taiwan, *Marine Geology, 137,* 305–319.

Mikumo T., Y. Yagi, S. K. Singh, and M. A. Santoyo (2002), Coseismic and postseismic stress changes in a subducting plate: possible stress interactions between large interplate thrust and intraplate normal-faulting earthquakes, *J. Geophys. Res., 107* (B1), 2023, doi:10.1029/2001JB000446.

McIntosh, K., Y. Nakamura, T.-K. Wang, R.-C. Shih, Allen Chen, and C.-S. Liu (2005), Crustal-scale seismic profiles across Taiwan and the western Philippine Sea, *Tectonophysics, 401*, 23–54.

Mortimer, N., A. J. Tulloch, R. N. Spark, N. W. Walker, E. Ladley, A. Allibone, and D. L. Kimbrough (1999), Overview of the Median Batholith, New Zealand: a new interpretation of the geology of the Median Tectonic Zone and adjacent rocks, *J. African Earth Sci., 29*, 257–268.

Mouthereau, F., and C. Petit (2003), Rheology and strength of the Eurasian continental lithosphere in the foreland of the Taiwan collision belt: constraints from seismicity, flexure, and structural styles, *J. Geophys. Res., 108*, 2512, doi:10.1029/2002JB002098.

Nakamura, M. (2004), Crustal deformation in the central and southern Ryukyu Arc estimated from GPS data, *Earth Planet. Sci. Letts., 217*, 389–398, doi:10.1016/S0012-821X(03)00604-6

Rau, R.-J., and F. Wu (1995), Tomographic imaging of lithospheric structures under Taiwan, *Earth Planet. Sci. Letts., 133*, 517–532.

Rau, R. J., W. T. Liang, H. Kao, and B. S. Huang (2000), Shear wave anisotropy beneath the Taiwan orogen, *Earth Planet. Sci. Letts., 177*, 177–192.

Reed, D. L., N. Lundberg, C. S. Liu, and B. Y. Kuo (1992), Structural relations along the margins of the offshore Taiwan accretionary wedge; implication for accretion and crustal kinematics, *Acta Geologica Taiwanica, 30*, 105–122.

Savage, M., M. Duclos, and K. Marson-Pidgeon (this volume), Seismic anisotropy in South Island, New Zealand.

Sibuet, J. C., and S. K. Hsu (2004), How was Taiwan created?, *Tectonophysics, 379*, 159–181.

Sibuet, J. C., S. K. Hsu, C. T. Shyu, and C. S. Liu (1995), Structural and kinematic evolutions of the Okinawa Trough backarc basin, in *Backarc Basins; Tectonics and Magmatism*, Plenum Press, New York, pp. 343–379.

Silver, P. G. (1996), Seismic anisotropy beneath the continents: probing the depths of geology, *Annu. Rev. Earth Planet. Sci., 24*, 385–432.

Stein, S., and M. Wysession (2003), *An Introduction to Seismology, Earthquakes, and Earth Structure*, Blackwell Publishing Ltd., Malden, MA.

Stern, T. A., P. Molnar, D. Okaya, and D. Eberhart-Phillips (2000), Teleseismic P-wave delays and modes of shortening the mantle beneath the South Island, New Zealand, *J. Geophys. Res., 105*, 21,615–21,631.

Stern T., D. Okaya, S. Kleffmann, M. Scherwath, S. Henrys, and F. Davey (this volume), Geophysical exploration and dynamics of the Alpine fault zone.

Suppe, J. (1985), *Principles of Structural Geology*, Prentice-Hall, Englewood Cliffs, NJ, 537 pp.

Suppe, J. (1987), The active Taiwan mountain belt, in *The Anatomy of Mountain Ranges*, Princeton Univ. Press, Princeton, NJ, pp. 277–293.

Tozer, R. S. J., R. W. H. Butler, and S. Corrado (2002), Comparing thin- and thick-skinned thrust tectonic models of the Central Apennines, Italy, EGU, *Stephan Mueller Spec. Publ. Ser., 1*, 181–194.

Tulloch, A. J., and D. L. Kimbrough (2003), Paleozoic plutonism in the New Zealand sector of Gondwana, *Geosci. Aust.*, 123–124.

Twiss, R. J., and E. Moores (1992), *Structural Geology*, W. H. Freeman, New York, 532 pp.

Walcott, R. I. (1998), Modes of oblique compression: late Cenozoic tectonics of the South Island, New Zealand, *Rev. Geophys., 36*, 1–26.

Wang. Z., D. Zhao, J. Wang, and H. Kao (2007), Tomographic evidence for the Eurasian lithosphere subducting beneath south Taiwan, *Geophys. Res. Letts., 33*, doi:10.1029/2006GL027166.

Wu, F., R. J. Rau, and D. Salzberg (1997), Taiwan orogeny: thin-skinned or lithospheric collision, *Tectonophysics, 274*, 191–220.

Wu F. T., L. C. Kuo, and H. Kuo-Chen (2007), *Deformation of Taiwan from Continuous GPS Monitoring*, AGU Fall Meeting, San Francisco, CA.

Seismogenic, Electrically Conductive, and Fluid Zones at Continental Plate Boundaries in New Zealand, Himalaya, and California, USA

George R. Jiracek[1], Victor M. Gonzalez[1], T. Grant Caldwell[2],
Philip E. Wannamaker[3], and Debi Kilb[4]

We explore the idea that fluid occurrence below the seismogenic zone plays an active role in the rupture process by examining how fluids spatially relate to seismicity at three continental plate boundaries: South Island of New Zealand, the Himalaya, and San Andreas fault, USA. With this objective, we project earthquake hypocenters onto magnetotelluric (MT) electrical resistivity cross-sections. MT detection of conductive zones in the crust containing low fractions of fluids (<1%) requires an interconnected network of fluid-filled porosity facilitated by shearing, fracturing, and/or grain-edge wetting. Mechanisms promoting fluid reservoirs in the ductile crust include: 1) stalling of upward propagating porosity waves, 2) tectonically induced neutral buoyancy, and 3) development of ductile shear zones. Distinct conductive horizons are detected at depth in the ductile crust in New Zealand and the Himalaya where the tectonic convergence is high. In the Parkfield segment of the San Andreas fault, where convergence is low, there is high conductivity in the ductile crust but it forms a sub-vertical corridor to the surface with no distinct top. The tops of sub-horizontal conductive zones are ~20 km depth in New Zealand and ~25–40 km in the Himalaya where the seismogenic crust extends only to 12 and 25 km depth, respectively. The deep conductive layer in New Zealand may have originated as a "water sill" facilitating water-weakening, localized deformation, and eventually becoming a water-rich, anisotropic, mylonized, ductile shear zone. Fluid exchange through the active Alpine fault may initiate or be initiated by fault rupture. Localized, unstable flow in deep fluidized zones detected by MT may trigger earthquakes above.

Supplemental Information is available at: http://eqinfo.ucsd. edu/~dkilb/Jiracek/Jiracek_web.html.

[1]Department of Geological Sciences, San Diego State University, San Diego, California.

[2]GNS Science, Lower Hutt, New Zealand.

[3]Energy & Geoscience Institute, University of Utah, Salt Lake City, Utah.

[4]Scripps Institution of Oceanography, University of California, San Diego, La Jolla, California.

A Continental Plate Boundary. Tectonics at South Island, New Zealand
Geophysical Monograph Series 175
Copyright 2007 by the American Geophysical Union.
10.1029/175GM18

1. INTRODUCTION

Understanding the dynamics of earthquake rupture remains a critically important role of science in human society. However, the physical-chemical processes that lead to earthquake nucleation and rupture remain elusive. There is increasing evidence that processes removed from the actual seismogenic zone may be very important. For example, in Japan *Iio and Kobayashi* [2002] suggested that accelerated aseismic slip on a shear zone in the lower ductile crust triggered a major earthquake on a steeply dipping fault in the upper brittle crust. *Rogers and Dragert* [2003] proposed that stress loading evidenced by deep non-earthquake signals monitored in the

Cascadia subduction zone could affect the locked portion of a fault, thereby triggering a large thrust earthquake. In California, USA, *Nadeau and Dolenc* [2005] discovered nonvolcanic tremor activity at 20 to 40 km below the San Andreas fault that correlates in space and time with local earthquake activity in the overlying, 15-km-thick, seismogenic crust. In all of these cases the authors have suggested fluid-related processes well below the seismogenic zone as the origin of stress increase in the upper brittle crust.

Before proceeding, we define three terms that will be used repeatedly. First, the brittle, seismogenic portion of the crust will mean the uppermost region where the localized frictional strength or resistance to shear increases linearly with depth in a manner often described as *Byerlee* [1978] behavior. Second, below the brittle section we define the ductile portion of the crust to be the increasingly weaker lower crust which undergoes aseismic, power law distributed creep. This depth profile of crustal rheology characterizes the famous "pine-tree plot" of crustal strength versus depth [e.g., *Sibson*, 1984]. Third, we use the common usage of the term brittle-ductile transition to be the transition between the upper seismogenic crust and the creeping section below.

The "standard" description of the crust of localized friction (brittle behavior) above to broadly distributed creep (ductile behavior) below is called the "older concept" by *Rice and Cocco* [2007]. They cite a "more recent concept" where the transition is from potentially unstable to inherently stable, but still localized, friction. The base of the seismogenic zone is the transition between the two modes of localized friction rather than a transition from localized friction to distributed creep. Since this model assumes localized deformation at all levels, *Rice and Cocco* [2007] pointed out a difficulty by suggesting that below the seismogenic zone there could be broadly distributed deformation during the interseismic period but localized deformation during deeply penetrating earthquake rupture. *Hobbs et al.* [2002] prefer the terms plastic (pressure sensitive/temperature insensitive) and viscous (pressure insensitive/temperature sensitive) instead of brittle and ductile, respectively. A viscous material is usually highly ductile but it can exhibit localized deformation, and a plastic material can be brittle or ductile depending on the conditions of deformation [*Hobbs et al.*, 2002]. Independent of what mechanism is used to describe the "brittle-ductile transition," there is a temperature limit for the depth of seismogenesis (e.g., ~350°C for wet granite) and continued interseismic creep occurs below the locked seismogenic zone [*Rice and Cocco*, 2007].

The San Andreas fault observatory at depth (SAFOD) experiment in California, USA seeks to characterize the uppermost portion of an active fault zone by drilling into it, sampling and monitoring it. The SAFOD program used an array of geophysical measurements including seismic reflec-

tion and tomography, gravity and magnetic surveys, microearthquake monitoring, and magnetotelluric (MT) soundings to locate the drill site. The ability of the MT method to image very low interconnected fractions (<1%) of aqueous fluid [e.g., *Wannamaker et al.*, 2002] proved valuable in evaluating the hydrologic conditions in uppermost brittle crust of the SAFOD site [*Unsworth and Bedrosian*, 2004a]. It has been accepted for some time that the redistribution of fluids in seismic fault zones can trigger shallow earthquakes [e.g., *Nur and Booker*, 1972; *Rice*, 1992; *Sibson*, 1992; *Bylerlee*, 1993; *Hickman et al.*, 1995; *Gratier et al.*, 2003] and there is evidence that interconnected pore fluid is ubiquitous throughout the seismogenic zone [e.g., *Fialko*, 2004]. However, the idea that fluid concentration and movement below the seismogenic zone plays an active role in the earthquake cycle is relatively new. In 1995 *Hickman et al.* [1995] posed the question "are fluids present in the sub-seismogenic crust?" Since then the evidence has been overwhelming leading *Ague* [2006] to state in an invited address at the American Geophysical Union's 2006 Spring Assembly that "the presence of fluids in deep settings is no longer in dispute." The presence of fluid in the deep crust influences the rheological properties; in particular, it can dramatically reduce the shear strength [e.g., *Cox*, 2005]. However, the response of the ductile crust to increased fluid content is not unique since increasing the geothermal gradient, changing the rock composition to more quartz-rich or less feldspar rich, or decreasing the strain rate all have the same effect on crustal strength versus depth [*Sibson*, 1984]. Finally, the fate of the fluid, especially how it affects the earthquake cycle, involves a complex, nonlinear coupling between thermal, chemical, mechanical, and hydrological processes.

Despite the recognition of the importance of fluids in the earthquake nucleation process, there are many critical unanswered questions such as their origin, transport, and storage. Another fundamental question is how does the spatial distribution of fluids relate to seismicity in active tectonic settings? We will attempt to answer this question for three major continental plate boundaries emphasizing the origin, movement, and accumulation at or below the brittle-ductile transition. A first order requirement is to locate and understand anomalous zones of aqueous fluid and/or partial melt concentrations in the ductile regions of active, earthquake prone areas. From the beginning we will assume, as have many before [e.g., *Gough*, 1986; *Jiracek et al.*, 1983, 1995; *Marquis and Hyndman*, 1992; *Wei et al.*, 2001; *Wannamaker et al.*, 2002; *Unsworth and Bedrosian*, 2004a, 2004b; *Ritter et al.*, 2005], that MT imaging of most crustal conductive features in active settings is mapping zones of interconnected, fluids (either aqueous or partial melt). The three regions we have chosen to study are locations of varying degrees of compressional, continental plate boundaries: 1) the Southern Alps in New Zealand, 2) the Himalaya, and 3) a "transitional" portion of the San Andreas fault. Distinct differences in tectonic style, particularly the rate and scale of convergence, influence fluid generation, migration, and occurrence of the electrically conductive features and their relationship to deformation processes. Ultimately we seek a dynamic coupling between high conductivity regions [fluids] and earthquake occurrences.

It is important to understand that not all features of deep conductive zones are well-resolved by the MT method. However, in the usual case where the MT wavelengths (or skin depths) are large compared to the thickness of such layers, there are two quantities that are well-resolved. These are [e.g., *Jiracek et al.*, 1995]: 1) the depth to the top of the conductor and 2) its vertically integrated conductivity or conductance, namely, the conductivity-thickness product (or, equivalently, the thickness over resistivity quotient since resistivity is the reciprocal of the conductivity). Consequently, when comparing crustal conductive zones, we will stress the depth to the top and their conductance. A requirement for MT detection of any layer-like conductive feature is that its conductance significantly exceeds the integrated conductance of the entire overlying crust.

We will review briefly the tectonic settings of the three study areas and compare and contrast the patterns of seismicity and conductivity, the latter inferring zones of fluid concentration in the ductile portion of the crust. Following the reviews we outline key results from the literature on the origin and fate of deep metamorphic fluids and the formation of fluid reservoirs below, or at, the brittle-ductile transition. In particular, fluid-filled fracture development and grain-edge wetting are discussed since fluids must be interconnected over tens of kilometers for detection by MT at mid-crustal depths. The requirement of sustained permeability in the ductile crust leads us to conclude that pervasive ductile shear zones may trap fluids and explain the high conductivity, an idea proposed decades ago by *Eaton* [1980]. Finally, possible links between earthquakes and the deep fluids (particularly aqueous) are proposed. Results from recent publications on crustal-scale fluid distribution in compressive tectonic environments also will be reviewed as they relate to our objectives. Details of the MT and earthquake data collection and modeling in the three study areas are not addressed herein since they are well-described in the literature cited below.

2. PLATE BOUNDARY STUDY AREAS: CONDUCTIVE AND SEISMOGENIC ZONES

2.1. New Zealand

South Island of New Zealand has been the focus of our group for over 10 years [*Wannamaker et al.*, 2002] so we present original MT results from this area along with a more complete analysis. Our MT work began with a pilot study of only 12 soundings in 1995 as part of an extensive multi-disciplinary, continental dynamics project [*Davey et al.*, 1998]. Few MT soundings had been recorded previously across the Southern Alps that dominate the striking topography of the central, 150 to 200-km-wide, portion of South Island. Here, the ~37 mm/yr displacement between the Pacific and Australian plates [*DeMets et al.*, 1994] is manifested by a 500-km-long mountain range with peaks rising to in excess of 3750 m. A convergent component of the orogen developed about 5-6 Ma ago and has resulted in ~70 km of shortening, crustal thickening, and exhumation of ~25-km-deep metamorphic rocks (Figure 1) along a ~50° southeast dipping, plate boundary fault called the Alpine fault [*Walcott*, 1998; *Norris*, 2004]. The Alpine fault accommodates most of the oblique plate motion between a strong Australian plate and a weaker Pacific plate (Figure 1). Convergence in the central segment of the Alpine fault is estimated to be about 10 mm/yr and there is ~35 mm/yr slip rate parallel to the fault [*Norris*, 2004; *Beavan et al.*, this volume]. We collected 54 MT soundings in 1995–1998 across the central segment including sites occupied over the 30-km-wide, rugged main divide of the Southern Alps using helicopter support. Detailed results of this effort appear in *Gonzalez* [2002] and *Wannamaker et al.* [2002].

For this study we use the 150-km-long two-dimensional (2-D) resistivity model (Plate 1) presented by *Gonzalez* [2002]. It differs slightly from that presented by *Wannamaker et al.* [2002] because *Gonzalez* [2002] applied near-surface distortion analysis [*Groom and Bailey*, 1989]. The model was derived using the 2-D MT inversion algorithm of *Rodi and Mackie* [2001] using the cross-strike, electric field response function (the transverse magnetic, TM mode). Inversion using both the TM and the along strike electric field response (the transverse electric, TE mode) produced a markedly different resistivity image due to partial violation of the 2-D assumption and significant anisotropy [*Gonzalez*, 2002; *Wannamaker et al.*, 2002]. However, the depth to the top of the deep conductor of ~20 km is in agreement. In three-dimensional (3-D) circumstances, a 2-D inversion using the TM mode alone gives results in better agreement with the actual resistivity cross-section perpendicular to the long axis of a buried 3-D conductive body [*Wannamaker et al.*, 1984; *Ledo*, 2005]. Such geometry has been established for the conductive body under the Southern Alps by additional MT soundings extending tens of km either side of the MT line marked in Figure 1. Resistivity values in Plate 1 span a large range from very highly resistive (10^4 ohm m and greater) to conductive (~10 ohm m). We stress that it is the depth to the top of the deep conductor (~20 km) and its conductance

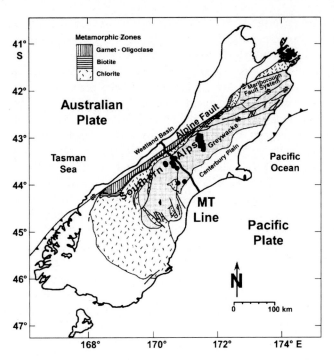

Figure 1. Location map of MT line across South Island of New Zealand. Metamorphic zones after *Vry et al.* [2001] and references therein. Epicenter locations from *Eberhart-Phillips and Bannister* [2002] (black dots) and *Bannister et al.* [2006] (black squares).

that are the best resolved quantities not its actual resistivity. Conductance calculations for the deep conductor in the ductile portion of the crust in Plate 1 range from 50 S in the cross-strike direction to 500 S in the along-strike direction. This electrical anisotropy is attributed to active, shearing in the fluidized zone of the ductile crust caused by plate motion [*Wannamaker et al.*, 2002].

Our focus in this paper is mainly on the conductive zone (imaged as ≤200 ohm m in the resistivity model) traced laterally from approximately 60 km southeast of the Alpine fault (Plate 1) to ~5–10 km southeast of the fault proper. Over this span, the conductive zone rises on both ends from a depth to its top of ~20 km under the eastern foothills of the Southern Alps. To the northwest it slopes sharply upward, nearly paralleling the depth projection of the Alpine fault, before it rises almost vertically from ~10 km depth (Plate 1). It intersects the surface at 5–10 km southeast of the fault trace. *Wannamaker et al.* [2002] note that 10 km depth is approximately at the brittle-ductile transition as established by others [e.g., *Beavan et al.*, 1999; *Leitner et al.*, 2001]. Where the conductor also rises to the surface ~60 km southeast of the Alpine fault the modeled conductivity, and therefore conductance, is more than an order of magnitude less

than that associated with the Alpine fault strand (Plate 1). A conductive outcrop and two upward extending conductive limbs at ~60 to 75 km distance in Plate 1 are coincident with mapped locations of backthrust faults associated with crustal shortening and uplift of the Southern Alps [*Cox and Findlay*, 1995]. *Wannamaker et al.* [2002] concluded that the origin of the fluids inferred to cause the high conductivity ~20 km beneath the Southern Alps lies mainly in the release of aqueous fluid from prograde metamorphism in a thickened crustal root. They estimated the amount of saline fluid to be ~0.02–0.2% where the resistivity is ~100 ohm m and 0.2 to 3% where the resistivity is lower (~10 ohm m). Since the top of the deep conductor is at ~20 km depth, it is well below the brittle-ductile transition of ~12 km [*Leitner et al.*, 2001] so the deepest fluids probably interconnect through grain-edge wetting [e.g., *Holness*, 1993; *Spear*, 1993] within a ductile shear zone [*Tullis et al.*, 1996] rather than in fractures.

There is a unique opportunity to evaluate the nature of the deep crustal conductive zone beneath the Southern Alps because exhumation along the hanging wall of the Alpine fault has exposed a sequence of rocks from over 20 km depth [*Norris*, 2004]. Near the fault, i.e., at the deepest part of the exhumed sequence, is a mylonite zone that shows very large shear strain, intense strain localization, and mineral assemblages that indicate deformation under metamorphic conditions at depths of 20–30 km [*Grapes and Watanabe*, 1994; *Vry et al.*, 2004]. *Norris* [2004] believes the mylonites formed during the last 5 Myr since plate convergence began and reactivation of an older discontinuity developed along the Alpine fault. Ductile shear recorded by the mylonites supports the projection of the Alpine fault via ductile creep to depths of the midcrustal conductor. Therefore, the preponderance of evidence lead us [*Wannamaker et al.*, 2002] to conclude that the ~20-km-deep conductor lies in an active, mylonized, ductile shear zone where deep-sourced fluids are trapped. Additional fluids may be released by strain-induced metamorphism [*Koons et al.*, 1998; *Upton et al.*, 2003; *Upton and Koons*, this volume] and shearing itself promotes deformation-enhanced grain boundary wetting and fluid interconnection [*Tullis et al.*, 1996; *Cox*, 2005]. Shear interconnection also explains the strong resistivity anisotropy below central South Island, because the greater degree of shear along strike promotes long-range "backbone" (connected) shears in that direction [*Cox*, 2005; *Wannamaker*, 2005]. The fluids in the ductile shear zone may have developed initially as a "water sill" (see Section 3).

The seismicity of the Alpine fault and the Southern Alps in New Zealand has been the focus of several recent efforts that have produced high quality hypocenter data [*Eberhart-Phillips*, 1995; *Leitner et al.*, 2001; *Eberhart-Phillips and Bannister*, 2002; *Bannister et al.*, 2006]. These data show

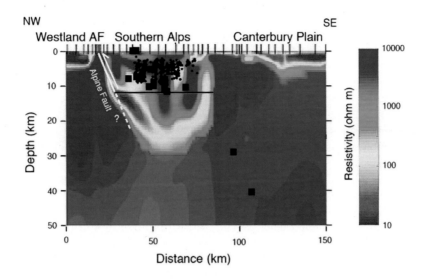

Plate 1. Electrical resistivity model of New Zealand MT line [*Gonzalez,* 2002]. Vertical ticks are locations of MT sites. AF denotes surface location of the Alpine fault; projection to depth assumes 50° SE dip. Black squares are earthquake hypocenters with magnitudes 2.0 to 3.9 occurring within ± 15 km of the MT profile [*Eberhart-Phillips and Bannister*, 2002] and black dots are locations of 1994 M$_w$ = 6.7 Arthur's Pass earthquake and 40 aftershocks [*Bannister et al.*, 2006] projected parallel to the Alpine fault (Figure 1). Horizontal black line at 12 km depth marks brittle-ductile transition [*Leitner et al.*, 2001].

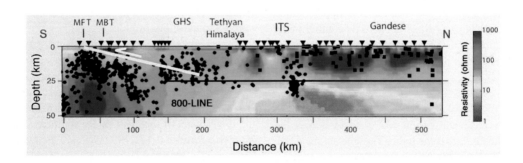

Plate 2. Electrical resistivity model for the Himalaya MT 800-line+Nepal from *Unsworth et al.* [2005]. Inverted triangles indicate MT stations. Abbreviations are: MFT (Main Frontal thrust), MBT (Main Boundary thrust), GHS (greater Himalaya sequence), and ITS (Indus–Tsangpo suture). Black circles are hypocenters from *Schulte-Pelkum et al.* [2005] and black squares are hypocenters from *Langin et al.* [2003] and *Monsalve et al.* [2006]. The extension of MFT corresponds to the shear zone presented by *Schulte-Pelkum et al.* [2005]. The black horizontal line at 25 km depth marks the brittle-ductile transition [*Langin et al.*, 2003].

that the base of the seismogenic zone is at 12 ± 2 km in the central part of the Southern Alps; ~10 km above the top of the midcrustal conductor shown in Plate 1. Here, the positions of nearby, accurately located hypocenters from these studies are projected parallel to strike onto the resistivity model. Shown are hypocenters from 17 earthquakes having local magnitudes ranging from 2.0 to 3.9, occurring during 1979–1996, all within 15 km of the MT profile [*Eberhart-Phillips and Bannister*, 2002], and the locations of the 1994 $M_w = 6.7$ Arthur's Pass earthquake and 40 aftershocks [*Bannister et al.*, 2006]. Epicenter locations for these earthquakes are presented in Figure 1. Although the 1994 sequence is ~70 km from the MT profile it is included here since the lateral and depth extent of the seismogenic zone is thought to be nearly uniform in this part of central South Island [*Leitner et al.*, 2001]. Moreover, this is the most reliable, detailed image of a moderate earthquake and its aftershock sequence associated with the Southern Alps. Nearly all of the seismic activity in this region occurs inboard (southeast) of the Alpine fault (Figure 1 and Plate 1) in a broad region of brittle deformation best modeled by numerous north-northeast trending reverse faults [*Leitner et al.*, 2001]. Earthquake focal mechanisms indicate significant components of dip slip (compression) and strike-slip, as expected. The comparatively low seismicity rate measured over the last 150 years indicates that only a small fraction of the strain accumulation has been released; this is interpreted to indicate the potential for large earthquakes along the Alpine fault [*Leitner et al.*, 2001].

A deeper northwest-dipping seismicity zone in this region of South Island defines a diffuse pattern that crosses the brittle-ductile transition and into the stiff upper mantle of the Pacific Plate [*Liu and Bird*, 2006]. The southernmost epicenters in Figure 1 and the corresponding hypocenters in Plate 1 at ~31 and 43 km depth are in this zone.

2.2. Himalaya

There have been several recent MT surveys across the Himalaya, specifically across the Indus-Tsangpo suture (ITS), also called the Yarlung-Tsangpo suture (YTS), in southern Tibet that divides rocks of Indian and Asian origin [e.g., *Wei et al.*, 2001; *Spratt et al.*, 2005, *Unsworth et al.*, 2005]. Underthrusting of the Indian continental lithosphere under Eurasia has resulted in the stacking of several crustal-scale thrust sheets [e.g., *Beaumont et al.*, 2001; *DeCelles et al.*, 2002] and a doubling of the crustal thickness to ~ 80 km [*Owens and Zandt*, 1997]. Convergence of India and Eurasia from global positioning system (GPS) measurements is 36–40 mm/yr; 15–20 mm/yr occurs within the Himalaya [*Zhang et al.*, 2004].

We have selected the MT 800-line+Nepal profile as representative of the MT results presented by *Unsworth et al.* [2005]. It traces a prominent crustal conductor for the greatest distance (>300 km) [Plate 2]. The Nepal portion of this profile (the southern ~170 km in Plate 2) was described earlier by *Lemonnier et al.* [1999]. They believe that the ~20-km-deep, ~30 ohm m conductive occurrence, and seismicity, near the downward terminus of the line representing the Main Frontal fault in Plate 2 is caused by fluids released from metamorphic reactions in the footwall of the thrust.

There is remarkable agreement of the salient geoelectric features from the four MT lines presented by *Unsworth et al.* [2005] even though they are spread over 1000 km distance. Location of these lines is mapped in Figure 2. Lowest resistivity zones (≤10 ohm m) are imaged north of the Tethyan Himalaya on each geoelectric profile with tops at ~25 to 40 km depth [e.g., Plate 2; *Unsworth et al.*, 2005]. The conductance of these zones is highest in the vicinity of the ITS, at ~20,000 S, and exceeds 1000 S for hundreds of km north [*Wei et al.*, 2001]. This is over an order of magnitude greater conductance than the deep conductor under the Southern Alps of New Zealand. *Unsworth et al.* [2005] interpreted this conductive zone as an interconnected, 5 to 14% partial melt fraction. *Makovsky and Klemperer* [1999] and *Li et al.* [2003] suggested that the nature of reflection seismic bright spots combined with the electrical conductivity favor an interpretation with both saline water and partial melt. The resulting conductive zone is estimated to have the equivalent of ~200 m of saline fluid above ~10 km thickness of ~10% partial melt. *Klemperer* [2006] believes this sequence best fits a picture of fluid rising from the subducting Indian subcontinent and triggering partial melting of granite at temperatures of ~650°C resulting from radiogenic heating. We would add that fluid production and possible melting would be enhanced by the crustal thickening itself in a fashion analogous to, but greater than, crustal development under the Southern Alps. *Klemperer* [2006] emphasizes that if either water or melt are present, the effect is a region of weak crust that would allow midcrustal flow that decouples the upper and lower parts of the crust under southern Tibet [*Nelson et al.*, 1996; *Royden et al.*, 1997; *Schulte-Pelkum et al.*, 2005]. *Unsworth et al.* [2005] estimate that the reduced viscosity calculated from the melt fraction required by the MT data is consistent with such midcrustal flow in southern Tibet.

Seismicity north of the ITS in central Tibet was evaluated by *Langin et al.* [2003] as part of the 1998–99 INDEPTH III program (International Deep Profiling of Tibet and Himalaya). Using a histogram of the focal depths (Figure 3) the authors concluded that 99% of the events are at depths shallower than 25 km in central Tibet. *Langin et al.* [2003] cited these data, along with MT data, as indicating that the crust in this region below the ~25 km level is undergoing ductile, aseismic deformation, and mid to deep crustal flow. The fo-

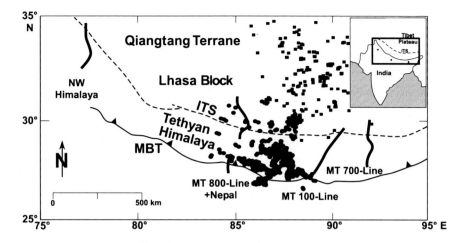

Figure 2. Location map of MT profiles crossing the Himalayan Indus-Tsangpo suture (ITS) and the Tibetan plateau. MT 800-line+Nepal is used in this paper. MBT is the Main Boundary thrust. Modified from *Unsworth et al.* [2005] and *Klemperer* [2006]. Epicenter locations from *Langin et al.* [2003], *Schulte-Pelkum et al.* [2005], and *Monsalve et al.* [2006].

cal mechanisms in the brittle 25-km-deep seismogenic crust are consistent with the north-south compression produced by India-Eurasia collision. The hypocenter data from *Langin et al.* [2003] are projected onto the MT 800-line+Nepal geoelectric section in Plate 2 as are the surface positions of the Main Boundary thrust (MBT) and the Main Frontal thrust (MFT) (with its projection downward). To provide additional coverage, particularly south of the ITS, we have included additional hypocenter data recently published by *Schulte-Pelkum et al.* [2005] and *Monsalve et al.* [2006] recorded in Nepal and Tibet in 2001–2003. Epicenter loca-

tions of all earthquake events used in Plate 2 are shown in Figure 2.

Chen and Yang [2004] showed that in western Himalaya and southern Tibet there are two distinct depth intervals of seismicity: 1) upper crustal events above 25 km (in agreement with *Langin et al.* [2003]) and 2) deeper events ~75–100 km deep, near or below the Moho discontinuity. The deep earthquakes are assumed to be in the strong uppermost mantle marking where subduction of oceanic lithosphere ceased tens of millions of years ago [*Chen and Yang*, 2004]. The region of deep (~75–100 km depth) seismicity as discussed by *Chen and Yang* [2004], and confirmed by *Schulte-Pelkum et al.* [2005], obviously would plot beneath the 50 km maximum depth in Plate 2.

Comparing all of the seismicity data with the 50-km-deep MT 800-line+Nepal resistivity section shows that on the northern two-thirds of the profile the seismogenic zone is shallow, above 25 km and above the electrical conductor (Plate 2). Earthquakes are present from the surface to 50 km depth at the southern end of the profile (Plate 2), where the deep conductive zone is absent, and the crust is relatively resistive. A local concentration of earthquakes extending from the surface downward into the conductive zone at ~ 25 km north of the ITS will be discussed later in concluding Section 7. The main observation in the context of our objectives is that the upper seismogenic zone is above a prominent crustal conductive zone in central and southern Tibet, similar to that observed under the Southern Alps of New Zealand. However, the conductance beneath Tibet is more than an order of magnitude greater reflecting the greater amount of crustal thickening and crustal melting that has occurred beneath this part of the Himalaya.

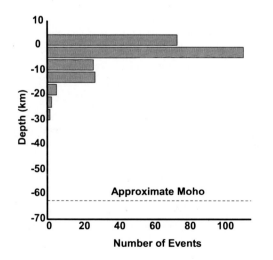

Figure 3. Histogram of depths of earthquakes in 1998–1999 from central Tibet [*Langin et al.*, 2003].

Published MT models for the Himalaya do not require significant resistivity anisotropy as deduced below the Southern Alps, presumably because under the Tibetan Himalaya there is a large melt fraction and less whole-crustal strike-slip shear.

2.3. San Andreas Fault

The 1200-km-long San Andreas fault in California, USA forms the boundary between the North American and Pacific plates (Figure 4). The central California portion of the fault zone is transpressive [Chery et al., 2001], the transpression existing for the last ~ 8 Ma of the 15 Ma old fault system [Atwater and Stock, 1998].

The Parkfield segment of the San Andreas fault, the subject of intense and ongoing research for decades, is located in a transition between a creeping portion of the fault to the northwest and a locked section to the southeast. Long-term strike-slip motion in the Parkfield segment (~35 mm/year) is mainly concentrated on the fault itself along a direction of ~N42°W with only a small (~3 mm/yr) convergent component [Argus and Gordon, 2001]. The Parkfield segment is also where the SAFOD drilling site (http://www.earthscope.org/safod/index.shtml) is located (Figure 4) and where a M=6.0, 7.8-km-deep earthquake occurred on September 28, 2004. Aftershocks from this event extended for over 30 km along the fault trace (Figure 4) with all hypocentral depths above 15 km (Thurber et al., 2006). The SAFOD drilling

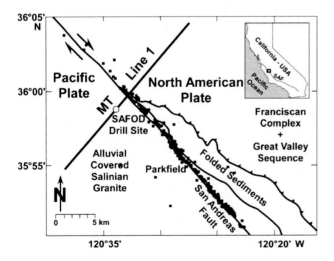

Figure 4. Location map of MT Line 1 in Parkfield, California vicinity of the San Andreas fault [modified from *Unsworth and Bedrosian*, 2004a]. Surface location of the San Andreas Fault Observatory at Depth (SAFOD) drill site is shown approximately 2 km from the fault. Epicenter locations are of M>2 aftershocks from the 2004 M=6.0 Parkfield earthquake [*Thurber et al.*, 2006].

program reached its phase 2 objective in the summer of 2005 when the deviated drill hole intersected an active section of the San Andreas fault (Plate 3) at a vertical depth of ~3 km.

There have been a series of papers describing the MT studies in the central portion of the San Andreas fault in California ranging from the creeping segment in the north near Hollister [Bedrosian et al., 2002], to the locked section in the south in the Carrizo Plain [Mackie et al., 1997; Unsworth et al., 1999]. MT results from the Parkfield-SAFOD area in the transitional segment of the fault are numerous [Unsworth et al., 1999; Park and Roberts, 2003; Unsworth and Bedrosian, 2004a, 2004b; Ritter et al. 2005; and Becken et al., 2006].

Although the MT data collected in the 1990s across the San Andreas fault were of unprecedented horizontal resolution for that time with site spacings of 100 m near the fault, the maximum depth of resolution was less than 10 km. This was because the profile lengths were short (e.g., <25-km-long for MT line 1 investigated by *Unsworth and Bedrosian* [2004a]; Figure 4) and the sounding periods were less than 1000 s. In their study of the Parkfield region *Unsworth and Bedrosian* [2004a, 2004b] performed a series of hypotheses tests using constrained inversions to evaluate the uniqueness of their models. For example, they concluded that the shallow fault zone conductor centered near the surface trace of the fault (SAF in Plate 3) possibly extends deeper than the 2 km they modeled. The very low resistivity (<2 ohm-m) in this zone is thought to be due primarily to saline fluids filling fractures with an overall porosity estimated using Archie's law of 8–30% [*Unsworth et al.*, 1997; *Unsworth and Bedrosian*, 2004a]. *Park and Roberts* [2003] suggested that thicker than expected, fluid-filled, conductive sedimentary rocks in the Parkfield syncline immediately (<1 km) east of the fault explains this shallow fault zone conductor. *Unsworth and Bedrosian* [2004a] cautioned that such shallow fluids in the Parkfield area may not be involved in the earthquake cycle. However, a small mantle component in the fluids sampled from groundwaters and faults adjacent to the San Andreas fault [*Kennedy et al.*, 1997] and in the SAFOD drill hole [*Thordsen et al.*, 2005], possibly associated with upwelling mantle [*Sass et al.* 1997; *Jove and Coleman*, 1998], provides evidence that there is a connection between the shallow and deeper aseismic crust.

Synthetic inversions of hypothesized structures extending to 40 km below the San Andreas fault by *Unsworth and Bedrosian* [2004b] concluded that lower crustal structure could be determined under "favorable conditions." Even so, a model containing a conductive, 5-km-wide vertical extension of the fault into the lower crust was imaged incorrectly as much wider at depth than in the model.

In the spring of 2005 *Becken et al.* [2006] recorded 45 combined long-period/broad-band MT soundings over a 50 by

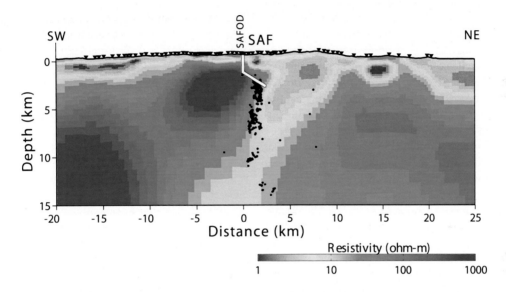

Plate 3. Electrical resistivity model for 2005 MT line from *Becken et al.* [2006; in press, 2007] in the Parkfield-SAFOD area of the San Andreas fault. Inverted triangles are 66 locations of MT sites used in the inversion. Black dots are aftershock hypocenters [*Thurber et al.*, 2006] within ±3 km of the MT line after projection along fault strike onto the plane of the MT profile.

50 km^2 area centered on Parkfield. They also collected 41 additional measurements along a previous 45-km-long seismic profile [*Hole et al.*, 2006]. Modeling results of *Becken et al.* [in press] are presented for the upper 15 km in Plate 3 where the upper 6 km of the resistivity profile closely agrees with the shallow section presented by *Unsworth and Bedrosian* [2004a, 2004b]. The prominent near-surface conductive zone within 10 km east of the fault (previously designated as the Eastern conductor by *Unsworth and Bedrosian* [2004a]) is thought to be caused by over-pressured fluid within the Franciscan complex and Great Valley sequence (Figure 4). A sub-vertical, ~10-km-wide corridor of high conductivity extends downward from the Eastern conductor through the entire brittle section of the crust where it links with the active seismic zone at a depth of ~5–15 km (Plate 3). Below the 15 km depth section shown in Plate 3 the conductive corridor penetrates vertically along the downward projection of the San Andreas fault through the underlying non-seismic (ductile) crust (15–25 km depth) [*Becken et al.*, 2006, in press]. *Becken et al.* [2006] hypothesized that this deep-reaching conductive anomaly marks a major pathway for deep crust or mantle fluids since it penetrates the entire ~25-km-thick crust.

The deep conductance presented by *Becken et al.* [2006, in press] integrated from 10 to 25 km depth has a maximum conductance of 1500 S and is greater than ~750 S within a zone ~5 km either side of the region of maximum conductance. The high conductance value of 1500 S in the ductile crust beneath the Parkfield segment of the San Andreas fault is intermediate between the high values in the Himalaya (~20,000 S) and New Zealand (~500 S). The lack of a clear depth to the top of the ~10-km-wide conductive zone in the ductile crust beneath the San Andreas fault means that the resistivity pattern is distinctly different than those at the other two continental plate boundaries (Plates 1 and 2). It is nearly centered on the downward projected surface trace of the fault and connects upward through the brittle crust sub-vertically where it is offset east of the surface trace. In New Zealand and the Himalaya the conductive zones in the ductile crust have sharply defined upper boundaries, are laterally extensive, and are situated down dip of the surface expression of the main faults that accommodate convergence. One explanation of the difference may relate to the low convergence rate across the Parkfield portion of the San Andreas fault (~3 mm/yr) compared to the Himalaya (15–20 mm/yr) and New Zealand (~10 mm/yr). In addition, since there is virtually no dip slip along the Parkfield segment of the San Andreas fault there is little opportunity for fluid production though crustal thickening and metamorphism.

Where the San Andreas fault zone is narrow, seismicity is localized (Figure 4 and Plate 3) within the near-vertical extensions of the fault itself, slightly offset from its surface trace, or along near-downward projections of steeping dipping auxiliary faults [*Thurber et al.*, 1997, 2006; *Bedrosian et al.*, 2004]. Whether the fault is locked, creeping, or has recently ruptured, more than 99% of modern and historic earthquakes occur in the upper 15 km for the San Andreas system as has been known for some time [*Brace and Byerlee*, 1970]. Plate 3 includes aftershock hypocenters from the 2004, M=6.0 Parkfield earthquake [*Thurber et al.*, 2006] that occurred within ±3 km of the *Becken et al.* [2006, in press] MT profile. This profile extends from both ends of MT Line 1 in Figure 4 where the epicenters of M>2 earthquakes from the 2004 event are plotted.

There is scant evidence of the exact nature of the ductile crust beneath the San Andreas fault. *Unsworth and Bedrosian* [2004b], citing the work of *Anderson et al.* [1983] on exhumed fault zones, suggest a narrow, vertical mylonite zone where fluid and shearing would enhance the electrical conductivity not unlike what we have proposed beneath the Southern Alps in New Zealand.

3. DEEP FLUIDIZED ZONES

Source regions of fluid release must feed each of the deep conductive zones. In the case of the Southern Alps the fluid is thought to originate from a thickened crust [*Koons et al.*, 1998; *Vry et al.*, 2001; *Wannamaker et al.*, 2002; *Upton et al.*, 2003; *Upton and Koons*, this volume] and in the Himalaya from within thickened underthrust Indian lithosphere [*Makovsky and Klemperer*, 1999; *DeCelles et al.*, 2002; *Klemperer*, 2006]. Below the San Andreas fault, fluid is hypothesized to be produced in the deep crust [*Rice*, 1992; *Pili et al.*, 1998; *Gratier et al.*, 2003] and from an upwelling mantle [*Zandt and Carrigan*, 1993; *Kennedy et al.*, 1997; *Sass et al.*, 1997; *Jove and Coleman*, 1998; *Thordsen et al.*, 2005]. In New Zealand and the Himalaya the conductive zones are distinct with upper bounds clearly beneath the seismogenic crust. In the ductile regime below the San Andreas fault, a sub-vertical conductive pattern is centered on the fault and it has no distinct top. To explore these relations we consult the extensive literature on the occurrence and fate of deep metamorphic fluids [e.g., *Ferry*, 1994; *Connolly*, 1997; *Manning and Ingebritsen*, 1999] and the formation of fluid reservoirs below and at the brittle-ductile transition [*Bailey*, 1994; *Connolly and Podladchikov*, 2004; *Hobbs et al.*, 2004; *Cox*, 2005].

Bailey [1994] described a sealing mechanism in the mid-crust due to the formation of immiscible CO_2-H_2O mixtures below which rising fluids could pond. He argued that the process would operate best just above the brittle-ductile transition and that such sealing would not be permanent below the transition. Hydraulic fracturing, induced by the continued arrival of high-pressure fluids from below, would be expected to be horizontal (perpendicular to the minimum principal stress which is vertical) in stable or compressional tectonic regimes. This would result in very thin fluid reservoirs compared to their horizontal extent as described earlier by *Bailey* [1990].

Connolly [1997] extended the work of *Connolly and Thompson* [1989] and *Thompson and Connolly* [1990] to study the migration of devolatilization-generated fluids by numerical simulations of heating at various rates at 25 km crustal depth. The accompanying metamorphic reactions released water exceeding 1% by weight in some cases. The main finding of Connolly's simulations was that metamorphic dehydration processes perturb the steady state hydrologic regime enough to produce shock waves of anomalous fluid pressure, porosity, and fluid flux. He found that compaction caused by the difference between the confining pressure and vertical fluid pressure gradients generated positive fluid pressure above the reaction (fluid release) front. This launched an upward propagating porosity wave leaving a connected pore network in its wake. Hydrofracture is possible at shallower depths as the upward strengthening of the crust slows the porosity waves and increases their wavelength and pressure amplitude. Upward slowing of porosity waves also allows newly launched waves to interact with earlier ones [*Connolly*, 1997]. As this occurs, the porosity of the wave continues to increase as fluid is transferred from deeper waves. This process leads to mechanical stability with a positive feedback system favoring the formation of high-porosity so-called "water sills" [*Fyfe et al.*, 1978; *Connolly*, 1997]. Such features have low shear strength so they would be unstable under deviatoric stress. The model calculations showed that water sills can form in rheologically continuous, upward strengthening media but the time required is long (>50 Myr). However, a crustal strength discontinuity greatly accelerates their formation. This means that a relatively impermeable or strong lithologic barrier below the brittle-ductile transition is required to arrest upward fluid migration in young, active tectonic regimes. For example, in a simulation presented by *Connolly* [1997], a stronger layer (by a factor of 20 times) at 17 km depth was enough to stall an upward propagating porosity wave within the time scale of 5–10 Myr after generation of the reaction front at 20 km depth.

Connolly and Podladchikov [2004] proposed another mechanism to trap fluid in the ductile crust in compressional tectonic settings. They showed that in such environments the absence of a vertical hydraulic gradient below the brittle-ductile transition produces a zone of fluid stagnation (Figure 5). This occurs because the buoyancy forces acting on a fluid are balanced by the vertical stress gradient in the rock [*Connolly and Podladchikov*, 2004]. This not only explains how upward migrating

fluids are arrested but also how fluids from above will flow downward into a reservoir below the brittle-ductile transition. Thus, wherever a stagnation zone is formed in the ductile crust there is a zone of negative hydraulic potential gradient above that drives fluid flow downward and a positive gradient below that drives flow upward (Figure 5). The vertical potential gradient in between is zero so the fluid is trapped. *Connolly and Podladchikov* [2004] called this tectonically induced neutral buoyancy because its formation depends on differential stress that develops in response to horizontal compressive stress.

Using ranges of estimated parameters, the authors calculated possible stagnant zones for aqueous fluid up to 12 km thick and 20 km below the brittle-ductile transition. The

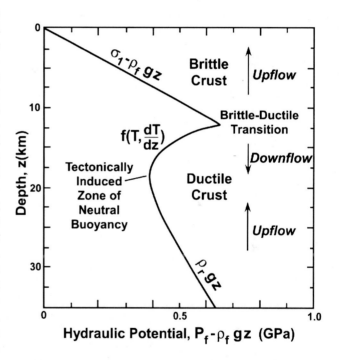

Figure 5. Profile of hydraulic potential versus depth for a 2–D pure shear, compressional model of the crust [modified after *Connolly and Podladchikov*, 2004]. Byerlee's law is assumed for the brittle crust using fluid density ρ_f. A differential stress ($\sigma_1 - \sigma_3$) term included in the fluid pressure, P_f calculation results in suprahydrostatic hydraulic potential, $\sigma_1 - \rho_f gz$ where g is the gravitational acceleration. Thermally activated ductile behavior below the brittle-ductile transition decays toward lithostatic conditions at depth for rock density ρ_r. A region of tectonically induced neutral buoyancy is developed below the brittle-ductile transition where the vertical hydraulic gradient is zero and water flows from above and below into a zone of fluid stagnation. This would result in a conductive zone in the ductile crust. Specific values of GPa versus depth were calculated using estimated crustal parameters including temperature, T and its vertical gradient [*Connolly and Podladchikov*, 2004].

concept of tectonically induced neutral buoyancy expands on the "Swiss cheese" model (Figure 6) of the crust [*Connolly and Podladchikov*, 2004] where fluid circulates freely in the brittle crust and is allowed only in self-propagating, isolated domains of various geometries in the ductile crust (Figure 6). Fluid flow in these domains is sensitive to tectonic forcing and can be accomplished through an individual fracture, a network of fluid-filled fractures, or via grain-edge porosity (discussed in Section 4). In either case, the vertical pressure gradient, $\partial P_f / \partial z$ in a spherical domain has the same mean stress gradient as the surrounding rocks. The gradient approaches the vertical gradient of the horizontal, σ_1 and vertical component, σ_3 of the far-field stress for vertically and horizontally elongate domains, respectively (Figure 6). *Connolly and Podladchikov* [2004] described the result that the brittle-ductile transition would act as a barrier to upward propagating domains (in the form of porosity waves as discussed above) with spherical or vertical geometries. However, far field compressive stress would promote a dynamic evolution of these geometries to spread laterally during upward ascent resulting in coalesced, quasi-static fluid bearing horizons. It is important to note that the fluid accumulation mechanism in the ductile crust described by *Connolly and Podladchikov* [2004] is valid for any fluid, including partial melt.

The work of *Hobbs et al.* [2004] independently confirmed Connolly and Podladchikov's prediction of regions of stag-

nant fluid flow in the ductile portion of the crust. In fact, where the geothermal gradient is high, multiple zones are predicted. Multiple zones are also predicted by *Connolly and Podladchikov* [2004] because numerous porosity waves would occur during episodic or continuous, deep metamorphic reactions. *Hobbs et al.* [2004] numerical models included a weak zone dipping at 45° in the brittle crust to simulate a fault extending to the brittle-ductile transition (Plate 4a). The 2-D numerical modeling assumed a crust initially saturated with fluid with pore pressure sufficient to hold pore space open without causing failure. The crust was then shortened horizontally at a given strain rate and patterns of dilatancy were computed. For example, Plate 4b shows the instantaneous volumetric strain rate where there was 1.5% shortening. Here, dilatant zones represent zones of increased porosity and, therefore, excess fluid. They are developed as near-horizontal, tabular features (Plate 4b) at the brittle-ductile transition and above along the hanging wall of the numerically inserted fault (Plate 4a) and *en echelon* to it along a diffuse backthrust shear zone. These features are the direct result of strain partitioning that produces a plastic wedge of material involving the entire brittle crust. This wedge shape is very similar to the conductive pattern below South Island of New Zealand (Plate 1) except for one major difference: the bottom of the conductive feature in New Zealand is not at the brittle-ductile transition, it is well below it in the ductile crust. *Hobbs et al.* [2004]

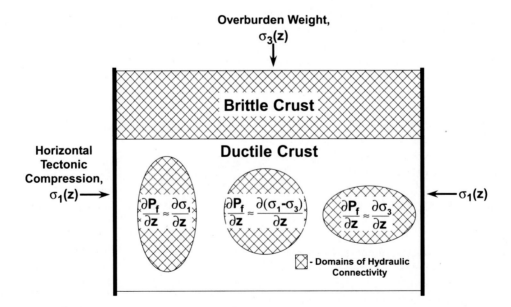

Figure 6. "Swiss cheese" model of the crust [modified after *Connolly and Podladchikov*, 2004] during horizontal compression, σ_1 in which fluid circulates freely in the brittle crust but occurs in isolated fluid-rich domains in the ductile crust. The shape of the isolated domains governs whether the vertical pressure gradient is dominated by the far field compressive stress, σ_1 (in vertically elongated domains), the mean stress $(\sigma_1 - \sigma_3)$ (in spherical domains), or the vertical overburden stress, σ_3 (in horizontally elongate domains).

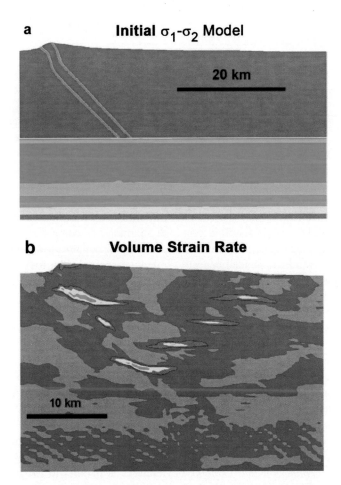

a   Initial σ_1-σ_2 Model

20 km

b   Volume Strain Rate

10 km

Plate 4. (a) Initial geometry of 2-D numerical model from *Hobbs et al.* [2004] shown by plot of $(\sigma_1 - \sigma_2)$ using a geothermal gradient of 20°C/km. Inserted fault is outlined in green. Red is plastic (brittle) portion of the crust corresponding to 0–500 MPa. Below is viscous (ductile) crust. Darkest blue is maximum $(\sigma_1 - \sigma_2)$ corresponding to 3 GPa and greater. Contour interval is 500 MPa. (b) Plot of instantaneous volumetric strain rate using geothermal gradient of 40°C/km. Dark green: 3.2×10^{-11}/s; yellow: 2.4×10^{-11}/s, purple: 1.6×10^{-11}/s, contour interval: 0.8×10^{-11}/s. Total shortening is 1.5%; from *Hobbs et al.* [2004].

numerical modeling did not include the necessary coupling between deformation and induced porosity production in the viscous (ductile) portion of the crust so zones of increased porosity were not modeled there as they were in the plastic (brittle) upper crust. Nevertheless, the result from *Hobbs et al.* [2004] illustrates the pattern of fluidized zones resulting from hydrofracturing in compressional settings and serves to encourage numerical modeling aimed at studying porosity evolution and its consequences in the ductile crust.

Hobbs et al. [2004] pointed out that there would be different mineral alteration patterns in fluid zones formed in the brittle crust compared to the stagnation zones in the ductile crust. This is because fluid flow above a fluid-stagnant zone is in the same direction as the geothermal gradient (downward) and it's in the opposite direction (upward) in the lower section. In the brittle crust, flow is opposite to the geothermal gradient and fluids ascend (Figure 5). The results would be asymmetric mineral alteration patterns for fluid stagnation zones originating in the lower ductile, crust and homogeneous alteration patterns in the upper crust, features that may be observable in exhumed midcrustal terranes.

The above considerations lead us to expect that fluid generated by deep episodic or near steady state metamorphic processes would be propagated upward and ponded beneath strong lithologic contrasts present below the brittle-ductile transition. Time scales required for the formation of such fluid-rich zones beneath inhomogeneities would be only ~5–10 Myr compared to >50 Myr in rheologically homogeneous, upward strengthening ductile crust. This process would occur independent of the tectonic setting [compressional or extensional; *Connolly and Podladchikov, 2004*]. Fluid-bearing horizons also would occur below the brittle-ductile transition in regions of neutral buoyancy in compressional environments. In either case the fluid is expected to spread horizontally. Water sills can be stable for long times (tens of Myr) even by geologic standards and even longer in compressional environments [*Thompson and Connolly*, 1990]. However, their low shear strength (due to high fluid content) would make them unstable under deviatoric stress [*Connolly*, 1997].

4. INTERCONNECTIVITY OF FLUIDIZED ZONES

If the proportion of aqueous fluid in the ductile crust is <1%, it must be interconnected to have resistivities lower than a few 100 ohm-m. This requires either a fluid-filled fracture network, grain-edge interconnection, or both (so-called dual porosity). Either possibility is allowed in the mechanisms discussed above [*Connolly and Podladchikov*, 2004]. Experimental work has shown that fracture networks with high crack connectivity and permeability can form in

ductile regimes at low stain rates provided the fluid pressure is elevated compared to the vertical stress [*Cox*, 2005]. Since the state of stress in a material is expressed by a second-order tensor, there are always three orthogonal vectors that define the principal stress directions. In compressional tectonics the maximum principal stress, σ_1 is horizontal and the least principal stress, σ_3 is vertical (Figure 6).

4.1. Fracture Connectivity

In the presence of fluids, pure extensional fractures are initiated perpendicular to the minimum principal stress (σ_3) if the fluid pressure equals or exceeds the sum of σ_3 and the tensile strength of the rock [*Cox*, 2005]. Stress on a fracture can be decomposed into a component in a direction parallel to the fracture plane (shear stress) and one perpendicular to it (normal stress). The effects of these components are highly coupled in the sense that deformation caused by a change in one is dependent on the magnitude of the other [*National Research Council*, 1996]. If an extensional fracture is already present but closed, the fluid pressure to reopen it is equal to the normal, i.e., compressive, stress across the fracture. For fractures to remain open, and therefore interconnected, the effective stress in the rock, generally taken as the normal stress on the fracture minus the fluid pressure therein [e.g., *National Research Council*, 1996] must remain positive. If the effective stress is zero the fractures close. The rock will undergo extensional hydrofracture if the effective stress is negative and satisfies a yield condition.

Because extensional cracks form perpendicular to the minimum compressive stress, they would be horizontal in compressive environments. However, shear fractures and hybrid extensional-shear fractures form at oblique angles to σ_3, typically between ±20° to 35° and 0° to ±25° to σ_1, respectively [*Cox*, 2005]. Pure extension fractures open perpendicular to the fracture wall, whereas, displacement in shear fractures is parallel to the fracture plane. Hybrid fractures are defined where there is displacement both parallel and perpendicular to the fracture plane. Pure extensional failure occurs at relatively small stress differences ($\sigma_1 - \sigma_3$) coupled with high fluid pressure. In contrast, if the stress difference is increased, shear fracture can result under conditions of low fluid pressure [*Cox*, 2005].

Maintenance of fracture porosity during interseismic periods is a complex function of space and time. Sustained elevated fracture permeability in the ductile crust usually requires continued deformation, high fluid pressure, and/or permeability enhancing reactions [*Cox*, 2005]. If such conditions are not met, fluids can become isolated and pressurized locally because cracks seal during episodes of low fluid pressure gradients [*Byerlee*, 1990].

4.2. Grain-Edge Connectivity

For grain-edge (intergranular) porosity, recrystallization before and during fluid ascent can lead to the creation of connected pore networks (permeability) that can be maintained over geologically significant times [*Brenan*, 1991]. Under conditions of both mechanical and chemical equilibrium, the distribution of intergranular fluid is one that minimizes the total interfacial energy of the system, leading to "textural equilibrium" [*Watson et al.*, 1900; *Spear*, 1993; *Holness*, 1997]. The interfacial energy depends on the mineral and fluid phases, the mineral major- and minor trace-element composition, temperature, pressure, and crystal orientations [*Watson et al.*, 1900]. Knowledge of the interfacial energy is usually unknown or is too imprecise to be useful [*Brenan*, 1991]. However, at textural equilibrium there are well-defined angular relationships between the fluid and grain boundaries that define the wetting characteristics. In a monomineralic, texturally equilibrated, isotropic material undergoing recrystallization, an angle, called the wetting angle or dihedral angle, θ determines the 3-D connectivity of the fluid-filled pores. The dihedral angle is the angle of intersection where mineral grains make contact with the fluid [Figure 7a; *Brenan*, 1991]. For low porosity rocks (< 1%), grain-edge tubes remain connected if $\theta \leq 60°$; the tubes pinch off and the fluid is isolated at grain-edge intersections if $\theta > 60°$ [Figure 7b; *Watson et al.*, 1990]. A stable, thick fluid would completely wet all grain boundaries at a dihedral angle of 0°. At $\theta = 180°$ the fluid is isolated in spherical cavities (Figure 7).

The majority of the experimental determinations of equilibrium fluid topologies have been obtained from monomineralic grain aggregates, but recent studies have demonstrated greatly increased pore fluid connectivity in systems with traces of other mineral phases [e.g., *Holness*, 1995; 1998; *Yoshino et al.*, 2002]. Earlier work by *Watson and Brenan* [1987] on aqueous fluids with monomineralic aggregates and synthetic mafic rocks (primarily quartz and dunite, respectively) discovered θ just below 60° for quartz (implying interconnectivity) but high dihedral angles for the mafic rocks. Salts (NaCl, KCl) added to the aqueous fluid significantly lowered θ (to ~40°) in the quartz-fluid system but had no affect on the high wetting angles in dunite (implying isolated pores). The addition of CO_2 increased θ to > 60° in all cases. Therefore, *Watson and Brenan* [1987] concluded that aqueous fluids in the ductile portion of the crust are likely to be interconnected especially if the water is saline and CO_2-poor, but aqueous fluid in the ultramafic uppermost mantle would exist in isolated pores.

For equilibrium wetting to occur, the rate at which equilibrium is reached must be faster than the strain rate of ductile deformation or of any other rate limited nonequilibrium processes [*Brenan*, 1991]. This means that the rate of equilibration is about equal to the rate of change of porosity so if fluid production (or loss) is too rapid, textural equilibrium is not achieved [*Holness and Siklos*, 2000]. *Tullis et al.* [1996] reported on laboratory experiments where the increased deformation of "water-added" feldspar aggregates at high temperature and pressure mimicked that of aggregates containing a fluid with a dihedral angle of 0° (total grain

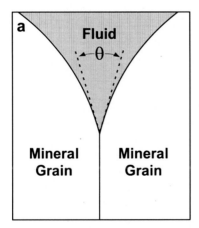 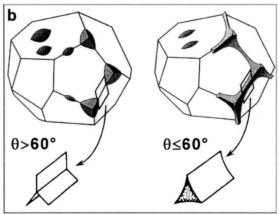

Figure 7. (a) Cross section through a fluid-filled grain-edge channel at the intersection of two minerals grains and the definition of the wetting or dihedral angle, θ; modified from *Brenan* [1991]. (b) 3-D fluid distribution along a mineral grain where fluid occurs in isolated pores, $\theta > 60°$ and where there is interconnectivity, $\theta \leq 60°$; modified from *Watson et al.* [1990].

wetting). The authors pointed out that this result is consistent with enhanced fluid permeability and material transport along fluid-filled grain boundaries in ductile shear zones in the mid-crust.

5. EARTHQUAKES AND FLUIDS IN THE DUCTILE CRUST

The question we explore now is in what ways fluidized zones at or below the brittle-ductile transition can play a role in the earthquake cycle. There seems to be no question that large earthquake ruptures in the brittle crust can propagate downward into the ductile region thereby releasing fluid trapped below the brittle-ductile transition [e.g., *Cox*, 2005]. Draining of these fluid reservoirs requires them to have enhanced permeability pathways as developed by deformation and/or high fluid pressure in ductile shear zones. Fluid-triggered aftershocks [e.g., *Nur and Booker*, 1972] occur in the brittle crust after the main shock of an earthquake due to upward migrating fluid-pressure pulses (*Sibson's* [1992] fault-valve concept), and from coseismic stress transfer [*Cox*, 2005]. A more uncertain scenario is how earthquake rupture in the brittle crust is initiated by permeable, overpressured fluidized zones in the ductile crust. Addressing this possibility requires reviewing the basics of how earthquake rupture occurs.

Failure initiating an earthquake occurs when the tectonic shear stress on the fault exceeds some critical value. The process leading to localized unstable stick-slip, or earthquake nucleation, can be expressed as a constitutive relation that depends on several factors including the slip rate, temperature, pressure, rock composition, and fluid state. When the relationship exhibits rate-weakening, rather than rate-strengthening, the localized deformation causes friction to decrease resulting in velocity weakening. Such a condition allows earthquake nucleation provided that a patch on the fault greater than some critical dimension becomes unstable and slips [*National Research Council*, 2003; *Rice and Cocco*, 2007]. The suggestion that earthquake nucleation can be triggered by fluid processes originating below the actual seismogenic zone is motivated by the spatial relationships between deep conductivity zones and seismicity in New Zealand (Plate 1) and the Himalaya (Plate 2). The idea has been independently bolstered by recent observations of nonvolcanic, episodic tremor and slow slip in the Cascadia subduction zone [*Rogers and Dragert*, 2003] and below the San Andreas fault [*Nadeau and Dolenc*, 2005]. The exact manner by which events tens of kilometers beneath the seismogenic zone connect to earthquakes in the upper seismogenic crust is unknown. However, the prevailing view is that the tremors are generated by processes associated with fluids [*Nadeau and Dolenc*, 2005; *Shelly et al.*, 2006].

Iio and Kobayashi [2002] proposed that seismic failure along an active fault in the upper, brittle crust can be triggered by localized, aseismic motion on a shallow-dipping shear zone in the ductile crust. *Hobbs et al.* [2004] investigated this possibility numerically; however, they concluded that to fully understand the proposal of *Iio and Kobayashi* [2002] requires a better understanding of crustal rheology than is presently available. Two of the questions posed by *Hobbs et al.* [2002] were: 1) what processes can produce the strain softening required to localize deformation in the lower, ductile crust and 2) are there mechanisms in the lower, ductile crust that could generate earthquakes?

Hobbs et al. [2002] cited evidence that localized deformation requires strain softening if the lower crust behaves in a simple viscous manner. Softening can occur in a viscous material via several processes including water-weakening, microcracking, and thermal softening. Deformation in a classical power-law, viscous material is always distributed unless there is such softening. *Hobbs et al.* [2002] also pointed out that localization would be ubiquitous for hybrid elasto-viscous-plastic behavior such as a transitional zone between the brittle and ductile portions of the crust. Such a transition zone would allow "ductile fracture" and a proposal that earthquakes could be generated by unstable sliding on the fractures. Velocity weakening and unstable sliding in materials under conditions where ductile behavior normally dominates is considered by *Hobbs et al.* [2002] to be an important mechanism for earthquake generation at midcrustal depths especially if spaces between asperities are filled with overpressured fluid.

Water-weakening would occur in the fluid stagnation zones described by *Connolly and Podladchikov* [2004] and *Hobbs et al.* [2004] if sufficient fluid-filled porosity exists. Combining the upward propagating porosity waves of *Connolly* [1997] with these stagnation zones is one mechanism that would produce such porosity. *Connolly and Podladchikov* [2004] discussed two mechanisms by which fluids in such overpressured zones could be propagated upward 5 to 7 km above the zone of stagnation thus breaching the brittle-ductile transition. According to *Connolly and Podladchikov* [2004] the zone in which fluids accumulate and are trapped may simply thicken upward eventually breaching the brittle-ductile transition releasing fluids into the overlying brittle crust. This process will occur if the fluid, accommodated by ductile deformation, accumulates faster than the ductile compaction rate. Another way to release fluids from the stagnation zone into the brittle crust is to invoke a relaxation of compressive tectonic stresses, thereby, changing the principal stress relations and causing upward propagating hydrofractures [*Connolly and Podladchikov*, 2004]. In either case, the release of fluid will increase fluid pressures in the upper crust and possibly initiate seismic rupture.

Recognizing the uncertain role of the ductile crust in earthquake generation, *Hobbs et al.* [2002] concluded that "there is a wealth of work to be done to understand the coupling between dilation, fluids, and constitutive relations for lower (ductile) crustal rocks and the controls on localization and unstable sliding." We agree that this remains a field of important future study.

6. DISCUSSION AND CONCLUSIONS

The role of fluids during the entire earthquake cycle may control the processes of fault rupture, propagation, and arrest [*Hickman et al.*, 1995]. We have attempted to answer the question of how the spatial distribution of aqueous fluids at or below the brittle-ductile transition relates to seismicity at three continental plate boundaries in New Zealand, the Himalaya, and along the Parkfield segment of the San Andreas fault. We assumed that MT imaging of conductive zones in these locations is mapping zones of interconnected fluid. Table 1 summarizes our observations.

The seismogenic zones in New Zealand and the Himalaya are distinctly above the midcrustal conductors. The situation is less clear for the Parkfield segment of the San Andreas fault where the most recent MT results show a near-vertical conductor throughout the ductile crust which extends upward to a near-surface conductive zone centered east of the fault. The conductance of the near-vertical feature is intermedi-ate in magnitude compared to the conductances of the sub-horizontal, low resistivity zones in the ductile crust of the Southern Alps and the Himalaya (Table 1). The Himalaya conductance exceeds that beneath the Southern Alps by more than an order of magnitude (Table 1). The low resistivity of the Himalaya deep conductor it is interpreted to be a thick, partial melt zone with aqueous fluid residing on top. We believe that the best explanation of the conductive zone under the Southern Alps is that of a mylonized, fluid-rich, active, ductile shear zone. The high conductance in the ductile crust beneath the San Andreas fault conductive zone may also mark a ductile shear zone that is vertically oriented as suggested by *Eberhart-Phillips et al.* [1995] and *Unsworth and Bedrosian.* [2004b]. Such interpretations are consistent with the results of *Tullis et al.* [1996] and *Famin and Nakashima* [2004] that microfracture permeability and/or grain-edge wetting is enhanced in active ductile shear zones.

Comparing conductive occurrences with seismicity in the three tectonic settings necessarily requires considering several factors. These include:

1. The compressional Himalaya has the longer history of convergence tectonics (~50 Myr), the width of the orogen (~400 km) and its convergence rate (15–25 mm/yr) are the largest, and the crustal thickening is greatest (>80 km).

2. The transpressional Southern Alps in New Zealand have a compressional history only for the last ~5 Myr with convergence averaging about 10 mm/yr, the width of the central

Table 1. Properties of Seismogenic, Electrically Conductive, and Fluid Zones in the Ductile Crust and Plate Motions at Three Continental Plate Boundaries (see text for sources of information).

	Alpine Fault, Southern Alps, New Zealand	Himalaya, Central and Southern Tibet	San Andreas Fault, California, USA
Depth Extent of Seismogenic Crust	~12 km	~ 25 km	~15 km
Depth to Top of Conductive Zone	~ 20 km	~25–40 km	No Distinct Top; Sub-Vertical Corridor Extends Through Entire Ductile Crust
Conductance of Conductive Zone	50 to 500 S	3000 to >20,000 S	750 to 1500 S
Fluid Content	≤3% Aqueous Fluid	~10% Aqueous Fluid Above Thicker, 5–14% Partial Melt	Uncertain % Aqueous Fluid
Fluid Origin	H_2O Released from Prograde Metamorphism in Thickened Crustal Root and Strain-Induced Metamorphism within a Ductile Shear Zone.	Heating of Double Thick Crust. H_2O Released from Melting, Prograde Metamorphism, and/or Dehydration of Subducted Indian Crust.	H_2O Released from Mantle and Strain-Induced Metamorphism within a Ductile Shear Zone.
Relative Plate Motion	Strike Slip: 35 mm/yr Convergence: 10 mm/yr	Convergence: 15–20 mm/yr	Strike Slip: 35 mm/yr Convergence: 3 mm/yr

portion of the orogen is ~100 km, and the maximum crustal thickness is ~45 km [*Scherwath et al.*, 2003; *van Avendonk et al.*, 2004].

3. In the central segment of the ~8 Myr old San Andreas fault there is very little convergence (~3 mm/yr) and, therefore, limited crustal thickening; the strike slip component of strain is large (~35 mm/yr), about the same as along the Alpine fault in New Zealand.

Given the above similarities and differences we now comment on their relevance:

1. High convergence in New Zealand and the Himalaya produces uplift, mountain building, crustal thickening, and the deep release of aqueous fluid through prograde metamorphism in a crustal root. Additionally, greater crustal thickening in the Himalaya has promoted partial melting and further release of aqueous fluid. Release of aqueous fluid through prograde metamorphism in a thickened crustal root would not occur below the low relief Parkfield area of the San Andreas fault.

2. Laterally extensive fluid reservoirs formed by the arrest of upward propagating porosity waves at permeability or strength discontinuities, or regions of neutral buoyancy, are favored in compressional settings [*Connolly*, 1997]. But low convergence rates, such as the 3 mm/yr across the San Andreas study area, may not allow their formation. The only place along the San Andreas fault where significantly higher compression occurs is in the Transverse Ranges where the convergence is 7.7 to 10.6 mm/yr [*Godfrey et al.*, 2002].

3. A sub-vertical conductive zone occurs in the ductile crust beneath the San Andreas fault in the Parkfield-SAFOD area and appears to continue uninterrupted to the surface. This may reflect insufficient permeability or lithologic heterogeneity needed to arrest and pond upward migrating fluid in such a young tectonic setting. For example, the time required to form water sills in rheologically homogeneous circumstances is estimated to exceed 50 Myr. However, in the simulations run by *Connolly* [1997], formation times decreased to ~5 Myr when there is an order of magnitude increase in rock strength. Therefore, with a sufficiently large strength change in the lower crust in the Southern Alps a water sill could have formed in the last ~5 Myr. This would have facilitated water-weakening and development, or increased development, of a near-horizontal, mylonized, ductile shear zone.

4. A horizontally limited conductor under the San Andreas fault may be a near-vertical, ductile shear zone below the brittle seismogenic zone. Shear strain strongly affects fluid interconnection enhancing electrical conductivity even though the amount of fluid may be small (<1%). Also, strike-slip deformation leads to the creation of long backbone shears

and increases the proportion of backbone to isolated, fluid-bearing elements [*Cox*, 2005].

5. Greater fluid interconnection and reduction of resistivity in the strike-slip direction is expected in highly transpressional orogens like the New Zealand Southern Alps [*Wannamaker*, 2005]. This would explain the strong resistivity anisotropy inferred there. The crustal conductor below the San Andreas fault may turn out to be anisotropic with greater conductivity along strike due to shearing. That the Himalaya exhibit relatively limited electrical anisotropy despite the pronounced high conductivity is attributed, in part, to it being a more purely compressional regime.

6. We note that there is no MT detection of through-going pathways from the surface to the deep conductor found in the Himalaya results of *Unsworth et al.* [2005] as there are in New Zealand and along the San Andreas fault. However, new MT modeling by *Arora et al.* [2007] along the NW Himalaya line (Figure 2) reveals that the ITS appears to mark a low resistivity connection to the deep conductor. This may relate to crustal extension in western Himalaya in the area [*Zhang et al.*, 2004]. The deep conductor in northwest Himalaya is also much less conductive (by a factor of 2 to 3) compared to that in Tibet possibly because there is no evidence there of continental magmatism [*Arora et al.*, 2007].

7. The occurrence of earthquakes crossing the upper boundary of the conductive layer in the Himalaya (at about 25 km north of the ITS in Plate 2) may have been initiated by fluid propagation downward along deep penetrating ruptures or by catastrophic upward breaching of the brittle-ductile transition by high pressure pore fluid as we have suggested. However, inaccuracies in focal depth estimates and the projection of hypocenters ~100 km from the MT profile (Figure 2 and Plate 2) make such assertions speculative.

Localized deformation in the ductile crust is possible where strain softening occurs. Since water-weakening is one mechanism to do this, the fluidized zones detected by MT soundings may be imaging zones of localized ductile shear as we surmise in New Zealand. Aseismic slip in these zones may couple directly upward along deeply penetrating faults or stress loading may be propagated above into the brittle crust. Rapid fluid expulsion could perturb the stress state enough to trigger earthquakes in the seismogenic crust. Fluid concentrations trapped under the seismogenic zones beneath the Southern Alps and the Himalaya are likely to be overpressured through continuous generation of fluids from below which upon release could initiate earthquakes. MT results from the Parkfield area of the San Andreas fault suggest that fluids are untapped and continuously raising which may facilitate lesser magnitude but more frequent seismicity compared to the Southern Alps and the Himalaya. This perhaps explains the short recurrence interval of tens of years for moderate earthquakes in the Parkfield area of the San Andreas fault [*Bakum and Lindh,* 1985].

Our study of three continental plate boundaries, and the appropriate literature, establishes experimental and numerical evidence for the occurrence of trapped fluidized zones in the ductile crust. Detectability of deep, interconnected fluidized zones by MT is unmatched below seismogenic depths, therefore, MT promises a unique contribution to understanding the earthquake cycle. MT images of the crust are providing well-constrained depths to fluid concentrations that may turn out to be tied to earthquake rupture above. We call for more high quality MT surveys in a variety of tectonic settings and laboratory conductivity measurements of continental rocks at midcrustal, fluid bearing conditions while undergoing strain. Ultimately, temporal changes in the measured MT response of large fluidized regions in the ductile crust and their extensions upward into the active seismogenic crust combined with continuously recorded GPS surface distortion data, tilt measurements, and tremor activity could reveal measurable earthquake precursors.

Acknowledgments. The MT program in New Zealand was supported by the U. S. National Science Foundation (NSF) grants EAR9 8530 and EAR97-25883 (Continental Dynamics) and by the New Zealand Foundation for Research Science and Technology. Kilb's contributions were supported by NSF award EAR 0545250. We are greatly indebted to many colleagues who furnished MT and earthquake data from the three study areas. Stephen Bannister and Donna Eberhart-Phillips provided seismic data from New Zealand. For the Himalaya, Martyn Unsworth furnished MT data and Larry Brown, Gaspar Monsalve, Jim Ni, Vera Schulte-Pelkum, Peter Shearer, and Francis Wu are all responsible for the seismic data that we received. Previous MT results for the San Andreas fault were furnished by Martyn Unsworth and Paul Bedrosian. Michael Becken, Steve Park, and Oliver Ritter graciously shared their new, unpublished MT data for the Parkfield area. Newly located hypocenters for the 2004 M=6.0 Parkfield earthquake and its aftershocks were provided by Jeanne Hardebeck. Discussions with Yuri Fialko and John Vidale proved to be very valuable. Hugh Bibby, Steve Day, Yuri Fialko, Rob Mellors, Steve Park, and Peter Shearer read early versions of the manuscript and made many constructive suggestions for improvement. Bruce Hobbs shared his recent 2-D geodynamic numerical models and answered several key questions regarding the results. We are solely responsible for any misunderstandings. We could not have handled the large earthquake data sets without full support from the University of California-San Diego, Scripps Institution of Oceanography's Visualization Center; Graham Kent and Tom Im answered the call. We also thank Tony Carrasco who worked tirelessly and cheerfully to provide many of the graphics. We greatly appreciated the efforts by reviewers Martyn Unsworth, James Connolly, and Fred Davey whose comments and suggestions improved the paper significantly. Finally, we thank the organizers of the Workshop on Geotectonic Investigation of a Modern Continent-Continent Collisional Oregon: Southern Alps, NZ held in Christchurch, New Zealand in June 2005 since this paper grew out of that workshop.

REFERENCES

Ague, J. J. (2006), Fluids in the deep crust and in subduction zones: Frontiers for research, *EOS Trans. AGU, 87*(36), Jt. Assem. Suppl. Abstract U23A-06.

Anderson, J. L., R. H. Osborne, and D. E. Palmer (1983), Cataclastic rocks of the San Gabriel fault–an expression of deformation at deeper crustal levels in the San Andreas fault zone, *Tectonophysics, 98*, 209–251.

Argus, D., and R. Gordon (2001), Present day motion across the Coast Range and San Andreas fault system in central California, *Geol. Soc. of Am. Bull., 113*, 1580–1592.

Arora, B., M. Unsworth, and G. Rawat (2007), Deep resistivity structure of the northwest Indian Himalaya and its tectonic implications, *Geophys. Res. Lett., 34*, doi:10.1029/2006GL029165.

Atwater, T., and J. M. Stock (1998), Pacific-North American plate tectonics of Neogene southwestern United States: An update: *Int. Geol. Rev., 40*(5), 375– 402.

Bailey, R. C. (1990), Trapping of aqueous fluids in the deep crust, *Geophys. Res. Lett., 17*(8),1129–1132.

Bailey, R. C. (1994), Fluid trapping in midcrustal reservoirs by H_2O-CO_2 mixtures, *Nature, 371*(6494), 238–240.

Bakum, W. H., and A. G. Lindh (1985), The Parkfield, California, earthquake prediction experiment, *Science, 229*, 619–624.

Bannister, S., C. Thurber, and J. Louie (2006), Detailed fault structure highlighted by finely relocated aftershocks, Arthur's Pass, New Zealand, *Geophys. Res. Lett., 33*, doi:10.1029/2006GL027462.

Beaumont, C. R., A. Jamieson, M. H. Nguyen, and B. Lee (2001), Himalayan tectonics explained by extrusion of a low-viscosity crustal channel coupled to focused surface denudation, *Nature, 414*, 738–742.

Beavan, J., M. Moore, C. Pearson, M. Henderson, B. Parsons, S. Bourne, P. England, D. Walcott, G. Blick, D. Darby, and K. Hodgkinson (1999), Crustal deformation during 1994–1998 due to oblique continental collision in the central Southern Alps, New Zealand and implications for seismic potential of the Alpine fault, *J. Geophys. Res., 104*, 25,233–25,255.

Beavan, J., S. Ellis, L. Wallace, and P. Denys (this volume), Kinematic constraints from GPS on oblique convergence of the Pacific and Australian plates, central South Island, New Zealand.

Becken, M., O. Ritter, S. K. Park, P. A. Bedrosian, U. Weckmann, and M. Weber (2006), A deep crustal fluid channel into the San Andreas fault system imaged by magnetotellurics, *EOS Trans. AGU, 87*(52), Fall Meet. Suppl., Abstract T21C-0436.

Becken, M., O. Ritter, S. K. Park, P. A. Bedrosian, U. Weckmann, and M. Weber (2007), A deep crustal fluid channel into the San Andreas fault system near Parkfield, California, *Geophys. J. Intl.,* in press.

Bedrosian, P. A., M. J. Unsworth, and G. Egbert (2002), Magnetotelluric imaging of the creeping segment of the San Andreas fault near Hollister, *Geophys. Res. Lett., 29,* doi: 2910.1029/2001GL014119.

Bedrosian, P. A., M. J. Unsworth, G. D. Egbert, and C. H. Thurber (2004), Geophysical images of the creeping segment of the San Andreas fault: implications for the role of crustal fluids in the earthquake process, *Tectonophysics, 385*, 137–158.

Brace, W. F., and J. D. Byerlee (1970), California earthquakes: Why only shallow focus? *Science, 168*(3939), 1573–1575.

Brenan, J. M. (1991), Development and maintenance of metamorphic permeability: implications for fluid transport, in *Contact Metamorphism, Rev. Mineral.*, vol. 26, edited by D.M. Kerrick, pp. 291–319, Mineral. Soc. of Am., Washington, D.C.

Byerlee, J. (1978), Friction of rocks, *Pure Appl. Geophys., 116,* 615–626.

Byerlee, J. (1990), Friction, overpressure and fault normal compression, *Geophys. Res. Lett., 17*, 2109–2112.

Byerlee, J. (1993), Model for episodic flow of high-pressure water in fault zones before earthquakes, *Geology, 21,* 303–306.

Chen, W.- P., and Z. Yang (2004), Earthquakes beneath the Himalayas and Tibet: Evidence for Strong lithospheric mantle, *Science, 304*, 1949–1952.

Chery, J., M. D. Zoback and R. Hassani (2001), An integrated mechanical model of the San Andreas fault in central and northern California, *J. Geophys. Res., 106*, 22,051– 22,066.

Connolly, J. A. D., and A. B. Thompson (1989), Fluid and enthalpy production during regional metamorphism, *Contrib. Mineral. Petrol., 102*, 346–366.

Connolly, J. A. D. (1997), Devolatization-generated fluid pressure and deformation-propagated fluid flow during metamorphism, *J. Geophys. Res., 102,* 18,149–18,173.

Connolly, J. A. D., and Y. Y. Podladchikov (2004), Fluid flow in compressive tectonic settings: implications for midcrustal seismic reflectors and downward fluid migration, *J. Geophys. Res., 109,* doi:10.1029/2003JB002822.

Cox , S. C., and R. H. Findlay (1995), The main divide fault zone and its role in the development of the Southern Alps, New Zealand, *N. Z. J. Geol. Geophys., 38*, 489–499.

Cox, S. F. (2005), Coupling between deformation, fluid pressures, and fluid flow in ore-producing hydrothermal systems at depth in the crust, in Economic Geology 100th Anniversary Volume, edited by J. W. Hedenquist, J. F. H. Thompson, R. J. Goldfarb, and J. P. Richards, *Econ. Geol.,* 39–76.

Davey, F. J., T. Henyey, W. S. Holbrook, D. Okaya, T. A. Stern, A. Melhuish, S. Henrys, H. Anderson, D. Eberhart-Phillips, T. McEvilly, R. Uhrhammer, F. Wu, G. R. Jiracek, P. E. Wannamaker, G. Caldwell, and N. Christenson (1998), Preliminary results from a geophysical study across a modern, continent-continent collisional plate boundary - the Southern Alps, New Zealand: *Tectonophysics, 288*, 221–235.

DeCelles, P. G., D. M. Robinson, and G. Zandt (2002), Implications of shortening in the Himalayan fold-thrust belt for uplift of the Tibetan Plateau, *Tectonics, 21*, 1062, doi: 10.1029/2001TC001322.

DeMets, C., G. Gordon, D. F. Argus, and S. Stein (1994), Effect of recent revisions of the geomagnetic time scale on estimates of current plate motions, *Geophys. Res. Lett.,21.* 2191–2194.

Eaton, G. P. (1980), Geophysical and geological characteristics of the crust of the Basin and Range province, in *Continental Tec-*

tonics, chm. B. C. Burchfiel, J. E. Oliver, and L. T. Silver, pp. 96–110, Nat. Res. Counc., Washington.

Eberhart-Phillips, D. (1995), Examination of seismicity in the central Alpine fault region, South Island, New Zealand, *N. Z. J. Geol. Geophys., 38*, 571–578.

Eberhart-Phillips, D., W. D. Stanley, B. D. Rodriguez, and W. J. Lutter (1995), Surface seismic and electrical methods to detect related faulting, *J. Geophys. Res., 100*, 12,919–12,936.

Eberhart-Phillips, D., and S. Bannister (2002), Three-dimensional crustal structure in the Southern Alps region of New Zealand from inversion of local earthquake and active source data, *J. Geophys. Res., 107*(B10), 2262, doi:10.1029/2001JB000567.

Famin. V., and S. Nakashima (2004), Fluid migration in fault zones and the evolution of detachments: The example of Tinos Island (Greece), in *Physicochemistry of Water in Geological and Biological Systems*, edited by S. Nakashima, C. J. Spiers, L. Mercury, P. A. Fenter, and M. F. Hochella Jr., Univ. Acad. Press, Tokyo, 189–209.

Ferry, J. M. (1994), A historical review of metamorphic fluid flow, *J. Geophys. Res., 99*, 15,487–15,498.

Fialko, Y. (2004), Evidence of fluid-filled upper crust from observations of postseismic deformation due to the 1992 M_w 7.3 Landers earthquake, *J. Geophys. Res., 109*, B08401, doi:10.1029/2004JB002985.

Fyfe, W. S., N. J. Price, and A. B. Thompson (1978), *Fluids in the Earth's crust*, 383 pp., Elsevier, Amsterdam.

Godfrey, N. J., G. S. Fuis, V. Langenheim, D. A. Okaya, and T. M. Brocher (2002), Lower crustal deformation beneath the central Transverse Ranges, southern California, *J. Geophys. Res., 107*(B7), doi:10.1029/2001JB000354.

Gonzalez. M. (2002), Magnetotelluric evidence for mid-crustal fluids in an active transpressive continental orogen, South Island, New Zealand , M.S. thesis, 151 pp., San Diego State Univ., San Diego, USA.

Gough, D. I. (1986), Seismic reflectors, conductivity, water, and stress in the continental crust, *Nature, 323*, 143–144.

Grapes, R. H., and T. Watanabe (1994), Mineral composition variation in Alpine Schist, Southern Alps, New Zealand: Implications for recrystallization and exhumation, *The Island Arc, 3*, 163–181.

Gratier, J. P., P. Favreau, and F. Renard (2003), Modeling fluid transfer along California faults when integrating pressure solution crack sealing and compaction processes, *J. Geophys. Res., 108*(B2), doi: 10.1029/2001JB000380.

Groom, R. W., and R. C. Bailey (1989), Decomposition of magnetotelluric impedance tensor in the presence of local three-dimensional galvanic distortion, *J. Geophys. Res., 94*, 1913–1925.

Hickman, S., R. Sibson, and R. Bruhn (1995), Introduction to special issue: mechanical involvement of fluids in faulting, *J. Geophys. Res. 100*, 12,831–12,840.

Hobbs, B. E., H. Tanaka, and Y. Iio (2002), Acceleration of slip motions in deep extensions of seismogenic faults in and below the seismogenic zone, *Earth Planets Space, 54*, 1195–1205.

Hobbs, B. E., A. Ord, K. Regenauer-Lieb, and B. Drummond (2004), Fluid reservoirs in the crust and mechanical coupling between the upper and lower crust, *Earth Planets Space, 56,* 1151–1161.

Hole, J. A., T. Ryberg, G. S. Fuis, F. Bleibinhaus, and A. K. Sharma (2006), Structure of the San Andreas fault zone at SAFOD from seismic refraction survey, *Geophys. Res. Lett., 33,* doi: 10.1029/2005GL025194.

Holness, M. B. (1993), Temperature and pressure dependence of quartz-aqueous fluid dihedral angles: The control of adsorbed H_2O on permeability of quartzites, *Earth Planet. Sci. Lett.,* 117, 363–377.

Holness, M. B. (1995), The effect of feldspar on quartz-H_2O-CO_2 dihedral angles at 4 kbar, with consequences for the behavior of aqueous fluids in migmatites, *Contributions to Mineralogy and Peterology,* 118, 356–364.

Holness, M. B. (1996), Surface chemical controls on pore-fluid connectivity in texturally equilibrated materials, in *Fluid Flow and Transport in Rocks: Mechanisms and Effects,* edited by B. D. Jamtveit and B.W.D. Yardley, pp. 149–170, Chapman and Hall, New York.

Holness, M. B. (1998), Contrasting rock permeability in the aureole of the Ballachulish igneous complex, Scottish Highlands: The influence of surface energy, *Contib. Mineral. Petrol., 131,* 86–94.

Holness, M. B., and S. T. C. Siklos (2000), The rates and extent of textural equilibrium in high-temperature fluid-bearing systems, *Chem. Geol., 162,* 137–153.

Iio, Y., and Y. Kobayashi (2002), A physical understanding of large intraplate earthquakes, *Earth Planets Space, 54,*1001–1004.

Jiracek, G. R., E. P. Gustafson, and P. S. Mitchell (1983), Magnetotelluric results opposing magma origin of crustal conductors in the Rio Grande rift, *Tectonophysics, 94,* 299–326.

Jiracek, G. R., V. Haak, and K. H. Olsen (1995), Practical magnetotellurics in a continental rift environment, in *Continental Rifts: Evolution, Structure and Tectonics,* edited by K. H. Olsen, Elsevier, New York, 103–129.

Jove, C. F., and R. G. Coleman (1998), Extension and mantle upwelling within the San Andreas fault zone, San Francisco Bay area, California, *Tectonics, 17,* 883–890.

Kennedy, B. M., Y. K. Kharaka, W. C. Evans, A. Allwood, D. J. DePaolo, J. Thordsen, G. Ambats, and R. H. Mariner (1997), Mantle fluids in the San Andreas fault system, *Science,* 1278–1281.

Klemperer, S. L. (2006), Crustal flow in Tibet: Geophysical evidence for the physical state of Tibetan lithosphere, and inferred patterns of active flow, in *Channel flow, ductile extrusion and exhumation in continental collision zones,* edited by R. D. Law, M. P. Searle and L. Godin,, *Geol. Soc. London Spec. Pub., 268,* 39–70.

Koons, P. O., D. Craw, S. C. Cox, P. Upton, A. S. Templeton, and C. P. Chamberlain (1998), Fluid flow during active oblique convergence: A Southern Alps model from mechanical and geochemical observations, *Geology, 26,* 159–162.

Langin, W. R., L. D. Brown, and E. A. Sandvol (2003), Seismicity of central Tibet from project INDEPTH III seismic recordings, *Bull. Seismol. Soc. Am., 93*(5), 2146–2159.

Ledo, J. (2005), 2-D versus 3-D magnetotelluric data interpretation, *Surveys Geophys., 26,* 511–543.

Leitner, B., D. Eberhart-Phillips, H. Andersen, and J. L. Nabelek (2001), A focused look at the Alpine fault, New Zealand: Seis-micity, focal mechanisms, and stress observations, *J. Geophys. Res., 106* (B2), 2193–2220.

Lemonnier, C., G. Marquis, F. Perrier, J.-P. Avouac, G. Chitrakar, B. Kafle, S. Sapkota, U. Gautam, D. Tiwari, and M. Bano (1999), Electrical structure of the Himalaya of central Nepal: High conductivity around the mid-crustal ramp along the MHT, *Geophys. Res. Lett., 26,* 3261–3264.

Li, S., M. J. Unsworth, J. R. Booker, W. Wei, H. Tan, and A. G. Jones (2003), Partial melt or aqueous fluid in the mid-crust of southern Tibet? Constraints from INDEPTH magnetotelluric data, *Geophys. J. Int., 153,* 289–304.

Liu, Z., and P. Bird (2006), Two-dimensional and three-dimensional finite element modeling of mantle processes beneath central South Island, New Zealand, *Geophys. J. Int., 165,* 1003–1028.

Mackie, R. L., D. W. Livelybrooks, T. R. Madden, and J. C. Larsen (1997), A magnetotelluric investigation of the San Andreas fault at Carrizo Plain California, *Geophys. Res. Lett., 24,* 1847–1850.

Makovsky, Y., and S. L. Klemperer (1999), Measuring the seismic properties Tibetan bright-spots: Free aqueous fluids in the Tibetan middle crust, *J. Geophys. Res. 104,* 10,795–10,825.

Marquis, G., and R. D. Hyndman (1992), Geophysical support for aqueous fluids in the deep crust: Seismic and electrical relationships, *Geophys. J. Int., 110,* 91–105.

Manning, C. E., and S. E. Ingebritsen (1999), Permeability of the continental crust: Implications of geothermal data and metamorphic systems, *Rev. Geophys., 37,* 127–150.

Monsalve, G., G., A. Sheehan, V. Schulte-Pelkum, S. Rajaure, M.R. Pandey, and F. Wu (2006), Seismicity and one-dimensional velocity structure of the Himalayan collision zone: Earthquakes in the crust and upper mantle, *J. Geophys. Res., 111,* doi:10.1029/2005JB004062.

Nadeau, R. M., and D. Dolenc (2005), Nonvolcanic tremors deep beneath the San Andreas fault, *Science, 307, 389.*

National Research Council (1996), *Rock fractures and fluid flow: Contemporary understanding and applications,* National Academies Press, Washington, D.C.

National Research Council (2003), *Living on an active Earth: Perspectives on earthquake science,* National Academies Press, Washington, D.C.

Nelson, K. D., W. Zhao, L. D. Brown, J. Kuo, J. Che, X. Liu, S. L. Klemperer, Y. Makovsky, R. Meissner, J. Mechie, R. Kind, F. Wenzel, J. Ni, J. Nabelek, C. Leshou, H. Tan, W. Wei, A. G. Jones, J. Booker, M. Unsworth, W. S. F. Kidd, M. Hauch, D. Alsdorf, A. Ross, M. Cogan, C. Wu., E. Sandvol, and M. Edwards (1996), Partially molten middle crust beneath southern Tibet: Synthesis of Project INDEPTH results, *Science, 274* (5293), 1684–1688.

Norris, R. J. (2004), Strain localization within ductile shear zones beneath active faults: Alpine fault contrasted with the adjacent Otago fault system, New Zealand, *Earth Planets Space, 56,* 1095–1101.

Nur, A. and J. R. Booker (1972), Aftershocks caused by pore fluid pressure? *Science, 175,* 885–877.

Owens, T. J., and G. Zandt (1997), Implications of crustal property variations for models of Tibetan plateau evolution, *Science, 387,* 37–43.

Park, S. K., and J. J. Roberts (2003), Conductivity structure of the San Andreas fault, Parkfield, revisited, *Geophys. Res. Lett., 30,* doi:10.1029/2003GL017689.

Pili, E., B. M. Kennedy, M. S. Conrad, and J. P. Gratier (1998), Isotope constraints on the involvement of fluids in the San Andreas fault, *Eos Trans. AGU, 79*(17), S229–S230, Spring Meet. Suppl.

Rice, J. R. (1992), Fault stress states, pore pressure distributions, and the weakness of the San Andreas fault, in *Fault Mechanics and Transport Properties of Rocks,* edited by B. Evans and T. F. Wong, pp. 475–503, Academic press, San Diego, CA.

Rice, J. R., and M. Cocco (2007), Seismic fault rheology and earthquake dynamics, in *Tectonic Faults,* edited by M. R. Handy, G. Hirth, and N. Hovius, The MIT press, Cambridge, MA, 99–138.

Ritter, O., A. Hoffmann-Rothe, P. A. Bedrosian, U. Weckmann, and V. Haak (2005), Electrical conductivity images of active and fossil fault zones, in *High Strain Zones: Structure and Physical Properties,* edited by D. Bruhn and L. Burini, *Geol. Soc., London, Spec. Publs., 245,* 165–186.

Rodi, W., and R. L. Mackie (2001), Nonlinear conjugate gradients algorithm for 2-D magnetotelluric inversion, *Geophysics, 66,* 174–187.

Rogers, G., and H. Dragert (2003), Episodic tremor and slip on the Cascadia subduction zone: The chatter of silent slip, *Science, 300,* 1942–1943.

Royden, L. H., B. C. Burchfiel, R. W. King, E. Wang, Z. Chen, F. Shen, and Y. Li (1997), *Science, 276,* 788–790.

Sass, J. H., C. F. Williams, A. H. Lachenbruch, S. P. Galanis Jr., and F. V. Grubb (1997), Thermal regime of the San Andreas fault near Parkfield, *J. Geophys. Res, 1029*(B12), 27,575–27,585.

Scherwath, M., T. Stern, F. J. Davey, D. Okaya, W. S. Holbrook, R. Davies, and S. Kleffmann (2003), Lithospheric structure across oblique continental collision in New Zealand from wide-angle P wave modeling, *J. Geophys. Res., 108(B12),* doi:10.1029/2002JB002286.

Schulte–Pelkum, V., M. Gaspar, A. Sheehan, M. R. Pandey, S. Sapkota, R. Biham, and F. Wu (2005), Imaging the Indian subcontinent beneath the Himalaya, *Nature,* doi:10.1038/nature03678.

Shelly, D. R., G. C. Beroza, I. Satoshi, and S. Nakamula (2006), Low-frequency earthquakes in Shikoku, Japan, and their relationship to episodic tremor and slip, *Nature, 442,* 188–191, doi:10.1038/nature04931.

Sibson, R. H. (1984), Roughness at the base of the seismogenic zone: Contributing factors, *J. Geophys. Res. 89*(B7), 5791–5799.

Sibson, R. H. (1992), Implications of fault-valve behavior for rupture nucleation and recurrence, *Tectonophysics, 211,* 283–293.

Spear, F. S. (1993), Metamorphic phase equilibria and pressure-temperature-time paths, *Miner. Soc. Amer. Monograph, 1,* Washington, D. C., 799 p.

Spratt, J. E., A. G. Jones, K. D. Nelson, M. J. Unsworth, and IN-DEPTH MT Team (2005), Crustal structure of the India-Asia collision zone, southern Tibet, from INDEPTH MT investigations, *Phys. Earth Planet. Inter., 150,* 227–237.

Thompson, A. B., and J. A. D. Connolly (1990), Metamorphic fluids and anomalous porosities in the lower crust, *Tectonophysics, 182,* 47–55.

Thordsen, J. J., W. C. Evans, Y. K. Kharaka, B. M. Kennedy, and M. van Soest (2005), Chemical and isotopic composition of water and gases from the SAFOD wells: Implications to the dynamics of the San Andreas fault near Parkfield, *EOS Trans. AGU, 86*(52), Fall Meet. Suppl., Abstract T23E-08.

Thurber, C., S. Roecker, W. Ellsworth, Y. Chen, W. Lutter, and R. Sessions (1997), Two-dimension image of the San Andreas fault in the Northern Gabilan Range, central California, *Geophys. Res. Lett., 24,* 1591–1594.

Thurber, C., H. Zhang, F. Waldhauser, J. Hardebeck, A. Michael, and D. Eberhart-Phillips (2006), Three-dimensional compressional wavespeed model, earthquake relocations, and focal mechanisms for the Parkfield, California, region, *Bull. Seism. Soc. Am., 96,* S38–S49.

Tullis, J., A. Yund, and J. Farver (1996), Deformation-enhanced fluid distribution in feldspar aggregates and implications for ductile shear zones, *Geology, 24,* 63–66.

Unsworth, M. J., P. E. Malin, G. D. Egbert, and J. R. Booker (1997), Internal structure of the San Andreas fault at Parkfield, California, *Geology, 25,* 359–362.

Unsworth, M., G. Egbert, and J. Booker (1999), High-resolution imaging of the San Andreas fault in central California, *J. Geophys. Res., 104*(B1), 1131–1150.

Unsworth, M., and P. A. Bedrosian (2004a), Electrical resistivity structure at the SAFOD site from magnetotelluric exploration, *Geophys. Res. Let., 31,* doi:10.1029/2003GL019405.

Unsworth, M., and P. A. Bedrosian (2004b), On the geoelectric structure of major strike-slip faults and shear zones, *Earth Planets Space, 56,* 1177–1184.

Unsworth. M. J., A. G. Jones, W. Wei, G. Marquis, S. Gokarn, J. E. Spratt and the INDEPTH-MT team (2005), Crustal rheology of the Himalaya and southern Tibet inferred from magnetotelluric data, *Nature, 438,* doi:10.1038/nature04154.

Upton, P., and P. O. Koons (this volume), Three-dimensional geodynamic framework for the Central Southern Alps, New Zealand: Integrating geology, geophysics and mechanical observations.

Upton, P., D. Craw, T. G. Caldwell, P. O. Koons, Z. James, P E. Wannamaker, G. R. Jiracek, and C. P. Chamberlain (2003), Upper crustal fluid flow in the outboard region of the Southern Alps, New Zealand, *Geofluids, 3,* 1–12.

van Avendonk, H. J. A., W. S. Holbrook, D. Okaya, J. K. Austin, F. Davey, and T. Stern (2004), Continental crust under compression: A seismic refraction study of South Island Geophysical Transect I, South Island, New Zealand, *J. Geophys. Res., 109*(B6), B06302.

Vry, J. K., R. Maas, T. A. Little, D. Phillips, R. Grapes, and M. Dixon (2004), Zoned (Cretaceous and Cenozoic) garnets and the timing of high-grade metamorphism, Southern Alps, New Zealand, *J. Metamorphic Geol., 22,* 137–157.

Vry, J. K., A. C. Storkey, and C. Harris (2001), Role of fluids in metamorphism of the Alpine fault zone, New Zealand, *J. Metamorphic Geol., 19,* 21–31.

Walcott, R. I. (1998), Modes of oblique compression: Late Cenozoic tectonics of the South Island of New Zealand, *Rev. Geophys., 36,* 1–26.

Wannamaker, P. E. (2005), Anisotropy versus heterogeneity in continental solid earth electromagnetic studies: fundamental response characteristics and implications for physicochemical state, *Surveys in Geophys., 26*, 733–765.

Wannamaker, P. E., G. W. Hohmann, and S. H. Ward (1984), Magnetotelluric responses of three-dimensional bodies in layered earths, *Geophysics, 49*, 1517–1534.

Wannamaker, P.E., G. R. Jiracek, J. A. Stodt, T. G. Caldwell, V. M. Gonzalez, J. D. McKnight, and A. D. Porter (2002), Fluid generation and pathways beneath an active compressional orogen, the New Zealand Southern Alps, inferred from magnetotelluric data, *J. Geophys. Res., 107*(B6), doi:10.1029/2001JB000186.

Watson, E. B., and J. M. Brenan (1987), Fluids in the lithosphere, 1, Experimentally determined wetting characteristics of CO_2-H_2O fluids and their implications for fluid transport, host rock physical properties, and fluid inclusion formation, *Earth Planet. Sci. Lett., 85*, 497–515.

Watson, E. B., J. M. Brenan, and D. R. Baker (1990), Distribution of fluids in the continental mantle, in *Continental Mantle*, edited by M. A. Menzies, pp. 111–125, Clarendon Press, Oxford.

Wei, W., M. Unsworth, A. Jones, J. Booker, H. Tan, D. Nelson, L. Chen, S. Li, K. Solon, P. Bedrosian, S. Jin, M. Deng, J. Ledo, D. Kay, and B. Roberts (2001), Detection of widespread fluids in the Tibetan crust by magnetotelluric studies, *Science, 292*, 716–718.

Yoshino, T., K. Mibe, A. Yasuda, and T. Fujii (2002), Wetting properties of anorthite aggregates: implications for fluid connectivity in continental lower crust, *J. Geophys. Res., 107*(B1), doi: 10.1029/2001JB000440.

Zandt, G., and C. R. Carigan (1993), Small-scale convectivity instability and upper mantle viscosity under California, *Science, 261*, 460–463.

Zhang, P.- Z., Z. Shen, M. Wang, W. Gan, R. Burgmann, P. Molnar, O. Wang, Z. Niu, J. Sun, J. Wu, S. Hanrong, and Y. Xinzhao (2004), Continuous deformation of the Tibetan Plateau from global position system data, *Geology, 32*(9), doi:10.1130/G20554.1, 808–812.

V. M. Gonzalez and G. R. Jiracek, Department of Geological Sciences, San Diego State University, San Diego, CA, USA.

T. G. Caldwel, GNS Science, Lower Hutt, New Zealand.

P. E. Wannamaker, Energy & Geoscience Institute, University of Utah, Salt Lake City, UT, USA.

D. Kilb, Scripps Institution of Oceanography, University of California, San Diego, La Jolla, CA, USA.